EARTH MATERIALS

EARTH MATERIALS

Components of a Diverse Planet

DEXTER PERKINS

Department of Geology and Geological Engineering, The University of North Dakota,
Grand Forks, USA

KEVIN R. HENKE

Retired, Center for Applied Energy Research, The University of Kentucky,
Lexington, USA

ADAM C. SIMON

Department of Earth and Environmental Sciences, The University of Michigan,
Ann Arbor, USA

LANCE D. YARBROUGH

Department of Geology and Geological Engineering, The University of Mississippi, Oxford, USA

CRC Press
Taylor & Francis Group
Boca Raton London New York Leiden

CRC Press is an imprint of the
Taylor & Francis Group, an **informa** business

A BALKEMA BOOK

Cover illustration: Courtesy of NASA Earth Observatory

CRC Press/Balkema is an imprint of the Taylor & Francis Group, an informa business

© 2019 Taylor & Francis Group, London, UK

Typeset by Apex CoVantage, LLC

Library of Congress Cataloging-in-Publication Data

Names: Perkins, Dexter, author. | Henke, Kevin R., author. | Simon, Adam C., author. | Yarbrough, Lance D., author.
Title: Earth materials : components of a diverse planet / Dexter Perkins (Department of Geology, University of North Dakota, Grand Forks, USA), Kevin R. Henke (emeritus professor, University of Kentucky, Lexington, USA), Adam C. Simon (University of Michigan, Ann Arbor, USA), Lance D. Yarbrough (The University of Mississippi, Oxford, USA).
Description: Leiden, The Netherlands : CRC Press/Balkema, [2019] | Includes bibliographical references and index.
Identifiers: LCCN 2019005125 (print) | LCCN 2019014605 (ebook) | ISBN 9780429197109 (ebook) | ISBN 9780367145651 (hardcover : alk. paper) | ISBN 9780367185947 (pbk. : alk. paper) | ISBN 9780429197109 (ebk.)
Subjects: LCSH: Earth sciences—Textbooks. | Geology—Textbooks.
Classification: LCC QE26.3 (ebook) | LCC QE26.3 .E235 2019 (print) | DDC 550—dc23
LC record available at https://lccn.loc.gov/2019005125

Published by: CRC Press/Balkema
       Schipholweg 107c, 2316 XC Leiden, The Netherlands
       e-mail: Pub.NL@taylorandfrancis.com
       www.crcpress.com—www.taylorandfrancis.com

ISBN: 978-0-367-14565-1 (Hbk)
ISBN: 978-0-367-18594-7 (Pbk)
ISBN: 978-0-429-19710-9 (eBook)

DOI: https://doi.org/10.1201/9780429197109

Contents

Part III
Surficial Geology and Resources

Preface

This book is about Earth materials. Although other textbooks include "Earth materials" in their titles, this is the only text that looks at all the diverse inorganic components that make up our planet. We have not considered Earth's atmosphere or biosphere in any detail, but we included everything else. So, this book covers many topics and includes a great deal of information. We doubt that any instructor will want to, or be able to, use it all during a single semester course.

Traditional geology (and other sciences) textbooks are focused, most dealing with small subdisciplines and replete with details. However, during the past several decades, it has become increasingly clear that we must think about Earth as a system if we are to understand and solve some important challenges we face. So, this book covers a spectrum of diverse materials that people encounter in their daily lives, and attempts to establish the relationships between all of them.

We start with considering Earth's origin, the overall structure and composition of Earth, and the nature of Earth processes and cycles. Subsequent chapters cover the fundamentals of mineralogy and crystallography and then igneous, sedimentary, and metamorphic petrology and stratigraphy. We have a chapter on water and the hydrosphere, a chapter on ore deposits, and a chapter on energy resources. We also have two chapters on soils—one focused on fundamental pedology and the other on soil mechanics, and we include a chapter on rock mechanics. We included geological engineering topics because we know that both science majors and engineers can benefit from a book such as ours. However, we fully recognize that different instructors may elect to omit some chapters, depending on the class they are teaching.

One problem with many textbooks is that they do not actively engage students. Many are well written, complete, and up to date, yet do a poor job of exciting students about science. We have done several things to overcome this problem.

First, we have written this book in what, we hope, is an engaging and conversational style. We always present information with a major goal of communicating why students might care about that information.

Second, although there is a great deal of information in this book, we have not tried to include everything about everything. Instead, our emphasis is on the most important and exciting things, and on fundamental ideas and skills that students will need to take with them for future professional development and to become informed citizens.

Third, we start every chapter with a vignette—a short and intriguing description of a geological event or topic that we hope will engage students' interest. Then we focus on smaller details and abstract details. The approach is from large pictures to small facts, because students most care about the bigger picture.

Fourth, although many textbooks, these days, include many graphic illustrations in lieu of photographs—we have gone the other direction. Our book contains many more photographs—perhaps twice as many—as in a traditional text. And, we have taken pains to track down and include spectacular photos. We drafted additional figures, too, making sure that they are most informative and comply with known best practices. Additionally, we have eschewed traditional lengthy figure captions and made sure that all figures are integrated with adjacent text material.

This is a different kind of textbook. In the end, its success or failure depends on the reactions of instructors and students. So, please feel free to tell us about your likes or dislikes, so that we may improve *Earth Materials: Components of a Diverse Planet* in the future. Contact details of the authors can be found in the author descriptions.

About the Authors

Dexter Perkins is Professor of Geology at the University of North Dakota, USA. He was an undergraduate at the University of Rochester before moving to Ann Arbor to attend law school at the University of Michigan, USA. He subsequently transferred to Michigan's Department of Geology and received a masters and a Ph.D. degree. He is presently in his 38th year at UND. Contact: dexter.perkins@und.edu.

Kevin R. Henke is a retired research scientist from the Center for Applied Energy Research (CAER) at the University of Kentucky, USA. He obtained his M.S. in Geology at the University of North Dakota in 1984, and his Ph.D. at the same university in 1997. His interests include metamorphic petrology, mineralogy, and geochemistry. Contact: kevin.r.henke@gmail.com.

Adam C. Simon is Professor of Earth and Environmental Sciences at the University of Michigan, USA. He obtained his Ph.D. at the University of Maryland, USA, in 2003. His fields of study are economic geology, igneous petrology, and geochemistry. Contact: simonac@umich.edu.

Lance D. Yarbrough is Assistant Professor of Geology and Geological Engineering at the University of Mississippi, USA, where he obtained his Ph.D. in 2006. His areas of expertise include engineering geology, remote sensing, and geotechnical engineering. Contact: ldyarbro@olemiss.edu.

Acknowledgments

The idea for this book stemmed from many discussions, most importantly discussions with Dave Mogk. Additional, but indirect, encouragement and motivation came from Steve Reynolds and Julia Johnson, who showed us that textbooks do not all have to be the same.

All mistakes in this book are attributable to the authors. However, reviews by Stephen Altaner, Eric Brevik, Josh Crowell, James Gardner, Elizabeth Goeke, Martha Growdon, Callum J. Hetherington, Jamey Jones, Erik Klemetti, Bill Ullman, and five anonymous reviewers were most helpful and led to significant improvements in the manuscript.

The many people who gave us permission to use their photos and Jamie Schod, who took several photos on demand, get our special thanks. We are especially grateful for all the people who have posted their photos and other images on Wikimedia Commons.

Adam C. Simon thanks Dr. Richard Graus for teaching an incredibly captivating Introduction to Geology course that motivated him to pursue geology as a career. Dexter Perkins thanks Eric Essene for teaching that a job is not over until it is completed and completed correctly.

And, most important, all the authors are especially thankful for the support provided by their spouses and other family members.

PART I
Introduction to Earth

1 The Origin of the Elements and Earth

1.1 Orion

Orion is shown in Figure 1.1. It is the brightest and most noticeable winter constellation in the Northern Hemisphere (and summer constellation in the Southern Hemisphere) and is distinctive because of its hourglass figure and the three bright stars that make up Orion's belt. We easily see it during winter months, but during the summer it does not rise until close to sunrise and so is difficult to discern. The constellation is named after the Greek hunter *Orion*. In one version of Greek mythology, Orion, a storied huntsman, was placed in the heavens by Zeus after he was killed by an arrow shot by *Artemis* (twin sister of Apollo and daughter of Zeus and Leto). An alternative legend is that Orion died because of the sting of a giant scorpion (that later became the constellation *Scorpius*). The myths are inconsistent and perhaps originally did not refer to the same individual.

The Orion constellation contains 8–10 bright stars (Fig. 1.2). Three noticeably aligned stars make up the narrow part of the hourglass, equivalent to Orion's belt. Two of the brightest stars in the sky, *Rigel* and *Betelgeuse*, mark Orion's left knee and right shoulder. Other bright stars distinguish the sword (or club depending on interpretation) and shield he is carrying and the dagger that is hanging from his belt.

The bright stars in Orion are mostly *supergiants*, and all of Orion's visible stars are more massive and brighter

Figure 1.1 Orion in the night sky.
Photo from Akira Fujii/David Malin Images. Inset from Joe Tucciarone.

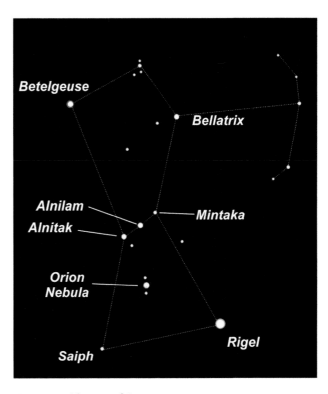

Figure 1.2 The stars of Orion.

than our Sun. They include some of the most distant stars that we can see without the aid of a telescope. Rigel and Betelgeuse are two of the ten brightest stars in the night sky. Rigel is a relatively young *blue-white supergiant* star, and Betelgeuse is an older *red supergiant* star. Blue and red refer to the most intense colors of light that the stars emit, but it sometimes takes good eyesight to discern these colors when looking at the stars in the sky. Red supergiants, like Betelgeuse, are the largest stars by volume, but not the most massive, because other stars are much denser.

However, what sets Orion apart from other constellations is that it includes the *Orion Nebula*, the brightest *nebula* in the sky. It is shown in Figure 1.3. We easily see the nebula with the naked eye. It appears as a single star in the dagger hanging from Orion's belt, but it is not a star. At 1500 light years (10^{16} kilometers) away, it is one of the closest nebulas to Earth and is the closest region associated with massive amounts of star formation. A closer look reveals this bright spot to be more diffuse and hazier than most stars and to have a reddish color. Viewing with binoculars or a telescope reveals many more fascinating features.

The Orion Nebula is a *nuclear furnace* made of a glowing cloud of dust, hydrogen, helium, and super-hot gas called *plasma*. Red and green hydrogen- and sulfur-rich gases, and some carbon molecules similar to components in car exhaust, surround the white-hot center of the nebula. All these gases are heated to extreme temperatures and blown about by winds generated during star formation. Four to six relatively energetic stars are forming near the center of the nebula, in the nebula's brightest white-yellow region. The photographs of the Orion Nebula (Fig. 1.4), taken with the Hubble telescope, show four relatively young protoplanetary disks with bright centers. The "hot spots" are the sites of the newly forming stars. At least 700 stars of lesser energy, in various stages of formation, lie within the nebula, and the nebula's total mass is 2000 times that of our sun. The stars may eventually be at the centers of planetary systems like our solar system.

Nebulas like Orion start as clouds of gaseous hydrogen and helium that collapse toward a focal point due to gravity. Gravitational and kinetic energy cause heating and, when hot enough, nuclear fusion begins. So, hydrogen and helium fuse to create heavier elements, and this process is occurring in many stars within the Orion Nebula. Elements have been created this way since the beginning of the universe, nearly 14 billion years ago, and this is one of only a few processes accounting for all the elements existing today.

Figure 1.3 The Orion Nebula.
Photo from NASA.

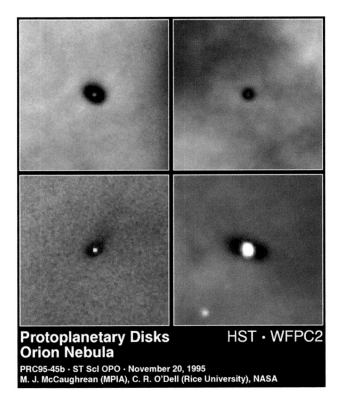

Figure 1.4   Protoplanetary disks in the Orion Nebula.
Photo from NASA.

1.2 The big picture

The *universe*, also called the *cosmos*, is everything. It is more than just planets, stars, and galaxies. It includes every known physical object, including our planet Earth, its life forms, and everything else that is on it, including you. The universe, estimated to be more than 93 billion *light years* (10^{24} kilometers) in diameter, is all of space and matter together. It has been expanding since it first formed about 13.8 billion years ago.

Gravity causes the matter of the universe to form and collect in large concentrations called *galaxies*. The word galaxy comes from the Greek *galaxias*, which means *milky*, in reference to our galaxy, the *Milky Way*. The Milky Way Galaxy, the galaxy that contains our solar system, shown in Figure 1.5, appears as a bar with spiral arms composed of giant stars that illuminate interstellar gas and dust. The Sun (labeled in the drawing), and the planets of our solar system, are in part of the galaxy called the *Orion Spur*; the radial lines and numbers in Figure 1.5 are the galactic longitude in relation to the Sun.

Every galaxy is a collection of *gas, dust, stars* and *star remnants*, and *dark matter*, all orbiting around a center point and held together by gravity. Galaxies range in size from *dwarfs* to *giants*. Small ones contain only a few billion stars; large ones may contain as many as 100 trillion stars. Spinning causes galaxies to flatten and become disk shaped, but they have diverse shapes—some are irregular, and others are elliptical, spiral, or spiral with bars extending across them (like our galaxy). Perhaps as many as 200 billion galaxies exist in our visible universe. They are separated by nearly empty space, estimated to contain less than one atom per cubic meter.

The Milky Way is about 100,000 light years (10^{18} kilometers) across and contains more than 100 billion stars. In the past, astronomers described the Milky Way as a simple spiral galaxy, like our closest galactic neighbor, the *Andromeda galaxy*. However, recent research suggests that it is a *bar-spiral* (Fig. 1.5). The Milky Way is in what astronomers call the *Local Group* of galaxies, which contains about 40 other galaxies.

In the Northern Hemisphere, on a clear summer night, the Milky Way stretches across the sky, appearing as a diffuse swath of light and stars (Fig. 1.6). When we look up and see the Milky Way, we are seeing an edge view of the entire galaxy. The galaxy's center is near the "Teapot" in the constellation Sagittarius, which appears on the southern horizon during the Northern Hemisphere summer. So, if you look just right and above Sagittarius, you are looking toward the center of our Milky Way galaxy. Our Sun is one star in the Milky Way's spiral arms about halfway between the center and outer edge of the galaxy (Fig. 1.5).

The planets, asteroids, and most other bodies of our solar system, shown in Figure 1.7, rotate around *Sol* (as in the adjective *solar*), more commonly called the *Sun*. Earth is the third planet from the sun after Mercury and Venus. It is tempting to think that our planet is special. It might be, but within the past 20 years, astronomers have found evidence of planets orbiting other stars. So, other planets and other solar systems exist where, perhaps, planets could be like ours. Yet, within our solar system, Earth is special because it has liquid water, an atmosphere of nitrogen and oxygen that shields Earth from harmful radiation, a constantly evolving outer crustal layer due to plate tectonics, and life.

A strong *solar wind* of charged particles, in many ways similar to the gases of the Orion Nebula, blows away from the Sun in all directions (Fig. 1.8). One consequence is that gases are swept away from the inner portions of our solar system. Consequently, the outer planets (Jupiter, Saturn, Uranus, and Neptune) are gaseous, mostly hydrogen and helium, but the inner

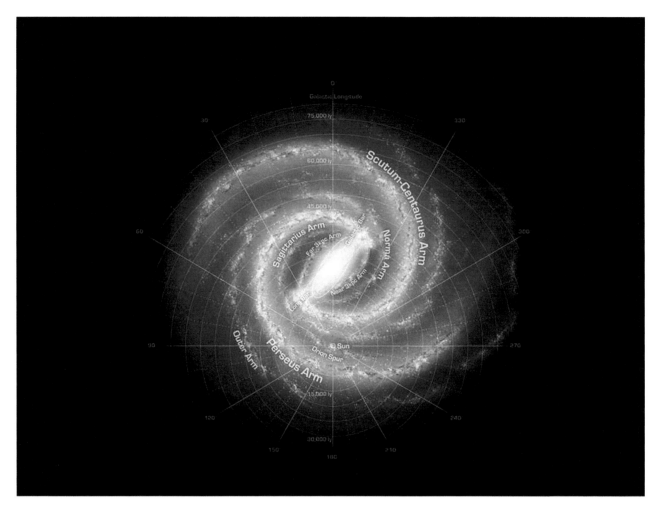

Figure 1.5   Conceptual view of our Milky Way galaxy.
Modified from a NASA drawing.

Figure 1.6   The Milky Way and two constellations: the Teapot (Sagittarius) and Scorpius.
Photo credit: Stellarium with additions by Bob King.

Figure 1.7 The planets and some other bodies of our solar system.
Drawing from NASA.

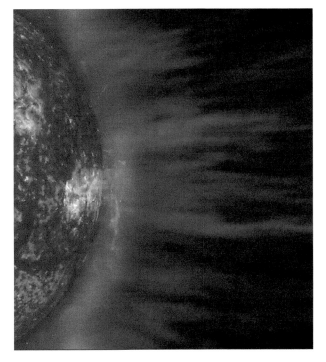

Figure 1.8 The Sun and its solar wind.
Photo from NASA.

planets (Mercury, Venus, Earth, and Mars) are mostly composed of solid material. The solid parts of all four inner planets have similar overall compositions. They also have a *differentiated* structure, meaning they contain layers that have different compositions.

The inner planets, unlike the outer planets, also have an evolved atmosphere—significantly different from their original atmospheres—that developed mainly by volcanism delivering gases from planetary interiors to exteriors and by comet impacts that added additional material. Earth's atmosphere has changed greatly over time and today is 78% nitrogen and 21% oxygen. In contrast, none of the other terrestrial planets contain more than a few percent nitrogen. Additionally, Mars and Venus atmospheres are about nearly all carbon dioxide, and Mercury's very thin atmosphere is heterogeneous but contains more than 40% oxygen and significant amounts of sodium. Earth also has the perfect temperature range that supports life, and this along with a life-supporting atmosphere separates Earth from the other terrestrial planets.

1.3 The beginning

1.3.1 The big bang

Just less than three-fourths of the mass of the universe is hydrogen. Most of the rest is helium, leaving less than 2% that consists of other elements. Although in low abundance in the universe, elements heavier than helium make up most of Earth and the life on it. They are one characteristic that makes us "special." So, where

did the elements that make up Earth come from, and how did Earth become so different from the other terrestrial planets and every other planetary body that we know of?

Currently, the most agreed upon explanation for the origin of the universe is the *Big Bang theory*. It is an explanation that is overwhelming and that pushes the limits of people's understandings of atomic physics, and some human notions about time and events. Although scientists do not fully understand or agree on all the details, the general model of the Big Bang is widely accepted. The theory is consistent with many observations, including Edwin Hubble's key observation that the universe is expanding today, and the universe age of 13.8 billion years is calculated, primarily, based on the rate of expansion. The theory also explains the overwhelming abundance of hydrogen and helium in the universe from its very beginning, and how, after the seminal event, the universe cooled enough for subatomic particles, and later atoms of many elements, to form.

During the *Big Bang*, about 13.8 billion years ago, all known matter and energy *inflated* from a *singular point*—a single point in space, at a single time—where everything that later became part of the universe was packed together. Contrary to popular misconceptions, the Big Bang was not a giant explosion in empty space. The Big Bang actually created space. What existed before the Big Bang, if anything, is unknown. Time may not have even existed before the Big Bang. That is, the Big Bang may have created time as well as space.

The initial temperature of the Big Bang was so high as to be hard to grasp (Fig. 1.9). Within a fraction of

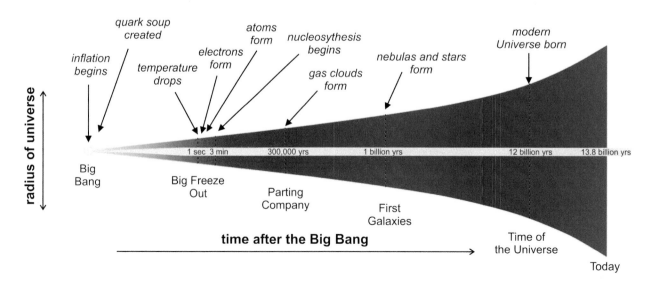

Figure 1.9   Timeline for the Big Bang.

a second after inflation, the temperature is estimated to have been 1000 billion K (degrees Kelvin). At these extremely high temperatures, the universe only consisted of a *quark soup* of *subatomic particles* including pairs of *quarks* and *antiquarks*, *electrons* and *protons*, other matching pairs of matter and antimatter, and enormous amounts of energy. No atoms, molecules, or compounds existed. As the matter and energy expanded, temperature fell dramatically during the *big freeze out*, and particle creation accelerated. All of the electrons in the universe formed within 5 seconds after the initiation of the Big Bang, at a temperature of 6 billion K. Perhaps 14 seconds after the Big Bang, temperatures had dropped to *only* 3 billion K. Protons, which constitute the entire nuclei of most hydrogen atoms, existed, and so hydrogen atoms were born. Fusion of hydrogen nuclei produced helium and other light atoms.

By 3 minutes after it started, the Big Bang had cooled to 1 billion K and the temperature was low enough that hydrogen atoms slowed and could interact even more. They began to fuse to create heavier elements. Initially, during this process, termed *nucleosynthesis*, deuterium (hydrogen with a neutron) formed. Helium and trace amounts of lithium then formed as well. Continued cooling and continued fusion may have led to formation of small amounts of other heavier elements, but most heavy elements had to wait for stars to be born.

The initial elements that formed were too hot to have any electrons. They were *plasma* or just bare nuclei. Electrons existed in the plasma, but because of the high temperatures, they moved at such high velocity that they were not associated with any nuclei. It took 300,000 to 400,000 years for the temperature of the universe to cool to 3000 K, cool enough for nuclei to pick up electrons, and only then were uncharged atoms born. Subsequently, matter could collect and large gas clouds could form that, eventually, parted company, separating into regions where matter clumped.

1.3.2 Stellar evolution

It took a long time, but about 1 billion years after its origin, the matter from the Big Bang began to clump into clouds of hydrogen and helium gas, called *nebulas*, or *nebulae* (plural for the Latin word, *nebula*, meaning cloud). Gravity caused the gaseous nebulas to collapse, their density increased, the temperature went up, and the first *stars* formed.

Stars of various sizes, including yellow stars like our Sun (an average star) and much larger blue giants (massive stars), condense from stellar nebulas of gas and dust (Fig. 1.10). Our Sun is about 4.6 billion years old and is about halfway through its life, but the life span of a star depends on its mass. The Sun and other *yellow stars* have life spans of about 9 billion years, but massive and hotter stars, called *blue giants*, rapidly consume their fuel and may die in less than 1 million years. Smaller stars are cooler, slowly consume their fuel, and have life spans of many billions of years. As the stars exhaust their fuel through nuclear fusion, they expand into *red giants* or *supergiants*. Eventually, reactions in the cores of the stars overcome gravity and the red giants and supergiants explode. Stars like our Sun will then become planetary nebulas composed of debris in a ring surrounding the remaining *white dwarf star*. Stars more massive than the Sun, eventually explode in what are called *supernovas*. The leftovers from supernovas may condense to form

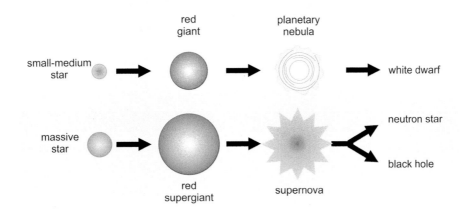

Figure 1.10 Life and death of stars.
Figure based in part on one from the Science Olympiad Student Center.

very dense, small radius (< 20 miles; 30 kilometers) *neutron stars* composed mostly of closely packed neutrons. The most massive supernovas, however, produce *black holes*. Black holes are so massive that not even light can escape their gravity.

When the universe was about a fifth of its present size, the stars were attracted to each other and formed the first *galaxies*. Although the initial element-forming part of the Big Bang had ended, elements continued to form, and by the time the universe had expanded to half its present size, nucleosynthesis was a continuous process in the hot interiors of stars. Hydrogen was fusing to produce heavier nuclides, first deuterium, then various isotopes of helium, beryllium, and carbon, and eventually heavier elements up to iron. Key reactions produced isotopes of carbon and nitrogen, critical to life on Earth today. Many fusion reactions yielded protons that then promoted other reactions, so chain reactions occurred, and the same chain reactions are occurring today in stars where new elements are constantly being born. Yet, during the earlier times of our universe, most matter consisted of very light elements, and without abundant heavier elements, Earth-like planets could not exist and orbit these first-generation stars.

Where did all the heaviest elements, such as gold or uranium, come from? In large part, they came from *supernovas*, and supernovas are creating them today. A crucial balance exists in stable stars. Gravity holds the star together and fusion in the center provides outward pressure. As a star runs out of fusible elements, it can collapse

and explode. When massive stars end their lives in *supernovas*, the short-lived (weeks or months) but huge explosions have immense energy. With a few exceptions, fusion in star interiors is incapable of producing elements heavier than iron. However, supernovas have enough energy to create and eject heavier elements into space—elements that may later become parts of new nebula and form new stars and solar systems. These new stars eventually become old and die, and the heavy elements are recycled. During the last 500–1000 years, astronomers have seen several supernovas (some with the naked eye) in our Milky Way and observed many (by telescope) in distant galaxies.

1.4 Origin of the solar system

1.4.1 What makes up the solar system?

Our solar system consists of the Sun, everything orbiting the Sun, and the space between (Fig. 1.11). The solar system is complex and incredibly diverse (Table 1.1). It includes planets, dwarf planets (also called planetoids), asteroids, meteoroids, comets, and dust. The planets of the inner solar system—Mercury, Venus, Earth, and Mars—are called the *terrestrial planets* because of their rocky compositions, and structures and sizes similar to Earth. The planets of the outer solar system—Jupiter, Saturn, Uranus, and Neptune—are called the *gaseous planets*, also called the *giant planets*.

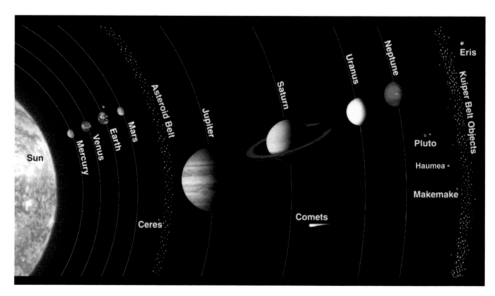

Figure 1.11   Some of the bodies in our solar system.
Image from NASA.

Table 1.1 Some properties of major bodies in the solar system.

Body	Classification	Diameter (km)	Average distance from Sun (x 10^6 km)	Rotation period at equator— Earth days and hours	Revolution period around Sun or parent planet in Earth years, days and years	Number of moons
Sun	star	1,395,000	—	25.4 days	—	—
Mercury	planet	4878	57.9	58.6	88.0 days	0
Venus	planet	12,100	108	243 days	225 days	0
Earth	planet	12,756	150	24 hours	365.26 days	1
Moon	moon	3476	150	29.5 days	27.3 days	—
Mars	planet	6786	228	24.6 hours	687 days	2
Ceres	dwarf planet	945	414	9 hours	4.6 years	0
Jupiter	planet	142,984	778	9.84 hours	11.9 years	67
Europa	Jupiter moon	3126	778	3.55 days	3.55 days	—
Ganymede	Jupiter moon	5276	778	7.15 days	7.15 days	—
Io	Jupiter moon	3630	778	1.77 days	1.77 days	—
Callisto	Jupiter moon	4800	778	16.7 days	16.7 days	—
Saturn	planet	120,536	1430	10.2 hours	29.5 years	62
Titan	Saturn moon	5150	1430	15.9 days	15.9 days	—
Uranus	planet	51,118	2870	17.9 hours	84.0 years	27
Neptune	planet	49,562	4500	19.2 hours	165 years	14
Pluto	dwarf planet	1187	5900	6.39 days	249 years	5

Data from Levy (1994), Vilas (1999), Abbott (2004), Spudis (1999), Faure (1998), Stern *et al.* (2015), and NASA.

The *Asteroid Belt* separates Mars from Jupiter. The dwarf planet Ceres is in the Asteroid Belt. The Kuiper Belt, a region in the outermost solar system that contains small bodies of frozen gases, is home to other known dwarf planets, sometimes called Kuiper Belt Objects, or KBOs: Pluto, Haumea, Makemake, and Eris. Additionally, countless *comets*, *asteroids*, *meteoroids*, and fine-grained dust, called *micrometeoroids*, are found throughout the solar system. Astronomers have also identified about 180 *moons* that orbit the planets and dwarf planets; some are listed in Table 1.1.

1.4.2 Solar nebula hypothesis

The universe has expanded since its inception and was about two-thirds of its present size when the solar system formed. Currently, most scientists accept the *Solar Nebular hypothesis* as the best explanation for the events that occurred at that time. Thus, the solar system condensed out of a *Solar Nebula* about 5 billion years ago (Fig. 1.12). The nebula consisted mostly of hydrogen and helium with smaller amounts of heavier elements produced by fusion in earlier massive stars and their supernovas. A nearby supernova may have initiated the gravitational collapse of the nebula.

As the cloud collapsed, it became denser and began to rotate. Gravity and the centrifugal force from the rotation caused the cloud to flatten into a spinning disk looking much like the disks we see associated with some young stars today (Figs. 1.13 and 1.14). The disks contain newly formed stars surrounded by a sea of dust and gas that eventually gave birth to today's planets. The flattened rotating disk explains why the orbits of the planets of our solar system occur in a plane and why the planets orbit the Sun in the same direction.

Eventually, gravitational pull caused most of the mass in the nebula to collect at the center, where it became denser and hotter. About 4.6 billion years ago, temperatures at the center of our solar system reached around 20 million K, initiating fusion of hydrogen into helium, and a star (our Sun) was born (Fig. 1.12). While the Sun formed at the center of the newly formed solar system, material collided and coalesced in regions away from the center (Figs. 1.12 and 1.14), producing larger bodies and eventually *planetesimals* (defined as bodies in the early solar system that are at least 10 meters across). They varied in size, but some planetesimals grew to be more than 1000 kilometers in diameter.

1.4.3 The planets

The largest planetesimals, which were really *protoplanets*, had enough gravity to attract additional material, growing larger as they swept up debris while orbiting

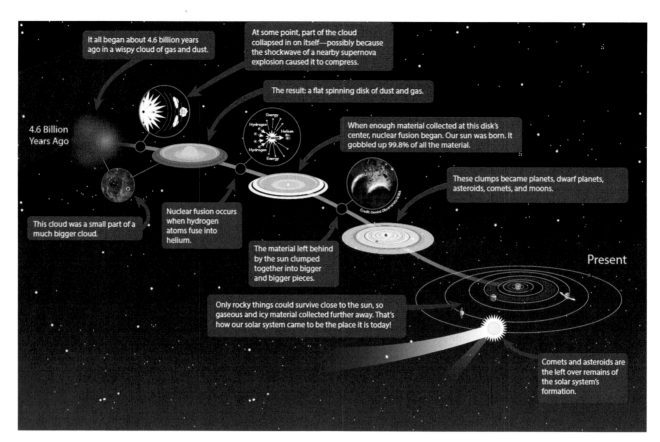

Figure 1.12   Stages in the formation of a solar system according to the Solar Nebula hypothesis.
Drawing from NASA.

Figure 1.13   Artist's conception of a newly formed star surrounded
by a swirling disk of protoplanets, dust, and gas.
Image from JPL.

Figure 1.14   A disk with likely forming planets around star HL
Tauri.
Taken by the Atacama Large Millimeter Array of telescopes in Chile. Photo
from NASA.

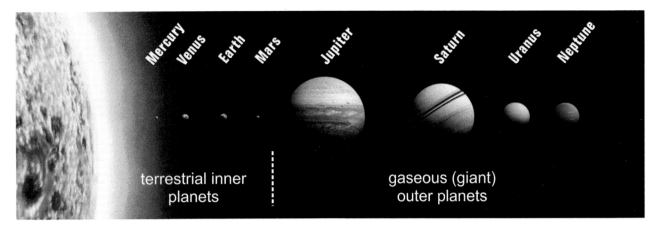

Figure 1.15 The planets of the solar system.
Image from NASA.

the early Sun. Some planetesimals, not all, combined to form today's planets (Fig. 1.15). The leftovers either were consumed by the Sun or became dwarf planets, asteroids, comets, meteoroids, or interplanetary dust. The leftover debris occurs throughout the solar system, but is especially abundant in the Asteroid Belt and in the outermost part of the solar system beyond the orbit of Neptune. Today, the same evolutionary process is taking place in many nebulas, where fine and coarse debris, planetesimals, and newly formed protoplanets rotate around young stars.

1.4.4 Dwarf planets

Astronomers define *planets* as being massive enough to be rounded by their own gravity but not massive enough to heat their cores to temperatures required for nucleosynthesis. Additionally, planets have sufficient gravity to attract all planetesimals and other debris from the regions around them. A *dwarf planet*, also called a *planetoid*, meets the first two-thirds of this definition but has insufficient gravity to clear the area around it of other space debris (Fig. 1.16). Pluto, discovered in 1930, was originally classified as a planet but now has been downgraded to dwarf planet status. The known and named dwarf planets, in increasing size, are Ceres, Makemake, Haumea, Eris, and Pluto. Ceres is the largest body of the Asteroid Belt but is the smallest dwarf planet with a diameter of about 945 kilometers (590 miles). Pluto (2370 kilometers diameter) and Eris (2330 kilometers diameter) are the largest dwarf planets. Some dwarf planets have moons; Pluto has Charon and four other moons. Most of the dwarf planets are much icier than Ceres because Ceres is in the Asteroid Belt (closer to the

Figure 1.16 Dwarf planet size compared to Earth.
Image from NASA.

sun) and the other four dwarf planets are outside the orbit of Neptune. Astronomers will likely discover other dwarf planets in the outermost solar system in the future.

1.4.5 Asteroids and meteoroids

Asteroids (Fig. 1.17) and related, but much smaller, *meteoroids* originate in the inner solar system. Both orbit the Sun. They are largely composed of metal alloys, crystalline minerals including sulfides, and silicates. The chemistry and mineralogy of asteroids are probably similar to meteoroids. That is, both asteroids and meteoroids include varieties that are rich in silicates, rich in alloys, or somewhere between.

Asteroids are irregularly shaped bodies. The largest known are 500 or so kilometers across, and the smallest perhaps only a few meters across. Vesta, the largest shown in Figure 1.17, is about 525 kilometers (325 miles) in diameter. Ceres has been identified, in the past, as the largest asteroid, but today it is considered a dwarf planet.

Most asteroids are in the *Asteroid Belt* between Mars and Jupiter. In the past, scientists hypothesized that these asteroids are the fragments of a planet that disintegrated in an encounter with Jupiter. Currently, most astronomers believe that the asteroids are debris created by collisions of dwarf planets and perhaps also materials that never coalesced into larger objects. A few asteroids have orbits outside the Asteroid Belt, and occasionally they collide with or pass near Earth. An asteroid impact is the current explanation for the extinction of dinosaurs and many other life forms about 65 million years ago.

Meteoroids are space debris smaller than an asteroid, commonly just a few millimeters across but sometimes up to several meters in longest dimension. Due to their small size, most meteoroids vaporize as they enter Earth's atmosphere. Although they do not reach Earth's surface, they may create light trails called *shooting stars*, or just *meteors*. Figure 1.18 shows what NASA called "Nature's light show" when it was photographed. A meteor trail from a *Geminid meteor*, combined with a view of the northern lights, makes quite a display. Geminid meteors, originating in the constellation Gemini, are thought to be pieces of an extinct comet. Earth runs into a cascade of Geminid meteors every year in mid-December.

If they make it to Earth's surface, meteoroids are termed *meteorites*. Scientists have collected about 40,000 meteorites from around the world. About 1000 were seen falling to Earth; the others fell at earlier, undetermined times. Overall, 500–1000 meteorites reach Earth each year, but the vast majority are never found. Most meteorites are harmless, but sometimes they create shock waves causing some damage at Earth's surface.

1.4.6 Classification and origin of meteorites

We commonly divide meteorites into *stony*, *iron*, and *stony-iron* varieties. Stony meteorites, which are low in nickel, iron, and other metals compared with the other two varieties, are further divided into achondrites and chondrites. Chondrites, by far the most common type of meteorite, contain abundant microscopic silicate glass spheres called *chondrules*; examples are shown in

Figure 1.18   A shooting star passes over the northern lights. The photo was taken in Norway. Photo from NASA.

Figure 1.17   Figure comparing the sizes of some asteroids. Photo from NASA.

Figure 1.19   10 cm wide chondrite meteorite. Photo from NASA.

Figure 1.19. Achondrites do not contain chondrules. Scientists think the chondrules are quenched liquid droplets that condensed from the Solar Nebula. Both achondrites and chondrites consist mostly of silicon, oxygen, magnesium, and some iron. They are mostly nonmetallic minerals, dominated by mafic silicates and sometimes carbonates. Before the differentiation of the core, mantle, and crusts, the overall chemical compositions of Earth and other planets of the inner solar system probably resembled the chemistry of chondrites. A subclass of chondrites, called *carbonaceous chondrites*, is meteorites that contain about 5% organic materials, including amino acids. Although amino acids are the building blocks of life, amino acids in meteorites are not thought to have had a biological origin.

Iron meteorites largely or entirely consist of nickel and iron and may be the remains of core materials from dwarf planets or planetesimals that existed in the early solar system. As the name implies, *stony-iron* meteorites have compositions between those of the stony and iron varieties. They are believed to represent materials from the mantle-core boundaries of planetesimals broken apart by impacts.

Most meteorites yield radiometric dates of around 4.5 billion years. Chondrites are generally 4.55 billion years old and are the oldest known materials in the solar system. However, a few meteorites are less than 4 billion years old, and their chemistry suggests that they probably originated from Mars and the Moon. Scientists believe that some early impacts on Mars and the Moon were so severe that they sent crustal debris into space, and that some of this debris later landed on Earth as meteorites. Some scientists have suggested that meteorites may have originated from Venus as well, but Venus has a much greater mass and escape velocity than Mars and the Moon. So, impacts on Venus are less likely to send meteorites to Earth. Additionally, meteorites, if ejected from Venus, would most likely be attracted by the gravitational pull of the Sun and never make it to Earth.

1.4.7 Comets

Comets consist mostly of rock, dust, ice, and frozen gases including methane, carbon monoxide, carbon dioxide, and ammonia. They come from the *Kuiper Belt*, which is just beyond the orbit of Neptune, and from the *Oort cloud*, which is even farther out. The inset in Figure 1.20 shows the Kuiper Belt and typical orbits

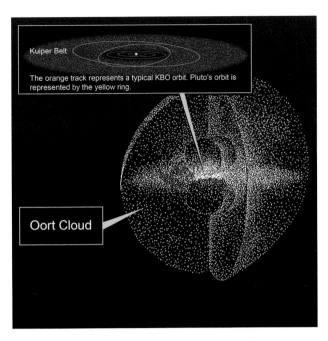

Figure 1.20   The Oort cloud and the orbits of Kuiper Belt Objects (KBOs).

Image from NASA.

for Pluto and other planetary bodies there. The larger image in Figure 1.20 shows the Oort cloud, the spherical cloud of icy planetesimals that encircles the entire disk-shaped solar system, including its outermost reaches. Scientists think that most comets formed similarly to the planets but were dispersed to the outermost parts of the solar system due to gravitational pull by the giant planets. Yet, a few comets enter new orbits that bring them into the inner solar system. After multiple orbits, they will eventually disintegrate from interaction with the *solar wind* (Fig. 1.8) or perhaps collide with a planet or the Sun. In July 1992, the comet Shoemaker-Levy 9 was torn apart in a close encounter with Jupiter, and in July 1994, the remains of the comet impacted Jupiter.

1.5   Evolution of the solar system

1.5.1 Planetesimals and formation of protoplanets

Shortly after their formation, the planetesimals of our solar system began to evolve. Collision of planetesimals led to heating, and the decay of radioactive elements added even more heat. Consequently, growing planetesimals and protoplanets experienced widespread melting on their surfaces, producing an early version of a crust. Besides this physical change, high-energy subatomic particles and radiation of the solar wind blew the interplanetary hydrogen, helium, and dust away from the inner solar system, leaving the larger and denser silicate and metallic materials to continue to coalesce. For the first 10 million years or so, sometimes called *the intense solar phase*, the Sun was especially luminous and solar winds were great. The growing terrestrial planets may have lost their entire atmospheres then.

So, the solar system became differentiated with the lightest material collecting in the outer regions and the densest material in the inner regions. The planetary bodies of the inner solar system became enriched in heavier and less volatile elements and depleted in hydrogen gas and helium. Elemental iron and nickel, as well as magnesium, and calcium compounds were most common in the planetesimals closest to the Sun that eventually coalesced to form Earth. Water ice, ammonia (NH_3), methane (CH_4), and other volatiles were blown away and largely concentrated in the outer solar system, or perhaps trapped in the deep interiors of the terrestrial planets. Hydrogen and helium gases concentrated in the Sun and the outer solar system, where solar winds were too weak to blow light elements away.

The planetesimals in the inner solar system collided and coalesced into a few larger protoplanets and eventually became the planets we have today, but because the distinction between planetesimals, protoplanets, and true planets is a hazy one, the timing of these events is hard to say with certainty. Scientists have estimated the age of the solar system by studying the ages of meteorites, because meteorites are believed to be the oldest material in the solar system. The oldest meteorites are about 4.55 billion years old, and consequently Earth's age is estimated at 4.55 billion years. However, the oldest rocks on Earth are only 3.8 billion years old, although some individual mineral grains are 4.4 billion years old.

1.5.2 Differentiation of the terrestrial planets and the moon

The terrestrial planets were, in a sense, lucky. Nucleosynthesis began with hydrogen, and then helium. Fusion created heavier elements but overall, the heavier the element, the less it was formed. Most of the lightest elements, however, were gaseous and blown away from the terrestrial planets by solar winds. Three other elements

(lithium, beryllium, and boron) were produced only in small amounts by nucleosynthesis. In fact, of these light elements, only oxygen existed in high amounts and stayed with the inner planets. Slightly heavier elements and transition metals, however, were relatively abundant and prone to bond with oxygen to form dense solids. So, the early Earth was rich in sodium, magnesium, aluminum, silicon, sulfur, potassium, and calcium—and especially iron.

One other singular event also was significant in determining the nature of early Earth. All evidence suggests that about 4.4 billion years ago, some 50–120 million years after Earth began forming from planetesimals, a large Mars-sized planet, called *Theia*, collided with Earth (Fig. 1.21). The collision entirely or almost entirely melted Earth. The iron from Theia was heavy and sank into the core of the molten Earth. The lighter silicates from Theia and Earth's mantle were thrown into orbit around Earth and coalesced into a relatively iron-poor and very dry body that has evolved to become our Moon today.

There is little doubt that the early versions of the terrestrial planets were largely molten due to heat from gravitational contraction, collisions, and radioactive decay. After Theia collided with Earth, it too melted. In the melts of the Moon and inner planets, elements separated by their densities—heavy elements sank toward the core and lighter elements remained in outer layers. So, iron, nickel, and similar elements concentrated in the core, while oxygen, silicon, uranium, and others built up in the mantle and crust. As heavy elements

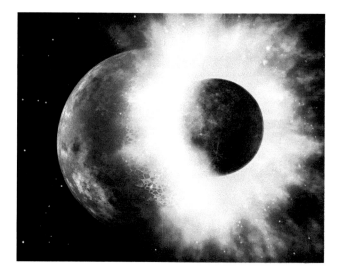

Figure 1.21   Theia colliding with Earth.
Image from NASA.

sank, the gravitational energy released by sinking was enough to cause more melting and produce denser melts that sank and released even more energy. The prevailing view is that a largely iron-rich (Fe-rich) core had developed within a few tens of millions of years after Earth formed.

As the core formed, silicates, containing the light elements silicon and oxygen, moved upward and over time produced a silica-rich (SiO_2-rich) crust that also contained much aluminum, sodium, potassium, calcium, magnesium, and some iron. Magnesium and iron silicates have moderate densities and today are mostly in the mantle. Differentiation, however, was not entirely due to density difference, because elements also have chemical properties that dictate whether they tend to stay in the core or move outward. Uranium and thorium, for example, are heavy elements that we might expect to build up in the core. However, they have an affinity for silicate minerals and so moved upward along with silicon and oxygen. Although their compositions vary, today each of the terrestrial planets probably has a core, mantle, and crust that formed before 4 billion years ago (Fig. 1.22). The Moon may or may not have a core.

During the early stages of solar system formation, there were many asteroids and meteorites. Craters probably pockmarked planetary surfaces, like they pockmark the Moon today. Eventually, the number and size of meteorites and asteroids decreased. Subsequently, on Earth, unlike on the Moon, erosion and tectonic forces largely erased the impact craters that once were present. Considering the massive bombardments and the eventual recycling (at least on Earth) of crust through plate tectonics, very little, if any, of the original crust may be preserved in planets of the inner solar system. However, although the Moon, too, was heavily impacted by meteorites and asteroids, some studies suggest that fragments of the primary crust may exist in the lunar highlands. Additionally, heat in planetary interiors has over time produced magmas that erupted as lavas on the surfaces. Most such lavas were basaltic, but at times recycling of earlier formed crustal material produced granitic magmas, and today granitic rocks dominate most of the continental portions of Earth.

1.5.3 Planetary atmospheres

Once the intense solar phase ceased, Earth and Mars regained water and other volatiles on their surfaces and in their atmospheres. Venus reestablished a dense atmosphere mostly of carbon dioxide. Geochemical studies suggest that Earth had a significant atmosphere within a few hundred million years after it formed 4.6 billion years ago. Venus and Mars, too, probably became massive and cool enough to form substantial atmospheres early in their histories. Water, carbon dioxide, methane, nitrogen, sulfur dioxide, and other gases either degassed out of the planets through volcanism or arrived with comets and meteorites. Degassing is a very efficient way to create an atmosphere, because an element like oxygen expands nearly 2000 times when it leaves a mineral and becomes a gas. Besides developing an atmosphere, early in their histories, Earth, Mars, and perhaps Venus also became cool enough at least temporarily to have oceans and atmospheric water. However, any oceans that once existed on Venus boiled away when the planet heated due to an intense greenhouse effect (Chapter 2).

One reason that Earth has preserved its atmosphere is that the outer core of Earth is still molten and generates Earth's magnetic field. The field protects Earth from intense solar winds because the magnetic field lines deflect solar particles around our planet (Fig. 1.23). Mars

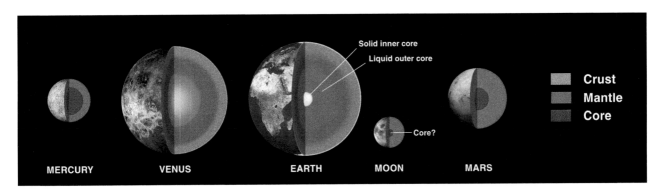

Figure 1.22 The likely interiors of the terrestrial planets showing their crusts, mantles, and cores. Image from NASA.

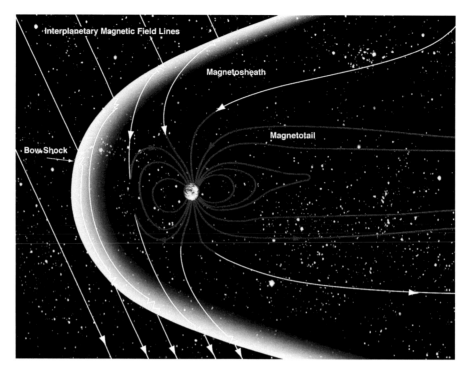

Figure 1.23   Solar wind interacting with Earth's magnetic field.
Image from NASA.

is a different story. Once the interior of Mars cooled and its protective magnetic field diminished, solar winds removed most of the atmosphere and surface water from the planet.

Earth's early atmosphere probably consisted mostly of carbon dioxide (CO_2) and nitrogen (N_2) along with water vapor emitted by volcanoes. Because carbon dioxide reacts readily with rocks, it slowly became incorporated in carbonate minerals, and so moved from the atmosphere to the solid part of Earth. Carbon dioxide also reacts with water and in part became carbonic acid (H_2CO_3) dissolved in water, moving from the atmosphere to Earth's hydrosphere. Nitrogen (N_2), however, is very inert. Except for lightning converting some of it into nitrogen oxides, or cosmic rays temporarily forming radioactive carbon-14, N_2 remained in the atmosphere and today it is the most abundant atmospheric gas. Besides nitrogen, atmospheric oxygen also increased over time. By 1.8 billion years ago, photosynthesis from blue-green algae had substantially consumed atmospheric carbon dioxide and produced substantial amounts of oxygen. Oxygen accumulated in Earth's atmosphere faster than oxidation of minerals could consume it, and today animal life survives on what was once a waste product of early terrestrial life.

1.6   What is Earth made of today?

The Milky Way, our galaxy, is 74% hydrogen and 24% helium, but Earth contains much less of these two elements (Fig. 1.24). Instead, Earth, like all the terrestrial planets, is dominated by silicate rocks and minerals. Subequal amounts of silicon, oxygen, iron, and magnesium make up more than 90% of Earth; significant but lesser amounts of aluminum, calcium, and other elements are present, and extremely small amounts of other elements are present too. As shown in Figure 1.24, these elements are not distributed uniformly because Earth over time has differentiated, and different elements have become concentrated in different Earth layers (crust, mantle, and core). The core is mostly iron, which explains why the composition of the entire Earth has so much iron compared with the crust and mantle.

Differentiation of Earth, and other planets into a crust, mantle, and core, was primarily the result of heating. Originally, the heat was mostly from radioactive decay of uranium, thorium, potassium, and other radioactive elements. These same elements produce heat today, but only about 20% as much as they did at the time of formation of the solar system, because they

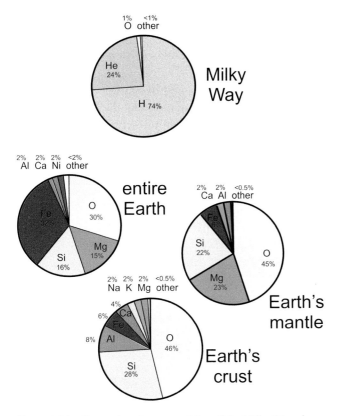

Figure 1.24 Comparing the composition of the Milky Way, the entire Earth, and Earth's crust and mantle.

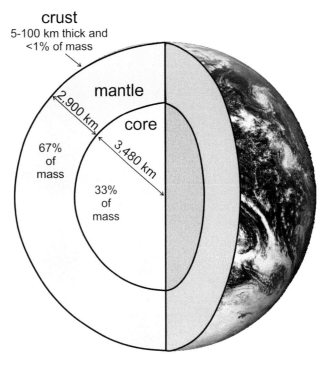

Figure 1.25 Earth's three principal layers have distinctly different compositions.

have been decaying and so disappearing for more than 4 billion years. Additional heat came from the ongoing gravitational contraction and from meteorite/asteroid impacts that converted kinetic energy to heat.

Because when first formed Earth was much hotter than it is today, and much of it was melted, the planet was squishy, not rigid. Over time, however, it cooled enough for rigid layers to form near the surface and this, along with continuing convection at depth, eventually led to the beginning of *plate tectonics* (see Chapter 2). Spreading began in the centers of ocean basins, subduction occurred at ocean margins, and Earth materials cycled from the surface to depth as Earth became further differentiated. It is unclear exactly when plate tectonics began, but much evidence suggests that ocean spreading and subduction were both happening at least 3.2 billion years ago.

Today we have a layered Earth akin to a hard-boiled egg (Fig. 1.25). The crust (egg shell) averages 30–50 kilometers (20–30 miles) thick in continental regions and 5–10 kilometers (3–6 miles) in oceanic regions. It is composed primarily of silicate rocks and minerals. The mantle (egg white) averages about 2900 kilometers

(1800 miles) thick and accounts for 84% of Earth's volume. It is mostly solid, but on a geological time scale it flows like a very viscous liquid. Like the crust, the mantle is composed mostly of silicates, but it contains less silicon and oxygen than the crust. The core (egg yolk) extends from the center to about 3480 kilometers (2100 miles). It is mostly iron (80%) and nickel (19%). The inner core is solid, and the outer core is molten.

Although Earth's mantle is about the same composition everywhere, this is not true of Earth's crust. Initially the crust may have had a single uniform composition, but after the development of continents and oceans, two distinctly different kinds of crust were present: *oceanic crust* and *continental crust* (Table 1.2). They have continued to evolve to the present day. The oceanic crust is relatively homogeneous and its composition is easily determined. The continental crust, however, is another story, because it varies in rock type both vertically and laterally. Estimates of the overall composition of the crust exposed at Earth's surface have been made by considering the distribution of different rock types and multiplying by their average compositions. Determining composition of the middle and lower crust is much more difficult. Some estimates have been made based on seismic velocities. Other estimates were based

Table 1.2   The compositions of the oceanic and continental crusts given in terms of chemical oxides.

	Oceanic crust	Continental crust
SiO_2	50.00	59.30
TiO_2	1.55	0.76
Al_2O_3	15.65	15.60
FeO^*	10.45	7.33
MgO	7.65	4.47
CaO	11.30	6.43
Na_2O	2.75	3.17
K_2O	0.18	1.79
total	99.53	98.85

*Total iron is reported as FeO. Data from Yanagi (2011).

on the types of magmas that come from the crust, and still others were based on rocks exposed in shield areas (which are assumed to represent the deep crust). Overall, the oceanic crust has the composition of a normal basalt. The continental crust is richer in silica and alkali oxides and poorer in iron and magnesium oxides. It has an overall composition that is equivalent to the volcanic rock *andesite* and the plutonic rock *diorite*. These and other igneous rock types are discussed in detail in Chapter 5.

Questions for thought—chapter 1

1. How can a smaller star by volume be more massive than a larger star by volume?
2. Why are many galaxies flat disks?
3. Most galaxies are moving away from us, but the Andromeda galaxy is not. If the Andromeda galaxy is moving toward us at about 402,000 km/hour and is currently 2,537,000 light years from us, predict when the Andromeda and Milky Way galaxies will collide. 1 light year is 9.461 x 1012 kilometers. Should you be worried?
4. Why is hydrogen the most abundant element in the Universe?
5. Why were electrons separate from atomic nuclei in the early Big Bang?
6. Why do massive stars have short life spans of millions of years?
7. Why will the Sun never become a black hole when it dies?
8. Why did the first stars that formed after the Big Bang not have any planets like the Earth?
9. How do gold and uranium form differently than carbon, nitrogen and iron?
10. Why are the orbits of the planets in our Solar System in one plane?
11. According to the solar nebula hypothesis, how does a planet form?
12. What are the major differences between a star, a planet and a dwarf planet?
13. Both Ceres and Eris are dwarf planets. Why is Eris icy, but not Ceres?
14. Why is Pluto no longer identified as a planet? Do you agree with that decision? Why or why not?
15. Why do Jupiter and the Sun have far more hydrogen gas than the Earth?
16. Why is iron more abundant in the core of the Earth than the mantle or crust?
17. Why does the Moon have more craters than the Earth when the Earth is a larger target with more gravity?
18. Originally, the Earth's atmosphere was probably mostly carbon dioxide with little or no oxygen just like the atmosphere of Venus is today. Now the Earth's atmosphere is 21% oxygen. What caused the concentration of carbon dioxide to decrease over the Earth's history and oxygen increase?
19. Oxygen is a very abundant element on the Moon and on planets that do not have oxygen-rich atmospheres. Where is the oxygen located and how is the oxygen associated with other elements?
20. Why is the Earth's interior still hot after 4.55 billion years?
21. How does the composition of the Earth's continental crust differ from the oceanic crust? And, how do both compositions differ from the composition of the mantle?

2   Earth Systems and Cycles

2.1   A Sand County Almanac

Aldo Leopold's essay "Odyssey" was first published in *Audubon Magazine* in 1942. It later became part of his book *A Sand County Almanac: And Sketches Here and There*, published in 1949, a year after Leopold passed away. Leopold's book of essays was little noticed when it first came out, but during the rapid rise of the environmental movement in the 1970s, it became a bestseller. Today, it is considered one of the seminal books of the American conservation movement. *Sand County Almanac* is sometimes compared with Henry David Thoreau's *Walden* and Rachel Carson's *Silent Spring*. "Odyssey" is one of the most compelling of Leopold's essays.

"Odyssey" is about an atom, named X, that starts in a rock outcrop (Fig. 2.1, top), where it had "marked time" in the limestone ledge since the Paleozoic seas covered the land. "Time to an atom locked in a rock," says Leopold, "does not pass." More than 250 million years after the rock formed, X began an epic journey, parts of which are shown in Figure 2.1, when a burr oak root cracked the limestone, and weathering and erosion freed the atom. During the next year, X became part of a flower, an acorn, a deer, and then a person. When the person died, X was interred in the ground until a bluestem root picked it up, and so X became part of a prairie. X became part of a leaf, then a mouse nest, then grama grass, and consequently was consumed by a buffalo before returning to the ground in a buffalo chip.

A prairie fire destroyed the prairie vegetation and reduced the buffalo chip to ash, and so erosion and runoff took their toll. X dallied for a while, but eventually was consumed by a gopher and later a fox, an eagle, and finally a beaver. When the beaver died and decayed, X was washed downhill into a stream (middle photo) and then into a river. After several detours and a prolonged journey, X ended up in the ocean (bottom photograph). Leopold concludes that "An atom at large in the biota is too free to know freedom; an atom back in the seas has

Figure 2.1   The beginning and the end of an atom's journey.
Photos from (top to bottom) Donvitocorleone and Ustill, Wikimedia Commons, and the National Park Service.

forgotten it. For every atom lost to the sea, the prairie pulls another out of the decaying rocks." Thus, according to Leopold, as complicated as it was, X's journey was not extraordinary. He only described one immense journey followed by a single atom, but (almost) all atoms on Earth have similar stories. The exceptions may be atoms of a few radioactive isotopes that decay quickly, but even they had a history before their demise, and many radioactive isotopes have very long lives.

Some natural materials persist unchanged for long times, but others do not. Organic compounds, for example, decompose rapidly when exposed to the elements. Inorganic materials, including most of the minerals that make up rocks, also decompose over time. When compounds, organic or inorganic, decompose or weather, the atoms that comprise them are released and recombine to form other Earth materials. So, everything in nature is eventually recycled. Rocks, water, gases in the air, and even people are part of endless natural cycles. People consume food, and their wastes reenter nature where they are broken down, used by other organisms, and perhaps eventually consumed again by people or other organisms. The water in your body, in a clay particle, in a stream in a forest, or in a cloud in the sky may once have been in the iceberg that sank the *Titanic*, perhaps drunk by President Abraham Lincoln, or maybe fallen as snow on the Himalaya.

Earth materials are diverse. They include many common elements, minerals, rocks of different kinds, soils, water, snow, and ice. These naturally occurring inorganic materials are resources used by all living things, and the materials are involved in many important and complicated processes. Some of the processes, such as the falling and subsequent melting of snow to produce runoff that later enters a river, are relatively fast and observable. Other natural processes, including the uplifting and erosion of mountains or the movement of continents, occur over millions and billions of years, so only their consequences can be observed by humans. This book is about Earth materials, and this chapter is about a few of the most significant processes that move atoms and other materials through natural "cycles" and from one setting to another.

2.2   The Earth system and my aquarium

2.2.1   Earth from space

Figure 2.2 shows two composite views of Earth, one of the eastern and one of the western hemisphere,

constructed by the National Aeronautics and Space Administration (NASA). Most of these images were created by combining data collected between 2001 and 2004 by the Moderate Resolution Imaging Spectroradiometer (MODIS) on NASA's Terra satellite. The city lights on the dark side of Earth come from the Defense Meteorological Satellite Program mission between 1994 and 1995, and the hints of topography are based on radar data collected by the space shuttle *Endeavour* in February 2000.

Figure 2.2   Composite views of Earth from space. Photos from NASA.

These images show a complex planet—and we are only looking at the outside. We see water in the form of clouds and oceans, as well as continents of different sorts. We also see vegetation in these renderings; we can't see the specific plants, but we can see different hues reflecting different kinds of land cover. If we had a very high-resolution telescope and could focus in, we would see individual plants, rocks, and minerals. We would see eroded geological materials that make up sediments and soils. Clouds, oceans, rocks, minerals, animals, plants, sediments, and soils—they all seem to be different and distinct entities, but they, and what is beneath the surface of the planet, are part of a single system—our planet Earth.

2.2.2 A simpler system

The Earth system is immense and complicated, but in my living room, I have a smaller system—an aquarium like the one shown in Figure 2.3. It contains about 30 gallons of water and, at last count, 26 fish of various kinds. Snails, several varieties of aquatic plants, and some algae make up the rest of the visible life forms. Unseen bacteria and other microorganisms are present, mostly in the gravel at the bottom of the tank. In large part, my aquarium operates independently of what is around it. I add food once a day and water when needed. Water evaporates, but otherwise nothing leaves the tank. Light and heat flow in, through the glass, but a significant amount of heat comes from a small aquarium heater in the water. The fish metabolize the food and excrete leftovers. The plants process carbon dioxide and, through photosynthesis, grow and release oxygen into the water, so the fish have something to breathe. The fish, returning the favor, add carbon dioxide to the water when they breathe. In the tank bottom, the microorganisms recycle fish and plant discards and return nutrients to the water for further use.

My aquarium is well balanced, overall, and has functioned the same way for several decades with no obvious changes. Yet, it is not static. The system components, both energy and matter, move between the water and organisms, changing form but never leaving completely. Individual atoms and compounds are part of continuous cycling processes that, as long as they are not disrupted, are unstoppable. These processes are vital characteristics of the aquarium, just as many Earth processes are vital characteristics of Earth. Some would say that the definition of a system includes not just its components or boundaries, but also the processes that take place within it.

2.3 Systems and scientific investigations

Humans' quest for knowledge is immense, but Earth's system is larger. Scientists cannot study everything

Figure 2.3 Model for a closed system: an aquarium.
Photo from indianlifestylez.com.

simultaneously and thus must focus investigations and study what is manageable. They put boundaries around the topic or region of study and perhaps develop expertise in a particular discipline. Investigators use the term *system* to refer to the part of the universe, or of Earth, that is the focus of their investigations. Things outside the system, things not considered, are called the *surroundings*. Generally, scientists choose system boundaries so there is some degree of isolation between what happens within and outside the system in the surroundings.

Systems range in size from atoms or single organs in an organism to an entire organism, and systems may be even larger—perhaps a rain forest ecosystem that contains different organisms, a continent, a planet, the solar system, or even the universe. It is tacit that people studying small systems focus on small details, and people studying larger systems often focus on larger features. Microbiologists may study viruses, fungi, parasites, or bacteria too small to see with the unaided human eye. Physicians may study human circulatory systems. Biologists may study a single kind of plant or perhaps a pond. Astronomers may study a remote planet or sometimes the entire solar system.

2.3.1   Classification of systems

The flow of energy and matter across natural system boundaries is highly variable, and scientists classify systems as *open*, *isolated*, or *closed*, according to whether and how the systems interact with their surroundings (Fig. 2.4). *Open systems* freely exchange both matter and energy with their surroundings. A living human being is an example of an open system. People get food, warmth, oxygen, and water from their surroundings and replace them with wastes and carbon dioxide. They also give off small amounts of heat that are absorbed by their surroundings.

An *isolated system* does not exchange matter and energy with its surroundings. Natural systems are never truly isolated, because some energy always flows in and out—by conduction, convection, or radiation—but some natural systems are more isolated than others. For example, in a laboratory, well-insulated calorimeters come very close to being isolated systems. In our homes, sealed thermos bottles are, perhaps, the best examples.

Closed systems exchange energy with their surroundings, but not matter. A well-sealed garbage landfill, when functioning properly, is an example of a closed

system. Small amounts of heat may pass in and out of a landfill, but the trash and waste stay there. If rain or groundwater percolate into a leaky landfill and escape into the surroundings, the system becomes open, and a potential environmental or health hazard. So, clay, plastic liners, and topsoil are used to keep the landfill a closed system.

Systems may have different degrees of openness and closedness. Over long times, Earth is an open system with spacecraft and small amounts of helium and hydrogen permanently leaving Earth, and dust and larger meteors being added. Overall, however, little matter flows in and out of the Earth system, keeping it almost closed. My aquarium, too, has some characteristics of

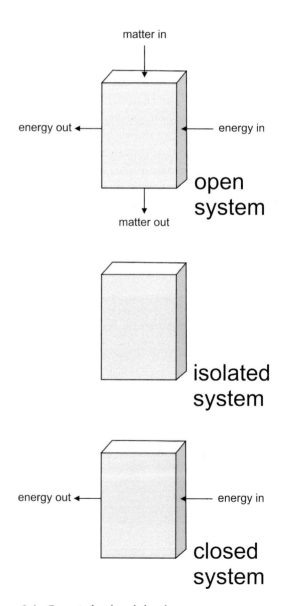

Figure 2.4   Open, isolated, and closed systems.

a closed system, but it could not survive without the addition of food every day, and probably not without the electricity that powers the heater, and thus its survival depends on being open.

Because Earth is an essentially closed system, our resources are limited. We can recycle water, aluminum, and some other commodities, but not petroleum and other fossil fuels. Once fossil fuel supplies are exhausted, it will be a long time before nature can replace them, too long even to contemplate. Other things, such as waste disposal, also pose problems for Earth's closed system. If we cannot recycle wastes, or effectively destroy them, we must store them somewhere on our planet. Today, launching garbage, or waste from nuclear reactors, into outer space is impractical. Thus, we have landfills, and we have potentially dangerous radioactive waste stored near every nuclear power plant.

Yet, although little matter moves between Earth and its surroundings, energy is a different story. Some heat derives from Earth's interior, but the Sun is our major energy source (Fig. 2.5). When solar radiation encounters Earth's atmosphere, some is reflected back into space by the atmosphere or by clouds. The remainder may be changed into heat and absorbed by the atmosphere or by Earth. Earth subsequently reemits some radiation, and that, along with incoming radiation that is reflected, may leave the Earth system and return to space.

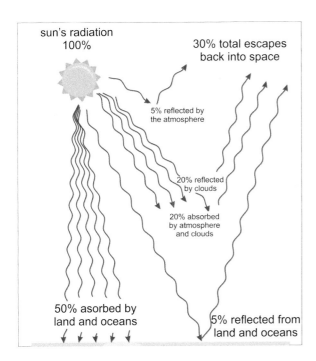

Figure 2.5   Earth's heat budget.

2.3.2 *Earth's spheres*

The most important Earth materials fall into one of four categories, or groups, of materials:
- water (and snow or ice)
- air
- living organisms
- rock and mineral materials (including sediments and soils)

Each of these four groups of materials is scattered in a spherical volume that is the same size and shape as Earth, and the groups are called the *hydrosphere*, *atmosphere*, *biosphere*, and the *geosphere*.

Sometimes scientists redefine, or subdivide, the spheres for particular purposes, perhaps so they can focus on some particular processes that have practical implications. Geologists, for example, separate the geosphere into the *lithosphere* (Earth's crust and uppermost mantle), the *asthenosphere* (a layer below the lithosphere), and the rest of Earth when they study plate tectonics (Fig. 2.6). Other scientists invoke the term *cryosphere* to refer to frozen matter, including ice, snow, glaciers, ice sheets, and frozen soils (permafrost), if they are concerned only about the effects of freezing. In agriculture and related disciplines, soils are usually identified as the *pedosphere*, elevating the importance of soils as parts of natural systems.

The spheres are distinctly different. The hydrosphere contains ice, water vapor, and liquid water. It is mostly H_2O; other compounds or elements are absent or present only in minor amounts. The lithosphere includes Earth's crust and the upper mantle and is the key player in the plate tectonic system that moves continents, and creates and destroys ocean crust. The lithosphere mostly consists of rocks and other materials that are dominated by the elements oxygen and silicon, with lesser amounts of aluminum, iron, calcium, sodium, and potassium. The mineralogy and chemistry of the crust and mantle are very heterogeneous. Continental crusts are more enriched in silicon, oxygen, sodium, and potassium, and depleted in magnesium, calcium, and iron, when compared with oceanic crusts. Beneath the crust, the mineralogy of the mantle varies with depth, mostly because different minerals are stable at different pressures. The upper mantle consists mostly of olivine, pyroxene, and garnet. At extremely high pressures, at depths of 400–1000 kilometers (250–620 miles), olivine transforms into a different mineral, spinel. And, as discussed later in this chapter, carbon, when present in the upper mantle, exists mostly as carbon dioxide or the

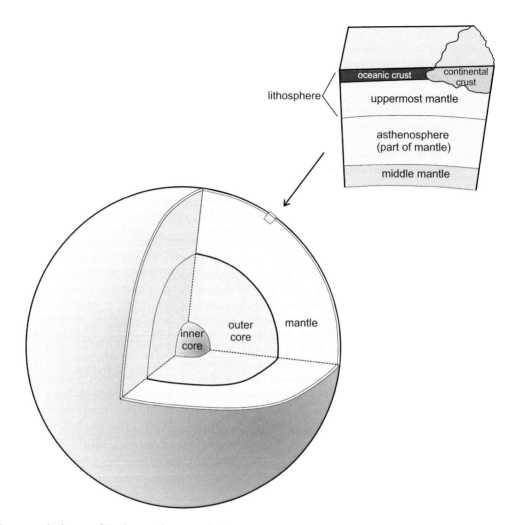

Figure 2.6   Some standard parts of Earth, according to geologists.

carbon mineral graphite, but at the high pressures of the deeper mantle, carbon is mostly in diamonds.

Volcanic eruptions bring some shallow mantle rocks to the surface where they may be collected and studied, but overall the chemistry of the mantle is poorly understood. Although the specifics are not well known, partial melting in the upper mantle probably produces chemical heterogeneities that are unlikely to occur in the lower unmelted mantle. Additionally, the lowest portion of the mantle near the iron-rich outer core is probably more enriched in iron than the rest of the mantle.

The pedosphere (the world's soils) contains the same elements as the crust, but also other elements that are essential for plant growth and sustaining life, including carbon and nitrogen. Carbon is the most important element in all organic matter, making it dominant in the biosphere. Oxygen, nitrogen, and hydrogen, too,

are essential components of living organisms and thus important in the biosphere. The atmosphere largely consists of nitrogen and oxygen gases, N_2 and O_2.

Hydrologists may study the hydrosphere, meteorologists the atmosphere, biologists parts of the biosphere, and mineralogists aspects of the geosphere. Yet, these four spheres are really not isolated. Matter and energy pass between them, and each system affects its surroundings. At Earth's surface, the solid part of our planet meets water, air, and life, and thus all the major spheres are intimately connected in this boundary region. Interactions between the hydrosphere and atmosphere control Earth's climate and weather, which affect the biosphere. Climate and weather also control erosion and deposition, thus maintaining or modifying environments occupied by living organisms. Changes in climate affect the sizes of glaciers and ice packs, which in turn affect sea level and coastal environments.

2.3.3 Broader views

For practical purposes, most scientists study systems of small size compared with the spheres mentioned above. Perhaps they study a single lake or pond, a volcano, a landslide zone, or a rock outcrop. Alternatively, they may broaden their views and study lakes, ponds, volcanoes, landslides, or outcrops of some particular type—larger systems that contain subsystems. The different subsystems are not identical; they vary in chemical and physical properties from others of the same type. No two forests, ponds, landslides, or outcrops are exactly alike, and although scientists tend to focus their investigations, what they learn in one study often has broader implications. Once specific bits of nature are adequately understood, they may be combined with other bits, leading to more holistic understandings.

Sometimes, studies of small systems are broadened and applied in different venues. As an example, a geologist may study certain types of rocks, sediments, or soils at a given location, such as granites in the Sierra Nevada Mountains of California. Once the chemistry, mineralogy, origins, and other characteristics of the granites are understood, the information may provide a guide for study of granites in other places. Subsequent studies may be broadened further to consider different kinds of rocks associated with the granites, or perhaps how granites weather and erode to produce sediment and soil. From there, a biologist may use the information about sediments and soils as a foundation for studying the nature and distribution of native plants in the Sierra Nevada Mountains. Thus, scientific studies are connected, and often one leads to another.

2.3.4 System reservoirs and fluxes

One important aspect of any system is the way in which different components are distributed within it. *Reservoirs* are locations or "compartments" in a system where energy and matter of any sort, including organisms, may reside temporarily or for the long term. Figure 2.7 shows the principal reservoirs for Earth's water: oceans, surface water or groundwater on land, and the atmosphere. The way this Figure is drawn is somewhat misleading, because the oceans contain 1400 million cubic kilometers of water—accounting for 97% of the water on Earth. The other water reservoirs are dinky in comparison.

Reservoirs for Earth materials include water bodies, rocks or magma, sediments or soil—any part of the Earth system that can hold matter or energy. Within a system, matter and energy flow between reservoirs, but the rates of flow are variable. Consider, for example, a water molecule in an ocean (reservoir). It may evaporate to become part of a cloud (a different reservoir) and return to the ocean if it rains. This process is rapid and commonplace. If the rain occurs on land, that same molecule may infiltrate the ground and become part of a groundwater reservoir, where it may remain for millions of years.

The rate of flow between reservoirs, and also between two systems, is termed the *flux*. Figure 2.7 has arrows showing the fluxes between different water reservoirs; the size of the arrows and numbers allow comparison of flux values. Common fluxes involving water include evaporation, precipitation, runoff, and infiltration. Most water simply cycles back and forth between the oceans and the atmosphere.

Fluxes, however, are not restricted to water. Carbon dioxide moves from Earth to the atmosphere and back again. Organic compounds and inorganic nutrients move through Earth's biological systems. Magmas flow within Earth and sometimes at Earth's surface as lava or some other volcanic material, moving inorganic materials from one reservoir to another and also moving heat. Additional heat flows between space, Earth's atmosphere, and the solid Earth.

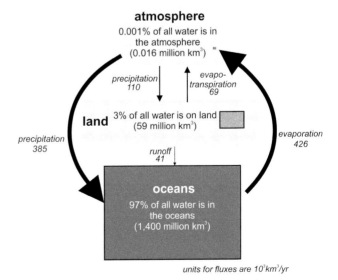

Figure 2.7 Reservoirs and fluxes for the main reservoirs of Earth's hydrosphere.

Data from The Woods Hole Oceanographic Institute.

Aldo Leopold's atom started its journey in limestone, where it resided for more than 250 million years. Subsequently it traveled to different reservoirs—including animals, plants, soils, and water. The average amount of time that a molecule, atom, or ion of one kind spends in a reservoir is called the *residence time*. We can calculate or measure residence times for any element or compound in the atmosphere, in an individual lake, and in any other reservoir. These times are important because they influence how stable a reservoir is and how fast a system can adapt or change. If we are concerned about a particular pollutant, we want to know how long it will remain in a particular place and to where it might move over time.

The residence times of different substances in different reservoirs vary, and not all substances will follow the same path. For example, the residence time of dissolved aluminum in seawater is estimated at about 100 years. Thus, *on average*, an aluminum atom, molecule, or ion remains dissolved in seawater for about 100 years before leaving the ocean by precipitating onto the ocean floor or by another process, including consumption by an aquatic animal. Unlike aluminum, sodium is very soluble and stable in seawater. It usually stays in solution for a long time and does not readily precipitate as a sodium compound. Therefore, it is not surprising that the residence time of sodium in seawater is great, about 50 million years on average. In contrast with aluminum and sodium, phosphorus in seawater is often involved with biological cycles and so moves seasonally.

Fluxes and residence times for many geological processes are poorly known, because processes that take place within Earth cannot be studied directly. Additionally, the times involved in many Earth processes are immense—often millions of years—and the fluxes may be very small compared with the total volume of material in a reservoir. However, fluxes and residence times involving Earth surface or near-Earth surface processes, especially those that are relatively rapid, can be measured or calculated. For example, although there is some variation, residence times for water in different reservoirs are reasonably well known (Table 2.1). These times range from days (clouds) to thousands of years (oceans, groundwater, ice). Scientists determine residence times, to get values such as those in Table 2.1, in many different ways. For instance, an intriguing study of groundwater in Maryland was based on the pollutants in water produced by natural springs. The conclusion was that average groundwater residence time in near-surface aquifers was about 20 years in the eastern

Table 2.1   Residence times for water in different reservoirs.

Reservoirs	Residence time
oceans	4000 years
lakes	10 years
swamps	1–10 years
rivers	2 weeks
soil pores	2 weeks to a year
groundwater	2 weeks to 10,000 years
ice caps	10–1000 years
atmosphere (clouds)	10 days
biosphere (organisms)	1 week

Data modified from *Groundwater* by Freeze and Cherry (1979).

part of Maryland, but groundwater in many other places is much older.

2.3.5 Cycles

All Earth materials move through Earth's systems over time. Some radioactive elements decay, forming new elements, but most elements have the potential to move from one reservoir to another and eventually return to where they started. The processes involved are called *geochemical cycles* or *biogeochemical cycles*, depending on whether living organisms are involved. Scientists have developed models for cycling involving many different elements, including carbon, calcium, nitrogen, sulfur, mercury, and lead.

Nitrogen provides a good example of a geochemical cycle. It makes up 78% of Earth's atmosphere, and small amounts of nitrogen can be found in soil, in plants and animals of many kinds, and dissolved in the oceans. In the nitrogen cycle, shown in Figure 2.8, nitrogen takes several forms. Most of the nitrogen is in the atmosphere in the form of N_2 gas, with lesser amounts of nitrous oxide (N_2O). Within Earth, nitrogen exists mostly as ammonium (NH_4^+), nitrate (NO_3^-), or nitrite (NO_2^-). It is also present in some organic molecules in both plants and animals. The cycle that relates all these different forms of nitrogen is complicated and very important, because nitrogen is a key nutrient essential for plant growth.

Although it dominates the atmosphere, the nitrogen in the air cannot be easily used by plants because it is N_2 gas, a very strongly bonded molecule of two atoms that plants cannot break apart. Through a process called *nitrogen fixation*, however, N_2 can be converted to ammonia or a nitrate, different nitrogen species with weaker bonds that can be used by plants. In nature,

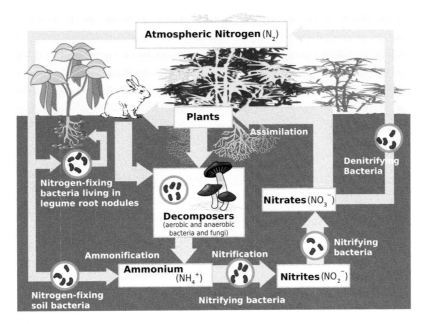

Figure 2.8 The nitrogen cycle.
Figure from NOAA, Wikimedia Commons.

nitrogen fixing is done mostly by bacteria in soil, commonly associated with roots of beans and other legumes, that produce nitrate. So, in natural systems, most plants get nitrogen from nitrate. Insufficient nitrate can limit plant growth and agricultural food production, but farmers can overcome this by putting ammonia-based fertilizers on their fields.

Cycling of compounds is more complicated than cycling of individual elements, because many compounds decompose or react to form different compounds over time. Some compounds, however, are more resilient than others. While water molecules are rarely created or destroyed in the hydrologic cycle, organic compounds, including plant and animal material and compounds such as methane or benzene, are unstable in the presence of oxygen and most organisms. They tend to decompose rapidly in nature. Additionally, some elements or compounds may become stalled in a particular place and form, not cycling at all. So, cycles are descriptions of how materials could move through Earth systems, but they are models that may not apply precisely to every atom or molecule in every system.

Earth processes include an uncountable number of different cycles. Some are fast and some are slow, but the billions of years of Earth history provide ample time for cycling to occur and for average rates to be determined.

Some of the more important cycles of geological interest include the water cycle, the carbon cycle, the rock cycle, and plate tectonics. These cycles are not completely separate, just as Earth's systems are not completely separate. The different cycles interact and share both energy and matter.

Understanding Earth's cycles is important for many reasons. The carbon cycle, for example, has great significance because it is at the heart of all living organisms. Additionally, today the human-modified carbon cycle has implications for climate change and global warming. Understanding plate tectonic cycles is important when we explore for ore deposits and because plate tectonics is responsible for destructive earthquakes and volcanic eruptions. Other cycles are important because they determine the nature and fate of pollutants and other toxics in the environment.

Aldo Leopold's atom started in a rock and, eventually, made its way to the ocean, and Leopold stopped the story there. Yet, there is no reason why his atom could not precipitate from the ocean water and become part of a new rock that eventually is uplifted and exposed on dry land—and then begin a new journey to the ocean. This kind of thing has happened countless times during Earth's lifetime, but the process is so slow that when we look at an outcrop, it does not cross our minds that it is still in the process of evolving.

2.4 Rocks and the Rock Cycle

Rocks consist of minerals or related solid materials, including volcanic glass, fossils, and sometimes organic compounds, compacted and cemented together. Some rocks consist only of a single mineral; for example, many sandstones are nearly 100% quartz. Other rocks, such as coal and glass-rich volcanic rocks, may contain very few minerals. Most rocks, however, contain several different mineral components. Consequently, unlike minerals, most rocks are heterogeneous.

Rocks fall into one of three fundamental groups: *igneous*, *sedimentary*, and *metamorphic*, discussed in more detail in later chapters. Figure 2.9 shows examples of each kind; all three specimens are about 12 centimeters (5 inches) across. The top rock is an igneous rock called *tonalite*. It contains the minerals black biotite, and light-colored quartz and plagioclase. The banded rock in the center is a deformed *gneiss*, a metamorphic rock that contains mostly biotite, quartz, garnet, and plagioclase. The bottom rock is a *conglomerate*, a type of sedimentary rock. It contains large easy-to-see quartz pebbles surrounded by finer quartz, feldspar, and mica.

Igneous rocks form from the cooling and solidification of magmas, most of which originate deep in Earth. Magmas, and thus igneous rocks, originate by whole or partial melting of a *parent rock*. The elements silicon and oxygen dominate almost all magmas. Depending on the depth in Earth, typical magmas have temperatures between 700 °C and 1300 °C (1300 °F to 2400 °F). If a magma cools and solidifies underground, the result is an *intrusive* (also called *plutonic*) igneous rock. If a magma reaches the surface, it becomes lava and will cool to form an *extrusive* (also called *volcanic*) rock. Some extrusive rocks form from lava *flows*, but if an eruption is very explosive, magmas may become airborne fragmental material, including fine ash and sometimes coarser material, which settles to produce *air-fall* volcanic rocks.

Weathering and erosion of any kind of rock can produce sediment composed of pieces of the original rock and, often, of new minerals produced during weathering. Additionally, some soluble components of the original rock may dissolve in water and subsequently be carried away. Some sediment is deposited close to where it forms, but often it is carried long distances by water, wind, ice, or gravity before deposition. Sediment may also be deposited if dissolved material precipitates from water or if remnants of aquatic animals settle to the bottom of a stream, lake, or ocean. Thus, sedimentary rocks are formed by deposition and subsequent cementation of particulate or dissolved material derived from parent rocks. Although sedimentary rocks are widespread on Earth's surface, covering about three quarters of all continents, they are only a relatively thin veneer on top of mostly igneous and metamorphic rocks.

Metamorphic rocks are igneous, sedimentary, or preexisting metamorphic rocks that have been squeezed, heated, or otherwise substantially altered without melting. Most metamorphic rocks that we see are formed over a range of pressures up to, perhaps, 10,000 atmospheres. Higher pressure rocks generally only form deep within Earth (>30 km) and, as a consequence, are rare at Earth's surface. Metamorphic temperatures are generally between 200 °C and 750 °C (400 °F to 1400 °F) because at higher temperatures, melting is likely to occur, creating a magma and thus an igneous rock. Heat and pressure, typically due to burial or mountain building, are the main agents of metamorphism. Chemically active fluids, too, may cause metamorphism. Less

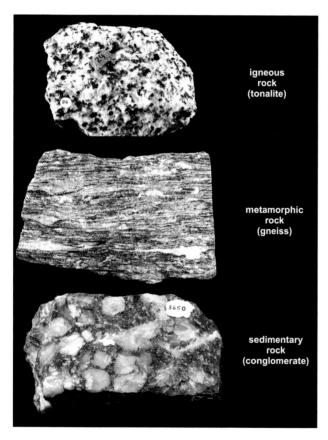

Figure 2.9 Examples of igneous, metamorphic, and sedimentary rocks.

igneous
rock
(tonalite)

metamorphic
rock
(gneiss)

sedimentary
rock
(conglomerate)

commonly, shearing associated with movement of active faults produces a special kind of metamorphic rock called a *cataclastic rock*.

Although some rocks may remain relatively unchanged for billions of years, many rocks, like other natural materials, undergo change over time. Sometimes, a preexisting rock melts to produce magma. In other cases, nature produces sediment when rocks weather and erode. So, as shown in Figure 2.10, igneous or sedimentary rocks may derive from preexisting rocks. There are other such processes, too, that change one kind of rock into another. In fact, any preexisting rock can be changed into a new rock if it

- melts, partially or entirely, to produce magma that ultimately becomes an igneous rock.
- weathers and erodes to produce material that later becomes sediment and, perhaps, a sedimentary rock.
- changes texture or mineralogy in reaction to heat and pressure, or chemically reactive fluids, to produce a metamorphic rock.

All these processes make up what geologists call the *rock cycle*, commonly depicted as shown in Figure 2.11. The depiction and name are both misleading because few rocks go through the entire cycle. Nonetheless, the underlying principles are valid: igneous rocks can become different igneous rocks or may become sedimentary or metamorphic rocks. Metamorphic rocks may change into different kinds of metamorphic rocks or into igneous or sedimentary rocks. And sedimentary rocks may weather and erode to produce new sedimentary rocks, may melt and become igneous rocks, or may be metamorphosed.

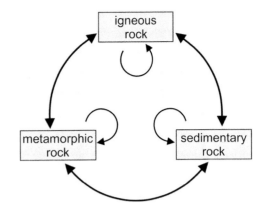

Figure 2.11 The rock cycle.

2.4.1 At Earth's surface

Minerals in metamorphic and igneous rocks generally form at high temperatures and under relatively dry conditions. Within Earth, they form under low-oxygen conditions. Consequently, in the presence of water, oxygen, and low temperatures at or near Earth's surface, most metamorphic and igneous minerals are unstable and weather rapidly. Over time, iron-bearing minerals oxidize. Olivines, pyroxenes, sulfides, other minerals, and volcanic glass react with oxygen and water and break down into iron oxides and hydroxides, magnesium carbonate, calcite, clay minerals, and other weathering products. Wind, running water, and ice physically break large rocks into smaller pieces, generally termed *clasts*. Erosion by water, ice, and wind transport the clasts to new locations, where they accumulate as *clastic sediments*.

At Earth's surface, physical weathering and erosion are the major contributors to the rock cycle. These processes are powered, in large part, by solar energy and gravity. The sun causes water to evaporate, which ultimately leads to precipitation, and water is the key catalyst causing rocks and minerals to weather and turn into different compounds. Differential heating by the sun also causes winds to blow, and blowing wind, flowing water, and gravity lead to erosion. Gravity causes water and glaciers to flow, causing more erosion that moves weathered materials from one setting to another until they are deposited as sediments.

Sedimentary deposits accumulate in many different environments, both *subaerial* (formed in open air at Earth's surface) and *subaqueous* (deposited under water). Subaerial deposits form, for example, on slopes beneath outcrops, in valley bottoms, or in dunes. Subaqueous

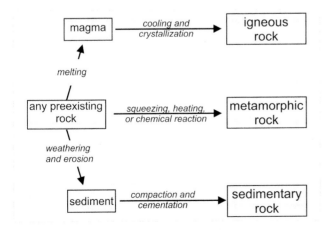

Figure 2.10 How rocks become new rocks.

Figure 2.12   Loose olivine crystals that have weathered from basalt in Hawaii.
Photo from Broken Inaglory, Wikimedia Commons.

deposits collect on river, lake, or ocean bottoms. Any of these different kinds of deposits may evolve and eventually become a sedimentary rock, with physical and mineralogical characteristics that reflect where it formed, and sometimes how it was transported.

Regardless of the nature of the parent rock, weathering commonly produces *detritus* (a general term for lose material produced by erosion) composed of particles of recognizable rock or of individual mineral grains. The photograph (Fig. 2.12) shown here is of a weathered surface of basalt (black). The green grains on top are individual crystals of olivine that were once in solid rock. Besides containing clasts of rock and mineral fragments, some sediments include biogenic debris, for example bones or shells, or occasionally vegetation litter. Other sediments contain calcite ($CaCO_3$), gypsum ($CaSO_4 \cdot 2H_2O$), or other compounds that dissolved in water during weathering and subsequently reprecipitated. Thus, sedimentary material is highly variable.

2.4.2   Lithification

Sediments of all sorts are deposited on dry land, in lakes or streams, or on ocean bottoms. As gravity deposits new sediment on top of old sediment, thick sedimentary layers, up to thousands of meters thick, may form. Silt, sand, mud, gravel, soil, and biogenic debris such as seashells on a beach are some more common constituents. However, the nature of a sediment depends on the original parent material and on climatic factors, including temperature and precipitation, and on

biological activity. Landscape and topography, too, affect the thickness and nature of sedimentary layers. The deposits remain sediments until they are consolidated and lithify (harden), at which point they become sedimentary rocks.

Some sediments, such as salt deposits or carbonate precipitates, may turn into sedimentary rocks (evaporites and limestones, respectively) at or near Earth's surface. However, most sediments, especially clastic sediments, are converted to rocks through deep burial during *lithification*, a process also called *diagenesis*. As lithification occurs, the weight of water and overlying sediments causes loose sediments to compact. At the same time, water flowing through pore spaces between grains can deposit mineral material, and individual clasts and grains under pressure may become cemented together. So, sand may turn into sandstone, and mud may turn into shale. These processes occur at depths in Earth as great as several kilometers and at temperatures up to about 200 °C.

2.4.3   Soil

Soil is a loosely defined term that refers to sediment and detrital organic matter that is exposed at Earth's surface. So, mud at the bottom of a lake is not soil, but if the lake dries up, the mud becomes soil. Soils consist of layers, or horizons, of fine-grained rock and mineral materials. Gases and liquids occupy spaces between grains, and living organisms are commonly present. Most soils contain organic material, but the amounts are variable, ranging from minor to dominant. The soil-making process is dynamic, in many settings being a continuous work in progress, during which once solid rock is transformed into a substrate that can support plant and animal communities. Most soils evolve over long times in spurts as they experience cycles of wetting, drying, and plant growth. Further discussion of soil is found in later chapters.

2.4.4   Deeper in Earth

As shown in Figure 2.13, both temperature and pressure increase with depth in Earth. The curves that depict temperature increase with depth are called *geotherms*. The temperature increase is due to heat flowing from Earth's center. Most of the heat is left over from the time of Earth's formation, especially the formation of the core, 4.5 billion years ago, but some derives from decay of radioactive elements (Chapter 1). Because heat flow is greater in oceanic regions compared with continental regions, temperatures increase faster with depth in ocean areas (blue geotherm in Fig. 2.13) compared with continental areas (black geotherm). Pressure increases with depth in Earth the same way that pressure increases when you swim to the bottom of a swimming pool—the deeper you go, the more mass there is above you pushing down.

Sediments and rocks that are deeply buried on continents, or that are carried down in subduction zones, eventually reach depths where temperatures and pressures are great enough for metamorphism to begin. At these temperatures and pressures, and especially in the presence of water, chemical reactions occur that produce new minerals and change rock textures. Thus, new metamorphic rocks are born.

At temperature greater than 650 °C, some minerals will melt; other minerals require temperatures significantly greater. The exact melting temperature depends on the particular mineral and the depth in Earth. The presence of water or other fluids can reduce melting temperature, but no matter the reason, when melting occurs, metamorphism gives way to igneous processes.

Sometimes a rock may melt completely, but often only partial melting occurs, because most rocks contain more than one mineral and different minerals melt at different temperatures. So, some rocks contain relict (leftover) minerals that did not melt and veins or patches of newly produced magma. When such rocks cool, the result is a *migmatite*, a rock that contains two different distinguishable components. Figure 2.14 shows a migmatite from western Norway. In migmatites, the minerals that did not melt are equivalent to metamorphic minerals, and the

Figure 2.14 A typical migmatite.
Photo from Simm Sepp, Wikimedia Commons.

minerals that formed from crystallization of partial melts are igneous minerals.

During tectonism (mountain building), portions of Earth's crust are uplifted, creating mountain belts with high relief. This process results in topography that is only temporary. Erosion and gravity move material from high elevations to low elevations and, in the process, expose rocks that were once deeply buried. The removal of large amounts of material by erosion causes Earth's crust to move upward, and thus even more rocks are exposed that will eventually erode. In this way, igneous and metamorphic rocks, which formed deep in Earth, become sediment that travels to lower elevations where it is deposited. Sometimes, due to topography, sediment accumulates on land, but ultimately, much of it makes its way to ocean basins where it is deposited and may eventually become a sedimentary rock. So, tectonism is a key process that drives the rock cycle.

2.5 From continental drift to plate tectonics

2.5.1 *Continental drift: A hypothesis that led to the theory of plate tectonics*

The hypothesis that continents move—like islands through Earth's crust—was first proposed by Dutch geographer Abraham Ortelius in 1596. Ortelius said that North America, Europe, and Africa were all once a single entity that later became separated by earthquakes

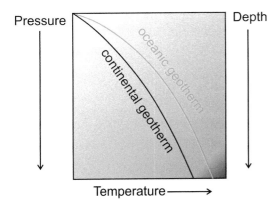

Figure 2.13 Continental and oceanic geotherms.

and floods. In the following 350 years, geographers, explorers, and others remarked on the similarity of coastline shapes on maps, in particular how Africa and South America seem to fit into each other like a jigsaw puzzle. While these other philosopher scientists echoed Ortelius's ideas, it was not until 1912 that a more comprehensive hypothesis, with much supporting evidence, was developed. At that time, Alfred Wegener, a German meteorologist, proposed a model for *continental drift.*

Wegener wrote two major articles about continental drift, summarizing evidence and concluding correctly that Europe, North America, South America, Asia, Africa, Australia, and Antarctica were once (200–300 million years ago) together as a single "super" continent, shown in Figure 2.15, that eventually broke up, creating the present-day continents and the Atlantic Ocean.

Some of the key evidence supporting this history included the following:

* the way continents fit together like a jigsaw puzzle
* the way different kinds of rock formations match on opposite sides of an ocean and would line up if the ocean was absent
* the similarity of fossils on different sides of the same ocean
* the way fossil organisms found on some continents are inconsistent with present-day climates

Wegener called the ancient supercontinent *Pangaea* (Greek for "all lands"), and the super ocean *Panthalassa*. Although Wegener published his ideas before radiometric dating was well developed, today, based on radiometric age dates, we know that Pangaea broke apart beginning about 250 million years ago.

Wegener had good evidence to support the idea of continental drift and the (former) existence of Pangaea, and the evidence he used is still valid today. However, Wegener could not explain how continents could actually move, and geophysicists, especially, questioned how continents could plow through ocean crust. It took about 50 years for most of the apparent problems and inconsistencies of continental drift to be resolved, and the result was the *theory of plate tectonics* that is accepted by all Earth scientists today.

The term *tectonics* refers to any significant and large-scale processes that change the structures of Earth's crust. *Plate tectonics* refers to those processes resulting from interactions between slabs of Earth's crust and mantle called *lithospheric plates*. Plate tectonics is the most significant recycling system on our planet and the main driving force for the rock cycle and many other important geochemical cycles.

2.5.2 Development of the theory of plate tectonics

2.5.2.1 Mapping the ocean floor

In the 1940s, scientists began to collect data that would eventually transform the hypothesis of continental drift into the theory of plate tectonics. In some ways, World War II and the Cold War are responsible for the key steps in the development of this theory.

During World War II, the U.S. Navy used sonar to find enemy submarines. Sonar waves would reflect off solid objects and return to the ships. By knowing the velocity of the sonar waves through seawater, the depth to the solid object could be calculated. Most often, the sonar waves simply reflected off the ocean bottom, providing a picture of submarine topography. Figure 2.16 shows the topography of the ocean floors, based on a great deal of data collecting.

With the coming of the Cold War, the Navy used this technology to map the ocean floor. They wanted information crucial for submarine navigation and for keeping track of where enemy submarines might hide. In the process, they found deep trenches in the Pacific Ocean—most near continental margins. They also found the Mid-Atlantic Ridge, a long underwater mountain chain that runs down the middle of the Atlantic Ocean (Fig. 2.16). Technology always gets better, and improvements in sonar technology today

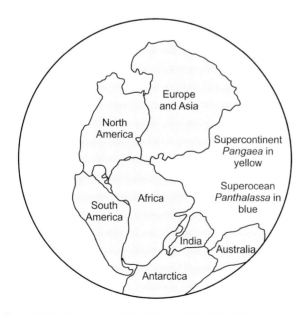

Figure 2.15   Pangaea and Panthalassa, 200–300 million years ago.

Figure 2.16 Continent and ocean floor topography.
Image from NOAA, Wikimedia Commons.

permit better mapping, including sometimes identification of rock and sediment layers in the ocean floor that may contain oil or natural gas.

2.5.2.2 Global seismicity

Beginning in the 1950s, scientists developed seismographs and seismic systems in order to monitor atomic weapons testing. Although hundreds of seismometers were already scattered around the globe, the technology was inconsistent and data recording incomplete. So, the World-Wide Standardized Seismographic Network (*WWSSN*) was developed to improve and standardize data collection and to use the data for scientific study of Earth. By 1967, the WWSSN had more than 120 stations in 60 different countries.

Besides keeping track of nuclear weapons development, the seismographs allowed geologists to plot the intensities, locations, and depths of earthquakes. Geologists soon realized that the distribution of earthquakes was not random. Earthquakes occur in linear zones, with depths that vary depending on the zone. The map in Figure 2.17 shows the locations (black dots) of all earthquakes with a magnitude of 4.5 or greater that occurred between 1963 and 1998. The linear features established by maps such as this one correlate with mid-ocean ridges (topographic highs) and subduction

zones (topographic lows) found during seafloor mapping. These features were eventually identified as being boundaries between *tectonic plates*, portions of the outermost Earth that move independently of each other.

2.5.2.3 Global volcanism

Geologists also studied volcanoes and found that volcanoes, like earthquakes, do not occur in random places. The red dots in Figure 2.18 show where volcanoes occurred in the recent geologic past. Most volcanoes occur in the same general places as earthquakes—in linear belts associated with tectonic plate boundaries. Many other volcanoes are not visible at the ocean surface but are known, based on studies of ocean bottom topography, to occur at plate boundaries beneath the oceans. Today, the vast majority of the volcanos that we see at Earth's surface are in the Ring of Fire that encircles the Pacific Ocean.

2.5.2.4 Earth's magnetic field and paleomagnetism

Long-term changes in Earth's magnetic field also provided important evidence for plate tectonics. Like a bar magnet, Earth has north and south magnetic poles (Fig. 2.19), although the magnetic poles do not coincide exactly with Earth's rotational axis. Earth's magnetic

Preliminary Determination of Epicenters
358,214 Events, 1963 - 1998

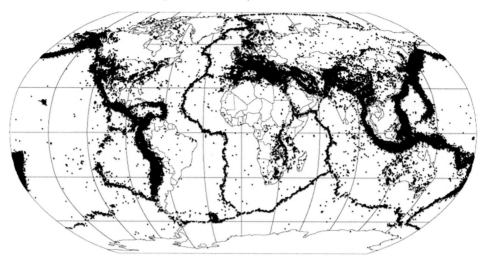

Figure 2.17   Locations of earthquake epicenters, 1963–1998. Map from NASA.

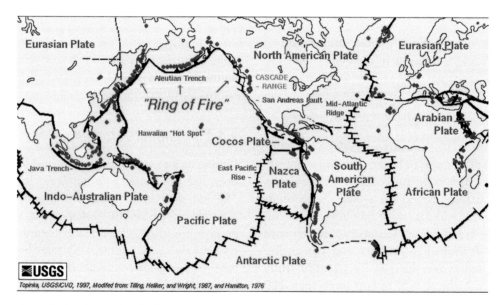

Figure 2.18   Locations of volcanic eruptions in the recent past. Map from the USGS.

north-south poles are currently inclined about 11.5° from the rotational axis, but the poles have shifted over time. These poles create a magnetic field (curving lines with arrows in Figure 2.19) that is parallel to Earth's surface near the equator and perpendicular near the poles.

Rocks that contain iron become slightly magnetized at the time they form. The oceanic crust largely consists of iron-rich volcanic rocks called *basalts*. During

eruptions, as iron-rich minerals crystallize and cool in basaltic lavas, they orient, like compass needles, with Earth's magnetic field. Once the lava entirely cools to basalt, the oriented iron minerals are frozen in place. Similarly, when continental lavas and magmas solidify, the iron minerals in them also orient with the current magnetic field of Earth and are frozen in place. Magnetic particles in ocean sediments may record similar

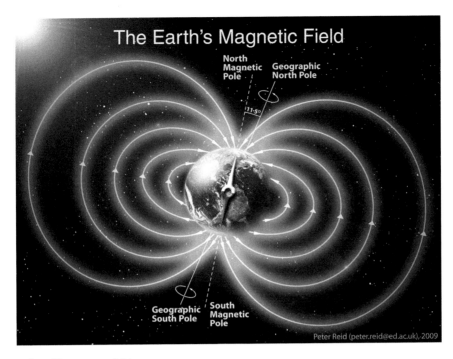

Figure 2.19　Depiction of Earth's magnetic field.
Drawing from NASA.

information. As magnetite and other magnetic minerals fall through ocean water and settle to the ocean bottom, they preferentially align with the magnetic field and leave a permanent magnetic record in the sediments. So geologists can collect rock samples from both continental and oceanic areas and measure the direction of magnetism, and thus determine the orientation Earth's magnetic field at the time the rocks formed.

The study of the magnetic fields recorded by rocks is called *paleomagnetism*. Paleomagnetic studies reveal not only the direction to the magnetic pole, but also the distance to the pole. So, by studying ancient rocks, scientists learn where Earth's north and south magnetic poles were at different times in the past. Geophysical considerations indicate that Earth can only have a single set of north and south poles at a given time, and paleomagnetic studies showed that the magnetic poles, while staying near the rotational axis, have wandered slightly (moved around some with respect to the rotational axis) over geological time. The change in pole positions is called *apparent polar wander*.

However, the first paleomagnetic studies, in the late 1950s and early 1960s, led to some contradictory results. Investigations of ancient rocks in Europe, North America, Africa, South America, and elsewhere did not all yield the same polar wander paths. For example, the left half of Figure 2.20 shows apparent polar wander paths for the north magnetic pole over the past 500 million years. Studies based on rocks in Europe and in North America do not give the same results. Although the *apparent polar wander* paths do not match, the inconsistencies were resolved in the mid-1960s when geologists recognized that continents can move independently of each other. If the continents are moved so that the pole paths overlap (Fig. 2.20, right side), the resulting continental assembly matches Pangaea, the supercontinent hypothesized by Wegener and others (Fig. 2.15). Thus, paleomagnetic studies confirmed the former existence of Pangaea.

Paleomagnetic investigations also produced other information of fundamental importance. Surveys of magnetism recorded by basaltic rocks in the ocean floors revealed that different fields were recorded by rocks of different ages and by rocks in different places. These variations occurred because Earth's magnetic field periodically changes polarity because of changes in the flow of molten iron-rich material in Earth's core. When this happens, Earth's north and south poles switch, and a compass needle that once pointed toward one pole will point toward the other. Thus, the magnetic fields recorded by rocks change. The last long-term magnetic field reversal occurred about 750,000 years ago.

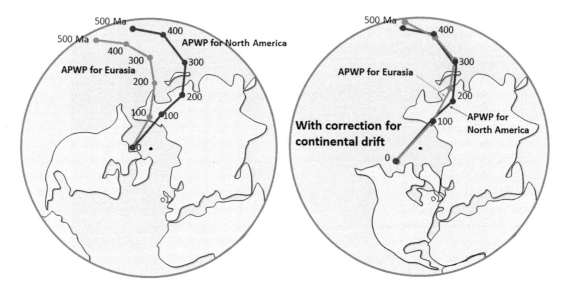

Figure 2.20   Apparent polar wander paths (APWP) and corrected polar wander paths (right). Graphic from Stephen Earle.

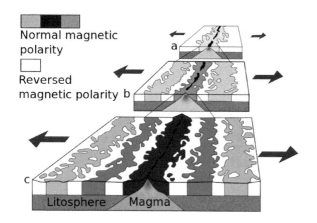

Figure 2.21   Magnetic anomalies near a mid-ocean ridge. From USGS.

Scientists discovered that the ocean floor contained many magnetic "stripes" caused by changes in the polarity of Earth's magnetic field over time (Fig. 2.21). The stripes are regions of ocean floor basalt that record normal (equivalent to today's) and reversed magnetic pole positions. They are parallel to, and symmetrical about, mid-ocean ridges, and are older farther from the spreading centers (Fig. 2.22). These stripes, called *magnetic anomalies*, developed because, as newly formed seafloor moved away from spreading centers, Earth's magnetic field alternated between normal and reversed polarity. As it cooled, the new seafloor recorded the direction of the field, and symmetrical magnetic stripes formed on either side of the mid-ocean ridge.

2.5.2.5   *Confirming evidence*

During the 1960s, two exploration ships, *Glomar Explorer* and *Glomar Challenger*, began drilling and collecting rock samples from the deep ocean floor, including regions near the Mid-Atlantic Ridge. Radiometric dating and microfossils showed that the rocks at spreading ridges were relatively young and that the ocean floor farther from the spreading centers was older. In the image shown in Figure 2.22, the youngest ocean crust is in red (0 years old at mid-ocean ridges) and the oldest in blue (180 to 200 million years old near some continents). The same studies also revealed that only thin layers of sediments covered basalts near mid-ocean ridges, while thicker layers covered those farther away. Rocks farther away had more time to be covered by the shells of microscopic organisms, ocean water precipitates, and sediments that settled through ocean waters. Thus, the Glomar studies of the ocean floor further supported Wegener's ideas.

2.5.2.6   *Developing a theory*

In science, a *theory* is not a hunch, guess, or tentative explanation. Those words define a *hypothesis*. A scientific theory is a set of well-supported explanations that have credence beyond a hypothetical level. Changing Alfred Wegener's hypothesis of continental drift into an acceptable theory and developing a comprehensive theory of plate tectonics was the work of many people. In

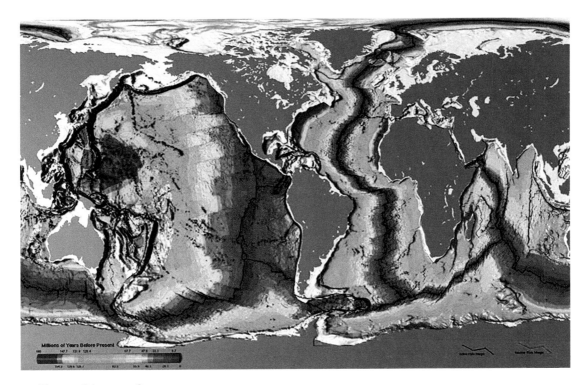

Figure 2.22 The age of the ocean floor.
Graphic from NOAA.

1960, Harry Hess, Robert Dietz, Bruce Heezen, Marie Tharp, and others began to synthesize all of the seismic, paleomagnetic, and other data into a single theory. They demonstrated conclusively that ocean crust was created at spreading centers, like the Mid-Atlantic Ridge, and subducted and destroyed at ocean trenches, like those that surround the Pacific Ocean. While the evidence used by Alfred Wegener to support continental drift also supports plate tectonics, the theory of plate tectonics is far more detailed and supported by much information that was unknown to Wegener.

2.6 The theory of plate tectonics

The outermost part of Earth consists of a brittle, rigid layer called the *lithosphere* (Fig. 2.23; see also Fig. 2.6). The top part of the lithosphere is Earth's crust, a layer with overall chemical composition similar to granite in continental regions and similar to basalt in oceanic regions. The lower part of the lithosphere is the uppermost part of the mantle. The lithosphere is thinner in oceanic regions compared with continental regions.

The oceanic lithosphere varies in thickness. It is just a few kilometers thick near mid-ocean ridges and up to 100 kilometers (60 miles) thick in places distant from the ridges. Continental lithosphere thickness ranges from about 40 kilometers (25 miles) near the margins of young continents to nearly 200 kilometers (125 miles) in the centers of old continents. The oceanic and continental lithospheres have marked differences in composition. The principal elemental components are the same, but oceanic lithosphere contains relatively more iron and magnesium and less silicon and oxygen than continental lithosphere does. Consequently, the oceanic lithosphere contains different kinds of minerals and rocks and has considerably greater density than its continental counterpart.

The *asthenosphere*, the layer of Earth's mantle just beneath the lithosphere, is so hot that it is partially melted in some places, notably near mid-ocean ridges. Additionally, because of its high temperature, it is less rigid than the lithosphere is and can flow in response to deformation or gravity. Although it is mostly solid, the asthenosphere behaves like a very viscous fluid or a deformable plastic; a good analogy is *Silly Putty*. The asthenosphere is not at the same depth, nor does it have the same thickness, everywhere. In most places it extends from about 100 kilometers (60 miles) to about 700 kilometers (450 miles) below Earth's surface (Fig. 2.23).

The lithosphere comprises seven principal tectonic plates and many smaller ones, shown in Figure 2.24, which can move independently of each other. The plates sort of "float" above the more plastic asthenosphere, which explains how continents appear to drift. Continents do not plow through the ocean crust, as envisioned by Wegener's critics. Instead, the rigid lithospheric plates, which include both continental and oceanic crusts, as well as portions of the upper mantle, move as units across Earth's surface. Continental movement has been measured in real time using GPS (global positioning systems) and can be calculated by studying the age of the seafloor at different distances from a mid-ocean ridge. Plates' velocities range from 1 to 20 centimeters/year. (In comparison, human hair grows at about 7.5 centimeters/year and fingernails at about 4 centimeters/year.)

The speed at which plates move seems slow compared with many other things, but geological time is long. Over the billions and millions of years of Earth history, plates that were once at the North or South Poles have migrated to the equator, and some have made a return trip. So, Earth materials are constantly moving, and rocks or other materials that form at one place may be thousands of kilometers from where they formed in a relatively short (geological) time.

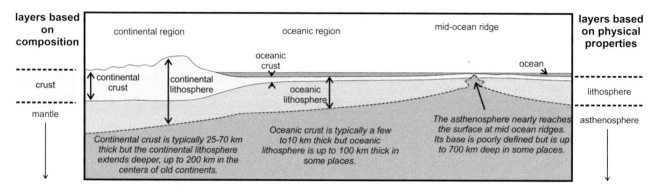

Figure 2.23   Earth's lithosphere and asthenosphere.

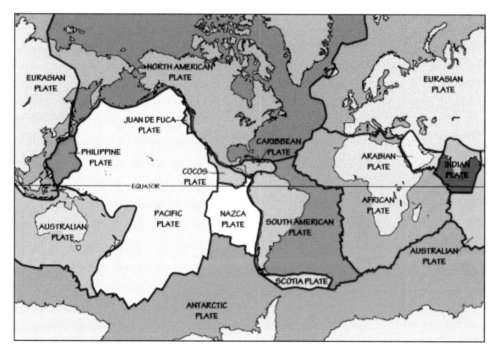

Figure 2.24   Earth's major lithospheric plates.
Figure from USGS.

The sizes and compositions of the lithospheric plates vary greatly. The largest, the Pacific Plate, is about 103 million square kilometers (40 million square miles) and is composed almost entirely of oceanic crust and mantle (Fig. 2.24). In contrast, the second largest plate, the North American Plate (76 million square kilometers; 30 million square miles), is about half continental material but also includes some lithosphere beneath the Atlantic and other oceans. The smallest of the seven major plates is the South American Plate (43 million square kilometers; 17 million square miles), but geologists have identified many smaller plates termed *minor plates* and even smaller ones called *microplates*. All of the plates, no matter their size, have the potential to move independently of each other.

Geologists do not agree on the exact number of lithospheric plates. In part, that number depends on whether plates that were once separate, but move as a single unit today, are lumped or split. For example, the boundary between North America and the Atlantic Ocean was once an active subduction zone. Today, however, the North American continent and the Atlantic Ocean, from the east coast of the United States to the Mid-Atlantic Ridge, move together as a single plate (see Fig. 2.24). The two formerly independent plates (North America and the western half of the North Atlantic) are said to be *welded* together, and the boundary between them is called a *passive margin* (because no subduction or other tectonic activity is occurring there). Together the two former plates make up most of the present-day North American Plate. Greenland, parts of the Caribbean, a small part of Russia, and a few islands in the Atlantic Ocean are also part of the North American Plate. Today the western edge of the North American Plate is an *active margin* characterized by subduction and volcanism

on the coasts of Washington and Oregon, the San Andreas transform fault in California, and more subduction and volcanism in Mexico. The eastern edge of the North American Plate is an active margin characterized by volcanism and a plate boundary at the Mid-Atlantic Ridge.

Other large plates, too, are amalgamations of smaller ones. The large Eurasian Plate, for example, is known to contain at least 15 microplates of various sizes. Recent studies suggest that the Eurasian Plate in China and eastern Siberia consists of several smaller plates. The microplates and other small plates moved independently of each other in the past. Additionally, geological investigations are ongoing today, and new plate boundaries are being discovered and characterized.

2.6.1 Dynamic Earth

Plate tectonic processes divide Earth into two types of regions, characterized by different amounts of tectonic activity. The boundary regions between plates are relatively narrow but often lengthy, dynamic zones characterized by deformation (folding and faulting), earthquakes, and often volcanoes (Table 2.2). In contrast, the vast intraplate regions, far from plate boundaries, are generally very stable and quiescent. The few exceptions are associated with *hot spots*, where plumes of magma rise to the crust from deep in the mantle. These hot spots may be near plate margins or in plate centers.

As shown in Figure 2.25, at plate boundaries, relative plate movement may be of three kinds: (1) plates may diverge (move apart), allowing magmas or solid material to rise from beneath and fill the gaps opened between the plates; (2) plates may converge (move toward each

Table 2.2 Dynamic regions associated with plate tectonics.

Relative movement	Process	Tectonic characteristics
plate divergence at mid-ocean ridges	seafloor spreading	faulting, earthquakes, and volcanoes
plate divergence in continental regions	continental rifting	faulting, earthquakes, and volcanoes
ocean plate–ocean plate convergence	subduction	folding, faulting, earthquakes, and volcanoes
ocean plate–continental plate convergence	subduction	folding, faulting, earthquakes, and volcanoes
continental plate–continental plate convergence	continental collision	folding, faulting, earthquakes, and only very rare volcanoes
plates sliding past each other in ocean regions	transform faults	faulting, earthquakes
plates sliding past each other in continental regions	transform faults	faulting, earthquakes

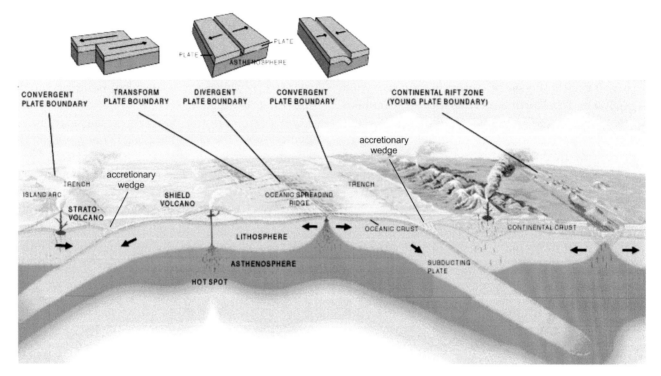

Figure 2.25   The three kinds of plate boundaries.
From USGS. Modified from USGS graphic.

other), which often leads to subduction as one plate slides beneath the other, creating linear subduction zones; and (3) plates may slide past each other along *transform faults*, with little associated deformation to either plate, except right along the transform fault zone.

Plate boundaries are further differentiated by where they are situated, on continents or in the oceans, and the nature of the plates involved. Table 2.2 summarizes the seven possible combinations, the processes that occur at each, and the nature of tectonism that is typical.

2.6.2 Divergent plate boundaries: spreading zones

2.6.2.1 Mid-ocean ridges

The world's major oceans contain spreading zones—long linear undersea ridges between two diverging tectonic plates. Figure 2.26 shows the most important of these ridges. The ridges are not straight but zigzag because they are offset in many places by countless transform faults. Most of the volcanism on Earth occurs at mid-ocean ridges as magma, derived from solid mantle rocks, rises to fill gaps between separating lithospheric plates. Older lithosphere on both sides of the spreading center moves away from the ridge, and the intruding

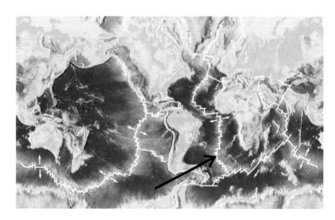

Figure 2.26   Mid-ocean ridges of the world.
Drawing from J. M. Watson, USGS.

magmas add new igneous rocks to both of the adjacent plates. The added rocks are responsible for the magnetic anomalies shown in Figure 2.21 and the ages of the ocean floor shown in Figure 2.22.

Figure 2.27 shows the most important process that occurs at mid-ocean ridges. Rising asthenosphere melts, producing magma that fills space created as ocean plates on either side of the ridge move apart. At fast-spreading centers, such as the major ridge in the eastern Pacific, spreading rates exceed 10 centimeters/year and magma rises quickly. Most mid-ocean ridges are only 2–3

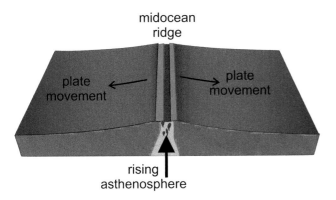

Figure 2.27 Topography at spreading centers.

kilometers wide, with smooth domed shapes, extend long distances, and are only a few hundred meters high. However, at slow-spreading centers (less than 4 centimeters/year), like the Mid-Atlantic Ridge, magma upwells slowly and deep rift valleys form at the centers of the spreading ridges.

At spreading centers, rising magmas create fractures that subsequently provide conduits for cold seawater to flow into the subsurface. As this seawater nears hot magma, the cold water is heated, producing hydrothermal water that extracts chemicals from surrounding rocks and sometimes from magma. The hydrothermal waters may eventually escape at the ocean floor, adding dissolved material to seawater and producing *black smokers*, like the one shown in Figure 2.28, that deposit sulfide minerals. Some of the world's most important ore deposits formed in this way at mid-ocean ridges. Additionally, hydrothermal conditions, such as those commonly found at spreading centers, are thought by many scientists to have been sites where life originated about 4 billion years ago. Studying these processes is challenging because of the depth at which they occur; this photograph was taken from inside the *Alvin*, a research submersible vessel, in 2009.

2.6.2.2 Continental rifts

Spreading centers are not restricted to oceanic regions. Although not as common as spreading at ocean ridges, spreading may develop in continental regions in linear zones called *continental rifts*. In the past few tens of millions of years, continental rifting has separated the Arabian Peninsula from Africa and created the Red Sea This can be seen in Figure 2.29. The Figurealso shows a related spreading rift, called the *East African Rift*, that extends south from the Red Sea and that is slowly splitting eastern Africa from central Africa. Figure 2.30

Figure 2.28 A black smoker on the Juan de Fuca Ridge, southwest of Vancouver, Canada.

From NOAA, Wikimedia Commons.

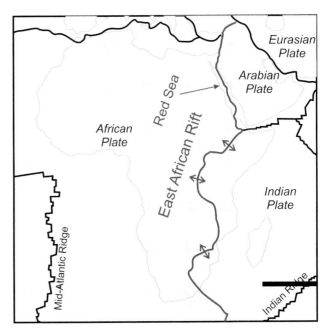

Figure 2.29 Spreading of the Red Sea and the East African Rift.

Figure 2.30   A portion of the East African Rift in Eritrea.
From Clay Gilliland, Wikimedia Commons.

shows a rift valley in Eritrea that was created as the African crust spread and thinned. This spreading is associated with earthquakes and volcanoes, similar to what happens at mid-ocean ridges. Note the people for scale in Figure 2.30.

When the East African Rift becomes deep enough and long enough to connect to the sea, water will enter and the rift will become like the Red Sea—a long narrow valley filled with salt water. With continued spreading, both the Red Sea and the East African Rift may eventually evolve to become new oceans, as seafloor spreading takes over from continental rifting. This process is how all of the world's major oceans formed—starting as rifts and evolving to ocean basins. For example, several different rifts, beginning about 200 million years ago, separated Pangaea into the continents that we see today (Fig. 2.15). As part of this process, the Atlantic Ocean formed about 130 million years ago when rifting and then seafloor spreading separated Africa from the Americas.

2.6.3 Convergent boundaries (plate collisions)

2.6.3.1 Subduction zones

Earth's diameter is not increasing, so if seafloor spreading creates new lithosphere, there must be places where older lithosphere sinks into Earth to be recycled. This happens at convergent boundaries—at subduction zones—where one lithospheric plate slides under another and into the mantle. Convection may involve

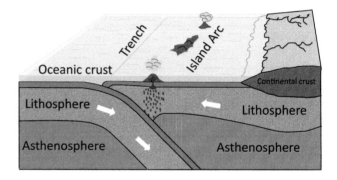

Figure 2.31   Subduction involving two oceanic plates.
Modified from an original USGS figure.

oceanic lithosphere subducting beneath oceanic lithosphere, as is shown in Figure 2.31, or oceanic lithosphere subducting beneath continental lithosphere. Thus, subduction always involves an oceanic plate that passes beneath another oceanic plate or beneath a continental plate. Continental plates do not subduct in any setting, in large part because they are much less dense than the mantle and thus buoyant. Rates of subduction range from about 2 centimeters to 8 centimeters per year along most subduction zones. Geophysical studies have detected the remains of subducted plates in the mantle at depths of 1500 kilometers.

2.6.3.2 Ocean-ocean subduction

Oceanic lithosphere cools after it forms at mid-ocean ridges, and cold rocks are denser than hot rocks. Consequently, when two oceanic plates converge, the older of the two plates is usually cooler and denser, and will subduct under the younger, warmer, and less dense plate (Fig. 2.31). Thus, rocks of the oceanic lithosphere are recycled into the mantle, where they may eventually be metamorphosed or melted. Subduction leads to melting of the mantle above the subducting slab, and as volcanism proceeds, the new mantle-derived magmas move material up and add it to the overlying crust. Subduction of one ocean plate under another is occurring, most notably, around the Pacific Ocean rim (in the Aleutian Islands, Indonesia, the Philippines, and Japan) but also is taking place in the Caribbean and elsewhere. This type of subduction is typified by volcanism, dominated by eruptions of ash and other airborne material, which produces curved island chains called *island arcs*. The island chains of Indonesia and the Aleutian Islands of Alaska are examples of island arcs. Most islands of the Aleutians, and of other island arcs, are small, sometimes no more than a single volcano. But continuing eruptions cause islands to grow larger, and originally separated islands may eventually be combined into one. This process explains why the Japanese islands of Honshu, Hokkaido, Shikoku, and Kyushu are larger than the islands of the Aleutians.

As plates subduct, valleys form and deep trenches may cut into the ocean floor. Trenches off the east coast of Japan are up to 10,554 meters (35,600 feet) deep in some places. The trench south of the Aleutian Islands is more than 3200 kilometers (2000 miles) long, 80 to 160 kilometers (50 to 100 miles) wide, and up to 7600 meters (25,000 feet) deep in some places. The deepest place in the oceans is the Marianas Trench in the eastern Pacific; it extends to 11,000 meters (36,000 feet) below sea level.

Trenches are natural places for sediment from continents or islands to collect. Additionally, by the time a plate enters a subduction zone, it may have as much as 500 to 1000 meters of sediment on top of it that accumulated as it drifted across the ocean basin. As subduction progresses, some sediment moves into the mantle with the subducting plate, along with subducted igneous and metamorphic rock of the oceanic lithosphere. Any sediment that does not subduct collects and forms a wedge of material, called an *accretionary wedge*, in the trench (Fig. 2.25).

2.6.3.3 *Ocean-continent subduction*

Oceanic lithosphere is denser than continental lithosphere, and when oceanic plates converge on a continent,

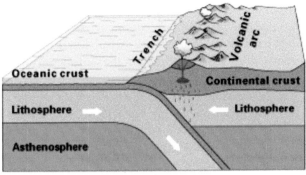

Figure 2.32   Subduction of an oceanic plate beneath a continental plate.
From the USGS.

the oceanic plates subduct beneath the continent, as shown in Figure 2.32. These kinds of subduction zones are also shown in Figure 2.25. Ocean-continent subduction shares many features with ocean-ocean subduction, including trenches, sediment accretionary prisms, earthquakes, and volcanoes. However, subduction under a continent produces long linear volcanic mountain chains on the margins of continents instead of island arcs.

On the west coast of South America, the Nazca plate subducts under the South American continental plate, producing the volcanoes and earthquakes associated with the Andes Mountains shown on the maps in Figures 2.17 and 2.18. The deep Peru-Chile Trench, just offshore, is another product of the subduction. Similar volcanic mountains, earthquakes, and trenches characterize other ocean-continent convergence zones on the eastern side of the Pacific. Subduction of the Cocos Plate beneath Central America is responsible for active volcanism in Costa Rica and other countries, and farther north, along the coasts of northern California, Oregon, and Washington, subduction of the Juan de Fuca Plate beneath North America causes earthquakes and the volcanoes of the Cascade Mountains (also shown in Figs. 2.17 and 2.18). In all these ocean-continent subduction zones, formerly mantle material melts, magmas move upwards, and new material is added to continents.

2.6.3.4 *Continent-continent collisions*

If ocean lithosphere subducts beneath a continent, unless new ocean lithosphere is being generated somewhere, the width of an ocean must decrease. Eventually, the ocean may disappear entirely, and once widely separated continents may collide with each other in a continental convergence zone. Because continent crusts have relatively low density compared to underlying mantle, and because all continents have about the same density, little subduction of either colliding plate occurs. Instead, the plates become crushed together, forming high mountain ranges with deep roots. This has happened in Asia as the Indian subcontinent moved north and collided with Tibet, beginning about 50 million years ago. The result is the Himalaya Mountains, seen as snow-capped peaks in Figure 2.33. The flat planes in the south (bottom of image) are part of India, the high mountains with snow are mostly in Nepal, and the high-elevation lakes are in Tibet. Because subduction is mostly absent, volcanism is rarely associated with continent-continent collisions. Continent-continent collision is how the Alps, the mountains that divide southern Europe from northern Europe, were created. The main alpine mountain building was 25–35 million years ago when the African Plate moved north, closing an ancient sea called the Tethys Sea, and collided with the European Plate. During that event, rocks were buckled, faulted, and folded on top of each other, but volcanism was scarce.

About 6100 kilometers (3800 miles) east of the European Alps, the Himalaya Mountains are rising today. The process leading to their formation began when India separated from Pangaea and moved northward.

Subduction of oceanic crust under Eurasia eventually brought India into contact with Eurasia about 10 million years ago (Fig. 2.34).

Prior to India hitting Eurasia, subduction produced volcanism, but once the continents met, volcanism ended and the continents were deformed and uplifted to create the Himalaya Mountains. The collision continues today and produces deadly earthquakes in India and Pakistan. This continent-continent collision between India and Eurasia is similar to the process that created the Ural Mountains 300 million years ago, in Russia and Kazakhstan, that divide Europe from Asia.

2.6.4   *Transform plate boundaries*

Some plate boundaries do not involve convergence or divergence, but instead involve horizontal movement as two plates slide past each other. These are *transform plate boundaries*. The boundary between the North American Plate and the Pacific Plate that runs north-northwest, parallel to the coast of California, is an example (Fig. 2.35). This boundary is the very active San Andreas Fault zone—really a network of faults—which extends from the Gulf of California in Mexico 1050 kilometers (650 miles) northwest before entering the Pacific Ocean near Point Reyes, a peninsula that juts out into

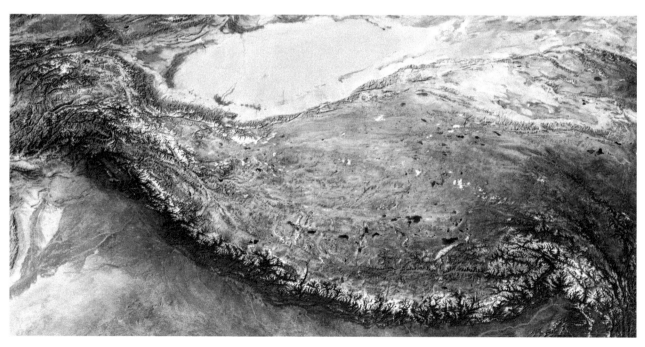

Figure 2.33   The Himalaya mountain chain.
From NASA, Wikimedia Commons.

the ocean 55 kilometers (35 miles) northwest of San Francisco.

Transform faults create only minor rock deformation, except within, or very close to, the fault zones. Additionally, because no subduction occurs, volcanism is absent. Consequently, transform boundaries are narrow features with undisturbed lithospheric plates on each side. Movement along transform faults is generally horizontal, sometimes by continuous slow movement called *creep* and sometimes in fits and spurts. So, contrary to what might be seen in some movies, California is not

going to fall into the Pacific Ocean, although many small earthquakes can be expected.

Today, the Pacific Plate (which includes a sliver of the California coast west of the San Andreas Fault from Baja California north to San Francisco) is moving north relative to the North American Plate. Figure 2.36 shows the San Andreas *fault scarp* (displacement of the ground

Figure 2.35   The San Andreas Fault in California.
From the USGS, https://pubs.usgs.gov/gip/dynamic/graphics/Fig35.gif.

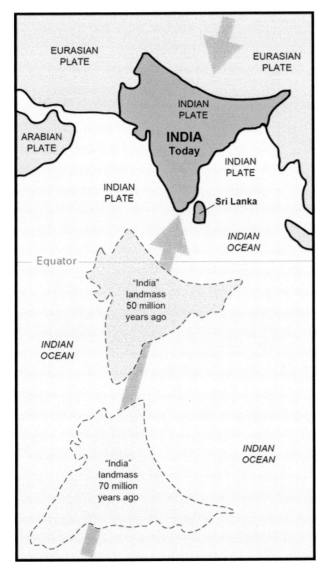

Figure 2.34   Indian subcontinent converging on Asia.
From USGS.

Figure 2.36   Aerial view of the San Andreas Fault on the Carrizo Plain, California.
From I. Kluft, Wikimedia Commons.

surface) in central California. Movement along the fault has left a visible trench along the plate boundary. Streams that cross the fault have been offset, moved to the right, due to fault movement. Movement continues along this fault today, and in the next 10 or 20 million years, Baja and southern California will become an island in the northeastern Pacific. The island may collide with Alaska in about 50 million years, and the granitic rocks that are seen east of the fault in present-day Joshua Tree National Park in southern California may be juxtaposed with subduction zone volcanic rocks in the Aleutian Peninsula.

The San Andreas is really a fault system, not just one fault (Fig. 2.35). In places, it consists of more than one major, but parallel, fault, and many smaller faults branch off from the main ones. Today, the Pacific Plate is moving northwest past the North American Plate at an average rate of about 6 centimeters (2.5 inches)/year, but movement along the San Andreas Fault accounts for only two-thirds of that. The rest occurs on other faults, such as the Hayward Fault, that are parallel to the San Andreas. Although often shown as lines on maps, many large faults are actually fault zones many meters across. Although the San Andreas fault system is responsible for hundreds of earthquakes each year, most are too small for people to feel. Throughout history, however, the fault system has been responsible for many, and sometimes powerful and devastating, earthquakes affecting California.

2.6.5   Hot spots

Most of the volcanism on Earth is associated with plate tectonics and, in particular, spreading centers and subduction zones. The other major sites of volcanism are *hot spots*. Many hot spots are found around the world, in both continental and oceanic regions of all ages. Figure 2.37 shows the locations for the most significant ones according to the United States Geological Survey.

Figure 2.38 shows a mantle plume at a hot spot. The causes of hot spots and plumes are debated, but the features at the surface are similar everywhere. In oceanic regions, hot spots may produce large submarine volcanoes called *seamounts* which, if exposed above water, form islands, such as the Hawaiian Islands. In continental regions, hot spots produce areas of volcanic and hydrothermal activity, such as the area around Yellowstone National Park. As shown in Figure 2.38, hot spots are places where large amounts of magma rise from deep in the mantle, carrying heat toward the surface. At

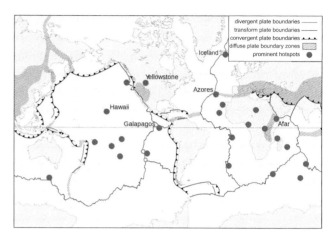

Figure 2.37   Locations of the most significant hot spots on Earth. Modified from graphic by the USGS, Wikimedia Commons.

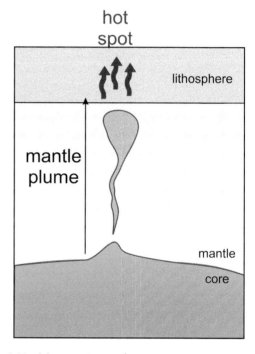

Figure 2.38   Magma rising at a hot spot.

least some of the magmas originate at the mantle-core boundary. So, hot spot magmas, which have far deeper origins than spreading and subduction zone magmas do, may be the most significant process moving Earth materials from the deep mantle toward the surface.

Lithospheric plates are decoupled from underlying mantle, and move independently, above hot spots. We can, therefore, think of hot spots as being motionless with plates moving above them. For example, the Pacific Plate has moved over the Hawaiian hot spot and left a chain of volcanoes that stretches from Siberia to present-day Hawaii (Fig. 2.39). Some of the older

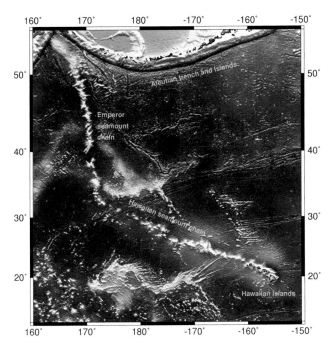

Figure 2.39 The Emperor Seamount Chain and the Hawaiian Ridge.

From Ingo Wölbern, Wikimedia Commons.

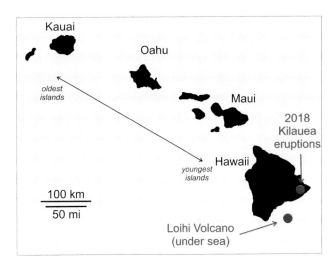

Figure 2.40 The Hawaiian Islands and recent volcanic activity.

volcanoes are seamounts—submarine mountains that do not extend above the sea surface—but the younger ones form islands. About 43 million years ago, the movement of the Pacific Plate changed direction from nearly due north to northwest, which explains the different orientation of the younger Emperor Seamount Chain compared with the Hawaiian Ridge and island chain (Fig. 2.39).

The oldest Hawaiian volcanoes are below sea level at the northwest end of the chain, and the youngest are to the southeast. The oldest volcanoes above sea level are on Kauai; they are 3.8 to 5.6 million years old. Other volcanoes can be seen on Oahu, Molokai, Maui, and the big island Hawaii (Fig. 2.40). Most of the Hawaiian volcanoes are extinct, but the big island of Hawaii contains several active volcanoes, and the relatively young underwater *Loihi Volcano* is active just offshore to the southeast of the big island (Fig. 2.40). In the next few hundreds of thousands of years, Loihi Volcano will rise above sea level and become the next Hawaiian Island. Yet, without the input of new lava, most of the islands are becoming smaller. They are being recycled as they cool and contract, slowly eroding to produce sediment deposited on the Pacific Ocean floor.

Hot spots are found in many diverse settings on Earth (Fig. 2.37). The Hawaiian Islands result from a hot spot in the middle of the oceanic Pacific Plate. A hot spot created Iceland near a plate boundary, the Mid-Atlantic Ridge, but the volcanoes of Iceland are fed by magmas that have a much deeper origin than those associated with normal spreading centers. Hot spots also occur in the centers of continents. In North America, westward movement of the North American Plate above the Yellowstone hot spot was, in the past, responsible for the Columbia Plateau flood basalts that erupted in Washington, Oregon, and Idaho 15–17 million years ago (Fig. 2.37). It was also responsible for several huge volcanic eruptions that changed global climate during the past 2.1 million years. Today, the Yellowstone hot spot provides the energy that powers geysers and other geothermal features of Yellowstone National Park, Wyoming. Yellowstone could, in the future, become more active, and the next eruption could be catastrophic, as Yellowstone has been in the past.

2.6.6 Orogenies

Orogeny is a general term used to describe large structural deformation of the continental lithosphere that produces long, relatively narrow belts called *orogenic belts*, or simply *orogens*. The process involved is termed *orogenesis*. Topographically, many of these belts are mountainous and are commonly called *mountain belts* or *mountain ranges*. Figure 2.41 shows, in blue, the most significant orogens active today or within the past several hundred million years.

Orogenies occur when lithospheric plates converge, and orogenic belts are characterized by combinations of deformation (folding and faulting), crustal thickening,

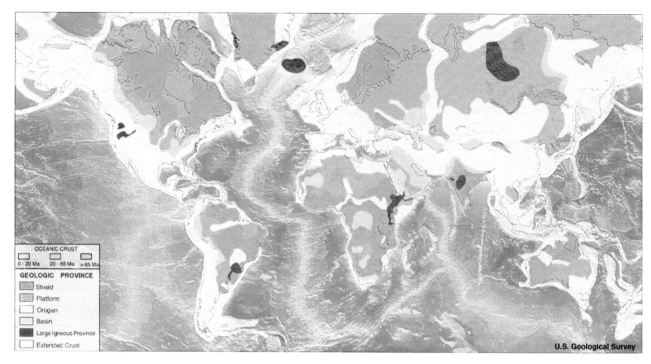

Figure 2.41   Orogens, shown in blue, and other geological provinces.
From the USGS.

metamorphism, melting, and magmatic activity. The specifics depend on the nature and strength of the continental lithosphere that is involved and how fast convergence occurs. In continent-continent collisions, deformation generally dominates over volcanism. In ocean-continent collisions, the opposite is true.

The European Alps and the Himalaya are orogenic belts caused by continent-continent collisions, and the entire chain of mountain ranges from the Pyrenees in Europe to the Himalaya in India is called the *Alpine-Himalayan Orogen* (Fig. 2.41). Other mountain ranges in this orogen include the Carpathian, Balkan, Caucasus, Hindu Kush, and Karakoram. All of these and some other smaller ranges grew when southern continents collided with Europe or Asia. The Andes and mountain ranges in Central America, western North America, and north all the way to Alaska are part of a single orogenic belt that is associated with subduction of Pacific Ocean lithosphere under the North and South American plates. The belt is called the *Cordillera*, or *Cordilleran Orogen*, in Central and North America and the *Andean Orogen* in South America.

Because continental lithosphere is too light to subduct, the primary process that recycles continental material is erosion. In mountain belts, when weathering and erosion occur, once deeply buried sedimentary,

igneous, or metamorphic materials are exposed at the surface and eventually removed by erosion. Removal of material lightens the lithosphere, causing it to rise upwards, which leads to further erosion. A common result is that, over time, orogens evolve into linear or slightly curved belts with old igneous and metamorphic rocks exposed as cores at the center and younger sedimentary layers overlapping along the sides.

Weathering and erosion in mountainous terrain is much greater than in flat regions, so mountain belts are the sources of most sediment that makes it into continental or ocean basins. The sedimentary material may subsequently be buried or subducted, and thus orogenic belts are focuses for much recycling of sedimentary material from Earth's continental lithosphere. Yet, despite potentially extensive erosion, continental crust sometimes survives for billions of years. In comparison, plate tectonics ensures that most oceanic lithosphere subducts and is recycled within 200 million years of when it formed (Fig. 2.22). The oldest large regions of oceanic crust are 180–200 million years old (western Pacific and northwest Atlantic), but some fragments of crust in the Mediterranean Sea are 270–340 million years old.

Today, active orogens are found next to all active convergent plate boundaries. Present-day orogens, such

Figure 2.42 View of the Himalaya from the north.
Photo from NASA.

as the Himalaya or Andes, maintain great topographic relief, and many earthquakes attest to ongoing tectonic activity. The Himalaya continue to rise at an average rate of about 1 centimeter/year, and the Andes at about 0.1 centimeter/year. Huge amounts of sediment are removed and transported by rivers to the oceans near both ranges, but still the mountains continue to grow taller.

Figure 2.42 shows a view of the Himalaya from the International Space Station. Although hard to pick out, the mountains in this photograph include Everest, Lhotse, Makalu, and Cho Oyu, all more than 8000 meters (26,000 feet) tall. Everest is the highest mountain in the world, and the other three rank 4, 5, and 6, respectively. Numbers 2 and 3, K2 and Kangchenjunga, are elsewhere in the Himalaya.

The Himalaya and Andes are growing ranges today, but other mountain ranges, including the Appalachians and Rocky Mountains of North America, are far from convergent boundaries and so are not growing. These ranges, which once might have been as tall as the Himalaya, are associated with no active volcanoes and very few earthquakes. Erosion continues and, over time, the Appalachians and Rocky Mountains may be entirely removed, leaving flat topography behind.

One consequence of orogenesis is that continents may become larger over time. The sizes of Earth's present-day continents have slowly increased since the Precambrian. In large part, this is because smaller continents have combined to produce larger ones over time. North America, for example, is composed of more than half a dozen different regions that formed separately before joining to produce what is North America today. Additionally, during continental collision, magmatism commonly moves material from Earth's mantle to the crust, thus increasing the overall volume of crustal material.

Because most tectonic activity takes place at continental margins, the oldest parts of continents are generally in their centers. All of the world's major continents contain regions, called *Precambrian shields*, where erosion has been so great that the ancient roots of once-present mountain chains are exposed in relatively flat terrain (orange regions in Figure 2.41). Figure 2.43 shows Precambrian rocks of the Canadian Shield. In the Canadian Shield, and other shields, the exposed rocks contain much evidence of deformation, metamorphism, and igneous activity of the past. They include igneous and metamorphic rocks from deep within Earth, and they contain folds and faults created during mountain building. Most continents contain regions adjacent to shields where Precambrian shield rocks are covered by relatively flat-lying Paleozoic sedimentary rocks. The sedimentary rocks are termed *platform rocks* (pink regions in Fig. 2.41). Together, the shield regions, and adjacent areas where sedimentary rocks cover shield rocks, create *cratons*—a name for the very stable interior portions of continents that have seen little tectonic activity since the Precambrian.

Figure 2.43   Outcrops of rocks in the Canadian Shield along the Missinaibi River, northern Ontario.
Photo from Lkovac, Wikimedia Commons.

2.6.7 Convection in the mantle

In 1929, shortly after Wegener proposed that continental drift occurred, Arthur Holmes suggested that convection, involving flow of generally solid material, occurs in the mantle. He proposed that hot materials are less dense than cold materials are and will naturally rise to Earth's surface where they will split the crust. Continued upward flow forces the crust to move apart and, when they cool, pieces of the crust will eventually descend back to depth again. According to Holmes, these two processes could drive the continental drift proposed by Wegener. It was not until the 1960s, however, that Holmes's ideas received much attention. Today, mantle convection is acknowledged as a key driver of plate tectonics. Although the mantle is mostly solid, because it is hot and somewhat elastic, it can flow like an extremely viscous liquid. Flow rates are slow, but over geological time, the distances can be great.

At mid-ocean ridges and, to a much lesser extent, at continental rifts, Earth's lithosphere splits apart as new material rises from the mantle. Newly formed lithosphere, including crust, moves away from the spreading center. At subduction zones, oceanic lithosphere sinks into the mantle as one plate slides beneath another. These two processes mean that, over time, Earth's mantle must be circulating—upwards at ocean ridges, down at subduction zones, and laterally from subduction zones to spreading centers at depth in Earth. This circulation involves convection, much as Holmes envisioned it. Because lateral convection in the mantle must balance with the rates of seafloor spreading and subduction, most geologists estimate that convection occurs at rates of centimeters/year. Thus, the rocks that make up Earth's mantle are moving and cycling just as the lithospheric plates do at Earth's surface.

Convection also occurs vertically in Earth, because Earth is cooling. The temperature in Earth's core is estimated to be 6000 °C (about 10,800 °F), much hotter than the mantle or crust, so heat is constantly flowing outward. Additional heat is generated by decay of radioactive elements—mostly uranium, potassium, and thorium—and must also flow to Earth's surface to dissipate. Mantle convection is the main mechanism by which this heat escapes from the interior of Earth, and the heat escapes primarily at mid-ocean ridges.

The combination of lateral and vertical convection means that the mantle contains convection cells that are in some ways similar to the cells that form when you pour cream into coffee or that develop when water is boiling in a saucepan. Figure 2.44 is a simple Figure showing only a few large cells descending to Earth's mantle-core boundary. Some geologists today debate the size of the cells and how deep they go. The relationship between hot spots and convection is also a topic of current research.

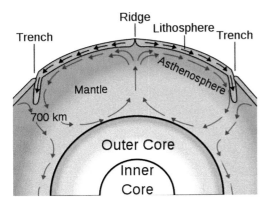

Figure 2.44 Convection in Earth.
Modified from an image from Surachit, Wikimedia Commons.

2.7 The water cycle

The abundance of liquid water on Earth is unique in the solar system, and all terrestrial life forms, including desert organisms, need water. Water also plays a prominent role in chemical and physical weathering, erosion, diagenesis, most magma formation, and metamorphism in the rock cycle. So, the hydrosphere greatly influences the atmosphere, the biosphere, the lithosphere, and the geosphere. These interactions have been going on for a long time; chemical evidence from ancient zircon mineral grains suggests that liquid water has existed on Earth for at least 4.4 billion years.

The characteristics of the hydrosphere, including the properties of water and the water cycle, are discussed in detail in Chapter 12, but that chapter focuses mostly on water at or near Earth's surface or in the atmosphere and the processes depicted in Figure 2.7 (this chapter). Although water cycles through Earth's atmosphere, surface, and the near subsurface, it also circulates at depth. Water is a key component in the rock cycle. It cycles between pure liquid form and being a component of magmas and minerals. It also cycles from great depth in Earth to the surface.

Water that originates from Earth's deep interior and that has never previously reached Earth's surface and atmosphere is called *juvenile water*. In contrast, *meteoric water* is, or was at one time, atmospheric precipitation in the form of snow, rain, or hail. Stable isotope studies indicate that water released by volcanoes and hot spots is mostly meteoric rather than juvenile. In fact, just about all the water at Earth's surface has a meteoric origin.

Interactions between the hydrologic cycle and the rock cycle, during plate tectonics, are important and affect magmatism, metamorphism, the formation of

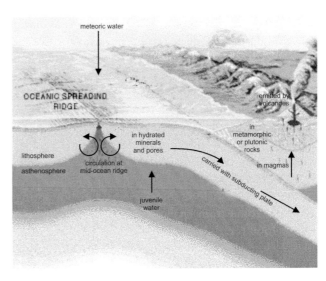

Figure 2.45 Water and plate tectonics.
Modified from a USGS drawing.

mineral resources, and ocean water chemistry. Figure 2.45 shows some key interactions. Escaping heat along mid-ocean ridges causes hot water to circulate through the crust in *hydrothermal systems*, sometimes producing black smokers such as the one shown in Figure 2.28. As seawater reacts with the hot rocks, its composition changes. Over the course of several million years, all of the water in the world's oceans cycles through the oceanic crust at mid-ocean ridges. Thus, hydrothermal circulation influences ocean composition. Additionally, as it circulates through the ocean crust, seawater becomes incorporated in, and alters, rocks of the ocean floor (Fig. 2.45). Thus, the basalt that makes up pristine ocean crust becomes hydrated, and hydrous minerals such as serpentine or chlorite may form. Rocks of the shallow ocean mantle can also incorporate ocean water. Eventually, many rocks of the ocean lithosphere are so altered to serpentine and other hydrous minerals that their original composition can only be known by inference.

Surface waters can infiltrate soils or sediments and percolate into Earth to become groundwater. Although some groundwater flow may extend to depths of thousands of feet, the pores in rocks at depth are mostly saturated with water and, consequently, have no room for additional water. Horizontal flow takes over and water moves through porous rocks or sediments to a river, lake, or ocean. Thus, infiltration is generally limited to relatively shallow depths.

During subduction, however, water in minerals can be carried to greater depths (Fig. 2.45). Hydrous minerals that form by alteration of the ocean lithosphere, or during metamorphism of subducted sediments, remain stable and may be carried to significant depth. At about 100 kilometers, pressures become so great that metamorphism produces new anhydrous minerals and water is released. Some recent studies of fluid inclusions in diamonds have shown that water exists at greater depths, but the exact amounts at different depths have yet to be determined.

Water from deep in Earth may return to the surface at volcanoes, geysers, and hot springs, but the bulk of the waters that feed these geysers and hot springs generally come from shallow depths, averaging 2000 meters (6600 feet). Water is, however, a key catalyst for rock melting that occurs deeper in Earth. Thus, water released by metamorphic dehydration promotes melting and can become incorporated in magmas at subduction zones, and so make its way back to the surface during volcanic eruptions or to the near-surface where it is incorporated into plutonic rocks (Fig. 2.45). In addition, during mountain building associated with orogenesis, water may become incorporated in metamorphic minerals and, with subsequent uplift and erosion, can then return to the surface.

2.8   Carbon and the carbon cycle

This chapter started with a summary of Aldo Leopold's description of an atom moving through natural cycles. Leopold never tells the reader what species (element) the atom is, but based on the hints he provides, it is most likely carbon. (It started life in a limestone, it was a key component of many life forms, and carbon dioxide has high solubility in water.) Carbon is an essential building block for all life on Earth and one of the

most widespread and mobile elements on our planet. It is also a common element in minerals, rocks, fossil fuels, plastics, and many other materials both natural and man-made.

Like water and many other substances, carbon cycles through the environment over time, and its residence time varies in different reservoirs. The carbon cycle (Fig. 2.46) describes the movement of carbon through the atmosphere, hydrosphere, Earth's interior, surface rocks, soils, fossil fuels, and biological organisms. The cycle is complicated, involving several overlapping processes. Much carbon cycling involves living organisms; that cycling occurs relatively quickly. Other cycling occurs much more slowly. It involves geological processes that move inorganic carbon between soil, permafrost, bedrock, fossil fuels, and other solid Earth materials, the atmosphere, and Earth's waters.

Carbon cycling is particularly important today because carbon dioxide (CO_2) and, to a much lesser extent, methane (CH_4) are the major carbon gases in Earth's atmosphere, and both are key *greenhouse gases* responsible for global climate change, including atmospheric warming, today. During much of the past, CO_2 consumed by plants during photosynthesis was just about balanced by CO_2 emitted during respiration and by wildfires. And, CO_2 exchange between the oceans and the atmosphere came out about even. So, the amount of CO_2 in the atmosphere did not vary much over time. But today, human activities (shown by red arrows in Fig. 2.46), including mostly burning fossil fuels but also cement manufacturing, land-use changes, and other things, are putting large amounts of extra CO_2 into the atmosphere. These anthropogenic contributions are shown by red arrows in Figure 2.46. Nature cannot currently keep up with these contributions, so atmospheric CO_2 is increasing and our planet is warming.

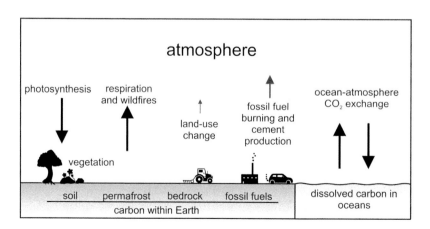

Figure 2.46   The carbon cycle, with arrows showing movement of CO_2 between reservoirs.

2.8.1 Isotopes of carbon

Carbon has an atomic number (*Z*) of 6, so every carbon atom has six protons in its nucleus. The number of neutrons varies, however, and carbon has many known isotopes; only three are natural and the rest are made synthetically in reactors. The three naturally occurring carbon isotopes are carbon-12, carbon-13, and carbon-14, abbreviated ^{12}C, ^{13}C, and ^{14}C (Fig. 2.47). They contain six protons and six, seven, or eight neutrons, respectively. ^{12}C and ^{13}C are stable (non-radioactive), always occurring in nature with proportions of about 99:1. ^{14}C is radioactive carbon that decays over time and would disappear completely if it were not replenished.

2.8.1.1 Carbon-14

^{14}C is a rare radioactive isotope with a half-life of 5700 years. It forms in the atmosphere when cosmic rays collide with ^{14}N atoms, converting that atom into ^{14}C (Fig. 2.48). The radioactive ^{14}C then behaves much like stable C until it decays back into ^{14}N. The radioactive carbon is incorporated in, and cycled through, the environment much as the more abundant, but stable, ^{12}C. Only living organisms can absorb ^{14}C. Consequently,

accumulation ceases when an organism dies. Plants that absorb ^{14}C pass it on to herbivores, and herbivores on to carnivores. So, ^{14}C concentrations can be used to obtain radiometric age dates for dead biological materials of many sorts. ^{14}C has a short half-life, compared with many radioactive isotopes, but for organisms between 150 and 50,000 years old, ^{14}C dating is practical. Thus, ^{14}C is used to date baskets, mummy wrappings in Egyptian tombs, and recent sedimentary deposits, but organic remains in dinosaur bones or coal are far too

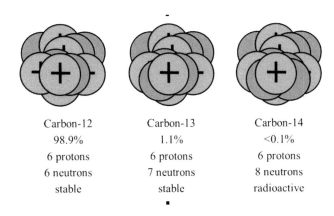

Figure 2.47 Naturally occurring carbon isotopes.

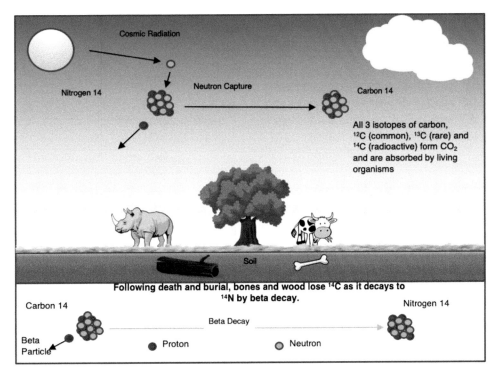

Figure 2.48 Carbon-14 pathways.
Drawing from Stewart Fallon.

old for this method because almost all the ^{14}C they once contained is gone.

2.8.1.2 Stable carbon isotopes

Most of the carbon in CO_2 is either ^{12}C or ^{13}C, both stable carbon. Because ^{12}C has one less neutron and is less massive than ^{13}C is, plants more readily incorporate ^{12}C during photosynthesis than they do ^{13}C. Other organic processes, too, concentrate ^{12}C. So, coal and other biological materials are enriched in ^{12}C, and scientists use the ratio of ^{12}C to ^{13}C to decide whether carbon in a specimen is of biological origin. Petroleum is also enriched in ^{12}C, suggesting that it has a biological origin. In contrast, carbonate minerals in meteorites are non-biological and are relatively enriched in ^{13}C.

^{12}C and ^{13}C ratios are also used to study the diets of ancient humans and other animals. Plants use two principal methods for metabolizing carbon: termed *C3* and *C4*. Most plants are C3 plants. These include temperate, or cool-season, plants such as wheat, rye, or oats. Tropical, or warm-season, plants such as millet, corn, or sorghum are C4 plants. The ratio of ^{12}C to ^{13}C is slightly higher in C3 plants, and the isotopic differences move through the food chain. Consequently, carbon isotope ratios help scientists learn what humans or other animals ate in the past.

2.8.2 Carbon in water

Carbon dioxide, carbonate minerals, and many different organic compounds affect the chemistry of water and the amounts of different carbon compounds in water. Although most carbonate minerals and organic compounds (including petroleum) are not significantly soluble, many organic compounds are soluble enough to pollute water. This pollution can change the chemistry of natural waters and harm organisms in many ways. For example, oil from leaking tanker ships and offshore petroleum wells adds carbon to ocean waters and produces slicks that are deadly to wildlife. Natural processes, too, such as the decay of dead organisms, will release organic compounds into the environment, which may be, depending on water acidity, significantly soluble in water.

Carbon dioxide gas is soluble in water, and some natural waters are carbonated, like carbonated beverages. Figure 2.49 shows an example of carbonated water flowing from an Idaho spring. If present, dissolved CO_2 reacts with water to form inorganic carbon species, including carbonic acid. Depending on how acidic the

water is, the acid may dissociate into variable amounts of hydrogen (H^+) and carbon-bearing ions. The amount of H^+ determines a water's pH (acidity), and so reactions of carbon species in water are of great importance.

The acidity of natural waters, including precipitation, affects the weathering of rocks, minerals, buildings, and statues. It also affects the behavior and toxicity of pollutants in water and the distribution of CO_2 between the atmosphere and oceans. Some toxic metals, including lead and cadmium, are quite soluble in acidic waters and, if present in high enough concentrations, may threaten humans, animals, and plants. Other toxins—including arsenic, selenium, and chromium—are commonly more soluble in alkaline waters.

Figure 2.50 shows where carbon is found in the oceans. Most carbon in natural waters is in the form of

Figure 2.49   Carbonated water at Soda Springs, Devils Postpile National Monument.

Photo from Thomas McGuire.

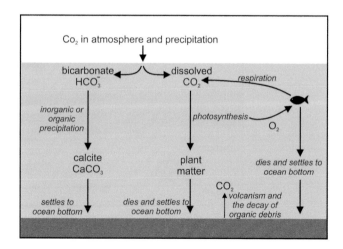

Figure 2.50   Carbon in the ocean.

carbon dioxide bubbles and bicarbonate ions derived from air, rain, other precipitation, terrestrial runoff, volcanism (especially at oceanic spreading centers), and respiration by animals (Fig. 2.50). Decay of dead organic material that settles to the ocean bottom and volcanic activity also release CO_2, although ocean circulation is slow, and it may take hundreds to thousand years for CO_2 at depth to rise to the ocean surface where the bulk of marine organisms flourish.

Photosynthesizing marine organisms remove dissolved carbon dioxide from seawater, but they do not remove bicarbonate ions. Bicarbonate is primarily removed from seawater when it reacts with dissolved calcium to form inorganic calcite that settles to the ocean floor (Fig. 2.50) or when it is converted to carbon dioxide, which is then taken up by plants or lost to the atmosphere. Additionally, clams, corals, foraminifera, and many other marine organisms remove bicarbonate from seawater to produce biogenic precipitates of calcite and aragonite in shells and skeletons. These biological remains may accumulate in sediments and eventually become fossilized.

In moderately deep ocean basins, carbon may be in *methane hydrates*, high concentrations of methane, derived from the decay of dead microorganisms trapped in ice. Such hydrates are stable at low temperatures and high pressures, but are absent from the deepest parts of the oceans where there are no microorganisms to decay. Some scientists are concerned that warming of the oceans will cause methane hydrates to decompose and release greenhouse methane.

2.8.3 Carbon in sediments, soils, and rocks

In the presence of oxygen, most organic compounds will oxidize to produce carbon dioxide, water, and other byproducts. If oxygen concentrations are low, however, oxidation may be slow or may not occur. Consequently, in swamps and some other wetlands, low oxygen levels in water hinder plant decay, and plant remains may accumulate as *peat*. Figure 2.51 shows peat that, when dried, can be burned as a heat source. If deeply buried and compressed, peat becomes coal, an even better source of energy. Similarly, in ocean basins and some lake waters, dead microorganisms may accumulate in the sediments. After deep burial, with a lack of oxygen, the organic matter may convert into *oil* and, if temperatures are high enough, *natural gas* (mostly methane) may form.

Figure 2.51 Peat deposits.
Photo by E. Freese, Wikimedia Commons.

Carbon in carbonate minerals, most importantly calcite, is stable and does not readily convert to carbon dioxide. Thus, carbonate minerals often accumulate in sediments and soils, tropical marine environments, and alkaline lake bottoms. Some sediments and soils contain fine disseminated calcite that is not easily seen. Other sediments and soils contain conspicuous layers or nodules of calcite called *caliche*. Figure 2.52 shows light-colored caliche nodules in red mud. Although most rocks contain only small amounts of carbonate minerals, some kinds of sedimentary and metamorphic rocks are dominated by carbonates. For example, limestones (sedimentary) and marbles (metamorphic) may be composed entirely, or nearly entirely, of carbonates including calcite or dolomite.

In geological settings with an abundance of organic matter, the organic matter and water, commonly aided by microbial action, may react to produce carbonic acid. Carbonate minerals are unstable in acidic waters, so mineral dissolution can occur. This explains why all major cave systems are in limestones. No other common rock type has the right solubility and the ability to react to produce carbonic acid.

Calcite is not the only mineral that dissolves in the presence of carbonic acid. Feldspars, arguably the most abundant minerals in Earth's crust, and some other minerals also react with acid to produce a variety of minerals. The transformation of calcium silicate minerals into calcite may be a very important mechanism for removing carbon dioxide from air over time. For example, CO_2 may have been removed from the atmosphere by the uplift of the Himalaya Mountains. The uplift, which is still ongoing today, has exposed many feldspars to weathering. Some investigators believe this weathering has removed considerable carbon dioxide from the atmosphere. It may have been responsible for global cooling that started 55 million years ago and, perhaps, was a contributor to the Pleistocene glaciations that began 2 million years ago.

Carbonate minerals and CO_2 reach the deep Earth when subduction carries them into Earth's mantle. Metamorphism associated with subduction, or during orogenesis, may cause new carbonate minerals to form, moving carbon from one mineral to another. The presence of calcite-rich veins in metamorphic rocks suggests that carbon may also move through the deep Earth when carbonate minerals dissolve in water and flow along cracks and fractures. Metamorphism may give way to igneous activity if temperatures get hot enough. When melting occurs, CO_2 enters melts and is carried back toward the surface to be recycled.

In the deepest parts of Earth's mantle, CO_2 is absent and carbon is present as graphite or diamond; both are elemental forms of carbon, but in the shallower mantle carbon dioxide is common. *Xenoliths* (from Greek for

"foreign rocks"), samples of the mantle carried to the surface by magma, often contain minerals with trapped bubbles of carbon dioxide gas and less commonly other carbon gases. Geophysical studies, too, confirm the presence of carbon dioxide in the mantle, and recent studies suggest that carbon dioxide may be responsible for significant amounts of mantle melting.

2.8.4 *Carbon in biological organisms*

Through photosynthesis, plants transform (gaseous) carbon dioxide into their (solid) organic parts, but the carbon does not remain in most plants long. Animals eat the plants and change the organic compounds into animal matter, nutrients, and wastes. Carnivores eat other animals and make new animal matter, nutrients, and wastes. Eventually, when animals die, their decomposing bodies release carbon dioxide and organic compounds into the air, water, sediments, and soils. These processes occur quickly compared with other parts of the carbon cycle, especially those parts involving geological cycles.

Some carbon atoms may remain in the wood of trees for hundreds or, less commonly, thousands of years. In fact, the oldest living organisms on Earth are bristlecone pine trees on the order of 5000 years old. Figure 2.53 shows bristlecones from the White Mountains of California, which is said to contain the oldest grove of trees on Earth. Eventually, however, all plant material, including ancient wood, burns or decays. During decay, fungi, bacteria, insects, and other organisms consume and recycle dead plant material. Decay releases, into the environment, carbon dioxide and other compounds

Figure 2.52   Caliche in mud.
Photo by M.C. Rygel, Wikimedia Commons.

Figure 2.53   Bristlecone pine.
Photo by M. Hedin, Flickr, Creative Commons License.

that can move on to become parts of other processes and organisms.

2.8.5 *Carbon in the atmosphere*

2.8.5.1 *Carbon dioxide*

Carbon in the atmosphere, mostly in the form of CO_2, is a key part of the carbon cycle. Natural processes including wildfires, volcanic eruptions, hot springs, animal respiration, plant transpiration, and decay of organic material add CO_2 to the atmosphere. However, photosynthesis, weathering of feldspars and other minerals to form calcite, reactions of CO_2 with rain and surface waters to form carbonic acid, and deposition of carbonate minerals to form limestone remove it. Most sources of CO_2 are natural and, for most of Earth's history, the natural flux of CO_2 into the atmosphere has been balanced by the flux removing it from the atmosphere. But, since the Industrial Revolution, *anthropogenic* (human) contributions have caused the carbon dioxide concentration in the atmosphere to increase.

For the past hundreds of thousands of years, glacial ice has trapped bubbles of atmospheric gases, and these bubbles provide a record of the composition of Earth's atmosphere. This record allows scientists to examine the changes in atmospheric carbon dioxide concentrations over time and to compare those changes with global temperatures.

Studies based on the content of air bubbles trapped in ancient glacial ice of Greenland and Antarctica reveal that, during the past 650,000 years, atmospheric carbon dioxide was as low as 180 ppm (parts per million) during glaciations and did not exceed 300 ppm until the Industrial Revolution. Figure 2.54 shows the record of atmospheric CO_2 for the past 400,000 years, and the inset shows the last 1000 years. Carbon dioxide concentrations are always highest (peaks on the graph) during warm interglacial periods and lowest during the glacial periods. Today, carbon dioxide concentration in the atmosphere is about 405 ppm—and is measurably increasing. This level of concentration is more than was in the atmosphere during any of the most recent warm periods between ice ages.

2.8.5.2 *Methane*

Like carbon dioxide, methane cycles in and out of the atmosphere. Most natural methane comes from wetlands, methane hydrates (compounds made of ice and

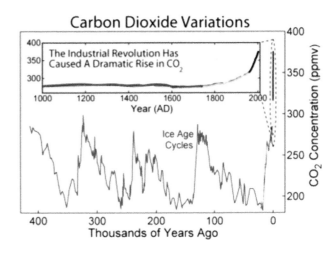

Figure 2.54 Carbon dioxide variations. From R.A. Rohde, Wikimedia Commons.

methane) in the oceans, and termites. More than half of the anthropogenic (human) methane comes from fossil fuel production and use, as well as livestock raising. Other sources include rice paddies, rotting food in landfills, and fires. Methane will react with oxygen in the atmosphere to produce carbon dioxide and water, but the process is very slow. Before the Industrial Revolution, natural methane sources were balanced by natural methane sinks, but today anthropogenic sources have upset this balance. Studies of ice cores from Greenland and Antarctica indicate that methane concentrations have increased since the Industrial Revolution.

Although usually in minor concentrations, methane is an important component in the carbon cycle because it is a major greenhouse gas and contributor to global climate change. Much methane is trapped in frozen tundra in polar regions, and many climatologists are concerned that, as Earth warms, the tundra will melt and release more methane into the atmosphere. This release would substantially accelerate the present-day global warming.

2.8.5.3 *Global climate change*

Temperatures at Earth's surface depend almost entirely on the amount of energy received from the sun and retained by Earth. Thus, the *greenhouse effect*—the ability of Earth's atmosphere to keep heat from escaping—has great influence on all living organisms. Greenhouse gases have been present in the atmosphere since early in Earth's history and are necessary for life to exist. If Earth's atmosphere had no greenhouse gases, average surface temperatures would be around –22 °C, Earth's surface would be entirely or almost entirely glaciated, and much of Earth would be devoid of life.

The most important greenhouse gases today are water vapor, carbon dioxide, and methane. They have always been present in Earth's atmosphere. As discussed previously, and depicted in Figure 2.46, a problem today is that human activities are removing carbon from inside Earth and releasing it to the atmosphere, where it adds to greenhouse gases. The most noticeable effect is global warming which, in turn, causes more water to enter the atmosphere, causing even more warming. Figure 2.55 shows mean global temperatures for the past century and a half. The zero line is the mean temperature for this time period. This graph shows that average temperature has increased by 1.1 °C (2 °F) since 1880, and all evidence suggests that the temperature will continue to increase in the future.

The number one cause of Earth warming is the burning of fossil fuels—coal, oil, and natural gas—but deforestation and some industrial processes contribute too. Fossil fuels formed over a long time by the slow accumulation of carbon, burial in Earth, and storage in the subsurface for millions of years. The extraction of these fuels (by humans) is millions of times faster than such formation and storage, and, consequently, natural balancing systems cannot keep up. The sudden release of large quantities of carbon into the atmosphere over the past 150 years, mostly as a result of extracting and burning fossil fuels, is causing profound environmental effects and global climate change.

Since the time of the Industrial Revolution, in about 1850, CO_2 levels have increased by more than 40%,

and methane has increased by more than 150%. During this time, Earth's average temperature has increased by about 1.1 °C, which seems small but may have drastic consequences. Figure 2.56 shows the increase in carbon dioxide levels in the atmosphere of the Northern Hemisphere since 1960. The measurements were made at high elevations in Hawaii. The temporary downward cycles resembling saw-blade teeth are due to plants in the Northern Hemisphere using carbon dioxide for photosynthesis in the spring and summer seasons.

Understanding global climate and predicting what will happen in the future is complicated because even a small amount of warming can cause changes that lead to the release of additional greenhouse gases into the atmosphere, making the situation even worse. There are also some mitigating effects that could work the other way. For example, more atmospheric carbon dioxide means that plant growth on land may increase, but only if there are enough nutrients (especially nitrogen and phosphorus) and water available, which does not seem to be the case today. Additionally, feldspars and other silicate minerals in sediments, soils, and rocks can neutralize carbonic acid and remove carbon dioxide from the atmosphere, but the process is too slow to compensate for the present rate of anthropogenic emissions. Yet, the cumulative effects of humans' disregard for natural cycles are clear today: Earth is warming, ice is melting, and sea levels are rising. There is substantial evidence, too, that normal weather patterns have been disrupted, precipitation has increased (on average) across the globe, hurricanes and other storms are stronger, and floods and droughts are more common than before the Industrial Revolution.

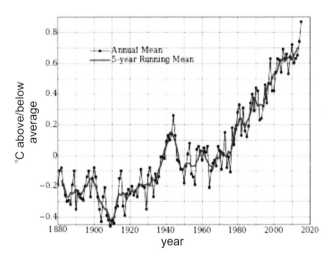

Figure 2.55   Temperatures during the past 140 years.
Modified from Wikipedia.

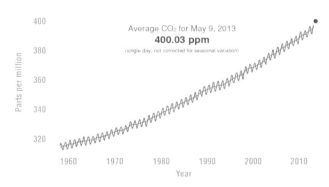

Figure 2.56   Increases in atmospheric carbon dioxide in the Northern Hemisphere since 1960.
Modified from a NASA image.

Questions for thought—chapter 2

1. Why would a silicon atom cycle move through the Earth's systems differently than a helium atom?
2. Why can life in a lake continue indefinitely without human assistance, but not life in an aquarium?
3. Why are landfills designed to be closed rather than isolated systems?
4. The Earth appears to be a closed system, but explain why it's technically open.
5. Why don't we depose of our garbage by launching it into space?
6. Explain how a water molecule can be in more than one sphere at once.
7. Why are igneous minerals often stable deep within the Earth, but not on the Earth's surface?
8. Why is the ocean crust younger in the middle of the Atlantic Ocean and older off the coasts of North America and Africa? How are geologists able to date ocean crusts and sediments?
9. Satellite studies can actually measure the rate at which southeastern Brazil and Angola, Africa are spreading apart. The speed is about 4 centimeters/year. Although the spreading rate has probably varied somewhat since the breakup of Pangaea and the formation of the southern Atlantic Ocean, we can use the current spreading rate to roughly estimate the amount of time that Brazil and Angola took to spread apart. To make this estimation, the distance between the edges of the continental shelves off the coasts of Campos, Brazil and southwestern Angola were measured. (The continents split apart at their shelves and not at present sea level locations.) The continental shelves are about 5500 kilometers apart. If Brazil and Angola have been constantly separating at 4 centimeters/year, how long ago were they together in Pangaea? How does your answer compare to the age of 135 million years ago predicted by fossils and radiometric dating?
10. In science, what is the difference between a hypothesis and a theory?
11. Name five pieces of evidence supporting the idea that plate tectonics is occurring and continents are drifting.
12. Why are the oldest oceanic crustal rocks in today's oceans only about 200 million years old when the Earth is 4.55 billion years old?
13. What determines why one piece of oceanic lithosphere subducts under another in an ocean-ocean subduction zone?
14. Would you guess that the volcanic rocks in Japan are older or younger than those of the Aleutian Islands of Alaska? Explain why.
15. Why are volcanoes rare in continent-continent collision zones?
16. Marine limestones are found on the summit of Mt. Everest. How did they get there?
17. The movie *Volcano* (1997) is a story about volcanoes erupting in Los Angeles. Why is this unlikely to happen?
18. The song *Day After Day (It's Slippin' Away")* by Shango (1969) describes California slipping off the North American continent and sinking into the ocean. What is actually expected to happen to southern California in 50 million years?
19. Consider Figure 2.40, why are the Hawaiian islands to the northwest smaller than the big island of Hawai'i?
20. Why does the ratio of carbon-12 to carbon-13 in petroleum indicate that it had a biological origin? What does the ratio tell us about the origin of meteorites?
21. Why do caves form in limestones but not in granites or sandstones?
22. Why could the formation of the Himalayas been partially responsible for the Pleistocene glaciations?
23. Why is carbon dioxide more abundant in the Earth's atmosphere than methane?
24. Why would the complete removal of greenhouse gases from the atmosphere be bad for life on Earth?

PART II
Fundamental Earth Materials

3 Minerals

3.1 Zeolites

What do water softeners, fish tanks, kitty litter, and people with radioactive poisoning have in common? The answer is zeolites. Zeolites, are a large group of related silicate minerals that contain alkali and alkali earth elements, along with aluminum, silicon, oxygen, and water. Zeolite compositions are essentially equivalent to feldspar minerals plus water, and many zeolites form in openings in feldspar-containing rocks. Most commonly, natural zeolites form in cavities and cracks in volcanic rocks. Some form by *diagenesis* (low-temperature alteration of already-existing sediment or rock) and others by *metamorphism* (higher temperature change in mineralogy or texture), but no sharp line divides the two processes.

Figure 3.1 shows orange chabazite crystals (up to 1 cm across) on top of grayish heulandite. Both of these zeolite minerals occur most commonly in openings within basalt and less commonly in metamorphic rocks or hydrothermal veins. This sample comes from a classic locality near Wasson Bluff, on the shore of the Bay of Fundy, Nova Scotia, where many outcrops of 200 million-year-old basalt contain cavities filled with spectacular mineral specimens.

Figure 3.2 shows the atomic arrangement in the zeolite *analcime*—it is similar in most ways to other zeolites. Zeolites contain *tetrahedra* (equal-sided pyramids shown in green and yellow) formed by aluminum and silicon cations surrounded by four oxygen anions. The oxygen bond to other tetrahedra and to large ions such as Na+ (purple) and to water molecules (blue). Analcime and other zeolites contain holes and channels that can hold or transmit large ions, such as sodium, calcium or potassium, water, or other molecules. This property sets them apart from most other minerals. Nearly 50 different natural zeolite species are known. Some more common varieties are listed and shown in Table 3.1.

Figure 3.1 Chabazite and heulandite.
Photo from R. Lavinsky, Wikimedia Commons.

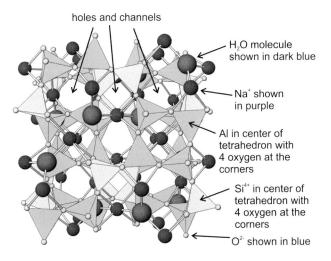

holes and channels

H_2O molecule shown in dark blue

Na^+ shown in purple

Al in center of tetrahedron with 4 oxygen at the corners

Si^{4+} in center of tetrahedron with 4 oxygen at the corners

O^{2-} shown in blue

Figure 3.2 Atomic arrangement in analcime.

Table 3.1 Some of the more common zeolite minerals.

Mineral	Formula	
analcime	$NaAlSi_2O_6 \cdot H_2O$	
chabazite	$(Ca,Na_2,K_2,Mg)Al_2Si_4O_{12} \cdot 6H_2O$	
clinoptilolite	$(Na,K,Ca)_{2-3}Al_3(Al,Si)_2Si_{13}O_{36} \cdot 12(H_2O)$	
heulandite	$(Ca,Na)_{2-3}Al_3(Al,Si)_2Si_{13}O_{36} \cdot 12H_2O$	

Mineral	Formula
natrolite	$Na_2Al_2Si_3O_{10} \cdot 2H_2O$

phillipsite	$(Ca,Na_2,K_2)_3Al_6Si_{10}O_{32} \cdot 12H_2O$

stilbite	$NaCa_4(Si_{27}Al_9)O_{72} \cdot 28(H_2O)$

Photo credits: analcime—Linnell, Wikimedia Commons; chabazite and stilbite—D. Perkins, GeoDIL; clinoptilolite—J. Sobolevski, Wikimedia Commons; heulandite, natrolite, phillipsite—R. Lavinsky, Wikimedia Commons.

Zeolites contain water that can be driven off by heat with the basic structure left intact. Removal of water leaves voids that can be filled, making zeolites useful for many purposes, including ion exchange, filtering, odor removal, chemical sieving, and gas absorption. The best-known use for zeolites is in water softeners. Because calcium in water causes it to be "hard," suppressing the effects of soap, forming scale, or creating other problems, it is often filtered through zeolite. Zeolite charged with sodium ions (generally from dissolved salt) allows water to pass through its structure while exchanging calcium for sodium, thus making the water less hard. This process is reversible, so the zeolite can be flushed and used again.

In a similar way, zeolites can absorb unwanted chemicals, including toxins. For example, zeolites may be components of kitty litter because they absorb cat urine and zeolites are added to livestock feed to absorb toxins that are damaging or fatal to animals. Additionally, people who have been exposed to radioactive elements ingest zeolites to help rid their system of those dangerous elements, and aquarium hobbyists use zeolites to remove ammonia and other toxins from fish tanks. Of great significance, in the highly developed countries of the world, most municipal water is processed through zeolites to remove large ions and organic contaminants before public consumption. Synthetic zeolites are easily grown in the laboratory, and scientists have synthesized more than 150 varieties. For many applications, synthetic zeolites have advantages over their natural cousins, because they can be manufactured to be pure and homogenous and to have unique properties not found in nature.

3.2 Minerals defined

Zeolites are but one of the many groups of minerals. Minerals and mineral groups are diverse, varying greatly in composition and atomic arrangement. Adding more complications, the term *mineral* is used in many ways. Dieticians use the term to refer to elements required by living organisms, such as calcium, iron, and magnesium. Miners use the term to refer to any geological resource that they can remove from the ground, including ore minerals, energy minerals, and building materials. Mineralogists use the term in a more specific way. James D. Dana (1813–1895), who developed the first widely used mineral classification system (which forms the basis of the one used today), defined a mineral as

"a naturally occurring solid chemical substance formed through biogeochemical processes, having characteristic chemical composition, highly ordered atomic structure, and specific physical properties."

More recently, the Commission on New Minerals and Mineral Names of the International Mineralogical Association (IMA) interpreted Dana's definition and promulgated guidelines for researchers as they investigate potential new minerals. In its introductory paragraph, the IMA definition says, "A *mineral* is a naturally occurring solid that has been formed by geological processes, either on Earth or in extraterrestrial bodies."

After this basic start, the IMA definition tackles some key issues from the Dana definition, including the meaning of the phrase "characteristic chemical composition," whether minerals can be wholly or partly *anthropogenic* (created by human activity), whether minerals may have biological origins, and whether they need be *crystalline* (have an ordered and repetitive atomic arrangement). The exact definition of a mineral is sometimes debated today, but for most purposes it is sufficient to say that minerals are natural crystalline solids, with a well-defined composition, and generally form by inorganic processes. The few exceptions are dealt with on a case-by-case basis by mineralogists.

3.3 Importance of minerals

What do you think minerals are used for? If you are like most people, you will think of gems or gold first. Demand for gems and gold, however, is small compared to demand for many other important mineral resources. Highways and buildings, fertilizers, cars, jewelry, computers and other electronic devices, kitchenware, salt, magazines, vitamins and medicines, and just about everything we use on a daily basis require mineral resources for production. Today's society could not function without minerals and related commodities.

The United States Geological Survey (USGS) and others have compiled lists of minerals essential to our daily lives (Table 3.2). Such lists include many "true" minerals but also include rock material and elements that we extract from minerals, as well as construction materials such as gravel and cement. So, even the USGS uses the term *mineral* in a general way to refer to any geological material that can be produced and used to benefit humans.

We use stone, sand, gravel, and other *construction materials* more than we do other mineral commodities.

Table 3.2   Some key mineral resources, lifetime needs per person, and primary sources (from the USGS).

Mineral commodities	Lifetime needs	Primary source
aluminum (bauxite)	5677 pounds	bauxite
cement	65,480 pounds	limestone or related rocks
clays	19,245 pounds	sedimentary deposits
copper	1309 pounds	copper sulfide minerals
gold	1576 ounces	native gold in hard rock or sediments
iron ore	29,608 pounds	magnetite and hematite
lead	928 pounds	galena
phosphate rock	19,815 pounds	apatite
stone, sand, and gravel	1.61 million pounds	sedimentary deposits
zinc	671 pounds	sphalerite

Construction materials are generally mixtures of different minerals and are prized for their overall properties, not the properties of the individual minerals. We also use large amounts of what are termed *industrial minerals*—resources valued for their mineralogical properties. Industrial minerals include limestone, clays, bentonite, silica, barite, gypsum, and talc.

Also important to our daily lives are *ore minerals*—minerals that are mined and processed for the elements they contain, not because of their mineralogical properties. Especially important are minerals that contain iron, aluminum, copper, zinc, and other metals associated with steel making and modern industry. The rock in Figure 3.3 contains blue azurite and green malachite. It comes from the Morenci Mine, by far the most productive of about a dozen copper mines in Arizona. Arizona produces more copper than any other U.S. state, but robust production statistics, however, do not reflect the true importance of some mining operations. Small amounts of rare elements, including gallium, indium, and selenium, derived from equally rare minerals and only produced in small quantities, are keys to fast computers, smartphones, and other cutting-edge devices.

Mining and processing minerals is a huge, global business. Mineral resources, however, are not distributed evenly, and some countries dominate export markets while others rely almost entirely on imports. South Africa is the world leader in mining and mining exports, having huge reserves of manganese, platinum, gold, diamonds, chrome, and vanadium. Figure 3.4 summarizes South Africa's production in 2009: gold and platinum-group elements, along with coal, accounted for three quarters of the mining revenue. In contrast with South Africa, the United States only produces a small amount of what it uses; we rely on imports to meet most demand. Platinum-group metals, rare earths, indium, manganese, and niobium are especially problematic

Figure 3.3   Copper ore from the Morenci Mine, Arizona.
Photo from R. Lavinsky, Wikimedia Commons.

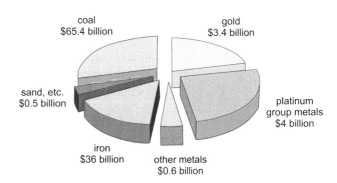

Figure 3.4   Mineral commodity production, South Africa, 2009.
Data from the Southern African Institute of Mining.

because we import them from sources and countries of uncertain reliability. The inequitable distribution of mineral resources has often led to conflict and trade issues. Wars have been fought to protect salt supplies in the East Indies and over diamonds in central Africa. Recent fighting in the Congo has been connected with gold and tin resources.

3.4 Studying minerals from the past to present

The use and processing of minerals goes back to at least 2900 BC, when Egyptians smelted gold and the Bronze Age began in Greece. The beginning of the field of mineralogy, however, is generally credited to the Greek philosopher Theophrastus, who lived ca. 372–287 BC and who wrote the first mineralogy book, *Concerning Stones*. Pliny (AD 23–79) expanded on the efforts of Theophrastus in his book *Natural History*, but major advances waited until the 19th century. Much of the delay in the development of mineralogy was because it is a hybrid science—mineralogy borrows from chemistry and physics, and the fundamental underpinnings of chemistry and physics were poorly understood until the late 1800s. In the mid-1600s, Robert Boyle, an Irish natural philosopher, chemist, and physicist, defined *elements*, and by 1800 many had been isolated. Between 1794 and 1812, French chemist Joseph Proust and others defined *compounds* and introduced the idea of chemical formulas. In 1848, Dana introduced a mineral classification scheme based on anions and anionic groups that is still used today. Thus, the stage was set, and by the end of the 19th century the fundamental understandings of mineral chemistry were well established.

Nineteenth-century mineralogists sought to create a complete list of minerals and their properties, especially diagnostic properties, so that minerals of different species could be distinguished. They collected detailed descriptions of minerals and used painstaking laboratory techniques, including *blowpipe analyses*, to measure mineral compositions. Similar studies, termed *descriptive mineralogy*, continue today, although the number of new minerals being discovered has slowed. Describing minerals can be tedious work and involves using a hand lens or microscope and making many different kinds of measurements and observations of properties, including composition, crystal shape, hardness, color, and density. Nonetheless, some mineralogists and amateur rockhounds travel to remote and exotic places to search for new minerals and new varieties of previously known ones.

William Conrad Roentgen's discovery of X-rays in 1895, and the subsequent discovery of radioactivity by Henri Becquerel and Marie Curie, set the stage for *X-ray diffraction studies*. In 1912 and 1913, multiple researchers, including Max Von Laue, William H. Bragg, and William L. Bragg, began collecting *X-ray diffraction patterns* for minerals and other crystalline compounds. Diffraction patterns, unique to any particular mineral, are produced when an X-ray beam enters a crystal and is re-emitted in multiple directions. X-ray diffractometers are common today, and diffraction patterns are now a key component of descriptive mineralogy. In fact, for nearly a century, diffraction patterns were the only definitive way to tell similar minerals apart. Roentgen, Von Laue, and the two Braggs all received Nobel Prizes for their pioneering X-ray work.

Figure 3.5 shows a powder diffractometer used for collecting X-ray patterns of minerals and other crystalline materials. Samples are ground to a fine powder for analysis and placed in a holder (center of photo). In the instrument shown, the X-ray beam comes from the source (X-ray tube), through slits, is diffracted from the sample, goes through another set of slits, and beam intensity is measured by the detector on the right. Measurements are taken as the detector rotates (black arrow) so that the X-ray hits it over a range of angles. The geometry sounds complicated, but X-ray analyses are very reproducible and reliable.

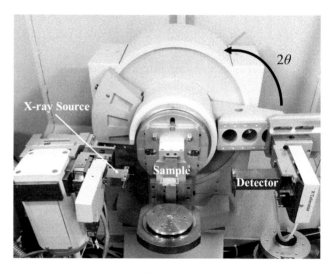

Figure 3.5   X-ray powder diffractometer.
Photo by K. Henke.

For routine mineral identification, mineralogists use powdered diffractometers to obtain diffraction patterns such as the one shown in Figure 3.6. This X-ray pattern is for vermiculite, a clay mineral. The horizontal axis is the angle between the detector and the impinging X-ray beam (Fig. 3.5). The vertical axis shows the intensity of diffraction in different directions.

For determining the positions of atoms within a crystal, mineralogists use a single crystal and a slightly different kind of diffractometer. A single mineral crystal may diffract X-rays in thousands of directions. The amount of information was overwhelming 100 years ago, when diffraction studies were first done. With modern X-ray diffractometers and fast computers, we can quickly measure the directions and intensities. We use the data to figure out how individual atoms are arranged and what they are bonded to, within a crystal. This approach is used by mineralogists who investigate *crystallography*—the study of how atoms bond together to make crystals—and allows us to make ball and stick drawings that depict atomic arrangement, such as the drawing of analcime in Figure 3.2.

Mineral analytical techniques have improved markedly over the last century, and today determining mineral composition is relatively straightforward. The most popular instrument for this purpose is an *electron microprobe*, also known as an *electron probe microanalyzer* (Fig. 3.7). A microprobe uses a narrowly focused

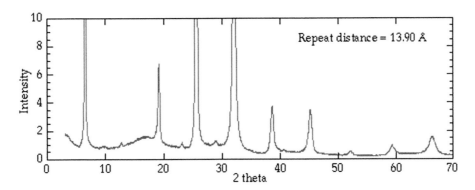

Figure 3.6   An X-ray powder diffraction pattern for vermiculite.
Courtesy of M. Stanley Whittingham.

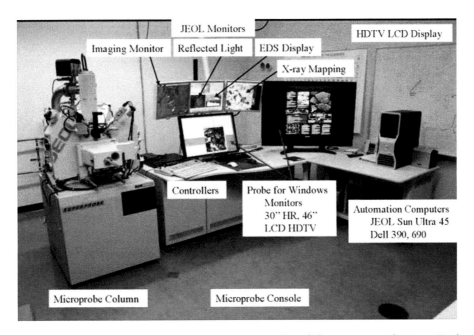

Figure 3.7   Electron microprobe and supporting equipment, Experimental Studies of Planetary Materials Group, Earth, and Planetary Sciences, Washington University at St. Louis.
https://espm.wustl.edu/research/

electron beam that strikes a sample. Electrons emitted when the beam strikes the sample allow imaging at very high magnifications. Additionally, the interaction between electrons and atoms in the sample causes emission of X-rays with wavelengths characteristic of the elements present. By measuring wavelengths and intensities, we can calculate mineral composition. The microprobe, and several related instruments, are a fundamental source of information for those who study *crystal chemistry*—the chemical nature and variability of mineral compositions.

3.5   Elements, minerals, and rocks

Figure 3.8 shows the relationships between elements (bottom), minerals (center), and rocks (top). Elements, singly or in combination, make up minerals. For example, some of the most common elements in Earth's crust make up

the minerals quartz, alkali feldspar, and biotite. Minerals, singly or in combination, make up rocks. For example, subequal amounts of quartz and alkali feldspar, sometimes with biotite and plagioclase, make up granite, a common crustal igneous rock (triangular diagram in Fig. 3.8).

All minerals, like all materials, consist of one or more elements, the building blocks of all matter (Fig. 3.8). Some minerals, diamond for example, contain a single element (carbon). Others contain many elements. Some minerals have compositions that vary little in nature. Quartz, for example, is always close to 100% silicon and oxygen in the atomic ratio 1:2. Other minerals incorporate elemental substitutions, so their compositions may vary a great deal from sample to sample. Biotite, for example, always contains potassium, magnesium, iron, aluminum, silicon, and oxygen. It generally also contains lesser amounts of manganese, sodium, titanium, and many other elements, so natural biotite compositions are quite variable.

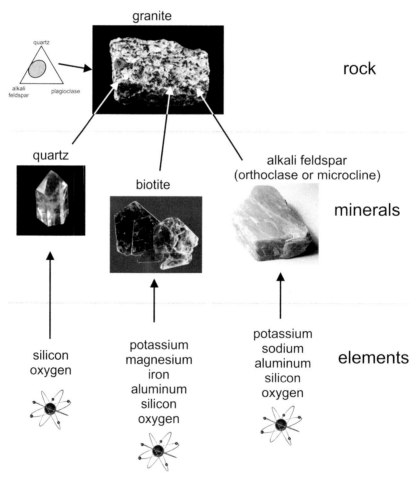

Figure 3.8   The relationships between elements (bottom), minerals (center), and rocks (top). Photos from D. Perkins, GeoDIL.

Rocks are aggregates of one or more minerals, mineral-like substances (termed *mineraloids*), and organic components. Rocks may form when minerals grow (*crystallize*) together, forming a *crystalline rock*, such as granite. They also can form when loose grains are cemented together, forming a *clastic rock*. Crystalline rocks may form from a magma (e.g., granite), may form by metamorphism (e.g., gneiss), or may form by precipitation from water (e.g., gypsum). Most clastic rocks form from consolidated sediments, but some form by volcanic processes.

Some rocks contain only one kind of mineral. Limestone (a sedimentary rock), for example, is often pure calcite (mineral). Dunite, an igneous rock that crystallizes from magma, is often nearly 100% olivine (mineral). Other rocks are composed of multiple minerals and generally named based on mineral proportions. Thus, all granite (rock), for example, contains subequal amounts of quartz (mineral) and alkali feldspar (mineral), often with lesser amounts of plagioclase (mineral).

3.6   Mineral formation

The formation of crystals involves the bringing together and ordering of constituent atoms; the atoms combine in an orderly repetitive arrangement (Fig. 3.9). For example, potassium and chlorine combine in an ordered way to form the mineral *sylvite*. This figure shows two different depictions of the atomic arrangement in sylvite (KCl). These are the typical kinds of drawings used to show atomic arrangements, depicting ions (incorrectly) as hard spheres of fixed radii.

Crystals grow from a small single molecule to their visible form. If the conditions are right, crystals may grow to be very large. In 2007, National Geographic reported some of the largest crystals in the world: gypsum crystals in Mexico's Cueva de los Cristales (Cave of Crystals, discussed more in Chapter 4) measure up to 10 meters long! Some mineral crystals—such as gypsum, garnet, beryl, and others—often grow to quite large size. Others do not. The reasons why some grow large and many do not are poorly understood.

3.6.1   *Igneous minerals*

One way that minerals form is when magma cools. In magma, *kinetic energy* ensures that atoms are always in motion. Some collide and may form bonds temporarily before breaking apart again. A balance is maintained between the formation of bonds and the rate at which they break apart. If bonds break as fast as they form, there will be no net crystallization. Kinetic energy is greater at higher temperatures, so magmas will be completely molten at high temperatures. When a magma cools sufficiently, however, atoms slow down and some bonds begin to persist, which is the beginning of the formation of crystals from a melt. The initial crystals form nuclei, many of which continue as the centers of continued crystal growth.

Figure 3.10 shows large light-colored quartz and green emerald (a variety of the mineral beryl) crystals. Many igneous crystals are so small that it takes a microscope to see them. Igneous mineral crystals in granites and other plutonic rocks may be sand- or pea-sized. But, some crystals in *pegmatites* (very coarse-grained igneous rocks such as the one shown) grow to be many centimeters or meters in their longest direction. Many

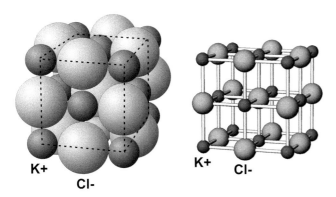

Figure 3.9   The atomic arrangement in sylvite (KCl).

Figure 3.10   Large beryl and quartz crystals in a pegmatite. Photo from Madereugeneandrew, Wikimedia Commons.

of the world's most spectacular mineral crystals come from pegmatites.

3.6.2   *Aqueous minerals*

Figure 3.11 shows *tufa* towers that precipitated underwater in Mono Lake, California. The tufa was exposed when the lake level dropped. The tufa (a variety of calcite) formed when hot spring waters rich in calcium encountered cool lake water rich in carbonate, resulting in the precipitation of calcium carbonate on the lake bottom. The tufa at Mono Lake is an example of minerals forming in an *aqueous system*, and the minerals that formed are termed *aqueous minerals*. Salt (halite) and gypsum crystals that commonly form in dry lake beds, evaporating ocean basins, or on salt flats are also examples of aqueous minerals.

The potential for formation of aqueous minerals is widespread. Atoms of different sorts may be present in water and, as long as the water is not saturated, no crystals will form. Atoms will bond temporarily, only to break apart and return to solution. Because most substances are more soluble in water at high temperature, a decrease in temperature may lead to oversaturation, nucleation, and precipitation (crystallization) of aqueous minerals.

Crystallization may also occur if composition changes. Suppose, for example, that seawater evaporates. The concentration of dissolved material in the remaining water will increase, leading to oversaturation and, eventually, precipitation of crystals. Besides changes in temperature and composition, changes in pressure, pH, or other things may also lead to the formation of aqueous or igneous crystals.

3.6.3   *Metamorphic minerals*

A third way minerals may crystallize is through *metamorphism*. Metamorphism often involves replacement of preexisting minerals by new ones. Bonds are broken and atoms migrate by *solid state diffusion* or are transported short distances by *intergranular fluids* to sites where new minerals crystallize and grow. Sometimes no new minerals form, but diffusion of atoms leads to *recrystallization* and a coarsening of mineral crystal size. In sedimentary rocks, a low-temperature form of metamorphism called *diagenesis* takes place.

Metamorphism may involve replacement of one mineral by another of equal composition. For example, calcite may become aragonite or vice versa. Both minerals are $CaCO_3$, but their atomic arrangements differ. Mineralogical changes due to metamorphism, however, usually involve several different minerals reacting together. Calcite ($CaCO_3$) and quartz (SiO_2) may react to form wollastonite ($CaSiO_3$) if a limestone containing quartz is metamorphosed at high temperature. Micas may react with quartz and other minerals to produce garnets in schists and other metamorphosed sedimentary rocks. Figure 3.12 shows an example of a rock in which this happened. The relatively large garnets can be seen sticking out of a sea of biotite. Sometimes, during metamorphism, a rock's overall composition may change significantly as fluids carry in or remove soluble materials. Such a process, called *metasomatism*, often leads to large, well-developed metamorphic minerals.

Figure 3.11   Tufa towers at Mono Lake, California.
Photo from D. Perkins, GeoDIL.

Figure 3.12   A typical garnet schist.
Photo from D. Perkins, GeoDIL.

3.6.4 *Mineraloids*

Some mineral-like substances, termed *mineraloids*, are partially or completely *amorphous*, which means they have a random atomic structure and thus are not crystalline. Some examples are shown in Figure 3.13. For example, natural volcanic glass, called *obsidian* (top photo),

obsidian

agate

opal

Figure 3.13 Three different mineraloids that are closely related to quartz.

Photos from D. Perkins, GeoDIL.

forms so quickly that atoms cannot arrange themselves in a regular, repetitive atomic structure. The process produces a glass that, unlike a mineral, is *noncrystalline*. So, obsidian is considered a mineraloid.

Agate (center photo) is a microcrystalline variety of silica (SiO_2) that is generally fine-grained and may have bright or distinctive coloration and, often, banding. The agate in Figure 3.13 is light-colored and rather drab compared with many agates. Although it has the same composition as quartz, strictly speaking, agate is not a mineral because it does not meet the definition of being crystalline; it does not have a homogeneous atomic arrangement and usually does not consist of a single mineral. Many mineralogists, however, consider agate, and the related mineraloid *chalcedony*, to be textural varieties of quartz.

Opal (bottom photo), another mineraloid, is an amorphous form of SiO_2 related to quartz (Fig. 3.13). It typically contains 6%–10% water, so it is not quite the same composition as true quartz. At an atomic level, precious opal is composed of spheres of silica, 150–300 nanometers in diameter, stacked together. The spheres cause light refraction and give opal its sought-after play of colors. Some opal has localized ordering of atoms and thus a small amount of crystalline character. The opal in Figure 3.13 has been rounded and polished as a gem stone.

3.6.5 *Natural and synthetic minerals*

Traditionally, only substances formed through natural processes are considered minerals, yet synthetic equivalents of zeolites and of many other minerals can be made in the laboratory (Fig. 3.14). Other crystalline substances can be created that have no natural analogs. Synthetic crystals, however, are not considered minerals because they are not formed by natural processes. Additionally, although some synthetic minerals may look like the real thing, they are not the same on a small scale. Mineralogists, for example, can often distinguish synthetic minerals from natural ones based on the number and type of minute *inclusions* they contain.

Photo a, in Figure 3.14, shows natural blue topaz. This is a rare color for topaz, which is typically clear or light-colored. So, most commercial blue topaz is synthetic. Photo b shows examples of synthetic blue topaz that have been cut as gemstones. The ruby on top of the rock in photo c is a natural gemmy red variety of the mineral corundum (aluminum oxide). The rock is a

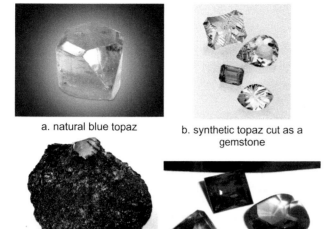

a. natural blue topaz

b. synthetic topaz cut as a gemstone

c. natural corundum crystal

d. synthetic corundum made in a laboratory

Figure 3.14   Natural and synthetic minerals. The photos of natural topaz, natural corundum, and synthetic ruby come from Roy Goldberg, Rob Lavinsky, and Ana Mladenovic, all at Wikipedia Commons.

The photo of synthetic blue topaz comes from the Gemological Institute of America.

Table 3.3   Some common examples of biominerals.

Mineral	Examples of organic origins
magnetite	bacteria, teeth of chitons, brains of vertebrates
calcite	shellfish, sponges, corals
quartz and related silica minerals	radiolarians, diatoms, sponges
apatite	vertebrate teeth, bone, conodonts

schist from Winza, Tanzania. The three synthetic rubies in photo d are also crystalline aluminum oxide, but they were made in a laboratory and then cut to produce gemstones. They are not, however, considered true minerals.

Many *synthetic gemstones* are sold today. They have essentially the same appearance and optical, physical, and chemical properties as their natural equivalents. Synthetic gemstones produced in the United States include alexandrite, coral, diamond, emerald, garnet, lapis lazuli, quartz, ruby, sapphire, spinel, and turquoise. Gem manufacturers also produce *simulants* that have an appearance similar to that of a natural gemstone but have different composition. Colored and colorless varieties of cubic zirconia, the most common simulants produced, may be hard to distinguish from real diamond by a layperson. Other simulants produced in the United States include coral, lapis lazuli, malachite, and turquoise. Additionally, certain colors of synthetic sapphire and spinel, used to represent other gemstones, are classed as simulants.

3.6.6 *Biominerals*

A strict definition of a mineral requires that it be inorganic. Yet minerals, or at least mineral-like substances,

are important components of many organisms, and some are created by combinations of organic and inorganic processes. So today, many mineralogists accept the existence of organic minerals. Table 3.3 lists some of the most common occurrences of biominerals.

Figure 3.15 shows examples of crystalline materials produced by organic processes (top row) and the same composition minerals produced by inorganic processes (bottom row). The materials in the top row today are referred to as *biominerals* by many mineralogists. Some of the most common examples of biominerals are seen here: carbonate shells produced by clams, phosphates produced by vertebrates (in teeth), and silica produced by diatoms,. *Biomineralization* is a widespread process involving many different types of organisms and, recently, a new field of specialization has developed (*geomicrobiology*) focused on minerals and mineraloids produced by bacteria and other microbes.

3.7   Common elements and the most common minerals

Mineralogists have described and named more than 4000 minerals, but most are extremely rare. Minerals containing just about all of the 115–120 known elements have been described, but as Figure 3.16 shows, fewer than 10 elements make up more than 99% of Earth's crust. Consequently, the most common minerals are restricted in composition. Most of the common rock-forming minerals are *silicates*—minerals with compositions rich in silicon and oxygen, because these two elements make up almost 75% of the crust.

Asking which are the most common minerals is really a trick question, because the answer depends on context and assumptions. Ice, olivine, plagioclase, quartz, or clays could be called the most common minerals (Fig. 3.17).

Ice, which meets all the requirements for being a mineral, is probably the most common mineral at Earth's

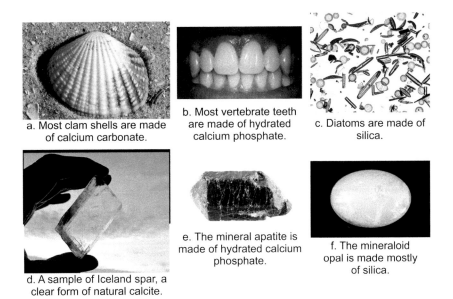

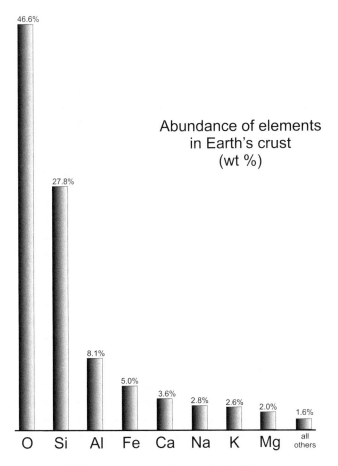

Figure 3.15 Examples of biominerals and their (inorganic) mineral equivalents.

Photo credits: a-Andrew Butko, b-dozenist, c-unknown, d-AmiEin, all at Wikimedia Commons. e and f - D. Perkins, GeoDIL.

**Abundance of elements
in Earth's crust
(wt %)**

46.6%

27.8%

8.1%

5.0%

3.6%

2.8%

2.6%

2.0%

1.6%

O   Si   Al   Fe   Ca   Na   K   Mg   all others

Figure 3.16 The most common elements in Earth's crust.

surface. Most mineralogists, however, ignore ice when talking about common minerals and consider only "traditional" minerals that make up rocks. Sediments and sedimentary rocks dominate the surface of continents, so the most often seen surface minerals are minerals typical of sedimentary rocks. This includes, especially, clay minerals and quartz because they are the most common products of weathering. Clays are the prime constituents of *shales*, the most common sedimentary rock type. Quartz is found in *sandstones*, *siltstones*, *conglomerates*, and in many igneous and metamorphic rocks exposed in outcrops.If we look deeper in Earth, but still in the crust, the most common minerals depend on where we look (Fig. 3.18). The oceanic crust is nearly entirely *basalt*, composed mostly of plagioclase. The oceans cover more than 70% of our planet's surface, so arguably plagioclase is the most abundant mineral in the crust, even if it is mostly underwater.

The average composition of the continental crust is more like *granite* than basalt. Quartz and the two feldspar groups dominate the mineralogy of the continental crust (Fig. 3.18). If we look deeper, into the uppermost mantle, we most commonly find *ultramafic rocks* dominated by olivine and pyroxenes. Different minerals are found at still greater depths in the mantle, but they are high-pressure minerals unstable at Earth's surface and so are almost never seen.

Ice, concentrated at polar regions, is arguably the most common mineral at Earth's sruface.

The green inclusion is a xenolith, a sample of the uppermost mantle dominated by olivine.

Basalt, containing pyroxene and olivine, makes up most of the oceanic crust.

Sandstone – mostly quartz – is widespread on continents

Shale -- the most common kind of sedimentary rock at Earth's surface – is mostly made of clays

Figure 3.17   The most common minerals on, and within, Earth.

Ice photo from Godot13, Wikimedia Commons; xenolith and sandstone photos from D. Perkins, GeoDIL; basalt photo from NOAA, Wikimedia Commons; shale photo from Pollinator, Wikimedia Commons.

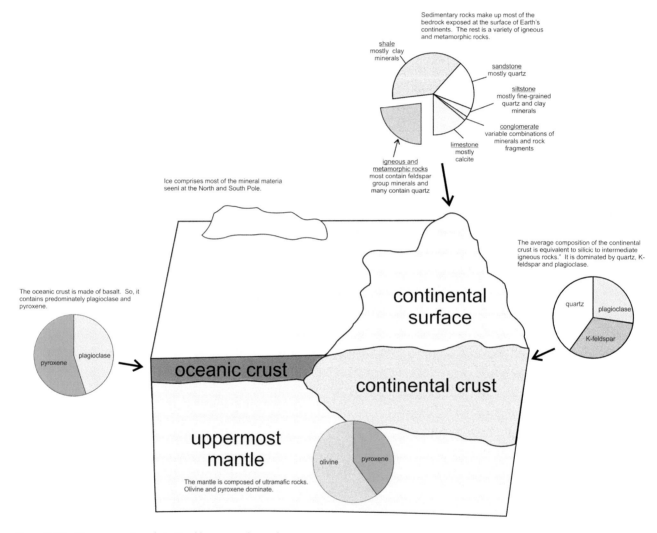

Figure 3.18   Common minerals in Earth's crust and mantle.

Table 3.4 lists some of the more common and important minerals. Included in this table are minerals, like quartz, that are widespread and found in abundance at Earth's surface. Others, including the feldspars (plagioclase and alkali feldspar), micas (biotite and muscovite), olivine, and several others, are listed because they are key minerals at depth in Earth. Still others are included because they are ore minerals, and so are important to society. This table provides a sampling only—there are other important

Table 3.4 Some of the most important minerals.

Example minerals	Most common occurrences
quartz	Quartz is one of the most common minerals in sedimentary rocks (because it is very stable at Earth's surface) and in many igneous and metamorphic rocks.
plagioclase and alkali feldspar	Plagioclase and alkali feldspar are the two major groups of feldspar minerals. They are common in many igneous and metamorphic rocks and make up more than half of Earth's crust.
biotite and muscovite	Biotite and muscovite are the two of the most common micas. They are found in many different kinds of igneous and metamorphic rocks.
hornblende and other amphibole group minerals	Amphibole group minerals are common in many types of igneous and metamorphic rocks. Hornblende is the most common amphibole.
augite	Pyroxene group minerals are common in mafic and ultramafic igneous rocks and in some high-temperature metamorphic rocks. Augite is the most common pyroxene.
olivine	Olivine occurs in mafic and ultramafic igneous rocks and in some metamorphic rocks. It makes up more than half the uppermost part of Earth's mantle.
almandine	Garnets are pretty and distinctive minerals found in many metamorphic rocks. Almandine, the most common garnet, occurs in metamorphosed sedimentary rocks.
kaolinite and other clay minerals	Minerals of the clay group are the most common kind of minerals in sedimentary rocks. Kaolinite, one kind of clay, is common in weathered rocks and soils.
calcite and dolomite	Calcite and dolomite are the two most common carbonate minerals. They are the dominant minerals in limestones and marbles.
hematite and magnetite	Hematite and magnetite are both iron oxides. They occur in rocks of many different kinds and are mined as iron ores.
pyrite and chalcopyrite	Pyrite and chalcopyrite are the two most common sulfide minerals. They occur as minor minerals in many kinds of rock. When concentrated, they are important ore minerals.
gypsum	Gypsum, the most common sulfate mineral, is common in sedimentary rocks, especially in thick beds associated with evaporating seas. It has many industrial uses.
sulfur	Sulfur is found as a deposit associated with hot springs, fumaroles, and other volcanic features. It has many industrial uses.
diamond	Diamonds originate at great depths in Earth. They are brought to the surface by explosive volcanic eruptions and are mined from igneous rock or associated sediments.
copper	Most copper is in multi-element minerals, but native copper is found associated with mafic volcanic rocks and, less commonly, sedimentary rocks.

minerals, too. A key thing to remember, however, is that only a relatively small number of the thousands of minerals that have been described are found in abundance or are used by people. Most mineral species are exceptionally rare, and most have little anthropocentric importance.

3.8 Mineral compositions

Diamond, graphite, copper, and a few other minerals contain a single element, but most minerals are compounds made of two or more elements, and we describe them using chemical formulas (Fig. 3.19). Formulas vary from short and simple to long and complex, depending on the minerals. When writing a mineral's formula, we use standard conventions to indicate proportions and relative abundances of elements and which elements occupy similar sites in the mineral's atomic arrangement. Some of these conventions are shown in Figure 3.19.

3.8.1 Solid solutions and mineral series

Minerals such as quartz (SiO_2), halite (NaCl), or fluorite (CaF_2) are generally close to the pure composition described by their formulas. They have compositions that vary little in nature, and only very minor amounts of other elements substitute for those in the ideal formula. Their formulas tell us the ratios of the constituent elements. Other minerals are *solid solutions*, and commas in formulas indicate elements that can substitute for each other (Fig. 3.19). Olivine, for example, may have any composition between Mg_2SiO_4 and Fe_2SiO_4, but the atomic ratio Mg+Fe:Si:O is always 2:1:4. So, olivine forms a solid solution and is a *binary mineral series* between two *end members*: Mg_2SiO_4 (called *forsterite*) and Fe_2SiO_4 (called *fayalite*). A general formula for olivine is $(Mg,Fe)_2SiO_4$ if it contains more Mg than Fe, and $(Fe,Mg)_2SiO_4$ if it contains more Fe than Mg. Subscripts may be used in the formula to reflect the ratio of Mg:Fe, if it is known. (e.g., $Mg_{1.82}Fe_{0.18}SiO_4$ describes olivine that is 1.81/2 = 81% forsterite and 9% fayalite.)

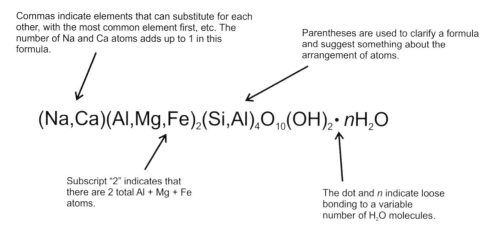

Figure 3.19   An approximate formula for montmorillonite, a clay mineral.

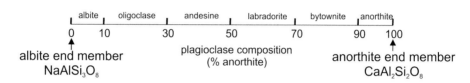

Figure 3.20   Range of compositions for plagioclase.

We can plot binary solid solutions on a line. For example, if we ignore minor components, Figure 3.20 shows the possible compositions of plagioclase, a binary solution of two end members, albite ($NaAlSi_3O_8$) and anorthite ($CaAl_2Si_2O_8$). Plagioclase can have any composition between pure end-member albite and pure anorthite, and a general formula is $(Na,Ca)(Si,Al)_4O_8$.

Depending on the composition, we give plagioclase a *variety name*: albite, oligoclase, andesine, labradorite, bytownite, or anorthite (Fig. 3.20). Adding some confusion, the name *albite* is used in two ways. It refers to both the end-member composition ($NaAlSi_3O_8$) and to any plagioclase that contains 90%–100% $NaAlSi_3O_8$. Similarly, the name *anorthite* refers to the end-member composition $CaAl_2Si_2O_8$ and to plagioclase that contains 90%–100% $CaAl_2Si_2O_8$.

Some minerals are solutions of more than two end members, and, as shown in Figure 3.21, triangular diagrams provide a way to plot compositions of ternary solutions (solutions of three end members). Compositions that are 100% A, B, or C plot at a corner on this diagram. Compositions that contain 0% of one end-member plot on an edge of the diagram. Other compositions plot somewhere in the middle. For example, the composition shown by the black dot on the left triangle contains 26% A, 12% B, and 62% C.

For a real example of minerals composed of three end members, consider the compositions of garnets in mantle samples from Arizona and New Mexico, shown on the right in Figure 3.21. This diagram shows that the mantle garnets are mostly > 50% pyrope end member, with subequal amounts of grossular and almandine.

3.8.2   Limited solid solutions

Some solid solutions are limited to specific composition ranges. Consider feldspars, a group of minerals that contain two series: the *alkali feldspar series* and the *plagioclase series* (Fig. 3.22). Alkali feldspars are generally solutions of end-member orthoclase and albite, perhaps containing minor anorthite. Plagioclase is generally a solution of albite and anorthite with only a limited amount of orthoclase. Compositions that would plot in the (white) middle of the feldspar triangular diagram are exceptionally rare in nature. The compositional gap (white region) is termed a *miscibility gap*.

The extent to which feldspars form solid solutions depends on temperature. The bottom diagram in Figure 3.22 shows, in tan color, compositions of stable plagioclase and alkali feldspars at 800 °C. Plagioclase may have any composition between albite and anorthite and

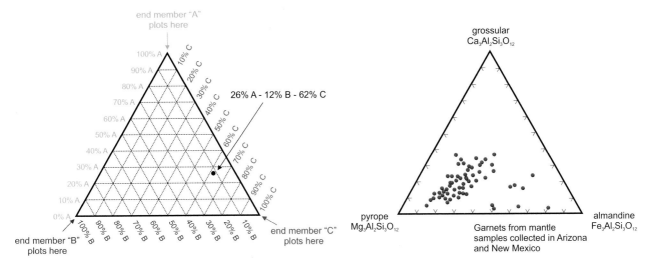

Figure 3.21   (Left) Plotting compositions on a triangular diagram. (Right) The composition of garnets in mantle xenoliths from New Mexico and Arizona.

Data from Switzer (1975).

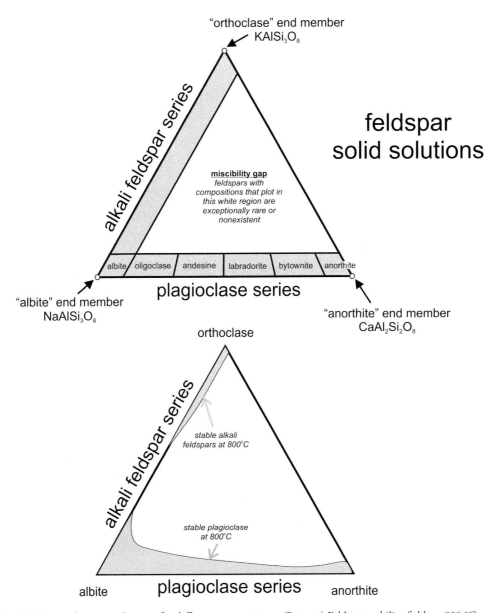

Figure 3.22   (Top) Feldspar solutions and names for different compositions. (Bottom) Feldspar stability fields at 800 °C.

may contain up to 10% orthoclase. Alkali feldspar compositions are restricted to a narrow range close to orthoclase. At higher temperatures, the miscibility gap grows smaller—which means more varied compositions are stable. At lower temperatures, the size of the gap expands as feldspar compositions become even more restricted than shown here. Many minerals exhibit *limited solid solutions* like the feldspars. At high temperatures, end members may mix freely, but at lower temperatures they unmix much like fat separates from chicken soup when it cools. This topic is further discussed in Chapter 4.

3.8.3   Mineral varieties

An individual mineral species, such as calcite, is defined by its composition and atomic arrangement. All calcite is mostly $CaCO_3$, with atoms arranged in the same way, no matter the size or shape of the sample. However, calcite, like many other minerals, has more than one named *variety*. Varieties are distinguished by crystal shape, composition, color, occurrence, or other characteristics. *Dogtooth spar*, for example, is a variety of calcite, found in some caves, that has a distinctive shape. And *Iceland spar* is a clear variety of calcite typically in cleavable rhomb shapes (Fig. 3.23).

Gemologists often name varieties based on color. Figure 3.24 shows three varieties of beryl. The mineral is called *aquamarine* if it has a light blue color, *emerald* if it is green, and *morganite* if it is pink. The different colors stem from very small compositional differences; all beryl is essentially $Be_3Al_2Si_6O_{18}$. Aquamarine, however, contains small amounts of Fe^{2+}, emerald contains small amounts of Fe^{2+} and Fe^{3+}, and morganite contains Mn^{2+}. Even small amounts of these transition metal ions can give minerals strong coloration.

Figure 3.23   Dogtooth spar and Iceland spar.
Photos from D. Perkins, GeoDIL.

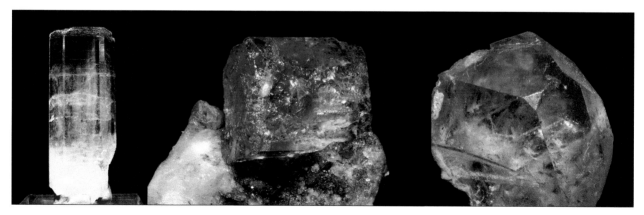

Figure 3.24   Aquamarine (left), emerald (center), and morganite (right).
Photos from Vassil, G. Parent, and P Géry, respectively, Wikimedia Commons.

3.8.4 Mineral groups

Figure 3.25 shows the atomic arrangement in spinel. The drawing on the right, with the atoms removed, emphasizes the fact that the arrangement is cubic. Mineralogists often group minerals that have the same basic atomic arrangements, even if they have different compositions. Sometimes, the group name is the same as one of the individual mineral species, adding some confusion. The mineral *spinel*, for example, has the specific composition $MgAl_2O_4$. The *spinel group*, however, consists of more than a dozen minerals, all with the general formula AB_2O_4. The A elements include Fe, Zn, Mn, Mg, or Ni; the B elements include Cr, Al, Fe, or Ti. No matter the elements present, the A atoms are bonded to four oxygen atoms, and the B atoms are bonded to six oxygen atoms. Angles between bonds and crystal shape are the same for all spinel group minerals, giving an overall cube-shaped structure, and the overall atomic arrangement is as shown in this figure.

3.9 Mineral stability

Although mineralogists have identified thousands of minerals, a relatively small number are common. One significant reason is that many minerals are unstable at normal Earth surface conditions and so react to form different minerals or decompose over time. As an example, let's consider the mineral ice, and related compounds water and water vapor.

Several different materials, or several different *phases*, have the composition H_2O. Hydrogen and oxygen can form a mineral: ice (H_2O), or they can exist as a liquid (water), or as a gas (water vapor). As we know from experience, ice forms at low temperature, water at intermediate temperature, and at even greater temperature water boils, producing steam (water vapor). The *Laws of Thermodynamics* control these relationships. Natural systems tend toward states of minimum energy (*stable states*), and the Laws of Thermodynamics tell us that, for a given chemical composition, the phase or phases with the lowest energy will be stable. In the H_2O example, the possible phases are ice, water, and water vapor. They each have their own *stability field*, and we can show the relationships between them on a *phase diagram* (Fig. 3.26).

There are many kinds of energy, but thermodynamics deals primarily with *Gibbs free energy*. Gibbs free energy is a value that describes how stable a particular phase is. At low temperatures, ice has a lower Gibbs free energy than water or water vapor. At intermediate temperatures, water has the lowest Gibbs free energy, and at high temperatures, water vapor has the lowest energy. On a phase diagram, along the boundary between two stability fields, two phases have equal energies. On the line separating the ice and water fields, for example, ice and water have equal energies, and both have less energy than water vapor.

The photographs in Figure 3.27 show diamond and graphite. They are *polymorphs*—both are made of carbon, but the carbon is bonded in different ways in the two minerals. Perfect diamond crystals may be octahedral (inset in Fig. 3.27); perfect graphite crystals are hexagonal tabs (inset). At normal Earth surface conditions, diamond is unstable, but fortunately, diamonds do not all turn into graphite and we have metastable diamonds to enjoy.

Thermodynamics and phase diagrams tell us what will form if a chemical system goes to *stable equilibrium*

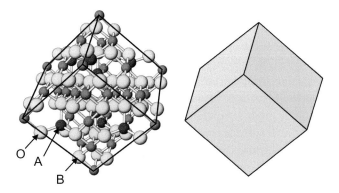

Figure 3.25 The cubic arrangement of atoms in spinel.

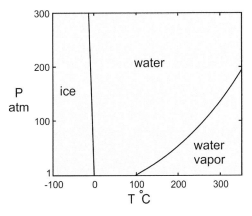

Figure 3.26 Phase diagram showing stability fields for ice, water, and water vapor.

Figure 3.27   Diamond and graphite.

Figure 3.28   Quartz and tridymite, two of the SiO_2 polymorphs.
Quartz photo from J. J. Harrison, tridymite photo from F. Kruijen, both Wikimedia Commons.

(lowest Gibbs free energy). Sometimes, for kinetic or other reasons, systems do not reach minimum energy. Diamond, for example, exists under normal Earth surface conditions even though graphite has a lower Gibbs free energy. Both consist of carbon, and graphite is the stable phase at room temperature and pressure. Yet diamond, like other phases termed *metastable*, can be found. Metastable minerals like diamond may persist at Earth surface temperatures for very long times, or even into perpetuity, because reaction rates are slow and the energy needed to start the change to stable minerals may be great. At higher temperatures most mineralogical systems reflect stable equilibria, so thermodynamics is a major tool for metamorphic and igneous mineralogists and petrologists, but not so useful to those who study low-temperature Earth surface processes.

For another example of polymorphs, we may consider minerals made of SiO_2. Quartz and tridymite, shown in Figure 3.28, are only two of many possible minerals of the same composition, and they have distinctive crystal shapes. Quartz is very common, but tridymite is only stable at high temperature and thus is rare under normal Earth surface conditions. In all, mineralogists have described more than a dozen crystalline forms of SiO_2. They are all polymorphs, meaning that they have the same composition but different arrangements of atoms, and are stable under different conditions.

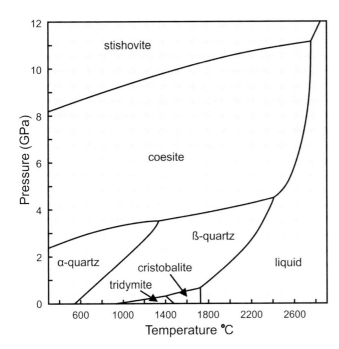

Figure 3.29 Phase diagram showing stability fields for some of the SiO$_2$ minerals.

This phase diagram (Fig. 3.29) shows stability fields for some of the SiO$_2$ polymorphs. At high temperature, SiO$_2$ melts to form a liquid; at lower temperatures, several different polymorphs are stable. α-quartz (commonly termed just quartz) is most common, but the other polymorphs (tridymite and cristobalite) occasionally exist in rocks that have not reacted to stable equilibrium.

Stishovite and coesite, rare dense varieties of SiO$_2$, are found in some rocks shocked by meteorite impacts. Less commonly, coesite is found in very high-pressure metamorphic rocks. Metastable cristobalite and tridymite occur in some glassy volcanic rocks, but over time may react to form α-quartz. β-quartz is never seen at normal temperatures because it reacts to form α-quartz instantaneously upon cooling. β-quartz is, however, stable at high temperature, and consequently is useful in some high-temperature applications.

The two phase diagrams shown above (Figs. 3.26 and 3.29) are relatively simple diagrams involving only a single component. The first depicts phases with composition H$_2$O, the second with composition SiO$_2$. Natural geological systems often involve many components, and the phase diagrams may be more complicated. The principles of thermodynamics, however, still apply. So, phase diagrams allow us to predict what will form under different conditions. They also allow us to interpret the conditions at which a particular mineral or mineral

assemblage must have equilibrated. If, for example, we find a rock containing coesite, we know it formed under very high-pressure conditions.

3.10 Mineral classification

Today, mineralogists group minerals into classes with chemical compositional similarity (Table 3.5). This system has not changed much since it was first created by James Dwight Dana in the mid-19th century. Most classes are based on the anions or anionic groups present in the minerals, but the native metal, sulfide, and sulfosalt classes include minerals in which the dominant type of bonding is not ionic and so must be classified in other ways. For some purposes, these main classes are further divided into mineral subclasses based on atomic arrangement, or mineral groups based on chemistry. Also, subgroups that contain only a few mineral species are often lumped. For example, the arsenates, vanadates, and phosphates are generally grouped together. The sulfosalts and sulfides, too, are generally grouped together.

Dividing minerals into classes based on anion or anionic group is convenient because we can determine class from chemical formula. However, this classification scheme makes sense for other reasons, too. The kinds of atomic arrangements and bonding are similar within a mineral class; thus, minerals within a class often have similar physical properties, making the classes useful in mineral identification. Such would not be the case if we divided minerals into groups based on cations. For example, lime (calcium oxide), wollastonite (calcium silicate), calcite (calcium carbonate), and fluorite (calcium fluoride) all contain Ca, but have few properties in common. In contrast, the carbonate group minerals, shown in Figure 3.30, share many properties. These carbonates include calcite (calcium carbonate), dolomite (calcium magnesium carbonate), rhodochrosite (manganese carbonate), smithsonite (zinc carbonate), and cerussite (lead carbonate) They are all quite soft and dissolve in acidic water. They have good cleavage, and, when they grow as flat-sided crystals, their crystals have similar shapes.

Another reason why classifying minerals based on anion content is useful is that minerals within a single class are often found together. The minerals that we most often see belong to the silicate class, but minerals of other classes are abundant, too. Silicates make up more than 99% of the minerals found in igneous rocks and account for more than 90% of Earth's crust and

Table 3.5　The different classes of minerals and examples.

Class	Chemical characteristics	Key element, anion, or anionic group	Example minerals
silicates	Silicates all contain Si^{4+} bonded to oxygen in a silicate (SiO_4) group.	$(SiO_4)^{2-}$	olivine $(Mg,Fe)_2SiO_4$ plagioclase $(Na,Ca)(Si,Al)_4O_8$ quartz SiO_2
native elements	Native elements are made of only a single element. Some rare, related minerals are alloys of two or more elements.	Generally a single metal	gold—Au copper—Cu diamond—C
sulfides	Sulfides contain sulfur unbonded to oxygen.	S	pyrite—FeS_2 chalcopyrite—$CuFeS_2$ galena—PbS
sulfosalts	In sulfosalts, a metallic element is bonded to a sulfur and a semimetal, such as or Sb	As or Sb with S	enargite—Cu_3AsS_4 pyrargyrite Ag_3SbS_3
oxides	In almost all oxide minerals, oxygen is bonded to metal cations only. Ice is the only exception because hydrogen is not generally considered a metal.	O^{2-}	ice—H_2O corundum—Al_2O_3 spinel—$MgAl_2O_4$
hydroxides	Hydroxides are closely related to oxides but contain $(OH)^-$ anionic groups.	$(OH)^-$	brucite—$Mg(OH)_2$ diaspore—$AlO(OH)$ gibbsite—$Al(OH)_3$
halides	Halide minerals have a halogen element as the principal anion.	F^-, Cl^-, Br^- and I^-	halite—NaCl fluorite—CaF_2 atacamite—$Cu_2Cl(OH)_3$
carbonates	Carbonate minerals contain carbon bonded to oxygen in a carbonate (CO_3) group.	$(CO_3)^{2-}$	calcite—$CaCO_3$ rhodochrosite—$MnCO_3$ dolomite—$CaMg(CO_3)_2$
nitrates	Nitrates contain nitrogen bonded to oxygen in a nitrate (NO_3) group.	$(NO_3)^-$	niter—KNO_3 nitratine—$NaNO_3$
phosphates	Phosphates contain phosphorous bonded to oxygen in a phosphate (PO_4) group.	$(PO_4)^{3-}$	apatite— $Ca_5(PO_4)_3(OH)$
sulfates	Sulfates contain sulfur bonded to oxygen in a sulfate (SO_4) group.	$(SO_4)^{2-}$	barite—$BaSO_4$ anhydrite—$CaSO_4$ gypsum—$CaSO_4 \cdot 2H_2O$
borates	Borates contain boron bonded to oxygen in a borate (BO_3) group.	$(BO_3)^{3-}$	kernite—$Na_2B_4O_6(OH)_3 \cdot 3(H_2O)$ borax—$Na_2B_4O_5(OH)_4 \cdot 8(H_2O)$ **boracite $Mg_3B_7O_{13}Cl$**
chromates	Chromates contain chrome bonded to oxygen in a chromate (CrO_4) group.	$(CrO_4)^{2-}$	crocoite—$PbCrO_4$
tungstates	Tungstates contain tungsten bonded to oxygen in a tungstate (WO_4) group.	$(WO_4)^{2-}$	wolframite— $(Fe,Mn)WO_4$ scheelite—$CaWO_4$
molybdates	Molybdates contain molybdenum bonded to oxygen in a molybdate (MoO_4) group.	$(MoO_4)^{2-}$	wulfenite—$PbMoO_4$
arsenates	Arsenates contain arsenic bonded to oxygen in an arsenate (AsO_4) group.	$(AsO_4)^{3-}$	erythrite—$Co_3(AsO_4)_2 \cdot 8H_2O$
vanadates	Vanadates contain vanadium bonded to oxygen in a vanadate (VO_4) group.	$(VO_4)^{3-}$	vanadinite—$Pb_5(VO_4)_3Cl$

mantle. Carbonates, the second most abundant mineral class, are similarly dominant in limestone, a common and widespread type of sedimentary rock. Sulfide minerals often concentrate in ore deposits, and hydroxide minerals typically form when a parent rock is weathered or altered in some other way.

Mineralogists have described and named about 4000 different mineral species, but most are exceedingly rare. Several dozen new minerals are described each year, and

a small number are discredited. The most complete listing of minerals *is Fleischer's Glossary of Mineral Species*, published by the Mineralogical Record every three or four years. Some mineral classes are more diverse than others. The silicate class, for example, contains the largest number of mineral species, with more than 800 known. Oxides and hydroxides together add up to 500 species. Sulfides and sulfosalts also make up about 500 species. Calcite accounts for most of the carbonates on

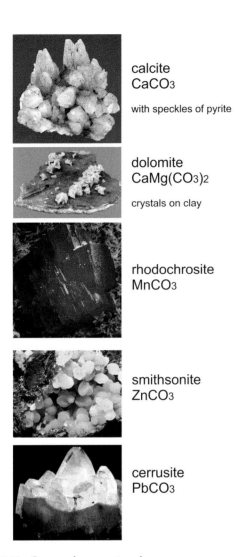

Figure 3.30 Some carbonate minerals.

Photo sources: calcite and dolomite—D. Perkins, GeoDIL; rhodochrosite—E. Hunt; smithsonite—P. Géry; cerussite—R. Lavinsky. The last three are from Wikimedia Commons.

calcite
$CaCO_3$

with speckles of pyrite

dolomite
$CaMg(CO_3)_2$

crystals on clay

rhodochrosite
$MnCO_3$

smithsonite
$ZnCO_3$

cerrusite
$PbCO_3$

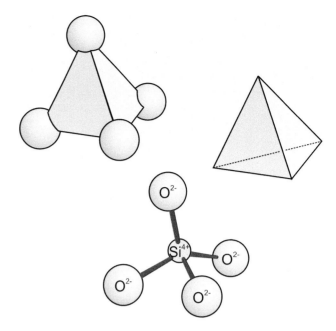

Figure 3.31 Three depictions of an SiO_4 tetrahedron.

oxygen anion is shared between two tetrahedra; the collection of tetrahedra make up a three-dimensional framework with an Si:O ratio of 1:2. Other subclasses have Si:O ratios between 1:4 and 1:2. They are discussed in further detail in Chapter 4. Some silicate minerals with complex structures do not fit conveniently into one of the six subclasses, but most common silicates do.

3.11 Mineral properties and identification

If you give a mineral specimen to nongeologists and ask them to describe it, generally color is mentioned first. However, as discussed later, color is often the least reliable property for identifying a mineral. With a little prodding, the nongeologist may go on to describe the shape and nature of visible crystals. For example, the pyrite shown in Figure 3.32 might be described as "metallic and forming cubic crystals." The piece of rose quartz in Figure 3.32 might be described as "glassy, transparent, and having a hexagonal shape." Metallic and glassy are terms describing *luster*. Transparent describes *diaphaneity*. Cubic and hexagonal describe *symmetry*, a property relating to *shape*. These three properties (luster, diaphaneity, and shape) are fundamental properties for mineral identification. Other properties—including *streak* (the color of a mineral when powdered), cleavage (the way a crystal breaks on planes), *parting* (the way a crystal

Earth, halite and fluorite for most of the fluorides, and apatite for most of the phosphates. Minerals of some groups, such as nitrates and molybdates and others at the bottom of Table 3.5, are exceptionally rare.

Because the silicate class contains many important minerals, we divide it into subclasses. In all silicates, except some very rare high-pressure minerals, four O^{2-} anions surround every Si^{4+} cation, forming a tetrahedron (Fig. 3.31), a pyramid shape with four identical faces. We name the subclasses according to how tetrahedra are linked (*polymerized*) in the atomic structure. In the *isolated tetrahedral silicates* (also called *island silicates*), tetrahedra are not polymerized; they are all separated (like a group of islands) with cations between them. The Si:O ratio for the structure is 1:4. In contrast, in *framework silicates*, each

Figure 3.32   Pyrite cubes on top of marl (left), and rose quartz (right).

Pyrite photo from C. Millan, quartz photo from L.M.B. Sánchez, both at Wikimedia Commons.

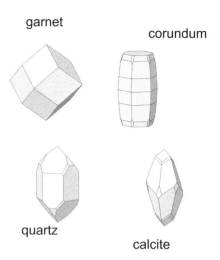

Figure 3.33   Drawings of perfectly formed crystals.

breaks due to structural weaknesses), and *hardness*—can be even more important.

Given a single property, for example luster, it is possible to divide minerals into groups. In the case of luster, we usually divide minerals into those that are *metallic* and those that are *nonmetallic*. There are, however, many metallic and many nonmetallic minerals; other properties must be considered if minerals are to be identified. Nonmetallic minerals, for instance, can be divided further based on more subtle luster differences. Other properties—including diaphaneity, crystal shape, cleavage, parting, fracture, and hardness—are also diagnostic. Ultimately, minerals can be identified by name or at least placed in small groups. It is tempting, then, to come up with a standard list of properties that should be evaluated when identifying minerals. However, most mineralogists know that, depending on the sample and circumstances, some properties are more important than others. Rather than going through a long list or filling out a standard table, experienced mineralogists focus on the properties that are most diagnostic. Sometimes a single property, such as strong effervescence by hydrochloric acid (diagnostic of calcite), may serve for mineral identification. At first, mineral identification may seem tedious, but with a little experience, it is possible to find shortcuts to make the process more efficient.

3.11.1 Crystal shape

To most people, a *crystal* is a sparkling gem-like solid with well-formed faces and a geometric shape. Figure 3.33 shows some examples. For many scientists, including all mineralogists, *crystal* and *crystalline* also refer to any solid compound having an ordered, repetitive, atomic structure, which may or may not result in crystal faces and a *gemmy* appearance. We use the term

"crystal" in both ways. When a mineralogist refers to a quartz crystal, the reference is usually to a six-sided shape with pyramid-like ends, perhaps like the one shown in Figure 3.33. On the other hand, petrologists and mineralogists may refer to crystals of quartz in a rock such as granite. The crystals in granite crystals rarely have smooth faces.

With just a few (debated) exceptions, all minerals are crystalline by definition, but perfectly formed crystals are rare. Yet, when faces on a mineral are fully or partially developed, *crystal shape* can be a powerful identification tool. When no faces are visible, we must rely on other properties to identify a mineral.

Mineral crystals always have an ordered arrangement of atoms within them, but they may not be geometrically shaped on the outside. All garnet crystals, for example, have the same highly ordered arrangement of atoms, as shown in Figure 3.34a, but only some crystals have visible crystal faces. The two photographs (Figs. 3.34b and c) show samples of natural garnet. Crystals, such as the well-formed garnet crystal in photo b (center), with well-developed faces, are termed *euhedral*. If no faces are visible, the crystal is *anhedral*, as in photo c. Those crystals that fall between euhedral and anhedral are called *subhedral*.

Compare the euhedral garnet crystal in Figure 3.34 with the drawing of garnet in Figure 3.33—the natural crystal is euhedral but not geometrically perfect. So, although all garnets have the same ordered atomic arrangement shown in the ball and stick drawing in Figure 3.34, perfectly formed crystals are rare. In fact, most natural garnet is anhedral and does not have flat crystal faces. The anhedral garnet shown in photo c is typical.

Figure 3.34 The arrangement of atoms in garnet, and two samples of natural garnet.
Drawing from D. Perkins. Photos b and c are from Didier Descouens and Teravolt, both at Wikimedia Commons.

Mineral *habit*, a property related to crystal shape, includes shape and also the way multiple crystals grow together. Some examples of different habits are shown in the photographs in Figure 3.35.

The most useful terms describing habit are self-explanatory (Table 3.6). Common ones used to describe the habit of single crystals include *equant* (equidimensional), *acicular* (needlelike), *tabular*, and *bladed*. For describing an assembly of multiple crystals, we use terms such as *massive*, *granular*, *radiating*, and *fibrous*. Compare the terms in Table 3.6 with the examples shown above.

Unfortunately, although museum specimens and pictures of minerals in textbooks often show distinctive shapes and habits, most mineral samples do not. Small anhedral crystals without flat faces, or massive aggregates, are typical, often rendering shape and habit of little use for identification. Additional complications arise because some minerals, for example calcite, have different crystal shapes or habits, depending on how they grow. Nonetheless, shape and habit reflect the internal arrangement of atoms in a crystal and, when visible, can be important diagnostic properties.

3.11.2 Mineral appearance

3.11.2.1 Luster

Luster refers to the sheen of a mineral, that is, to the way it reflects light; some examples are shown in Figure 3.36. For example, minerals that have the shiny appearance of polished metal are said to have a *metallic luster*. Most minerals, however, do not have a metallic luster; they have a *nonmetallic luster*. Mineralogists use many terms to describe nonmetallic minerals. We call those that appear only partially metallic *submetallic*. *Vitreous* minerals, like quartz, have a glassy appearance. *Adamantine* minerals sparkle or appear brilliant like diamonds. Some not-so-spectacular minerals do not have bright or flashy lusters; we may describe them as *earthy*, *dull*, or *resinous*. Some mineral specimens, such as the sphalerite shown in Figure 3.36, may have more than one luster or color depending on how they were formed. Table 3.7 lists some of the more common terms used to describe luster.

3.11.2.2 Diaphaneity

Diaphaneity refers to a mineral's ability to have light pass through it. Some mineral specimens, such as the calcite shown in the bottom of Figure 3.37 are *transparent* and, even when thick, light passes relatively freely through them. Quartz is another example of a mineral that is often transparent. Minerals that do not transmit light as well as clear quartz may be *translucent*. Although it is not possible to see through them as with transparent minerals, translucent minerals, if thin enough, transmit light. The quartz shown in Figure 3.37 is a good example. Still a third type of mineral, called an *opaque* mineral, does not transmit light unless the mineral is exceptionally thin and, perhaps, not even then. Most opaque minerals, like the molybdenite shown in the top of Figure 3.37, have metallic lusters. Pyrite and many other sulfide and oxide minerals are good examples. Their opaqueness sets them apart from most other minerals.

Figure 3.35 Examples of different crystal habits. Halite, gypsum, natrolite, actinolite, and cerussite.
Photos from D. Descouens, gypsum photo from Y. Chen, both Wikimedia Commons.

Table 3.6   Terms that describe crystal shape and habit

Shapes of Individual Crystals (With Example Minerals)

equant	the same dimensions in all directions (garnet, spinel)
blocky	equant with a nearly square cross sections (halite, galena)
acicular	needlelike (actinolite, sillimanite)
tabular or platy	appearing to be a plates or a thick sheet (gypsum, graphite)
capillary or filiform	hair-like or threadlike (serpentine, millerite)
bladed	elongated and flattened in one direction (kyanite, wollastonite)
prismatic or columnar	elongated with identical faces parallel to a common direction (apatite, beryl)
foliated or micaceous	easily split into sheets (muscovite, biotite)

Properties of Crystal Aggregates

massive	appearing as a solid mass with no distinguishing features
granular	composed of many individual grains
radiating or divergent	containing crystals emanating from a common point
fibrous	composed of fibers
stalactitic	appearing stalactite shaped
lamellar or tabular	appearing like flat plates or slabs growing together
stellated	containing an aggregate of crystals giving a star-like appearance
plumose	having feathery appearance
arborescent or dendritic	appearing like a branching tree or plant
reticulated or lattice-like	netlike, composed of slender crystals forming a lattice pattern
colloform or globular	composed of spherical or hemispherical shapes made of radiating crystals
botryoidal	having an appearance similar to a bunch of grapes
reniform	having a kidney-shaped appearance
mammillary	having breast-like shape
drusy	having surfaces covered with fine crystals
elliptic or pisolitic	composed of very small or small spheres

metallic luster

galena (silvery) on
top of barite

vitreous luster

rose quartz

adamantine luster

cerussite

submetallic luster

sphalerite

silky luster

satin spar
(a variety of gypsum)

pearly luster

talc

resinous luster

sphalerite (red) on
top of quartz

greasy luster

moss opal

dull luster

kaolinite

Figure 3.36 Some of the more common terms used to describe mineral luster, with example minerals.

Photos: metallic and pearly—R. Lavinsky; vitreous and adamantine—D. Descouens; silky and greasy, Ra'ike; dull—Minerals in Your World; submetallic—Andreas Fruh; all at Wikimedia Commons.

Table 3.7   Some of the more useful terms used to describe mineral luster

Luster	Meaning	Example minerals that may exhibit luster
metallic	having a shiny reflective appearance similar to polished metal	pyrite, galena
submetallic	having a somewhat metallic, but duller and less reflective, luster	sphalerite, cinnabar
vitreous	having a glassy appearance	quartz, tourmaline
resinous	having the appearance of resin	sphalerite, sulfur
greasy	reflecting light to give a play of colors; similar to oil on water	chlorite, nepheline
silky	having surfaces that appear to be composed of fine fibers	chrysotile (asbestos), gypsum
adamantine	having a bright, shiny, brilliant appearance similar to that of diamonds	diamond, cerussite
pearly	appearing iridescent, similar to pearls or some seashells	muscovite, talc
earthy	having a dull, granular, earthy appearance	kaolinite (clay)
dull	not reflecting significant amounts of light or showing any play of colors	kaolinite (clay), niter

Figure 3.37   Transparent calcite rhomb (bottom), translucent quartz with opaque molybdenite on top (top).
Quartz/molybdenite photo from R. Lavinsky, calcite photo from P. Géry, both at Wikimedia Commons.

3.11.2.3   *Color*

Color is often used for quick identification of minerals. Occasionally, it can be diagnostic, but often it is ambiguous or even misleading. For example, the deep red color of rubies may seem distinctive but ruby is just one variety of the mineral corundum. Corundum may also be blue or yellow sapphire, and a few other less common colors. Figure 3.38 shows three examples of corundum that have different colors. To add to the confusion, other minerals, such as spinel or garnet, may have the same deep red color as ruby. Color is ambiguous because many things can give a mineral its color. You can compare the shapes of the crystals in Figure 3.38 with the ideal shape depicted in Figure 3.33. Shape is often a better property for identifying corundum than color.

Color is one of the most misunderstood mineral properties. It is easy to look at a ruby illuminated by white light and say that the mineral has a red color. However, if the ruby is illuminated by light of a different color, it may not appear red. Color, then, is not a property of a mineral. It is instead the result we observe when light and a mineral interact. When we see that something has

Figure 3.38 Three samples of corundum: ruby on quartz (top), sapphire (below left), and a rare orange-pink variety. All photos from R. Lavinsky, Wikimedia Commons.

color, what we are really observing is the color of the light that is reflected or transmitted to our eye. Normal light, called *white light*, includes many different colors. When white light strikes a mineral surface, if all of the colors are reflected back to our eyes, the mineral will appear white. If none of the colors is reflected back to our eye, the mineral will appear black. Most minerals, like ruby, appear to have color because only one or a few wavelengths make it back to our eye. The other wavelengths of light are scattered in other directions or are absorbed or transmitted by the mineral in some way.

Metallic minerals, especially sulfides, tend to be constant in their coloration, so mineralogists use color as a key tool for sulfide identification. (However, metallic minerals easily tarnish, so we need a fresh surface to see the true color.) But, color is a poor property to use for identifying most *nonmetallic* minerals because it is so variable. Quartz may be colorless, rosy (rose quartz), yellow (citrine), purple (amethyst), milky, smoky, or black. Corundum may be just about any color, including colorless, black, brown, pink, yellow, blue, purple, or red.

We name gem varieties of corundum according to their colors, but that is mostly for commercial purposes.

The most significant control on color is a mineral's chemical composition. We call elements that give a mineral its color *chromophores*. If the elements controlling the selective reflection of certain wavelengths are major components in a mineral, the mineral is *idiochromatic*, or "self-coloring." Sphalerite, $(Zn,Fe)S$, for example, is an idiochromatic mineral. It changes from white to yellow to brown to black as its composition changes from pure ZnS to a mixture of ZnS and FeS. Many copper minerals are green or blue, while many manganese minerals are pinkish. These colors derive from selective absorption of certain colors by copper and manganese. Complicating things, however, is the fact that idiochromatic elements may have different effects in different minerals. Malachite is green and azurite is blue, but in both minerals the color is due to copper.

It does not take large amounts of chromophores to color a mineral. Minor amounts, less than 0.1 wt%, of transition metals, such as iron and copper, may control a mineral's color because electrons in the d-orbitals of transition metals are extremely efficient at absorbing certain visible wavelengths of light. The remaining wavelengths are reflected and give minerals their color. Ruby and sapphire are examples of *allochromatic* varieties of corundum. In allochromatic minerals, minor or trace elements determine the color. Very small amounts of iron and titanium give sapphire a deep blue color. Small amounts of chrome give ruby and some other gemstones deep red colors. The effects of allochromatic elements may be different in different minerals. Allochromatic chrome is also responsible for the striking green color of emerald (a variety of the mineral beryl), chrome diopside, and some tourmalines.

Structural defects in minerals may also give them color. Radiation damage gives quartz, for example, a purple, smoky, or black color. The purple color of many fluorites is also caused by defects. Other causes of coloration include the oxidation or reduction of certain elements (especially iron) and the presence of minute inclusions of other minerals.

3.11.2.4 Streak

Although it would never occur to many people to check a mineral's *streak*, which refers to the color it has when finely powdered, streak is sometimes a key diagnostic property (Fig. 3.39). As shown in Figure 3.39, hematite (the silvery metallic mineral on quartz in this

Figure 3.39   Examples of streak: hematite (top) and sulfur (bottom). Hematite photo from K. Hagen, sulfur photo from Ra'ike, both at Wikimedia Commons.

Table 3.8   Examples of minerals with colored streaks

magnetite, ilmenite	black
galena	lead gray
rutile	pale or light brown
sphalerite	light brown or yellow to white
sodalite	very pale blue to white
malachite	pale green
sulfur	light yellow to white
realgar	orange or reddish yellow
cinnabar	dark red to scarlet
hematite	rust red/brown to blood red

For example, a specimen of hematite may appear red, gray, or black and may or may not have a metallic luster. It always, however, has a diagnostic red streak.

3.11.2.5 Luminescence

Some minerals will emit light when they are activated by an energy form other than visible light. We call such an effect *luminescence*. Examples of luminescence include *fluorescence*, *phosphorescence*, and *thermoluminescence*. Fluorescent minerals give off visible light when they are struck by energy with wavelength less than that of visible light. The invisible radiation from ultraviolet lamps, for example, may cause scheelite, willemite, or fluorite to appear to glow in the dark. If the visible emission continues after the energy source is turned off (that is, the mineral "glows in the dark"), the mineral is *phosphorescent*, pectolite for example. Thermoluminescent minerals, such as some tourmalines, give off visible light in response to heating. Some varieties of fluorite, calcite, and apatite also have this property.

3.11.2.6 Play of colors

Some minerals appear to have multiple colors, similar in some ways to pearls or some clam shells, or to have different colors depending on how they are viewed. Figure 3.40 shows several examples of *play of colors*—a property that gives minerals a different color depending on how they are viewed. The photographs include opal displaying *opalescence*, moonstone (a feldspar variety) displaying *adularescence*, labradorite displaying *labradorescence*, and limonite displaying a weak purplish *iridescence*.

Opalescence, adularescence, labradorescence, and iridescence are examples of four of the many related varieties of play of colors. The different varieties vary slightly in their causes and the colors that result. The colorful effects are mostly due to light scattering caused by very fine particles or compositional layering in the minerals.

photograph) leaves a rusty red streak when scraped across a streak plate. The bottom photo shows sulfur which, when scraped across a dark-colored streak plate, leaves a characteristic light yellow trace.

For mineral identification, streak is much more reliable than mineral color and is easy to determine. The usual method of determining streak is to rub the mineral against a ceramic streak plate as shown in Figure 3.39. For example, quartz comes in many colors, but its streak is always white (clear). The same is true of calcite. Gold and pyrite (*fool's gold*) may have similar appearing colors, but their streaks are always different. Pyrite has a dark-colored streak; gold has a yellowish streak.

Mineralogists use streak routinely both in the laboratory and in the field. Table 3.8 lists minerals for which streak is particularly diagnostic. Streak is especially useful for identifying oxide and sulfide minerals, and streak can be extremely useful for differentiating dark-colored minerals.

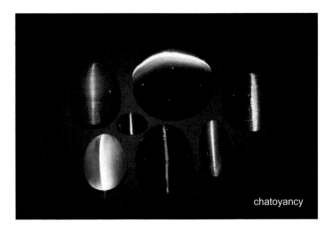

Figure 3.40 Four different examples of a play of color.
Photos from Dpulitzer, Didier Descouens, Shaddack, and Robert M. Lavinsky, all at Wikimedia Commons.

The included particles and layers separate white light into individual wavelengths of varying intensities that are emitted in different directions.

Chatoyancy and *asterism* are two other light-scattering effects, most easily seen in gemmy minerals, especially minerals polished to produce *cabochons* (stones with smooth rounded surfaces), as shown in Figure 3.41. These top photo shows seven minerals displaying chatoyancy and the bottom photo shows one displaying asterism. These stones are part of the mineral collection at the Smithsonian Institute of Washington. The chatoyant minerals show a bright band of scattered light, perpendicular to a long direction of the crystal. Such minerals are sometimes said to have a cat's-eye or tiger's-eye appearance. The satin spar variety of gypsum in Figure 3.36 is another example of a chatoyant mineral. Closely packed parallel fibers or inclusions of other minerals within a mineral cause chatoyancy. *Asterism*—a property sometimes visible in rubies, sapphires, garnets, and some other gems—refers to scattered light appearing as a "star." Like chatoyancy, asterism results from the scattering of light by small inclusions of a different mineral within a larger host. The bottom photograph in Figure 3.41 is of the Star of Asia, a famous sapphire that includes extremely small needle-like inclusions of rutile (titanium oxide) that scatter light.

3.11.3 Strength and breaking of minerals

The color and shape of minerals are obvious to anyone, but mineralogists notice other, more subtle, properties

Figure 3.41 Chatoyancy and asterism.
Images from the Smithsonian Institute.

too. Several relate to the strengths of bonds that hold crystals together. These properties are especially reliable for mineral identification, because chemical impurities or defects in crystal structure do not affect them.

3.11.3.1 Tenacity

The term *tenacity* refers to a mineral's toughness and its resistance to breaking or deformation. Those that break, bend, or deform easily have little tenacity. In contrast, strong unbreakable minerals have great tenacity. Figure 3.42 shows samples of the gemstone jade that have been shaped and polished to produce a figurine and a cabochon. Gemmy jade may be composed of either of two minerals: jadeite (a pyroxene) or nephrite (an amphibole). In either case, jade is one of the most tenacious natural materials known. It does not easily break or deform, even when under extreme stress. That is one of the reasons, besides beauty, that it is prized as a gemstone.

Figure 3.42   Jade figurine and cabochon.
Image from Gemological Institute of America.

The tenacity of a mineral is controlled by the nature of its chemical bonds. Ionic bonding often leads to rigid, *brittle* minerals. Halite is an excellent example of a brittle mineral. It shatters into many small pieces when struck. Quartz, too, is brittle, although the bonding in quartz is only about half ionic. Many metallically bonded minerals, such as native copper, are *malleable*, meaning they can be hammered into different shapes. Other minerals, such as gypsum, are *sectile*; they can be cut into shavings with a knife. Some minerals, including talc and chlorite, are *flexible* due to weak bonds holding well-bonded layers of atoms together. When force is applied, slippage between layers allows bending. When pressure is released, they do not return to their original shape. Still other minerals, notably the micas, are *elastic*. They may bend but resume their original shape after pressure is released if they were not too badly deformed. In micas and other elastic minerals, the bonds holding layers together are stronger than those in chlorite or clays.

3.11.3.2   *Fracture, cleavage, and parting*

Fracture is a general term used to describe the way a mineral breaks or cracks. Because atomic structure is not the same in all directions and chemical bonds are not all the same strength, most crystals break along preferred directions. The orientation and manner of breaking are important clues to crystal structure, and thus valuable and reliable tools for mineral identification.

Mineralogists have developed a vocabulary to describe the way minerals break (Table 3.9), and most of the terms are self-explanatory.

If a mineral fractures to produce multiple planar and smooth surfaces, the mineral has good *cleavage*. If a mineral cleaves along one particular plane, a nearly infinite number of parallel planes are equally prone to cleavage. This is due to the repetitive arrangement of atoms in atomic structures. The spacing between planes is the repeat distance of the atomic structure, on the order of angstroms (Å). The whole set of parallel planes, collectively referred to as a *cleavage*, represents planes of weak bonding in the crystal structure. Micas are the best examples of minerals with one excellent cleavage—they break easily into sheets of very small thickness. Minerals that have more than one direction of weakness will have more than one cleavage direction. The direction and angular relationships between cleavages, therefore, give valuable hints about atomic structure.

Cleavage is an excellent property for mineral identification because it depends on how atoms are arranged. Often the quality and number of cleavages can be seen quite easily. Sometimes a *hand lens* is needed to see the set of fine parallel cracks that suggest a cleavage too poorly developed to see with the naked eye. Angles between cleavages may be estimated or, if accurate

Table 3.9   Terms used to describe fracture and cleavage (and example minerals)

Terms Used to Describe Fracture

even	breaking to produce smooth planar surfaces (halite)
uneven or irregular	breaking to produce rough and irregular surfaces (rhodonite)
hackly	fracturing to produce jagged surfaces and sharp edges (copper)
splintery	forming sharp splinters (kyanite, pectolite)
fibrous	forming fibrous material (chrysotile, crocidolite)
conchoidal	breaking with curved surfaces as in the manner of glass (quartz)

Terms Used to Describe Cleavage

basal	well-developed planar cleavage in one direction only; also sometimes called "platy" (micas)
cubic	three cleavages at 90° to each other (galena)
octahedral	four cleavages that produce eight-sided cleavage fragments (fluorite)
prismatic	multiple directions of good cleavage all parallel to one direction in the crystal (tremolite)

angular measurements are needed, they can be measured using techniques involving a *petrographic microscope* or a device called a *goniometer*.

Minerals that are equally strong in all directions, such as quartz and olivine, fracture to form irregular surfaces. Quartz is one of only a few minerals that breaks only along curved surfaces to form *conchoidal fractures* (similar to what happens when a glass breaks), but most minerals exhibit cleavage.

Figure 3.43 shows examples of some different ways that minerals may cleave. Minerals with only one direction of weakness, such as the molybdenite and muscovite, have one direction of cleavage and usually break to form thick slabs or sheets. We say they have *basal cleavage*. Plagioclase (Fig. 3.43) and some other minerals have two cleavages. Some, such as kyanite and anthophyllite, easily break into splintery shapes. Other minerals, including halite and calcite, may have three directions of cleavage. In halite, the cleavages are perpendicular, in calcite they are not (Fig. 3.43). And still other minerals, such as fluorite, may have as many as six, or more, cleavages. We use geometric terms such as cubic, octahedral, or prismatic to describe cleavage when appropriate.

The ease with which a mineral cleaves is not the same for all minerals or for all the cleavages in a particular mineral. Mineralogists describe the quality of a particular cleavage with qualitative terms: perfect, good, distinct, indistinct, and poor. Quartz has poor cleavage in all directions, while micas have one perfect cleavage. Feldspars, like the plagioclase shown in Figure 3.43, have two cleavages, but one is very good and the other is better described as being distinct. The angle between the two feldspar cleavages is about 90°, which explains why the specimen shown in Figure 3.43 appears to have a square cross section. In all minerals, cleavages represent planes of relatively weak atomic bonding.

Crystal faces and cleavage surfaces may be difficult to tell apart. In some minerals, principal cleavage directions are parallel to crystal faces, but in most they are not. A set of parallel fractures suggests a cleavage, but if only one flat surface is visible, there can be ambiguity. However, this problem is sometimes mitigated because crystal faces often display subtle effects of crystal growth. *Twinning* (oriented intergrowths of multiple crystals) and other *striations* (parallel lines on a face), *growth rings* or layers, *pitting*, and other imperfections make a faceless smooth than a cleavage plane and give it lower reflectivity and a drabber luster.

Some minerals exhibit *parting*, a type of breaking that is often quite similar to cleavage. Parting occurs when a

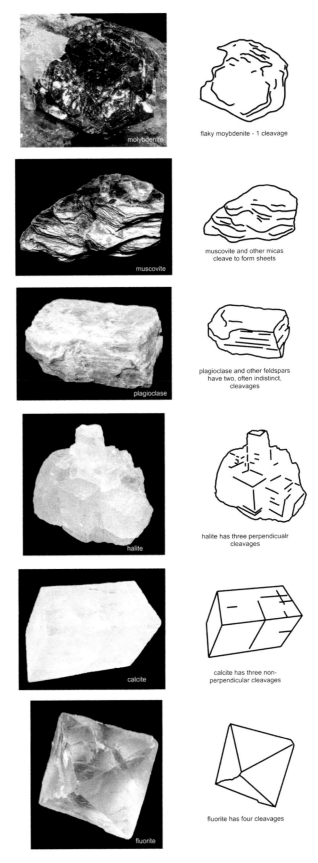

Figure 3.43 Six examples of the way that minerals can cleave. Photos and drawings from D. Perkins, GeoDIL.

mineral breaks along structural planes but, unlike cleavage, parting is not found in all samples of a particular mineral and does not repeat to form many parallel planes that are only a few angstroms apart. Parting can be induced by several things, perhaps most commonly when multiple crystals grow together (twinning). Distinguishing parting from cleavage can sometimes be problematic.

3.11.3.3 Hardness

Hardness, symbolized by *H*, is a mineral's resistance to abrasion or scratching. Some minerals, such as talc, molybdenite, graphite, and gypsum, are very soft; others are very hard, including topaz, chrysoberyl, corundum, and diamond. Most fall somewhere between these extremes. We determine *relative hardness* by trying to scratch a surface of one mineral with an edge or corner of another. If a scratch or abrasion results, the first mineral is the softer. On rare occasions, mineralogists determine *absolute values* of hardness in one of several ways. The easiest is to use an indenting tool similar to ones used to measure the hardness of steel. The indenting tool measures the force necessary to produce a permanent indentation in a flat surface.

Table 3.10 gives the relative hardness scale used by mineralogists. Based on 10 well-known minerals, it is called the *Mohs hardness scale*, named after Austrian mineralogist Friedrich Mohs who developed it in 1812. Compared with absolute hardness, the Mohs scale is not linear; it is close to exponential. The differences in hardness between talc and gypsum, and between gypsum and calcite, are small. The differences in hardness between topaz and corundum, and between corundum and diamond, are much greater. Differences for the intermediate minerals fall in between.

We can estimate mineral hardness by conducting scratch tests to compare the hardness of an unknown mineral to samples of the minerals in the Mohs hardness scale. Alternatively, we can approximate hardness by comparing mineral hardness to a fingernail, penny, pocketknife, glass, or several other common objects (see Table 3.10). Figure 3.44 shows gypsum being scratched by a fingernail. Gypsum, one of the softest minerals known, has a hardness of 2 on the Mohs hardness scale; fingernails have a hardness of about 2½.

Testing for a mineral's hardness may seem straightforward, but there can be complications. Mineral specimens may be too small or too valuable to scratch. Large samples may contain many grains loosely cemented together so that scratch tests are not possible. Others may cleave or fracture when we do tests. In still other cases, the results of scratch tests may be ambiguous.

The hardness of a mineral relates to its weakest bond strength. For example, graphite has a hardness of 1 and diamond has a hardness of 10, yet both are made of carbon. In diamond, carbon atoms are uniformly spaced and tightly bonded together, while in graphite bonding is very weak in one direction. Because bonds are usually not identical in all directions in crystals, hardness may vary depending on the direction a mineral is scratched. In kyanite, for example, hardness varies from 5.5 to 7 depending on the direction of the scratch test but, in most minerals, hardness is about the same in all directions. While the general relationship between hardness and bond strength is known, mineralogists have difficulty predicting hardness for complex atomic

Table 3.10   The mohs hardness scale

Reference mineral	Mohs hardness	Mohs hardness	Reference nonmineral
talc	1		
gypsum	2	2½	fingernail
calcite	3	3½	penny
fluorite	4		
apatite	5	5½	pocketknife or glass
feldspar	6	6½	steel file, streak plate
quartz	7		
topaz	8		
corundum	9		
diamond	10		

Figure 3.44   Scratching gypsum with a fingernail. Photo from Jamie Schod.

structures. For some simple ionic compounds, however, theoretical calculations match measurements well. Minerals with high density, highly charged ions, small ions, or covalent bonding tend to be hardest.

3.11.4 Density and specific gravity

The Greek letter ρ (rho) symbolizes *density*. We usually give the density of a mineral in units of grams/cubic centimeter (gm/cm^3). Density varies slightly depending on pressure or temperature, but most minerals have values between 2 and 8 gm/cm^3. Although there are exceptions, borates, halides, and sulfates usually have lower densities than do silicates or carbonates, which are of moderate density. Native metals, oxides, and sulfides, in contrast, are denser than most other minerals.

The density of a mineral depends on its constituent atoms and how closely packed they are. The polymorphs diamond and graphite are both made of carbon (C), but due to differences in atomic arrangements, diamond has specific gravity of 3.5, while graphite's is 2.2. Graphite forms under Earth surface conditions, but diamond, with its high specific gravity, only forms deep in Earth where pressures are great. The Laws of Thermodynamics tell us that high pressures favor dense minerals, which makes sense because at high pressure things are squeezed together.

Accurate determination of density can be difficult or impossible because it requires knowing the volume of a crystal, which can be difficult to measure with accuracy. A related property, *specific gravity (G)*, is often used instead. *Specific gravity* (unitless) is the ratio of the mass of a mineral to the mass of an equal volume of water at 1 atm and, because mass and weight are proportional, we normally determine specific gravity by comparing weights. If a mineral is at normal Earth surface conditions, density and specific gravity have about the same values.

Because mineral specific gravity varies greatly between minerals, we can easily distinguish minerals with high, moderate, or low specific gravity simply by picking them up. We use the term *heft* for estimations of *G* made by holding hand specimens; heft can be useful in mineral identification. For example, the mineral barite sometimes exists as massive white material that is easy to confuse with feldspars. However, its great heft, easily discerned by picking it up, helps identify it. Similarly, we can distinguish cerussite (lead carbonate) from other carbonate minerals by its heft.

Density differences can also help in the separation of minerals. In the laboratory, researchers separate crushed rock into mineral components by "floating" samples in liquids of different densities. In these *heavy liquids*, which are much denser than water, minerals separate as some float and others sink according to their specific gravities. In mining operations, ore minerals are often separated from valueless minerals by using gravity separation techniques that depend on density differences. This occurs in natural systems, too. Placer gold deposits form when gold from weathered rock, because of its high specific gravity, concentrates in streambeds.

3.11.5 Magnetism of minerals

The photograph in Figure 3.45 shows a sample of magnetite (from the Harvard Museum of Natural History) with nails stuck on its side. Magnetite is the only common mineral that is strongly magnetic. *Magnetism* derives from a property of electrons (the *magnetic moment*) that results from their spinning and orbiting motions. The sum of all the magnetic moments of all the atoms in a mineral gives it magnetism. We classify minerals as *ferromagnetic*, *diamagnetic*, or *paramagnetic*. If the moments of a mineral's atoms interact in a constructive way, the mineral will have properties similar to those of a magnet, such as the sample shown here.

A few minerals, including magnetite and pyrrhotite, exhibit marked magnetism. Magnetite, pyrrhotite, and other similar magnetic minerals are called *ferromagnetic* because they have the same magnetic properties as metallic iron. Most minerals exhibit little magnetic

Figure 3.45 Magnetite.
Harvard Museum of Natural History, DerHexer, Wikimedia Commons.

character but may be weakly repelled by a strong magnetic field; they are *diamagnetic*. Pure feldspars, halite, and quartz all exhibit weak diamagnetism. An impure feldspar, however, may contain iron, which results in *paramagnetism* (attraction to a strong magnet). Other paramagnetic minerals include garnet, hornblende, and many pyroxenes.

For mineral identification, magnetism is generally only useful for distinguishing magnetite from other dark heavy minerals. In the field, we can identify rocks containing magnetite because they will attract a magnet to them even if the magnetite grains are too small to see. In the laboratory, we use subtle differences in the magnetic properties of minerals to separate different minerals in crushed rock samples.

3.11.6 Electrical properties

Some minerals may conduct electricity. Electrical conduction occurs when a mineral's electrons can move throughout its structure. Such will be the case in structures containing metallic bonds. The native metals, such as copper, are the best examples. Small amounts of electrical conduction may also occur in minerals with defects and other imperfections in their structures. Other minerals, while being unable to conduct electricity, may hold static charges for brief times. They may be charged by exposure to a strong electric field, a change in temperature, or an application of pressure. A mineral charged by temperature change is *pyroelectric*; a mineral charged by pressure change is *piezoelectric*. Because they are difficult to measure, electrical properties are not often used for mineral identification.

3.11.7 Reaction to dilute hydrochloric acid

One chemical property, the reaction of minerals to dilute (5%) hydrochloric acid (HCl), is diagnostic for calcite, one of the most common minerals of the Earth's crust. As shown in Figure 3.46, drops of acid on calcite cause obvious bubbling or fizzing, termed *effervescence*. This occurs because the acid causes the calcite to give up the CO_2 that it contains, producing bubbles like those that are in carbonated soft drinks.

Dolomite, a closely related carbonate mineral that may be confused with calcite, effervesces when finely powdered but not when coarse, because powdering the sample means more surface area is exposed to the

Figure 3.46 Acid effervescing on calcite. Photograph from the Kentucky Geological Survey.

acid. Other carbonate minerals (smithsonite, aragonite, strontianite) effervesce to such a very small extent that they are best distinguished by crystal form, color, and other properties. Although acid tests have limited use, most mineralogy labs are equipped with small bottles of HCl and eyedroppers to aid in carbonate identification. Many geologists carry a small bottle of dilute hydrochloric acid when they go in the field so they may distinguish between rocks that contain calcite and rocks that do not.

3.11.8 Other properties

Minerals possess many other properties (for example, radioactivity and thermal conductivity). Because they are usually of little use for mineral identification, we will not discuss them individually here.

Questions for thought—chapter 3

1. Why would most geologists consider malachite rust on a bronze statue *not* to be a mineral?
2. Why is X-ray diffraction useful in distinguishing calcite from quartz, but not rose quartz from smoky quartz?
3. Minerals may form by igneous, aqueous, and metamorphic processes. Describe each of these processes and name one mineral that forms for each.
4. Magmas may cool underground, or the may erupt to become lavas that cool above ground. Which

scenario is most likely to produce large crystals? Why?

5. Name two mineraloids. Why are they not true minerals?

6. Some mineralogists study *biominerals*. What are biominerals – list several examples. How does each form.

7. Why are silicates the most common group of minerals on Earth?

8. Quartz and clays are the most common minerals that we see. Why are they so common and widespread on Earth's surface?

9. Is ice in an iceberg a mineral? Why? Is ice in your freezer a mineral or not? Why?

10. Some minerals, such as quartz or fluorite, are just about the same composition no matter where they are found. Other minerals are *solid solutions*. What does this mean?

11. A general formula for garnet is $(Ca,Fe,Mg,Mn)_3$ $(Fe,Al)_2Si_3O_{12}$. What is the significance of the elemental symbols separated by commas in this formula.?

12. Olivine is primarily a mixture of two end members. What are those end members (name them), and what are their formulas. Write a general formula for olivine that indicates that it is a solid solution of those two end members.

13. Why may an albite crystal not be 100% $NaAlSi_3O_8$, but still be correctly identified as albite?

14. Why are anions or anionic groups used to classify minerals rather than the cations?

15. What do all silicate minerals have in common?

16. Why is galena often covered with anglesite? (Hint: to start, look up the chemical formulas of each. You may have to go to the internet for this.)

17. Both diamond and graphite are made of carbon. Graphite is stable at Earth's surface, and diamond is stable deep within Earth but is metastable at the surface. What does *metastable* mean? And, some minerals are unstable at Earth's surface. What does this mean about them? How do they contrast with metastable minerals?

18. Are anhedral minerals crystalline? Explain why or why not. Why is crystal shape not a useful property for identifying a mineral sample that is anhedral?

19. Explain why color is one of the least reliable properties in the identification of quartz and many other non-metallic minerals.

20. Beryl is typically some shade of blue. But, it may be dark blue or light blue (aquamarine), and it is sometimes green (emerald), or pink (morganite), golden (helidor) or colorless (goshenite). Why all the variety – what causes the color differences? And, is color a good property for identifying beryl?

21. Why do silicate minerals generally not produce a streak? Hint - see Table 3.10.

22. Why is the ability of a mineral to scratch glass not a good enough criterion for identifying a diamond?

23. Why do minerals develop cleavages? For example, all micas cleave into infinitely thin sheets. What does this tell you about the arrangement and bonding of atoms within the mineral? Quartz is one of very few minerals that has no cleavage. What does this tell you about the arrangement and bonding of atoms within quartz?

24. Cleaning powders such as Ajax and Comet always have some powdered minerals in them to act as an abrasive. In the past, quartz was sometimes used but today it is pretty much all calcite. Why should cleaning products with calcite (instead of quartz) be used on mirrors?

25. How does dolomite react differently to dilute hydrochloric acid than calcite?

4 Mineral Crystals

4.1 Cuevo de los Cristales

The photograph in Figure 4.1 shows a geologist in Mexico's *Cueva de los Cristales* (Cave of Crystals). The cave contains some of the largest mineral crystals in the world—up to 12 meters long. These crystals are *selenite*, a transparent form of *gypsum*, which has the composition hydrated calcium sulfate ($CaSO_4 \cdot 2H_2O$). Gypsum is moderately water-soluble, so it is one of a relatively small number of common minerals that precipitate from natural water, often redissolve, and later reprecipitate somewhere else. Most natural gypsum crystals are centimeters in size or smaller. However, extremely large crystals, such as those shown here, exist in several places around the world.

Cueva de los Cristales is one of many caves associated with the Naica Mine in Chihuahua State, Mexico. The mine produces silver, lead, and zinc from deposits in a limestone host rock. Miners, exploring for new and deeper ore zones, drilled into the crystal cave in 2000. It was full of water that, when pumped out, revealed a wondrous crystal-lined natural chamber, irregularly shaped and about 30 meters in its longest dimension. The cave is more than 300 meters beneath the surface, and is very warm (> 49 °C; > 120 °F) and wet (99% humidity), which hinders both mining and geological investigations.

Large crystals had been found in other caves in the Naica Mine, but the discovery of Cueva de los Cristales was much more spectacular. Large blocky crystals, meters across, cover the floor, and huge crystal "beams," seemingly lying in random orientations, extend from floor to ceiling. The beams, large enough for people to use as bridges or walkways, measure up to 12 meters long and 4 meters in diameter and weigh as much as 55 tons.

Figure 4.1   Cuevo de los Cristales, Chihuahua, Mexico.
Photo from Alexander van Driessche, Wikimedia Commons.

The region around Naica is underlain by magma, left over from volcanic activity that began about 26 million years ago. Heat from the magma warmed groundwater, causing anhydrite ($CaSO_4$) in bedrock to dissolve. The result was hot, gypsum-saturated water that filled underground chambers. Subsequently, large gypsum crystals precipitated, one atom at a time, from the sometimes-supersaturated solution, at temperatures of around 50 °C (122 °F). The largest crystals may contain countless individual atoms, and crystal growth is estimated to have taken more than half a million years. Unfortunately, today, visiting Cueva de los Cristales is no longer possible. As of 2017, the Naica Mine ceased operations and the cave is now, once more, flooded with water.

4.2   Crystallography and crystal chemistry

Crystals, such as the huge ones in Cuevo de los Cristales, are solids with a fixed ratio of atoms held together by chemical bonds and arranged in an orderly and repeating way. The study of crystals and their formation is termed *crystallography*. Crystals may consist of a single element but, more commonly, they are *compounds* of two or more. The compositions of minerals, and how atoms bond together to create minerals, comprise the field of *crystal chemistry*. Because many natural materials are crystalline, crystallographic and crystal chemical studies today are directed at many kinds of materials—not just minerals. Among other applications, crystallography has been part of the semiconductor industry and the development of vitamins and drugs, and has allowed scientists to model the structure of DNA.

The first crystallographic investigations occurred in the 1600s. Johannes Kepler studied the symmetry of snowflakes and concluded that the hexagonal shapes were due to water particles packed together in hexagonal patterns. In 1669, Nicolas Steno showed that the angles between crystal faces were the same for all samples of any particular mineral. Eventually, in 1784, René Hauy deduced that crystal faces result from the stacking together of fundamental building blocks in regular ways. Subsequent studies concluded that crystals contain an ordered arrangement of atoms that repeat indefinitely in three directions that need not be perpendicular. Because the repeat distance in crystals is on the order of angstroms, we can think of crystals as many identical *unit cells* of angstrom scale stacked together

to create the entire crystal. Consequently, crystals of a given compound have the same composition and same relative arrangement of atoms no matter their sizes. The existence of a unit cell and the repetitive nature of the overall structure set crystals apart from other solid materials.

As described in the previous chapter, mineral crystals may be euhedral, subhedral, or anhedral. Euhedral crystalline materials have visible crystal shapes with flat crystal faces oriented at specific angles to each other. The properties and orientations of the faces are controlled by atomic arrangement in unit cells and the way unit cells stack together to make the entire crystal. Some mineral crystals are subhedral—they show a few, generally imperfect crystal faces. Most mineral crystals, however, are anhedral and do not show crystal faces at all. Nonetheless, they must contain unit cells of fixed composition with ordered atomic arrangements.

Many natural materials are noncrystalline, also termed *amorphous*. They contain atoms arranged in random patterns, schematically shown in the top part of Figure 4.2. Window glass and naturally occurring obsidian are good examples of amorphous solids. In contrast, in

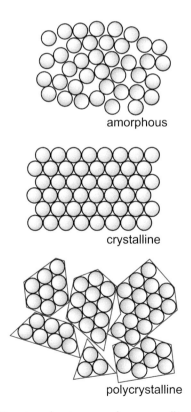

amorphous

crystalline

polycrystalline

Figure 4.2   Drawings showing amorphous, crystalline, and polycrystalline arrangements of atoms.

crystalline materials, the constituent atoms are arranged in repetitive patterns. Often the pattern appears to have symmetry, such as the hexagonal symmetry shown in the middle drawing here. All minerals, by definition, must be crystalline.

Other solids, including some minerals, grow to form *polycrystals* composed of multiple domains aggregated together (Figure 4.2, bottom). Each domain is crystalline, but the atoms in one domain are oriented randomly with respect to atoms in other domains. Many synthetic and natural materials are polycrystalline. Although minerals of all sorts are crystalline, most other inorganic solids are polycrystalline or amorphous.

Early crystallographic studies involved detailed studies of the shapes of crystals and angles between crystal faces. The discovery of *X-ray diffraction* in 1912, by Walter Friedrich and Paul Knipping, was the key that opened the door to modern studies. Friedrich and Knipping found that the atoms in copper sulfate crystals would cause a focused X-ray beam to be *diffracted*—redirected in many directions. The directions and intensities of diffracted beams provide information that mineralogists and others use to determine the arrangement of atoms in unit cells. X-ray diffraction techniques have improved markedly since 1912, and modern crystallographers study crystals using other kinds of diffraction besides X-ray (e.g., *electron diffraction* or *neutron diffraction*). Diffraction studies have allowed scientists to determine atomic arrangements in crystals of many different compositions and, today, diffraction is a major part of most crystallographic investigations.

4.3 The process of crystallization

As shown in the drawing in Figure 4.3, crystals start small and tend to grow larger with time. Whether in a magma or aqueous solution, or even in a metamorphic rock, initial crystallization usually involves many *nuclei* composed of a few atoms that form the initial focus of crystal growth. The first crystals to form are small.

Nucleation may be a slow process. Once crystallization starts, however, the process speeds up, and crystal growth spreads outward from the nuclei. Crystals grow larger as atoms add to the growing crystal in an orderly way, and in constant proportions, sometimes producing smooth planar crystal faces.

Nucleation is a kinetic process. Small nuclei composed of just a few atoms form relatively quickly compared with larger crystals. Larger crystals are, however,

more stable because molecules in the interior of crystals are less reactive and have lower energy than those on the outside. Larger crystals have greater volume-to-surface-area ratio and lower relative surface energy. Consequently, with time, energetics trumps kinetics, and molecules on the outside of small crystals diffuse and add to the outside of larger ones. So, over time, larger crystals form at the expense of smaller ones, and regions around large crystals become depleted in small crystals. We call this process *Ostwald ripening*. Ostwald ripening explains, for example, why ice crystals form over time in initially smooth ice cream (making old ice cream crunchy) and why large crystals surrounded by a sea of small crystals may form in some volcanic rocks.

The most important factors controlling crystal size and perfection are temperature, time, abundance of necessary elements, and the presence or absence of a flux, such as water, to carry atoms to growing crystals (Fig. 4.3). These factors explain why gypsum crystals in Cuevo de los Cristales grew so large: they formed at moderately warm temperatures; they had many thousands or millions of years to form; calcium sulfate existed in unlimited amounts in nearby country rocks; and the hot water rapidly delivered atoms to growing crystal faces.

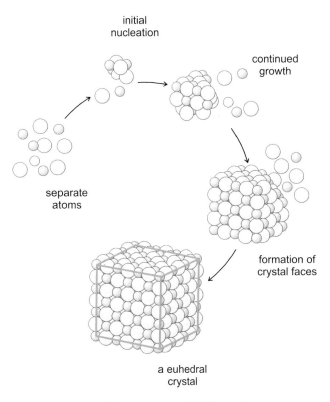

Figure 4.3 From a few atoms to a euhedral crystal.

The three photographs in Figure 4.4 show examples of mineral crystals formed by igneous, metamorphic, and sedimentary processes. The top photo shows large light-colored feldspar crystals in the Mont Blanc granite, France, with a 1 euro coin for scale. This rock and the crystals it contains formed from a high-temperature magma that crystallized slowly underground.

The middle photo shows blades of blue kyanite, with lesser amounts of brown staurolite, in a mica schist from St. Gotthard, Switzerland. The kyanite and staurolite crystals grew at high temperature over a very long time during metamorphism.

The bottom photo shows amethyst, a purple variety of quartz, that precipitated from warm water flowing through a cavity in a rock. The water acted as a flux, delivering silicon and oxygen to the growing quartz crystals. Because the crystals were growing into open space, they could grow large and euhedral. We call crystal masses that form this way *geodes*.

Temperature is a key factor affecting crystal growth because at elevated temperatures atoms are very mobile, which allows crystals to grow efficiently and often leads to large and well-formed crystals. *Principles of thermodynamics* tell us that crystals forming at higher temperatures have more complicated atomic structures than those forming at lower temperatures, which may relate to their tendency to be large and well ordered.

Time is important, because if a crystal has a long time to grow, it will naturally be larger and better ordered than one that grows more quickly. More atoms will have time to migrate to the growing crystal and to order themselves in a regular way. (Although we can make *synthetic minerals* in the laboratory, we cannot mimic the magnitude of geologic time. Consequently, when synthetic minerals are viewed at an atomic scale, it is often found that they are not as well ordered as their natural equivalents.) Thus, *intrusive igneous rocks*, such as the one shown in Figure 4.4, which cool slowly underground, contain larger crystals than *extrusive igneous rocks* of the same compositions. Some extrusive rocks, such as *obsidian*, cool so quickly that they solidify as amorphous *glass* containing a random arrangement of atoms in a solid form.

Some crystallization occurs quickly, some very slowly. For example, lava, which is molten rock on Earth's surface, can cool and crystallize in minutes, but magma that is deep underground crystallizes very slowly. An igneous mineral, therefore, may crystallize in minutes or in millions of years. Sedimentary minerals also grow over fast or slow times. Calcite that precipitates as marl in streambeds may dissolve and precipitate on a 24-hour cycle, but calcite that crystallizes in caves to form *stalactites* or *stalagmites* does so quite slowly, perhaps adding a millimeter of new calcite each year. Metamorphic minerals crystals may be the slowest to grow, sometimes requiring many millions of years in nature, although, in the laboratory, we can make some metamorphic minerals in just a few minutes.

Whatever the time and temperature, crystals cannot grow large if their component elements are scarce. A dozen elements or fewer account for 99% of the compositions of most rocks. Crystals composed of those abundant elements will usually be larger than those composed of rarer elements. However, even if time, temperature, and atoms are right, crystals may not grow large. Diffusion of atoms through solids is slow, and

Feldspar crystals in an igneous rock

Blue kyanite and brown staurolite crystals in a metamorphic rock

Amethyst (purple quartz) crystals in a geode.

Figure 4.4   Igneous, metamorphic, and sedimentary mineral crystals. Images from anonymous, Der Hexer, and Tracy Holman, all at Wikimedia Commons.

atoms may not be able to migrate to spots where crystals are growing. Yet, if a *fluid* such as *interstitial water* or a *magma* is present it may act as a *flux*, transporting atoms to growing crystals—elements may be carried long distances to sites of mineral growth, and even minerals composed of rare elements may grow to be large. This explains, in part, why some minerals of unusual composition are large in *pegmatites*, coarse-grained rocks that crystallize from water-rich melts left over after most of a magma has crystallized.

4.4 Ionic crystals

Atoms may bond together in different ways. The three most significant kinds of bonds in minerals are ionic, covalent, and metallic. Figure 4.5 shows the combinations of different kinds of bonding in some typical minerals. Most minerals are neither 100% ionic, 100% covalent, nor 100% metallic bonds, but some come close. Others are predominantly combinations of ionic and covalent bonding, or combinations of covalent and metallic bonding.

In *ionic crystals*, predominately ionic bonds hold atoms together. Halides and many oxide minerals fall into this category. Although bonding in many silicate minerals is only partially ionic, the properties of ions and ionic bonding have a strong influence on the atomic arrangement in all silicates. Other mineral groups, notably native elements and sulfides contain mostly covalent or metallic bonds, or combinations of both. However,

we focus here on ionic crystals before talking about crystal chemistry and crystallography in general, because the principles of ionic bonding are more easily analyzed than are those of covalent and metallic bonding, and because most lessons learned from the study of ionic crystals apply equally to crystals of different natures.

4.4.1 Ionic size

When atoms ionize, some lose valence electrons to become positively charged *cations*. Others gain valence electrons and become negatively charged *anions*. Figure 4.6 shows the relative ionic sizes of some common anions (blue) and cations (yellow). Overall, the anions are larger than the cations.

The size of an ion depends on the number of protons in the nucleus and the number of orbiting electrons. Un-ionized calcium, for example, contains 20 protons and an equal number of electrons; its atomic radius is about 1.74 Å. When ionized, calcium loses two electrons to become Ca^{2+}. The 20 protons in the nucleus pull strongly on the remaining 18 electrons, causing the size of the ion to shrink to 1.14 Å. This is true of all cations; cations are smaller than un-ionized forms of the same element because excess positive charge pulls the remaining electrons toward the nucleus. Furthermore, cations of high charge are generally smaller than cations of low charge. For opposite reasons, anions are always larger than un-ionized forms of the same element, and anions of high charge tend to be larger than those of low charge, as shown in Figure 4.6.

Because like charges repel and opposite charges attract, as atoms come together to form ionic crystals, cations bond to, and are surrounded by, anions, and anions bond to, and are surrounded by, cations. This occurs even if bonding is not entirely ionic. So, ionic and partially ionic crystal structures consist of alternating cations and anions in three dimensions. The number one control on how they fit together is ionic size.

4.4.2 Closest packing

Because oxygen and other common anions in minerals are generally larger than the common cations, we can think of ionic crystals as having large anions stacked together with smaller cations occupying holes between. As an analogy, consider a layer of equal-sized marbles. As shown in Figure 4.7, they naturally fit together in

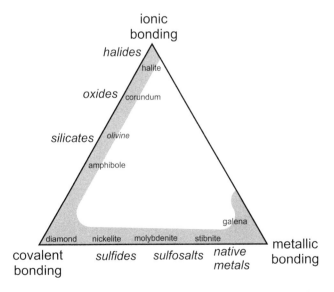

Figure 4.5   Different kinds of bonding in some different minerals.

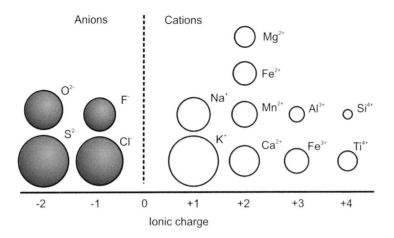

Figure 4.6 The relative sizes of some common anions and cations.

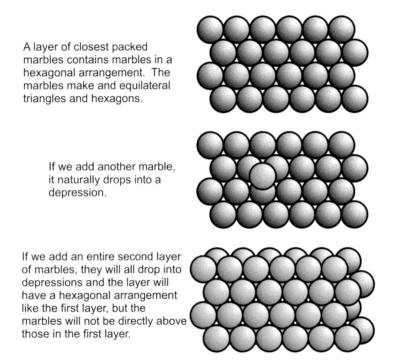

A layer of closest packed marbles contains marbles in a hexagonal arrangement. The marbles make and equilateral triangles and hexagons.

If we add another marble, it naturally drops into a depression.

If we add an entire second layer of marbles, they will all drop into depressions and the layer will have a hexagonal arrangement like the first layer, but the marbles will not be directly above those in the first layer.

if we add a third layer (shown in blue), it may be directly above the first layer, or it may not. If we add a fourth layer, however, it must align with one of the bottom two layers.

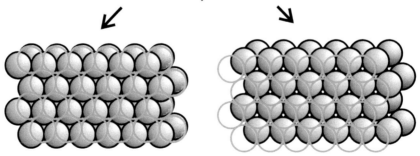

Figure 4.7 Closest packing of equal-sized spheres.

a pattern involving marbles arranged in hexagons and equilateral triangles. Because no other two-dimensional arrangement allows marbles to be closer together, we say the hexagonal arrangement is *closest packed*. Each marble touches six others, the maximum possible in a single layer of marbles.

Figure 4.7 shows what happens if we put another marble on top of the first layer—it will naturally fit in a low spot between three marbles in the bottom layer. If we continue to add marbles, we will get a second layer (with marbles arranged in the same pattern as the bottom layer) that is slightly offset from the bottom layer. As shown in Figure 4.7, if we add a third layer, it may be directly above the bottom layer or it may not. This is because each layer contains twice as many positions (low spots) for marbles to occupy as there are marbles, so each additional layer may be placed on the one below in either of two ways (shown by blue circles in the two drawings at the bottom). When we add a fourth layer, however, we run out of choices, and the fourth layer must lie directly above one of the first two. No matter how the layers repeat, we will end up with marbles that are closest packed and, in the three-dimensional arrangement, every anion is in contact with 12 others.

In ionic crystals, cations occupy holes between anions. A close look at Figure 4.7 reveals that there are two kinds of holes. As seen in Figure 4.8 (left side), one involves three anions forming an equilateral triangle with a fourth anion sitting directly above the center of the bottom three The other involves three anions forming an equilateral triangle, with three

anions in the next layer forming a triangle pointing in the opposite direction (Fig. 4.8, right side). If we draw lines connecting the centers of the anions, we get a *tetrahedron* in the first case and an *octahedron* in the second. So, cations occupying the spaces in the centers of the two arrangements are said to be in *tetrahedral coordination* (also called 4-fold coordination because bonds go to four anions) or in *octahedral coordination* (also called 6-fold coordination because bonds go to six anions).

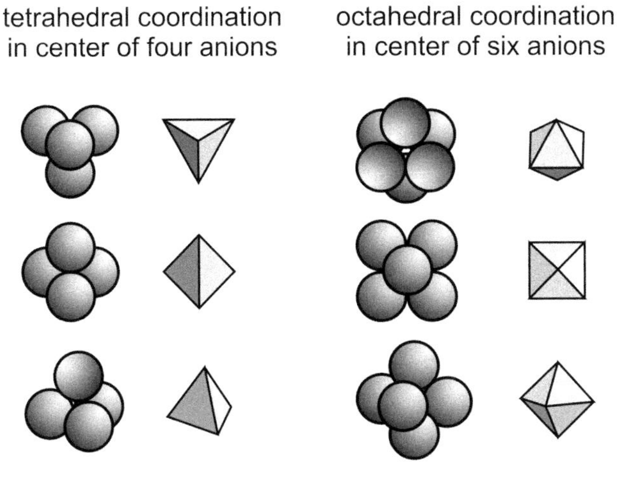

Figure 4.9 shows the atomic arrangements in sphalerite and wurtzite, both with composition ZnS. The sulfur atoms (yellow) in both minerals are nearly closest packed, but have been separated in these drawing so that the structure is more easily seen. The zinc atoms (red) are smaller than the closest-packed sulfur atoms around them. Both zinc and sulfur are in tetrahedral coordination; each sulfur atom bonds to four zinc atoms and vice versa. Although similar in many ways, sphalerite and wurtzite do not have identical atomic arrangements. They are *polymorphs*, meaning they have the same composition but different arrangements of atoms. In wurtzite, alternate layers of sulfur lie directly above or below each other; identical layers are labeled A and B in the drawing. In sphalerite, it takes three layers (A-B-C) before the pattern repeats.

4.4.3 Exceptions to closest packing

Sphalerite, wurtzite, halite, and some native metals (including gold, silver, platinum, and copper) are all examples of closest-packed, or almost closest-packed, minerals. Most other mineral structures are, however, not closest-packed. Some have complicated structures with mixed stacking sequences, and many have arrangements that are not closest-packed at all. Overall, dense minerals with few large cations tend to be closely packed. In contrast, minerals containing (larger) alkali or alkaline earth elements cannot be closest-packed because alkalis and alkaline earths are too large to fit in tetrahedral or octahedral sites. The closest-packed model also fails for other minerals in which small anions are between large cations, in some metals in which each atom is bonded to eight others, and for some minerals in which the dominant kind of bonding is not ionic. Still, the general principles of closest packing are useful guides to atomic arrangement in just about all crystalline structures.

tetrahedral coordination
in center of four anions

octahedral coordination
in center of six anions

Figure 4.8 Cation sites in closest-packed structures. Three views of tetrahedral coordination (left) and of octahedral coordination (right).

wurtzite

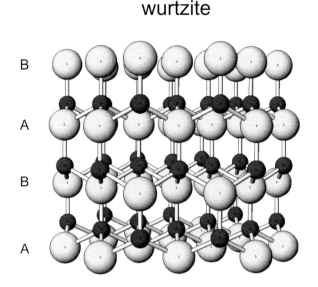

sphalerite

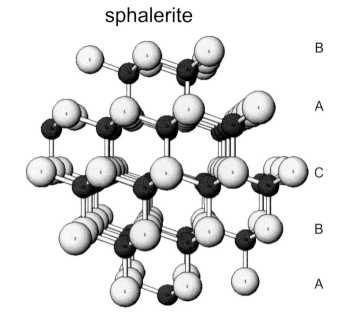

Figure 4.9   Atomic arrangements in wurtzite and sphalerite.

4.4.4 *Coordination number*

In closest-packed structures, cations can only be in tetrahedral or octahedral coordination, but arrangements that are not closest packed may contain atoms in other coordinations. Small cations bond to a small number of anions because the anions can pack together closely and still leave room for the cation between them. In contrast, large cations bond to a relatively large number of anions because more space is needed to accommodate the larger cation. The number of anions surrounding a cation is the cation's *coordination number*. In minerals, cation coordination numbers typically range from 3 to about 10, depending on ionic size.

As seen in Figure 4.10, we name the different kinds of coordination based on the geometric shape of the *coordinating polyhedra* obtained when the centers of coordinating anions are connected. So, we have triangular (3-fold), tetrahedral (4-fold), octahedral (6-fold), cubic (8-fold), and cubeoctahedral (12-fold) coordinations. Triangles, tetrahedra, octahedra, cubes, and cubeoctahedra are the only possible kinds of *regular polyhedra*—polyhedra with all bonds the same length. Many minerals contain predominantly regular coordination polyhedra, but *irregular polyhedra* (distorted versions of those shown in this drawing) and other coordinations (e.g., 5-fold, 7-fold) are also possible. When these less common coordinations

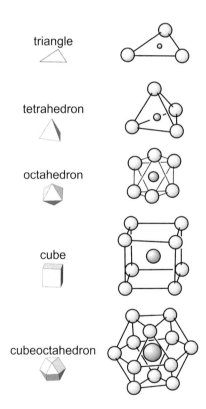

Figure 4.10   Coordination and coordinating polyhedra.

are present, it is often because bonds are not entirely ionic and, therefore, may be stronger in some directions than others, or because of the presence of large anionic groups such as $(CO_3)^{2-}$ instead of individual anions.

4.4.5 Pauling's rules

In 1929, Linus Pauling formalized some fundamental principles of ionic and covalent crystal structures, now called *Pauling's Rules*. The five rules, which have been little modified since first promulgated, provide a sound basis for understanding the orderly way in which atoms combine to form crystals.

4.4.5.1 Rule 1: the radius ratio principle

Pauling's first rule is the *Radius Ratio Principle*. Pauling observed that we can think of an ion as a sphere with a fixed radius, and we can think of ionic bonds as two touching spheres. So, the distance between a cation bonded to an anion is the sum of their ionic radii. Furthermore, Pauling noted, the coordination number of a

cation can be predicted by the ratio of the cation radius to coordinating anion radius (R_c/R_a). Larger cations, then, have greater coordination numbers than smaller cations. Ionic radii vary somewhat, depending on several things, including the arrangement of atoms and the degree to which bonds are ionic. The most commonly used list of *average* values was developed by Shannon and Prewitt (1969).

Figure 4.11 shows ideal situations for 3-fold and 6-fold coordination—where the cation fits perfectly snugly between surrounding anions. The ratios of cation radius to anion radius, such that the cation fits perfectly between the anions, are 0.155 (3-fold, triangular coordination) and 0.414 (6-fold, octahedral coordination). Similar geometric calculations can be done for 4-, 8-, and 12-fold coordination (Table 4.1). Pauling observed that these ratios can be stretched—cations a bit too large can still be in 3-fold (or 4-fold) coordination. If, however, cations are too small, they will rattle around and not maintain contact with surrounding anions and

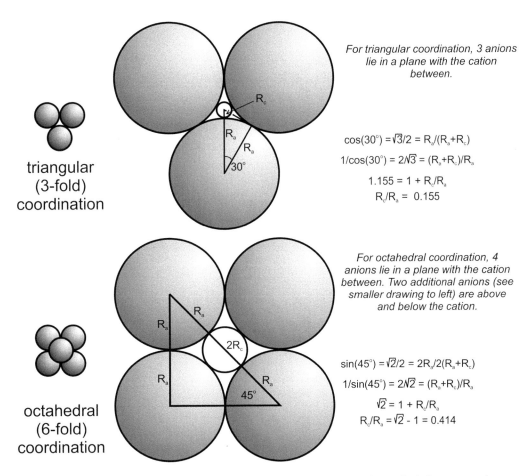

triangular (3-fold) coordination

For triangular coordination, 3 anions lie in a plane with the cation between.

$$\cos(30°) = \sqrt{3}/2 = R_a/(R_a+R_c)$$
$$1/\cos(30°) = 2/\sqrt{3} = (R_a+R_c)/R_a$$
$$1.155 = 1 + R_c/R_a$$
$$R_c/R_a = 0.155$$

octahedral (6-fold) coordination

For octahedral coordination, 4 anions lie in a plane with the cation between. Two additional anions (see smaller drawing to left) are above and below the cation.

$$\sin(45°) = \sqrt{2}/2 = 2R_a/2(R_a+R_c)$$
$$1/\sin(45°) = 2/\sqrt{2} = (R_a+R_c)/R_a$$
$$\sqrt{2} = 1 + R_c/R_a$$
$$R_c/R_a = \sqrt{2} - 1 = 0.414$$

Figure 4.11 Calculating ideal fits for cations in triangular and octahedral coordination. In the octahedral case, two other anions (not shown) are directly above and directly below the cation at the center.

Table 4.1 R_c/R_a and the predicted coordination of cations

R_c/R_a	Predicted coordination of cation	C.N.
< 0.15	2-fold coordination	2
0.15	ideal triangular coordination	3
0.15–0.22	triangular coordination	3
0.22	ideal tetrahedral coordination	4
0.22–0.41	tetrahedral coordination	4
0.41	ideal octahedral coordination	6
0.41–0.73	octahedral coordination	6
0.73	ideal cubic coordination	8
0.73–1.0	cubic coordination	8
1.0	ideal dodecahedral coordination	12
> 1.0	dodecahedral coordination	12

will prefer a site with lower coordination (Table 4.1). So, overall, cations will bond to as many anions as possible since they can stay in contact with the anions.

O^{2-} is the dominant anion in most minerals, and has an ionic radius of 1.27–1.34 Å, and we typically use an average value of 1.32 Å when calculating radius ratios. The most common cations include Si^{4+}, Al^{3+}, Fe^{2+}, Fe^{3+}, Ca^{2+}, Na^+, K^+, Mg^{2+}, Ti^{4+}, Mn^{2+}, P^{5+}, and C^{4+}. Table 4.2 lists ionic radii, typical coordination numbers (when coordinated with O^{2-}), and example minerals for each. Coordination numbers range from 3 to 12 and agree well with Pauling's rule, except for the smallest cations (P^{5+} and C^{4+}), which bond with little ionic character.

Some cations have different coordinations in different minerals. These differences in coordination arise, in part, because minerals may contain anions other than

Table 4.2 Ionic radii, typical coordination numbers with oxygen, and example minerals for some common cations.

Ion	Ionic radius (Å)	Predicted C.N.	Typical C.N. with oxygen	C.N.	Example minerals and formulas
K^+	1.59–1.68	12	8–12 (cubic or dodecahedral)	9	nepheline, $(Na,K)AlSiO_4$
					orthoclase, $KAlSi_3O_8$
				12	leucite, $KAlSi_2O_6$
					muscovite, $KAl_2(AlSi_3O_{10})(OH)_2$
Na^+	1.10–1.24	8	6–8 (octahedral or cubic)	6	pectolite, $NaCa_2Si_3O_8(OH)$
					albite, $NaAlSi_3O_8$
				7	sodalite, $Na_4(Al_3Si_3O_{12})•NaCl$
				8	nepheline, $(Na,K)AlSiO_4$
Ca^{2+}	1.08–1.20	8	6–8 (octahedral or cubic)	6	wollastonite, $CaSiO_3$
					pectolite, $NaCa_2Si_3O_8(OH)$
				7	plagioclase, $(Ca,Na)(Si,Al)_4O_8$
					titanite, $CaTiSiO_5$
				8	diopside, $CaMgSi_2O_6$
					garnet, $(Mg,Fe,Ca,Mn)_3Al_2Si_3O_{12}$
Mn^{2+}	0.83–1.01	6–8	6–8 (octahedral or cubic)	6	rhodonite, $MnSiO_3$
				8	garnet, $(Mg,Fe,Ca,Mn)_3Al_2Si_3O_{12}$
Mg^{2+}	0.80–0.97	6	6–8 (octahedral or cubic)	6	diopside, $CaMgSi_2O_6$
					olivine, $(Mg,Fe)_2SiO_4$
				8	garnet, $(Mg,Fe,Ca,Mn)_3Al_2Si_3O_{12}$
Fe^{2+}	0.71–0.77	6	6–8 (octahedral or cubic)	4	staurolite, $Fe_2Al_9Si_4O_{23}(OH)$
				6	biotite, $K(Mg,Fe)_3(AlSi_3O_{10})(OH)_2$
					olivine, $(Mg,Fe)_2SiO_4$
				8	garnet, $(Mg,Fe,Ca,Mn)_3Al_2Si_3O_{12}$
Ti^{4+}	0.69	6	6 (octahedral)	6	titanite, $CaTiSiO_5$; rutile, TiO_2
Fe^{3+}	0.57–0.68	4–6	4–6 (tetrahedral or octahedral)	6	epidote, $Ca_2(Al,Fe)_3Si_3O_{12}(OH)$
Al^{3+}	0.47–0.61	4–6	4–6 (tetrahedral or octahedral)	4	muscovite, $KAl_2(AlSi_3O_{10})(OH)_2$
					orthoclase, $KAlSi_3O_8$
				5	andalusite, Al_2SiO_5
				6	muscovite, $KAl_2(AlSi_3O_{10})(OH)_2$
					beryl, $Be_3Al_2Si_6O_{18}$
Si^{4+}	0.34–0.48	4	4 (tetrahedral)	4	quartz, SiO_2
					tremolite, $Ca_2Mg_5Si_8O_{22}(OH)_2$
				6	stishovite, SiO_2
P^{5+}	0.25	2	4 (tetrahedral)	4	apatite, $Ca_5(PO_4)_3(OH)$
C^{4+}	0.16	2	3 (triangular)	3	calcite, $CaCO_3$
					malachite, $Cu_2(CO_3)(OH)_2$

O^{2-}, and because ions are not truly spheres with fixed radii. (Their radii vary somewhat depending on coordination). Additionally, ionic radius varies depending on bond type, and bonds in minerals vary in character. The largest common cations, the alkalis and alkali earths, have radii close to that of O^{2-}. They may be in 6- to 12-fold coordination, and often occupy irregularly shaped sites with some bond lengths shorter than others. Coordinating polyhedra for the smaller cations are more predictable and generally have bonds of approximately equal length. Despite the complications, Pauling's Radius Ratio Principle is an excellent rule for just about all ionic and covalent crystalline materials.

4.4.5.2 *Rule 2: bond strength*

Pauling's second rule, the *Electrostatic Valency Principle*, deals with bond strength and defines *electrostatic valency* as the charge of an ion divided by its coordination number. The electrostatic valency is a measure of bond strength, and the total strength of the bonds that reach an anion from all neighboring cations is equal to the charge of the anion. Rules 1 and 2 are the most significant of Pauling's five rules. The best way to explore them is to look at examples.

4.4.5.2.1 Halite (NaCl)

Figure 4.12 shows the atomic arrangement in halite. Na^+ (red) and Cl^- (green) alternate in a cubic pattern. The radii of the cation (Na^+) and anion (Cl^-) are 1.10–1.24 Å and 1.67–1.72 Å, respectively. R_c/R_a, therefore, has a value of around 0.7, so Na^+ is in octahedral (6-fold) coordination. Bonds around Na^+ have a valence of 1/6, total charge ÷ number of bonds. The valence of

each bond around Cl^- is 1/6 as well. Six bonds of charge 1/6 add up to 1, the total charge of each ion.

4.4.5.2.2 Rutile (TiO₂)

Figure 4.13 shows the atomic arrangement in rutile. Ti^{4+} (red) and O^{2-} (blue) are in a rectangular arrangement with twice as many O^{2-} as Ti^{4+}. The radii of Ti^{4+} and O^{2-} are 0.69 Å and 1.32 Å, respectively. So, $R_c/R_a = 0.69/1.23 = 0.52$, confirming that Ti^{4+} is in octahedral (6-fold) coordination. The bonds around Ti^{4+} have a valence of $+4/6$ = total charge ÷ number of bonds = $+2/3$. Each O^{2-} bonds to three Ti^{4+}. So, the valence of bonds around oxygen is also $+2/3$.

4.4.5.2.3 Fluorite (CaF₂)

We can use Pauling's first two rules to analyze a more complicated mineral, fluorite (CaF_2), shown in Figure 4.14. The ions are arranged in a cubic pattern, and there are twice as many F^- (purple) as Ca^{2+} (blue). The radii of Ca^{2+} and F^- are 1.12 Å and 1.31 Å, respectively.

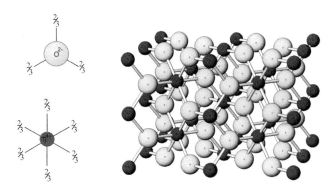

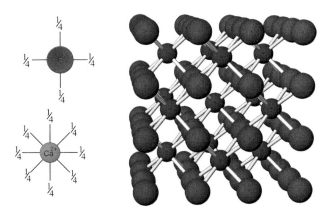

Figure 4.13   Arrangement of ions in rutile (TiO_2).

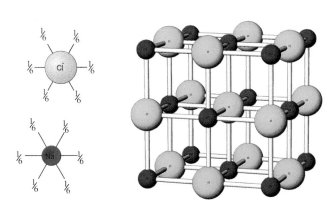

Figure 4.12   Arrangement of ions in halite (NaCl).

Figure 4.14   Arrangement of ions in fluorite (CaF_2).

So, R_c/R_a is 0.85 and, as predicted by Rule 1, Ca^{2+} is in 8-fold (cubic) coordination. Each bond around Ca^{2+} has a strength of 2/8 = 1/4 (ionic charge ÷ number of bonds). Since each F^- has a total charge of –1, it must be bonded to four Ca^{2+} to satisfy Rule 2. So, F^- is in tetrahedral coordination.

4.4.5.2.4   Anhydrite ($CaSO_4$)

Finally, let's consider anhydrite ($CaSO_4$), shown in Figure 4.15. In anhydrite, S^{6+} is in tetrahedral (4-fold) coordination with oxygen. So, bond strength between sulfur and oxygen is 6/4 = 1½. The $(SO_4)^{2-}$ tetrahedra are shown in yellow. Besides bonds to sulfur, the oxygen ions at the corners of the tetrahedra are also bonded to two calcium ions (blue). Since the oxygen anions have a total charge of 2, and 1½ is used bonding to sulfur, each O-Ca bond has valence of only ¼, much less than the valence of O-S bonds. Consequently, $(SO_4)^{2-}$ groups are tight units within the crystal structure.

Crystals such as anhydrite, which contain bonds with differing electrostatic valences, are termed *anisodesmic*. They contrast with *isodesmic* compounds such as rutile, halite, and fluorite, in which all bonds are the same strength. All sulfates, carbonates, nitrates, and other anisodesmic compounds have tightly bonded anion groups. So, we often think of $(SO_4)^{2-}$, $(CO_3)^{2-}$, $(NO_3)^-$ as single anionic units within crystals. In atomic drawings, we often depict the groups as polyhedra and do not show the cations at their centers (e.g., the sulfate group in Figure 4.15). When many sulfates, carbonates, and nitrates dissolve in water, the anionic units often do not completely disassociate, so the water may contain sulfate, carbonate, and nitrate groups as dissolved species.

4.4.5.3   Rules 3 and 4: sharing of anions

Figure 4.16 shows some of the ways that two cations (red) may share anions (blue). Pauling's Rule 3 says that two cations may bond to the same anion (left-hand drawing), but it is less likely that they will share two anions (middle drawing), and even more unlikely that they will share three (right-hand drawing). The dimensions (numbers) in these drawings are the relative distance between cations; the more sharing, the closer the cations become.

When cations share a single anion, their polyhedra have a common apex; when they share two anions, the polyhedra have a common edge; and when they share three anions, their polyhedra share a face. Rule 3 says that sharing edges, and particularly sharing faces, of two anion polyhedra in a crystal structure decreases its stability. The stability is decreased because the more anion sharing, the closer the cations must be, and because like charges repel. Rule 4 is a corollary to Rule 3 and says that in a crystal containing different cations, those of high valency and small coordination number tend not to share polyhedron elements with one another.

4.4.5.4   Rule 5: the principle of parsimony

Rule 5 says that the number of essentially different kinds of constituents in a crystal tends to be small. Silicon and oxygen, for example, bond the same way in most minerals because atoms of silicon and oxygen are most stable that way. A crystal may contain several types of polyhedra (e.g., tetrahedra, octahedra, dodecahedral) that contain cations, but there will not be many different types, and the elements that occupy the sites are predictable and often the same in different minerals.

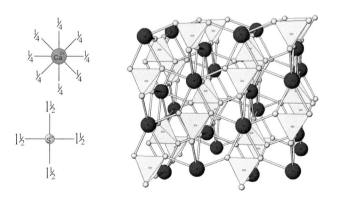

Figure 4.15   Arrangement of ions in anhydrite ($CaSO_4$).

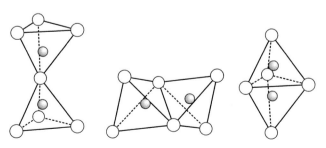

Figure 4.16   Corner sharing, edge sharing, and face sharing of anions (blue) around cations (red).

4.5 Silicate minerals

In minerals, Si^{4+} is always surrounded by, and bonded to, four anions, forming tetrahedra such as those shown in Figure 4.16. Although an ionic model works quite well for describing the bonding holding the tetrahedra together, the bonds between Si^{4+} and O^{2-} have a large covalent character, and because covalent bonds are relatively strong, SiO_4 tetrahedra act, in many ways, as single anionic units, like carbonate, nitrate, and phosphate groups. Because oxygen and silicon are the two most abundant elements in the Earth's crust, and because the $(SiO_4)^{4+}$ tetrahedron is such a strongly bonded group, silicate minerals are extremely stable and abundant in crustal rocks and sediments. They dominate igneous and metamorphic rocks as well as many sedimentary rocks.

Silicate minerals belong to a special group of compounds called *mesodesmic* compounds. If a bond distribution is mesodesmic, cation-anion bond strength equals exactly half the charge of the anion. In silicates, Si^{4+} is in tetrahedral coordination, and each Si:O bond has a strength of 1, exactly half the charge of O^{2-}. Consequently, the oxygen may coordinate to another cation just as strongly as to its coordinating Si^{4+}. The "other" cation may be another Si^{4+}, so two silica tetrahedra may share one oxygen. This is why silica tetrahedra, $(SiO_4)^{4+}$, can *polymerize* to form pairs, chains, rings, sheets, or networks.

Figure 4.17 shows some of the different ways that polymerization of silicon tetrahedra can occur. The Si^{4+} cation is not shown but lies in the center of every tetrahedron formed by four O^{2-}. The center columns shows the arrangement of oxygen anions; the right-hand column shows the resulting geometry of the arrangement.

The extent of polymerization varies between different mineral species. In quartz and most of its polymorphs,

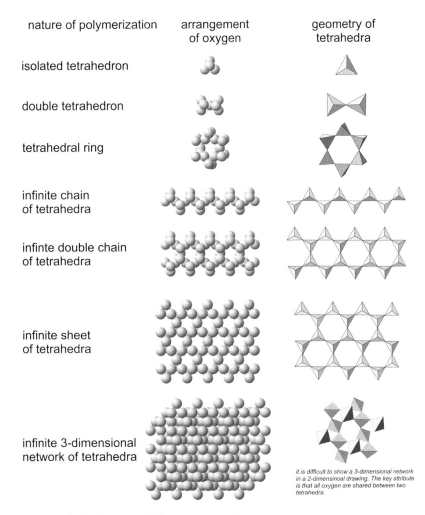

nature of polymerization	arrangement of oxygen	geometry of tetrahedra
isolated tetrahedron		
double tetrahedron		
tetrahedral ring		
infinite chain of tetrahedra		
infinte double chain of tetrahedra		
infinite sheet of tetrahedra		
infinite 3-dimensional network of tetrahedra		it is difficult to show a 3-dimensional network in a 2-dimensinoal drawing. The key attribute is that all oxygen are shared between two tetrahedra.

Figure 4.17   The different ways in which silicon tetrahedra may polymerize.

$(SiO_4)^{4+}$ groups share every oxygen with another $(SiO_4)^{4+}$ group, forming an infinite tetrahedral network. This is depicted in the bottom row of Figure 4.17. The strength of each Si-O bond is 1; each Si^{4+} bonds to four O^{2-}, and each O^{2-} to two Si^{4+}, so charge balance is maintained and the overall formula is SiO_2. Quartz has the maximum amount of polymerization possible.

In quartz and feldspars, polymerization results in a tightly bonded network. In other minerals, the silicon tetrahedra may be connected in pairs, chains, or sheets. Polymerization of silicon tetrahedra is absent in some silicates, such as olivine $(Mg,Fe)_2SiO_4$, that contain isolated silicon tetrahedra (top row in Fig. 4.17). In olivine, single tetrahedra bond to cations which bond to other tetrahedra, so bonds to cations between tetrahedra hold the structure together. In many silicates other than quartz and olivine, a combination of some oxygen sharing between silicon tetrahedra, and the presence of additional cations between tetrahedra, leads to charge balance and holds the structure together. With more oxygen sharing, fewer additional cations are needed.

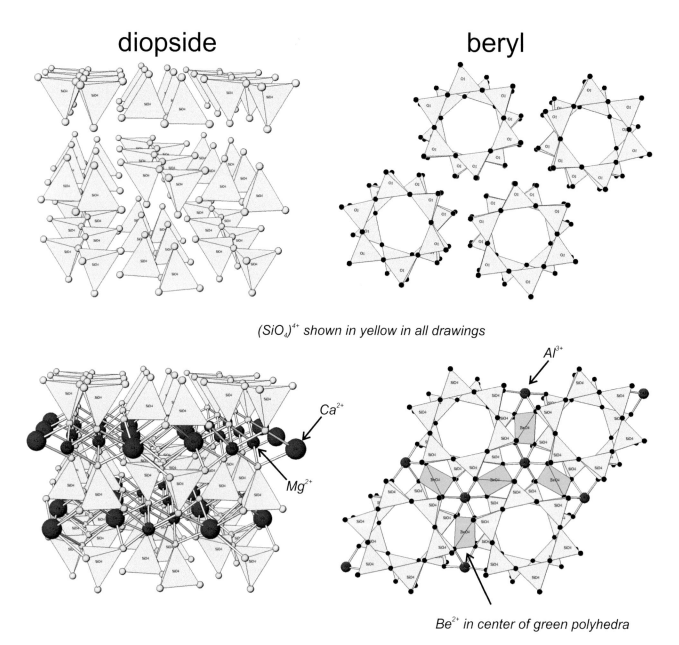

Figure 4.18   Polymerization of silicon tetrahedral in diopside and beryl.

In some silicates, Al^{3+} replaces some Si^{4+}, creating $(AlO_4)^{5+}$ tetrahedra that replace $(SiO_4)^{4+}$ tetrahedra. In these minerals, additional cations must be present to make up for the charge difference between Al^{3+} and Si^{4+}. In albite, for example, Al^{3+} replaces one-fourth of the tetrahedral Si^{4+}. Na^+ cations between tetrahedra make up for the missing charge. Albite's formula is $NaAlSi_3O_8$, which we may write as $Na(AlSi_3)O_8$ to emphasize that both Al^{3+} and Si^{4+} occupy the same structural sites. In anorthite, another feldspar, Al^{3+} replaces half of the Si^{4+}, resulting in the formula $Ca(Al_2Si_2)O_8$; the Ca^{2+} cation maintains the charge balance when two Al^{3+} replace two Si^{4+}. Besides feldspars, tetrahedral aluminum is common in micas, amphiboles, and, to a lesser extent, pyroxenes.

The orderly way silica (or alumina) tetrahedra polymerize leads naturally to the division of silicate minerals into the *subclasses* discussed in the previous chapter. We call silicates, such as olivine, in which tetrahedra share no O^{2-}, *isolated tetrahedral silicates*. Silicates in which pairs of tetrahedra share oxygen are *paired tetrahedral silicates*. If two oxygen on each tetrahedron link to other tetrahedra, we get *single-chain silicates* or *ring silicates*.

Figure 4.18 shows polymerization of silicon tetrahedral in diopside (chain silicate) and beryl (ring silicate). In the top images, all octahedral cations have been omitted so the polymerization is more easily seen. In the bottom images, cations and bonds are included to show how the entire structure is bonded together.

If some oxygen atoms are shared between two tetrahedra, and some between three, we get *double-chain silicates* (Fig. 4.17). If three oxygen atoms on each tetrahedron link to other tetrahedra to form tetrahedral planes, we get *sheet silicates*, and if all oxygen atoms are shared between tetrahedra, we get *framework silicates* (Fig. 4.17). The ratio of Si:O, then, indicates *silicate subclass* because different ratios result from different amounts of oxygen sharing (Table 4.3). In minerals containing tetrahedral aluminum, the ratio of total tetrahedral Al and Si to O, which we can abbreviate T:O, reflects the silicate subclass. If the only tetrahedral cation is silicon, isolated tetrahedral silicates are often characterized by SiO_4 in their formulas, paired tetrahedral silicates by Si_2O_7, single-chain silicates by SiO_3 or Si_2O_6, ring silicates by Si_6O_{18}, double-chain silicates by Si_4O_{11}, sheet silicates by Si_2O_5 or Si_4O_{10}, and framework silicates by SiO_2.

The chemistries of silicates correlate, in a general way, with the subclasses to which they belong (Table 4.3). This correlation reflects Si:O ratios and the way in which silica polymerization controls atomic structures. There are many variables, but we can make some generalizations. Isolated tetrahedral silicates and chain silicates include minerals rich in Fe^{2+} and Mg^{2+}, but framework silicates do not. The three-dimensional polymerization of framework silicates generally lacks sufficient anionic charge and the small crystallographic sites necessary for small highly charged cations. For opposite reasons, Na^+ and K^+ enjoy highly polymerized structures, which have large sites that easily accommodate monovalent cations.

Table 4.3 Silicate mineral subclasses.

Silicate subclass	Example minerals	Mineral formulas	Si:O or (Si,Al):O	Number of shared oxygen in tetrahedra
isolated tetrahedral silicate	olivine	$(Mg,Fe)_2SiO_4$	1:4	0
	almandine	$Fe_3Al_2Si_3O_{12}$		
paired tetrahedral silicates	lawsonite	$CaAl_2Si_2O_7(OH)_2 \cdot H_2O$	2:7 (1:3.5)	1
	akermanite	$Ca_2MgSi_2O_7$		
single-chain silicates	diopside	$CaMgSi_2O_6$	1:3	2
	wollastonite	$CaSiO_3$		
ring silicates	tourmaline	(Na,Ca) $(Fe,Mg,Al,Li)_3Al_6(BO_3)_3Si_6O_{18}(OH)_4$	1:3	2
	beryl	$Be_3Al_2Si_6O_{18}$		
double-chain silicates	anthophyllite	$Mg_7Si_8O_{22}(OH)_2$	4:11 (1:2.75)	2 or 3
	tremolite	$Ca_2Mg_5Si_8O_{22}(OH)_2$		
	talc	$Mg_3Si_4O_{10}(OH)_2$	4:10 (1:2.5)	3
sheet silicates	biotite	$K(Mg,Fe)_3(AlSi_3O_{10})(OH)_2$		
	kaolinite	$Al_2Si_2O_5(OH)_4$		
framework silicates	quartz	SiO_2	1:2	4
	microcline	$KAlSi_3O_8$		

Silicate crystal structures may be complex. Many silicates contain anions or anionic groups other than O^{2-}. Muscovite, $KAl_2(AlSi_3O_{10})(OH)_2$, and other micas, for example, contain $(OH)^-$. Other silicates, such as kyanite, $Al_2(SiO_4)O$, and titanite, $CaTi(SiO_4)O$, contain O^{2-} anions unassociated with the $(SiO_4)^{4+}$ tetrahedra. In muscovite and many other minerals, aluminum is in both tetrahedral and octahedral coordination. Still other silicates do not fit neatly into a subclass. Zoisite, $Ca_2Al_3O(SiO_4)(Si_2O_7)(OH)$, for example, contains both isolated tetrahedra and paired (double) tetrahedra (Fig. 4.17). Some texts and reference books separate elements and include extra parentheses in mineral formulas (as has been done in this paragraph) to emphasize crystal structure, but often we omit such niceties for brevity. In shorter form, we can write muscovite's formula as $KAl_3Si_3O_{10}(OH)_2$, kyanite's as Al_2SiO_5, titanite's as $CaTiSiO_5$, and zoisite's as $Ca_2Al_3Si_3O_{12}(OH)$. Though shorter and, perhaps, easier to write, these formulas give little hint of crystal structure.

4.6   Elemental substitutions in mineral crystals

Natural quartz is generally at least 99.9% pure SiO_2, fluorite is generally at least 99.9% CaF_2, and many other minerals always have compositions close to ideal formulas. Most minerals, however, have variable compositions due to *elemental substitutions*. They are *solid solutions* of two or more end members, discussed in Chapter 3. Natural feldspars, for example, are solid solutions of three end members—orthoclase ($KAlSi_3O_8$), albite ($NaAlSi_3O_8$), and anorthite ($CaAl_2Si_2O_8$). Pauling's Rules tell us that ionic size and charge control the extent to which one element may substitute for another. Elements with identical charges and similar sizes usually readily substitute for each other in crystals; those that differ greatly in charge and size may not.

Figure 4.19 compares size and charge for the common ions listed in Table 4.2. Fe^{3+} and Al^{3+} plot closely together and thus readily substitute for each other in minerals including garnets and spinel. Fe^{2+}, Mg^{2+}, and Mn^{2+} also plot closely together; they readily substitute for each other in many minerals including garnets and olivines. Ca^{2+} replaces Mn^{2+}, Fe^{2+}, and Mg^{2+} in some minerals but not in all minerals, because Ca^{2+} is larger than the other three cations and may not easily fit into the same site in some crystal structures.

For example, garnet solutions, $(Fe,Mg,Mn,Ca)_3Al_2Si_3O_{12}$, may have almost any composition that is a mixture of the four end members:
- almandine $Fe_3Al_2Si_3O_{12}$
- pyrope $Mg_3Al_2Si_3O_{12}$
- spessartine $Mn_3Al_2Si_3O_{12}$
- grossular $Ca_3Al_2Si_3O_{12}$

In contrast, olivine solutions, $(Fe,Mg,Mn)_2SiO_4$, contain only extremely small amounts of Ca^{2+} and are generally solutions of
- fayalite Fe_2SiO_4
- forsterite Mg_2SiO_4
- tephroite Mn_2SiO_4

4.6.1   Simple substitutions and coupled substitutions

Less obvious from Figure 4.19 is that Na^+, K^+, and Ca^{2+} substitute for each other in some minerals. K^+ is 25% larger than Na^+ and Ca^{2+} but is still able to replace them if the site in the crystal is large enough. Ca^{2+} has greater charge than the alkali cations (Na^+ or K^+) but may replace them if the extra charge is mitigated in some way.

Consider the feldspar solutions discussed previously. Feldspar compositions involve three end members:
- orthoclase $KAlSi_3O_8$
- albite $NaAlSi_3O_8$
- anorthite $CaAl_2Si_2O_8$

These three end members combine to form two kinds of feldspar: plagioclase and alkali feldspar. Figure 4.20 shows possible feldspar compositions at 800 °C and 1

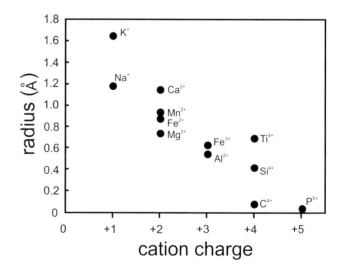

Figure 4.19   Cation radii and charge for some common cations.

atmosphere pressure. Alkali feldspars are predominantly solutions of orthoclase and albite, although not all compositions between are stable except at very high temperatures.

As shown in Figure 4.20, alkali feldspar solutions may contain small amounts of anorthite but are predominantly $(K,Na)AlSi_3O_8$, involving the *simple substitution* of K^+ for Na^+. Plagioclase solutions may include small amounts of orthoclase but are predominantly $(Na,Ca)(Al,Si)_4O_8$, involving the *coupled substitution* of $Ca^{2+}Al^{3+}$ for Na^+Si^{4+}. The coupled substitution maintains charge balance by replacing two ions simultaneously.

4.6.2 Limited and complete solid solutions

Plagioclase forms a *complete solid solution* between albite and anorthite, and any composition between is possible (Fig. 4.20). It may also form a *limited solid solution* with orthoclase; most plagioclase contains less than 5% orthoclase. In contrast, except at the highest temperatures, alkali feldspar solutions are always limited to compositions close to orthoclase and to albite, perhaps also containing a small amount of anorthite, so alkali feldspar forms a *limited solid solution*. Figure 4.20 shows that feldspars with compositions halfway between albite and orthoclase are unstable at 800 °C. Such feldspars may form at very high temperatures (> 800 °C) but will *exsolve* as they cool, which means that the original homogeneous feldspar grain will *unmix* to produce zones of two separate compositions (much like chicken soup separates into fat and broth on cooling).

Figure 4.21 shows a sample of alkali feldspar from the Black Hills, South Dakota. It originally crystallized as one homogeneous crystal at high temperature. Upon cooling, it unmixed into two compositions—one Na-rich and one K-rich—due to the limits on solid solutions between albite and orthoclase at lower temperatures. This produced *exsolution lamellae*, visible as near-vertical gray and white stripes in this photo. The lamellae are zones of different compositions within a single host crystal.

The limits on feldspar solutions relate directly to ion size. The radii of Ca^{2+} and Na^+ are similar and, because of this, they readily replace each other in plagioclase crystals and any composition between albite and anorthite is possible. The Ca^{2+}/Na^+ site, however, is too small to easily accommodate K^+, so plagioclase solutions can only incorporate a small amount of orthoclase. In alkali feldspar, Na^+ and K^+ occupy the same atomic sites, but the two cations have different sizes. If a feldspar is dominated by K^+, the sites are stretched and somewhat too large to contain Na^+; if a feldspar is dominated by Na^+, the sites are somewhat too small for K^+. So, the end members albite and orthoclase only mix to a limited extent. Alkali feldspar also cannot contain much Ca^{2+} because Ca^{2+} is too small to fit stably into the K^+/Na^+ site.

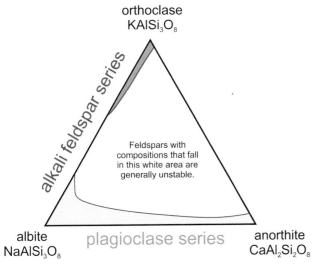

Figure 4.20 The feldspar series at 800 °C and 1 atm. pressure.

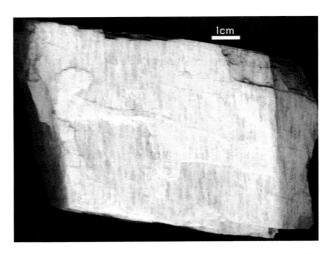

Figure 4.21 Exsolution lamellae in alkali feldspar from the Black Hills, South Dakota.

Photo from Jstudy, Wikimedia Commons.

4.7 The arrangement of atoms in crystalline solids

If a crystal is a perfect cube with six identical crystal faces at 90° to each other, the atoms that make up the crystal must be arranged in "cubic" pattern. The orderly arrangement of atoms must be the same in three mutually perpendicular directions. Similarly, if a mineral, for example quartz, forms minerals with hexagonal character, the atoms within must be arranged in a hexagonal pattern. This logic allowed 19th-century crystallographers to infer the nature of atomic arrangements even though they had no way to look directly at those arrangements. Furthermore, they realized that if an atomic arrangement in a crystal is hexagonal (or cubic), the unit cell—the basic structural unit—must also be arranged in a hexagonal (or cubic) pattern.

Cubic and *hexagonal* are terms that describe *symmetry*, a term closely related to shape. The symmetry of a crystal's unit cell limits the symmetry that the crystal can have. Beryl, for example, often grows as *hexagonal prisms* (crystals with a long direction and hexagonal cross section). It follows that the arrangement of atoms in beryl has a hexagonal pattern and the unit cell of beryl has a hexagonal shape.

Similarly, because fluorite and galena often form perfect cubic crystals, their atomic arrangements and unit cell shapes must also be cubic. The top part of Figure 4.22 shows the atomic arrangements in fluorite and galena and compares them to examples of natural crystals.

The bottom part of Figure 4.22 shows garnet and spinel, two other minerals that have cubic arrangements of atoms within them. For some minerals, such as fluorite and galena, the relationship between crystal shape and unit cell shape seems obvious. However, garnet also has a cubic unit cell, but euhedral garnet crystals do not have cubic shapes; instead, they typically have dodecahedral (12-sided) shapes. Similarly, spinel has a cubic unit cell, but crystals are generally octahedral (8-sided). What do a cube, a dodecahedron, and an octahedron have in common? The answer is that they have the same symmetry, even though they have different shapes. All crystals that have cubic symmetry—no matter what their shape—must have atoms arranged in a cubic

Figure 4.22   Fluorite, galena, garnet, and spinel crystals.
Drawings by D. Perkins. Garnet photo from an anonymous source, other photos from R. Lavinsky, both at Wikimedia Commons.

pattern. It is worth thinking, then, about what symmetry is and about the kinds of symmetry that may relate atoms to other atoms in a unit cell. It is easiest to do this by first considering only two dimensions.

4.7.1 Unit cells, atoms, and symmetry in two dimensions

Crystals consist of a near-infinite number of unit cells, and the unit cells must fit together without gaps between them. In two dimensions, the number of unique unit cell shapes is surprisingly small. In three dimensions, the number of possibilities, as we shall see, is not much greater.

In 2-d, some possibilities are shown in Figure 4.23: square, rectangle, diamond, parallelogram, diamond, rhomb, and hexagon. Floor tiles often come in these shapes. Although other more complicated shapes, such as an L shape, could work to tile a floor, any floor pattern that can be made with L-shaped tiles can be made with one of the simpler shapes in Figure 4.23.

4.7.1.1 Point symmetry

The unit cell shapes in Figure 4.23 demonstrate several kinds of symmetry. The top four, for example, have vertical *mirror planes* of symmetry (shown as dashed lines). The right side is a reflection of the left side; if we observed these unit cells in a mirror, they would appear the same.

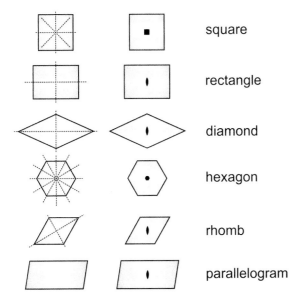

Figure 4.23 Possible two-dimensional unit cell shapes. The rhombohedral cell is a special kind of diamond with angles of 60° and 120°.

The top halves of the square, rectangle, diamond, and hexagon unit cells also reflect to the bottom halves. So, these four examples have both vertical and horizontal mirror planes. Additionally, the square, hexagon, and rhombus have diagonal mirror planes. The parallelogram, however, does not have any mirrors of symmetry at all.

The unit cells in Figure 4.23 also have *rotational symmetry*. The square cell has *4-fold symmetry*; it can be rotated 90° without changing appearance. Four rotations, and it is back where it started. The hexagonal cell has *6-fold symmetry*; when rotated 60°, it does not change appearance. The other unit cells have *2-fold symmetry*; rotation of 180° leaves them the same, and two rotations return them to where they started. Although not shown by the shapes of these unit cells, some mineral atomic arrangements (and mineral crystals) have *3-fold* symmetry. A small lens shape, diamond, triangle, square, or hexagon is used to show rotation axes perpendicular to the page; all except the triangle can be seen in one of the examples in Figure 4.23. Additionally, the six shapes shown have another kind of symmetry, termed *inversion symmetry*. A shape or an atom arrangement with inversion symmetry has a point called an *inversion center* at their centers. Features on one side can be *inverted* through the center and appear on the other side rotated 180°.

Figure 4.24 shows the seven fundamental *symmetry operators* needed to describe two-dimensional shapes and atomic arrangements. Drawings in the third column show iconic shapes with symmetry; drawings in the fourth column show atomic arrangements with the same symmetry. For completeness, a 1-fold rotation axis is included, although it leaves shapes and patterns unchanged. The kind of symmetry depicted in Figure 4.24, called *point symmetry*, involves symmetrical relationships associated with a single object or pattern. All except the mirror plane are symmetry around a point at the center of the object/pattern.

The 1-fold, 2-fold, 3-fold, 4-fold, and 6-fold rotation axes are associated with incremental rotations of 360°, 180°, 120°, 90°, and 60°, respectively. Repeating the increment 1, 2, 3, 4, or 6 times gets the image back to where it started. We need not consider 5-fold symmetry or symmetry greater than 6-fold; shapes with those symmetries cannot be unit cells because they will not fit together without leaving gaps.

The single example of a mirror plane shown in Figure 4.24 is oriented vertically, but mirror planes may have any orientation, and shapes may contain more than one mirror. The inversion center in Figure 4.24

shows how corners of a shape, or atoms in a drawing, invert through the center (dashed lines) and appear on the opposite side. In the 2-d drawings shown here, the inversion center appears to be the same as a 2-fold rotation axis, but in three dimensions inversion and rotation are distinctly different.

symmetry operator	symbol	example of a two-dimensional shape	example of an atom arrangement
1-fold rotation axis	1		
2-fold rotation axis	2		
3-fold rotation axis	3		
4-fold rotation axis	4		
6-fold rotation axis	6		
mirror plane	m		
inversion center	i		

Figure 4.24   The fundamental two-dimensional point symmetry operators.

4.7.1.2 Translational symmetry

Translational symmetry, a second type of symmetry, is the kind of symmetry generally seen in wallpaper. A *motif* of some sort repeats many times by *translation*, producing the overall wallpaper pattern. In crystals, a unit cell repeats the same way to produce the overall atomic arrangement of the crystal.

The drawing shown in Figure 4.25 contains a pattern, composed of five circles, that repeats many times. Think of this as a snapshot of part of a wall covered by wallpaper. There are four small greens, and one larger blue circle in each repeating unit. The five circles comprise the motif, and the motif repeats to give the overall wallpaper-like design. If the motif is replaced by a single point, the result is as shown in the right-hand part of the figure. Each red dot, called a *lattice point*, represents where one motif was situated. The collection of all such points is a *lattice*. Every lattice point is related to points around it by vectors, depicted by black arrows. In this example, the vectors are of equal length and are perpendicular. The overall pattern has a square arrangement, but this need not be the case. Rectangular arrangements or lattices involving non-90° angles are common.

So, vectors describe translational symmetry. They describe how far, and in what direction, one repeated motif is from another. Thus, they describe how far, and in what direction, one lattice point is from another. The vectors may be *orthogonal* (perpendicular to each other) or not. They may be of equal magnitude or not, giving four possibilities (equal magnitude, orthogonal; unequal magnitude, orthogonal; equal magnitude, not orthogonal; unequal magnitude, not orthogonal). Additionally, we consider a special case involving translations of equal magnitude at 60° to each other.

So we get five unique *plane lattices*, as shown in the drawing in Figure 4.26. The center column in this figure shows a unit cell that can repeat, according to the plane

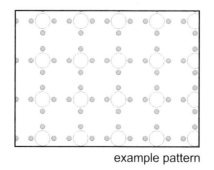

example pattern

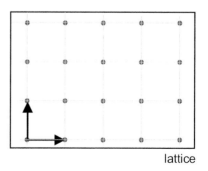

lattice

Figure 4.25   A repetitive pattern of five atoms and a lattice that depicts the repetition.

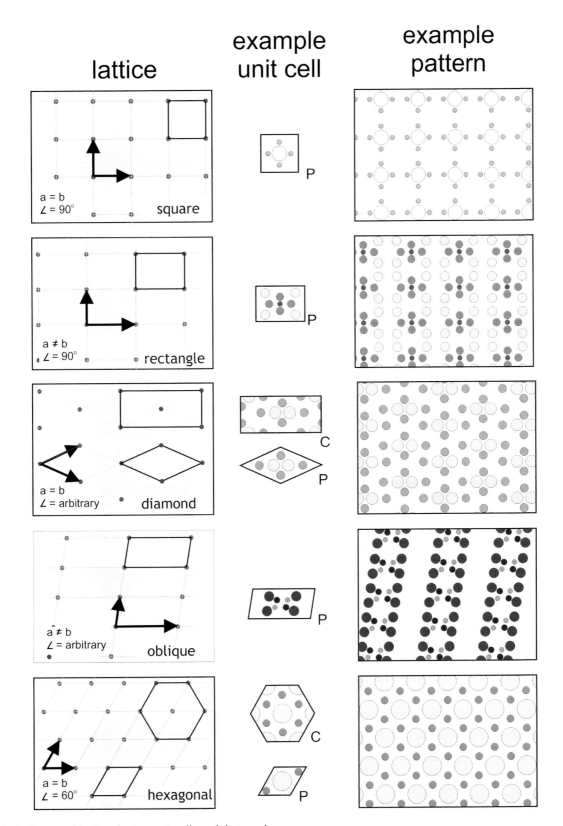

Figure 4.26 Five possible plane lattices, unit cells, and their product patterns.

lattice, to give the pattern in the right-hand column. In the simplest case, four lattice points outline a *primitive unit cell*—a unit cell that contains the minimum number of atoms needed to make a motif that can be used to make the entire atomic arrangement. Primitive unit cells are labeled with the letter *P* in Figure 4.26.

For some purposes, it is convenient to choose a *non-primitive* unit cell. Non-primitive cells contain one or more extra lattice points, and consequently one or more extra motifs. This is often done to emphasize a particular shape or symmetry. Non-primitive cells are designated with the letter *C* in Figure 4.26. The *C* stands for *centered* because the non-primitive unit cells have an extra motif at their centers. The unit cell shapes shown here are the same as those in Figure 4.23, confirming that using vectors to explore possible unit cells gives us a consistent result.

The right-hand column in Figure 4.26 shows how the hypothetical unit cell, containing multiple atoms, can be combined with each of the five lattices to generate an overall pattern. The five patterns all have different symmetries. The top pattern contains both 4-fold and 2-fold axes of symmetry perpendicular to the page. The three in the middle contain 2-fold axes perpendicular to the page; the bottom one contains both 6-fold and 3-fold symmetry perpendicular to the page. The top pattern has vertical, horizontal, and diagonal mirror planes, the next two have vertical and horizontal mirror planes, the fourth one has no mirror planes, and the hexagonal pattern has many mirror planes at 60° to each other.

In two dimensions, one additional symmetry operator must be considered if we are to consider all possible symmetry. Termed a *glide plane*, it is a combination of reflection and translation (Fig. 4.27). If a glide plane is present, a pattern, or arrangement of atoms, repeats by gliding in one direction and then reflecting across a mirror.

So, a motif composed of a group of atoms may have any of the point symmetries depicted in Figure 4.24. The atomic motifs repeat to produce an entire crystal in a translational pattern described by vectors. The combination of point symmetry and translational symmetry means that crystals may also contain glide planes of symmetry. Of particular importance is that translational symmetry, including standard translations and glide planes, differs from point symmetry because point symmetry is finite. For example, four applications of a 4-fold rotation axis return a motif to its original position. In contrast, translational symmetry is limitless; translation may be applied an infinite number of times without getting back to the original starting position.

The overall symmetry of an atomic arrangement depends both on the symmetry of the lattice (which defines the shape and symmetry of the unit cell) and on the symmetry of the atoms within the unit cell. There are 17 unique combinations, called the 17 *plane groups*. Patterns depicting each are shown in Figure 4.28. The labels for each group follow crystallographic conventions: *p* or *c* indicate whether the unit cell is primitive (*p*) or contains extra motifs (*c*). The presence of 1-fold, 2-fold, 3-fold, 4-fold, or 6-fold rotation axes is indicated by 1, 2, 3, 4, or 6. The letter *m* refers to a mirror plane, and *g* refers to a glide plane. By convention, non-primitive unit cells are chosen for two of the plane groups (*c2mm* and *cm*) so they have a rectangular shape instead of a diamond shape, which emphasizes their 90° angles. Additionally, a hexagonal unit cell is generally chosen for patterns with 6-fold symmetry (even though a smaller rhomb-shaped cell could be chosen), so that the 6-fold symmetry is clear. The symmetries of the 17 groups range from no symmetry (other than a 1-fold rotation axis) to complicated patterns with higher-order rotation axes, mirror planes, and glide planes. Unit cells may be *oblique* (shaped like a parallelogram), rhomb-shaped, *square*, *rectangular*, or *hexagonal*, and the 17 groups represent all possible combinations of symmetry in two dimensions.

4.7.2   Unit cells and symmetry in three dimensions

4.7.2.1   Unit cells

As shown in Figure 4.29, if we start with any of the two-dimensional plane lattices depicted in Figure 4.26 and define a third vector (black arrow) that describes translation in the third dimension, we can stack the two-dimensional lattices on top of each other to produce three-dimensional *space lattices*. Thus, we get a three-dimensional unit cell; the one shown in this figure has the shape of a distorted shoe box. The third translation may be perpendicular to one or both of the first

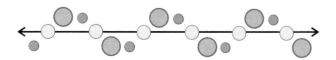

Figure 4.27   A glide plane in two dimensions.

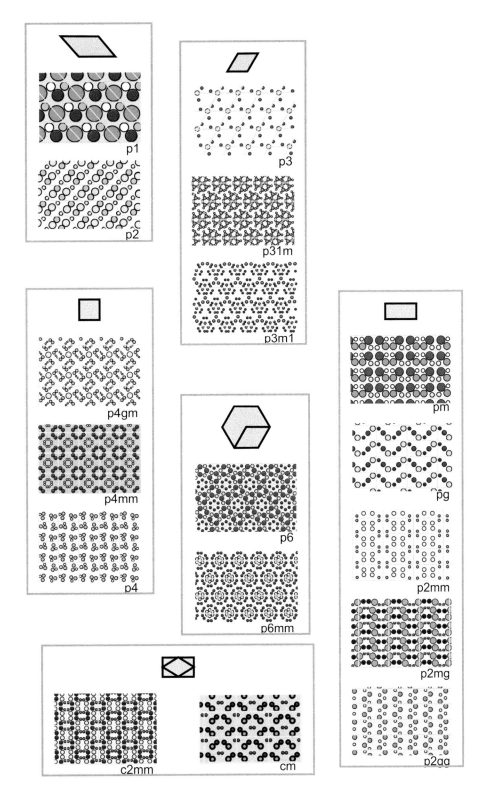

Figure 4.28 The 17 possible plane groups of symmetry.

two, but it need not be. Its magnitude may be the same as one or both of the first two, or not.

In total, there are 14 possible space lattices, first described by Auguste Bravais in 1850 and called the *Bravais lattices* today (Fig. 4.30). These 14 are the only unique combinations of symmetry (due to translation) possible. The 14 lattices involve unit cells of only six shapes. Six of the 14 lattices are associated with primitive

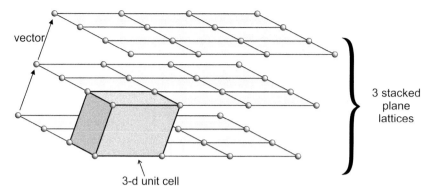

Figure 4.29   Stacking plane lattices to produce 3-d space lattices and 3-d unit cells.

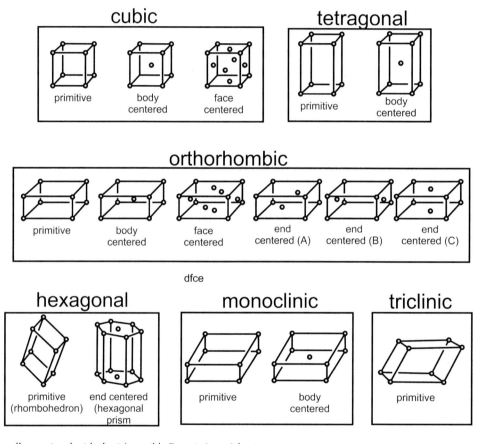

Figure 4.30   Unit cells associated with the 14 possible Bravais (space) lattices.

unit cells, and the others produce unit cells that contain extra lattice points (and, therefore, extra motifs). Note that the drawings in Figure 4.30 include three different examples of end centered lattices; they are all, however, equivalent.

So, all crystals are made of unit cells that can have any of the six fundamental shapes shown in Figure 4.31. This allows us to divide crystals into the *cubic, tetragonal, orthorhombic, hexagonal, monoclinic,* or *triclinic* crystal systems based on their unit cells. By convention, the variables *a*, *b*, and *c* describe the length of the unit cell edges, and the Greek letters α, β, and γ are the angles between the edges (Fig. 4.31). For some unit cells, angles are known to be 90° or 120°. And, for some, two or more of the edges have the same length. So, often we do not have to give values for all six variables to describe a unit cell's size and shape.

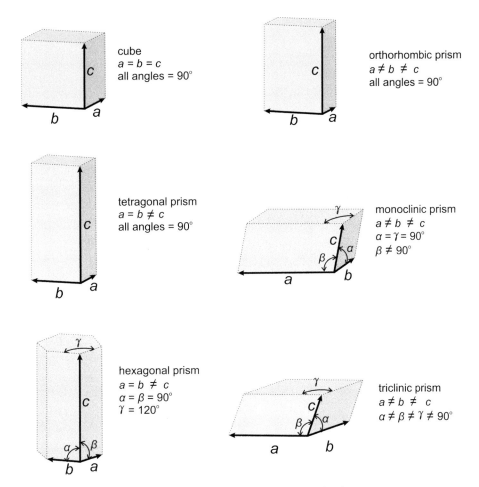

Figure 4.31 The six basic unit cell shapes and the dimensions and angles used to describe them.

The triclinic system includes only one possible (primitive) unit cell. The other systems include one or more non-primitive unit cells, which brings the total to 14. When talking about two-dimensional unit cells, we drew an analogy to possible shapes for floor tiles. Three-dimensional unit cells are akin to possible shapes that could be used to make a brick wall without leaving any gaps. No other fundamental shapes are possible, and more complicated brick shapes can be shown to be equivalent to one of those shown in Figure 4.31.

4.7.2.2   *Symmetry in three dimensions*

In three dimensions, point symmetry operators include the ones previously discussed for two-dimensional shapes plus several new ones termed *rotoinversion axes*. Rotoinversion axes involve the simultaneous application of rotation and inversion. They are symbolized by numbers with bars above them (Table 4.4).

Table 4.4   Conventional symmetry operators used to describe crystal symmetry.

Symmetry operator	Description	Equivalent operators
m	mirror plane	$1/m = \bar{2}$
2	2-fold rotation axis	
3	3-fold rotation axis	
4	4-fold rotation axis	
6	6-fold rotation axis	
2/m	2-fold rotation axis with perpendicular mirror	
4/m	4-fold rotation axis with perpendicular mirror	
6/m	6-fold rotation axis with perpendicular mirror	
i	inversion center	$\bar{1}$
$\bar{3}$	3-fold rotoinversion axis	
$\bar{4}$	4-fold rotoinversion axis	
$\bar{6}$	6-fold rotoinversion axis	3/m

A 1-fold rotoinversion axis is a combination of a 360° rotation plus inversion. A 2-fold rotoinversion axis is a combination of 180° rotation plus inversion. A 3-fold rotoinversion axis is a combination of 120° rotation plus inversion. A 4-fold rotoinversion axis is a combination of 90° rotation plus inversion. A 6-fold rotoinversion axis is a combination of 60° rotation plus inversion. Additionally, we use the symbols 1/m, 2/m, 3/m, 4/m, and 6/m to indicate a (normal) rotation axis with a mirror plane perpendicular to it.

Figure 4.32 shows examples of all the different kinds of symmetry operators that may be present in crystals. The many possible operators include some redundancies. For example, $\bar{2}$ is the same as a mirror plane. So, by convention, crystallographers only use the operators listed in the left-hand column of Table 4.4.

4.7.2.3 Symmetry of unit cells

As shown in Figure 4.33, each of the unit cells associated with each of the six crystal systems has different combinations of symmetry. The symmetry operators to the left of each drawing describe the symmetry present. If an operator, for example 2/m, is listed more than once, it means that there are 2/m axes in more than one orientation in the cell. The list of operators tells the different kinds of symmetry that are present but does not tell how many of each.

All six unit cells have inversion centers, and all but the triclinic cell have one or more mirror planes, and one or more 2-fold axes. The cubic and tetragonal cells have 4-fold symmetry; the hexagonal unit cell has 6-fold symmetry; the cubic unit cell has $\bar{3}$ symmetry. In these unit cells, 6-fold, 4-fold, and 2-fold rotation axes are always accompanied by mirrors perpendicular to them.

The triclinic cell has the least symmetry; it contains only an inversion center. In contrast, the cube contains the most symmetry; its many different symmetry operators are shown in Figure 4.34. They include:

- an inversion center
- nine different mirror planes (down the center of faces and cutting diagonally across faces), each perpendicular to a rotation axis
- six different 2-fold rotation axes perpendicular to mirrors (middle of edge to middle of edge)
- four different $\bar{3}$ axes (corner to corner)
- three different 4-fold rotation axes perpendicular to mirrors (perpendicular to faces)

4.7.2.4 Symmetry of atomic arrangements and the symmetry of crystals

In three dimensions, atomic arrangements may have any of the point symmetries in Figure 4.32. They also have translational symmetry and may contain glide planes. Additionally, some atomic arrangements contain symmetry elements, called *screw axes*, that can only exist in three dimensions. Screw axes are symmetry elements formed by a combination of rotation and translation, similar in some ways to glide planes that are combinations of reflection and rotation. The two are compared in Figure 4.35.

Some atomic arrangements contain more than one glide plane or screw axis, but many contain none. In Figure 4.35, the glide plane and screw axis are vertical but in crystals they can be in any direction. In these drawings, only one atom is shown repeating; in crystals all atoms are affected when these symmetries are present.

X-ray diffraction is the most common technique used to study the arrangement of atoms in crystals. Diffraction studies can reveal a crystal's lattice, unit cell dimensions, symmetry, bond types and orientations, and where different species of atoms are found in a unit cell. Such studies, termed *crystal structure refinements*, have been carried out for thousands of minerals and other crystalline substances and are used to construct models of crystals such as those seen in this chapter.

All crystals contain an ordered arrangement of atoms, and in many crystals, the atomic arrangement has a high degree of symmetry. As in two dimensions, in three dimensions the overall symmetry of a crystal's atomic arrangement depends both on the shape of the unit cell and on the symmetry of the atoms within the unit cell. The unit cell shapes and symmetry operators may combine in many ways; there are 230 possibilities, called the 230 *space groups*. Although all are theoretically possible, many of them correspond to no known natural examples.

The drawings shown in Figure 4.36 depict the arrangements of atoms in manganite, ilmenite, fluorite, and quartz. These atomic arrangements contain lots of symmetry, but symmetry is very difficult to see in two-dimensional drawings. Manganite's structure contains a 2-fold screw axis perpendicular to the page and a glide plane parallel to the page. Ilmenite contains only $\bar{3}$ symmetry, although it cannot be seen clearly here. Fluorite has lots of symmetry—a 4-fold axis and several mirror planes are easily seen in this drawing, and

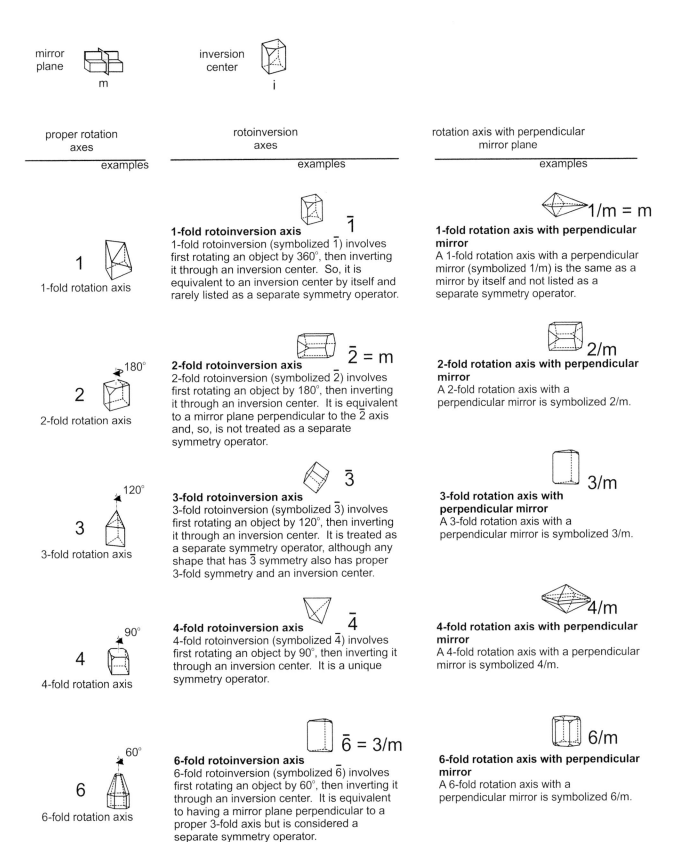

mirror plane

m

inversion center

i

proper rotation axes

rotoinversion axes

rotation axis with perpendicular mirror plane

examples

examples

examples

1

1-fold rotation axis

1-fold rotoinversion axis $\bar{1}$
1-fold rotoinversion (symbolized $\bar{1}$) involves first rotating an object by 360°, then inverting it through an inversion center. So, it is equivalent to an inversion center by itself and rarely listed as a separate symmetry operator.

$1/m = m$
1-fold rotation axis with perpendicular mirror
A 1-fold rotation axis with a perpendicular mirror (symbolized 1/m) is the same as a mirror by itself and not listed as a separate symmetry operator.

180°

2

2-fold rotation axis

2-fold rotoinversion axis $\bar{2} = m$
2-fold rotoinversion (symbolized $\bar{2}$) involves first rotating an object by 180°, then inverting it through an inversion center. It is equivalent to a mirror plane perpendicular to the $\bar{2}$ axis and, so, is not treated as a separate symmetry operator.

2/m
2-fold rotation axis with perpendicular mirror
A 2-fold rotation axis with a perpendicular mirror is symbolized 2/m.

120°

3

3-fold rotation axis

3-fold rotoinversion axis $\bar{3}$
3-fold rotoinversion (symbolized $\bar{3}$) involves first rotating an object by 120°, then inverting it through an inversion center. It is treated as a separate symmetry operator, although any shape that has $\bar{3}$ symmetry also has proper 3-fold symmetry and an inversion center.

3/m
3-fold rotation axis with perpendicular mirror
A 3-fold rotation axis with a perpendicular mirror is symbolized 3/m.

90°

4

4-fold rotation axis

4-fold rotoinversion axis $\bar{4}$
4-fold rotoinversion (symbolized $\bar{4}$) involves first rotating an object by 90°, then inverting it through an inversion center. It is a unique symmetry operator.

4/m
4-fold rotation axis with perpendicular mirror
A 4-fold rotation axis with a perpendicular mirror is symbolized 4/m.

60°

6

6-fold rotation axis

6-fold rotoinversion axis $\bar{6} = 3/m$
6-fold rotoinversion (symbolized $\bar{6}$) involves first rotating an object by 60°, then inverting it through an inversion center. It is equivalent to having a mirror plane perpendicular to a proper 3-fold axis but is considered a separate symmetry operator.

6/m
6-fold rotation axis with perpendicular mirror
A 6-fold rotation axis with a perpendicular mirror is symbolized 6/m.

Figure 4.32 Point symmetry operators used to describe crystal symmetry.

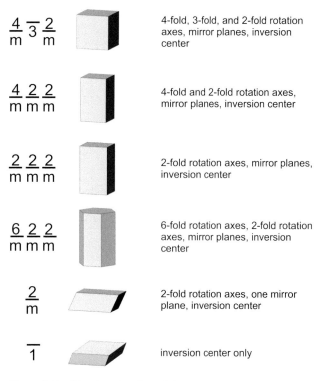

$\dfrac{4}{m}\ \overline{3}\ \dfrac{2}{m}$ 4-fold, 3-fold, and 2-fold rotation axes, mirror planes, inversion center

$\dfrac{4}{m}\ \dfrac{2}{m}\ \dfrac{2}{m}$ 4-fold and 2-fold rotation axes, mirror planes, inversion center

$\dfrac{2}{m}\ \dfrac{2}{m}\ \dfrac{2}{m}$ 2-fold rotation axes, mirror planes, inversion center

$\dfrac{6}{m}\ \dfrac{2}{m}\ \dfrac{2}{m}$ 6-fold rotation axes, 2-fold rotation axes, mirror planes, inversion center

$\dfrac{2}{m}$ 2-fold rotation axes, one mirror plane, inversion center

$\overline{1}$ inversion center only

Figure 4.33 The six unit cells and their symmetry.

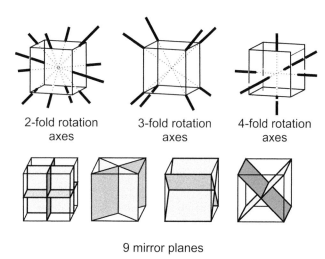

2-fold rotation axes   3-fold rotation axes   4-fold rotation axes

9 mirror planes

Figure 4.34 The symmetry of a cube.

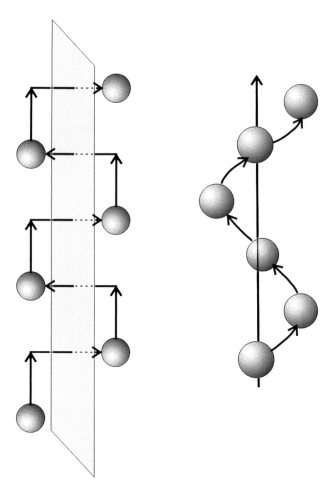

Figure 4.35 A glide plane (left) and a screw axis (right).

the arrangement also contains 3-fold symmetry at an oblique angle to this page. In the drawing of quartz, the (SiO$_4$) tetrahedra are shaded—picture a Si^{4+} ion at their centers. Quartz has 6-fold symmetry, roughly perpendicular to the page, and 2-fold symmetry in the plane of the page. Manganite, ilmenite, fluorite, and quartz represent four of the possible 230 space groups.

If we consider the symmetry of individual crystal shapes, rather than of atomic arrangements, we need only consider point symmetry and not symmetry involving translation. Consequently, the number of possibilities is much less. There are only 32 possible symmetry combinations, referred to as the *32 crystal classes*, or the *32 point groups*. These 32 are divided into the crystal systems, previously discussed, that are based on the unit cell shapes shown in Figure 4.33.

Crystallographers use conventional symbols, termed Hermann-Mauguin symbols, to label the different point groups. The notation includes all symmetry present for crystals of a given point group but does not say how many of each kind of symmetry operator are present; this information can be deduced from the crystal system.

The chart in Figure 4.37 lists all 32 point groups, separated by crystal system. Hermann-Mauguin symbols are given for each point group. Because the symmetry of a crystal, and thus of every point group, requires specific symmetry of a unit cell, each of the

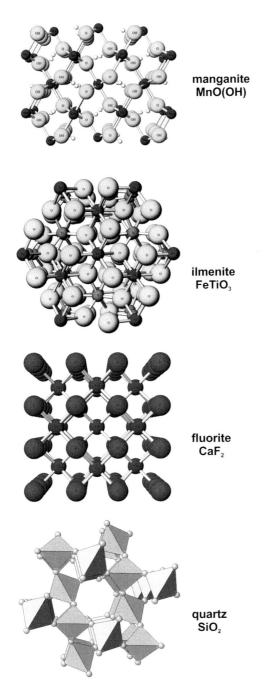

Figure 4.36   Atomic arrangements in manganite, ilmenite, fluorite, and quartz.

32 point groups can be unambiguously placed in a crystal system.

4.7.2.5   Crystal shapes and symmetry

It is important to remember that symmetry is a property. Crystals with different shapes or a different number of faces may have identical symmetry. All crystals shown in

Figure 4.38, for example, have the same symmetry as a cube. They have three 4-fold axes of symmetry, four axes of symmetry, six 2-fold axes of symmetry, nine mirror planes, and an inversion center. The Hermann-Mauguin symbol for all eight crystals is $4/m\bar{3}2/m$. Some contain more faces than others do, and some contain faces of differing shapes. Still, they all have the symmetry shown in Figure 4.34.

Most mineral samples are *anhedral* or *subhedral*—they do not show crystal faces, so we cannot easily see symmetry. For *euhedral* crystals, however, the shape of a crystal tells us the shape of the unit cell because of constraints placed by symmetry. A cubic unit cell is the only unit cell that has more than one 3-fold axis of symmetry. It follows, therefore, that all crystals that have more than a single 3-fold axis of symmetry must have a cubic unit cell. There are an infinite number of possible shapes that require a cubic unit cell.

Figure 4.38 shows some crystals that belong to the cubic system and their Hermann-Mauguin symmetry symbols. Figure 4.39 shows a few additional examples. Although none of the crystals shown in Figure 4.39 have symmetry equivalent to that of a cube, they all have four 3-fold axes of symmetry and so must have a cubic unit cell. Each drawing is labeled with the appropriate Hermann-Mauguin symbol listing its symmetry.

Similarly, if a crystal has a 6-fold axis of symmetry, it must belong to the hexagonal system and have a hexagonal unit cell—no other unit cell has 6-fold symmetry. Additionally, if a crystal has *a single* 3-fold axis of symmetry, it must also belong to the hexagonal system and have a hexagonal unit cell. So, crystal symmetry tells us about unit cell symmetry. However, the opposite need not be true. If a crystal's unit cell has a hexagonal shape, the unit cells may not combine to produce a crystal with visible hexagonal symmetry. Often, unit cells combine to produce crystals with no visible symmetry at all or with symmetry less than that of the unit cell.

How can different-shaped crystals be made from the same-shaped unit cell? The answer is that it depends on how the unit cells are stacked together. Figure 4.40 shows some examples using a cubic unit cell. The crystals all have different shapes, and some have different symmetries. Cubic unit cells may be stacked so that all cubic symmetry is preserved, but this need not be the case. In this figure, Hermann-Mauguin symmetry symbols are listed for each of the six drawings; they are not all the same. When unit cells are stacked, only some

System and Unit Cell			Hermann-Mauguin symbol for Point Group	Symmetry
cubic system		A cubic unit cell is shaped like a cube.	23 $2/m\overline{3}$ 432 $\overline{4}3m$ $4/m\overline{3}\,2/m$	Crystals with any of these symmetries belong to the cubic system. They must have a cubic unit cell because they all have four 3-fold axes of symmetry. Only a cubic unit cell has that symmetry.
hexagonal system	 *hexagonal division*	Hexagonal plane lattices may be stacked to produce a non-primitive hexagonal unit cell shaped like a hexagonal prism; it has hexagon-shaped ends with six identical faces perpendicular to the ends.	$\overline{6}$ 6/m 622 6mm $\overline{6}2m$ $6/m\,2/m\,2/m$	Crystals with any of these symmetries belong to the hexagonal system. They must have a hexagonal unit cell because they have a 6-fold axis of symmetry.
	 *trigonal division*	Hexagonal plane lattices may be stacked so that 3-fold symmetry is preserved but not 6-fold. This yields a non-primitive unit cell that is shaped like a rhombic prism (60° rhombs on the ends with four perpendicular faces) or a primitive cell that is a rhombohedron (all edges are the same length and all faces the same shape).	6 3 $\overline{3}$ 32 3m $\overline{3}\,2/m$	Crystals with any of these symmetries belong to the hexagonal system. They have a single 3-fold axis of symmetry, and only a rhombohedral-shaped unit cell has this symmetry.
tetragonal system		A tetragonal unit cell is shaped like a tetragonal prism; it has square ends and four identical faces perpendicular to the ends.	4 $\overline{4}$ 4/m 422 4mm $\overline{4}2m$ $4/m\,2/m\,2/m$	Crystals with any of these symmetries belong to the tetragonal system. They have a single 4-fold axis of symmetry, and only a tetragonal unit cell has this symmetry.
orthorhombic system		An orthorhombic unit cell has a shoe box shape.	222 mm2 $2/m\,2/m\,2/m$	Crystals with any of these symmetries belong to the orthorhombic system. They have three 2-fold axes or mirrors of symmetry. Only a shoe box-shaped unit cell has that symmetry.
monoclinic system		A monoclinic unit cell is shaped like a shoe box that has been crushed in one direction so two of the six faces are no longer rectangles.	2 m 2/m	Crystals with any of these symmetries belong to the monoclinic system. They have a 2-fold axis of symmetry, a mirror, or both. Only a monoclinic unit cell has that symmetry.
triclinic system		A triclinic unit cell is shaped like a shoe box that has been crushed so no angles are 90°.	1 $\overline{1}$	Crystals with any of these symmetries belong to the triclinic system. They have, at most, an inversion center, and only a triclinic unit cell is restricted to that symmetry.

Figure 4.37 Crystal systems and the 32 point groups.

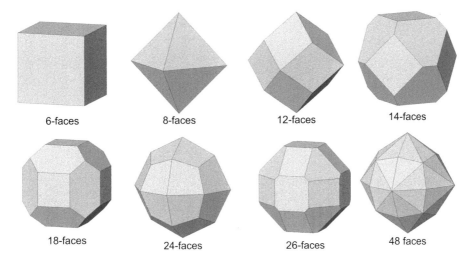

6-faces   8-faces   12-faces   14-faces

18-faces   24-faces   26-faces   48 faces

Figure 4.38   Examples of crystals with $4/m\bar{3}2/m$ symmetry.

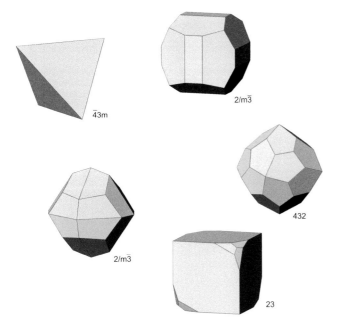

$\bar{4}3m$

$2/m\bar{3}$

432

$2/m\bar{3}$

23

Figure 4.39   Some crystals of the cubic system that do not have the symmetry of a cube.

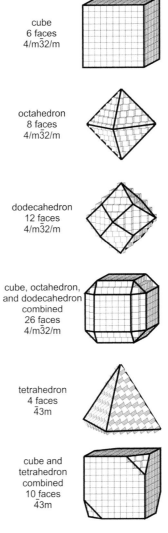

cube
6 faces
$4/m\bar{3}2/m$

octahedron
8 faces
$4/m\bar{3}2/m$

dodecahedron
12 faces
$4/m\bar{3}2/m$

cube, octahedron,
and dodecahedron
combined
26 faces
$4/m\bar{3}2/m$

tetrahedron
4 faces
$\bar{4}3m$

cube and
tetrahedron
combined
10 faces
$\bar{4}3m$

Figure 4.40   Crystals of different symmetry made by stacking cubic unit cells.

symmetry may preserved, or, in the case of anhedral crystals, no symmetry may be present at all.

Crystals belonging to each of the different crystal systems share some general shape characteristics (Fig. 4.41). Cubic crystals, because their unit cells have the same dimensions in all three directions, often grow as *equant* (equidimensional crystals). Hexagonal and tetragonal crystals are often long in one direction with *prism faces* (multiple faces all parallel to one direction) parallel to that direction. Orthorhombic and monoclinic crystals are often *tabular* (thin in one dimension), and triclinic

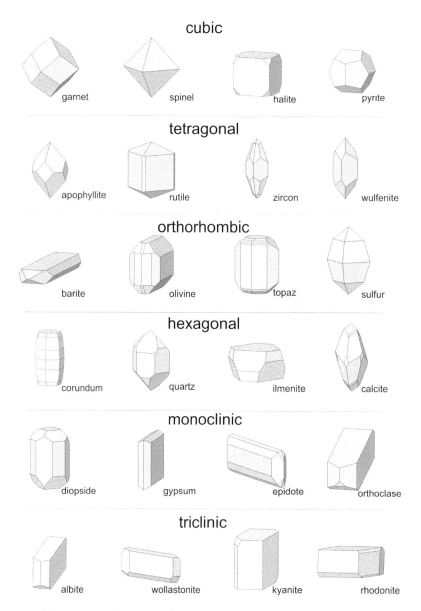

Figure 4.41   Examples of minerals belonging to the six crystal systems. Figure from *Mineralogy* (Perkins).

crystals may have many small different-shaped faces. Therefore, amazingly, experienced crystallographers can often infer the general nature of atomic ordering with a quick glance at a mineral crystal's shape and symmetry!

Questions for thought—chapter 4

1. Why did gypsum precipitate in the cave instead of anhydrite?
2. What is the difference between an amorphous and a polycrystalline substance?
3. Why is obsidian commonly present in volcanic rocks, but not intrusive igneous rocks?

4. What is the difference between anions and cations? Why is a sulfide anion larger than the sulfur atom in sulfate?
5. At one time it was believed that ionic crystals contained closest packed anions with cations between. Today we know that is incorrect for some minerals. Is it correct for any minerals? Explain.
6. What is an ion's *coordination number*? Do small cations tend to have high or low coordination numbers? Why?
7. Anhydrite ($CaSO_4$) is an anisodesmic mineral. Considering the strengths of the Ca-O and S-O bonds, which of the two bonds is most likely to

break if anhydrite is dissolved in water? That is, are the ions released by anhydrite in water: Ca^{2+} and SO_4^{2-} or S^{+6} and $4\ O^{2-}$?

8. What is polymerization?

9. What is *polymerization*? What are the different ways that silicon tetrahedra can polymerize in crystals. List them and list one example mineral for each.

10. Some ions can substitute for each other in mineral crystals. What must the ions have in common in order for this to occur?

11. Contrast and compare minerals that exhibit complete solid solution with those that have only limited solid solutions.

12. The atomic arrangements in mineral crystals have symmetry. There is two kinds of symmetry, point symmetry and translational symmetry. What is the difference between the two? Name one real-world example of each that is NOT a mineral.

13. All minerals are characterized by unit cells. What is a unit cell? These unit cell shapes allow crystals to be divided into crystal systems. What are the six (or seven if you count two different hexagonal cells) shapes that unit cells may have? So, what are the six crystal systems.

14. Mineral crystals may have lots of symmetry or not very much. Name one mineral that distinguishes lots of symmetry. To what crystal system does it belong? Name one mineral that exhibits only minimal symmetry. To what crystal system does it belong.

15. Overall, mineral crystals with the most symmetry belong to the cubic system. What is the most symmetry a crystal can have? And, mineral crystals with the least symmetry belong to the triclinic system. What is the minimal amount of symmetry that a crystal can have?

16. Mineral crystals may have any of 32 possible overall symmetries. What is the Hermann-Mauguin symmetry symbol for crystals with the maximum possible symmetry? What does each element in the symbol mean?

17. Explain how two mineral crystals may have the same symmetry but different shapes and appearances.

18. What don't minerals have five-fold symmetry? Can you cover a floor with pentagon-shaped tiles without leaving any gaps?

5 Igneous Petrology and the Nature of Magmas

5.1 Volcanism in Yellowstone National Park

Geologists generally use the term *magma* to refer to hot molten or partially molten rock material within Earth. Once magma reaches the surface, it may become *lava* that flows across land (or ocean floor), or the magma may become *pyroclastic material*—ash or coarser material that is ejected into the air—that subsequently falls and accumulates on the ground. Both scenarios, after cooling, produce *volcanic rocks*. Although Earth's volcanic rocks have variable compositions and contain many elements, silicon and oxygen dominate most magmas. The most common volcanic rocks range from *rhyolite*, typically containing 70–77 wt% silica (SiO_2, silicon oxide), to *basalt*, typified by 45–55 wt% silica. Yellowstone, America's first national park, contains both high-silica and lower-silica igneous rocks. Most of the Yellowstone eruptions yielded ashfall rhyolite, but lesser amounts of basaltic lava, containing significantly less SiO_2, are also present.

Figure 5.1 shows volcanic rocks exposed in Yellowstone National Park's Grand Canyon. The Yellowstone volcanic rocks accumulated during three major periods of volcanism that included some of the largest eruptions ever on Earth. The first of these eruptions occurred 2.1 million years ago; the last significant eruptions ended about 70,000 years ago. These huge, violent events threw much debris into the air; ash settled in layers over much of the western United States and circled the globe. In the Yellowstone region, soft ash, pumice, and other rock fragment material accumulated and ultimately became harder rocks. Nowhere are these rocks better exposed than in the Grand Canyon.

The canyon, shown in Figure 5.1, is a deeply eroded chasm exposing fresh volcanic rocks formed by volcanic ash deposits and lava flows about 500,000 years ago. The lava and ash formed from a type of magma called *rhyolite*. Yellowstone is an active geothermal area, so the original igneous rocks have been intensely altered by warm flowing groundwater. Many of their original features have been obscured, and oxidation during the alteration has produced colorful landscapes that thrill visitors and inspire photographers and artists.

Besides the rhyolites exposed in the Grand Canyon, Yellowstone volcanic activity has produced a significant amount of basalt, such as the basalt exposed in Sheepeater Cliffs seen in Figure 5.2. The basaltic eruptions

Figure 5.1 The Grand Canyon of the Yellowstone displaying rock formed from volcanic debris.
Photo from the USGS.

Figure 5.2 Sheepeater Cliffs in Yellowstone National Park.
Photo from the USGS.

produced horizontal flows that solidified; they, today, appear as hard, jointed cliffs with vertical columns. The magma that produced this basalt had a distinctly different composition from the more common rhyolitic rocks found in the region. The columns formed as lava flows cooled and shrank, much the same way that mudcracks form when a muddy puddle of water dries up. Although most well-formed columns are associated with basalt flows, in Yellowstone some columns also developed in rhyolite ash deposits.

The occurrence of rhyolite and basalt, without rocks of intermediate composition, is termed *bimodal volcanism*. Bimodal volcanism is often associated with *rifts* in Earth's continental crust which, given enough time, may become the sites of new oceans. Bimodal volcanism is also commonly associated with *hot spots* where upwelling heat rises from the mantle as a thermal plume that can cause melting and so produce volcanism. The Yellowstone region is characterized by both rifts and a major hot spot.

Although found together, Yellowstone's rhyolite and basalt formed from magmas coming from two different parts of Earth—thus explaining their different compositions—and, although related, they formed during different kinds of melting events. As seen in Figure 5.3, the basaltic magmas originated in the mantle, and the rhyolitic magmas originated in the crust.

Volcanism in the Yellowstone region was due to the presence of an underlying hot spot and an associated rift in the crust. The combination of extra heat and upwelling mantle caused melting to occur and create basaltic magma, a small amount of which reached the surface and produced lava flows. The basalt originated by *partial melting* of the uppermost portions of underlying mantle—at least 30 kilometers beneath the surface. Melting occurred, in part, because of extra heat delivered by the hot spot. It also occurred because pressure decreased as mantle material moved toward the surface, and lower pressure makes it easier for geological materials to melt. However, most magmas that originate in the mantle do not make it to the surface. Instead, they solidify underground or remain molten in deep magma chambers. The small amounts that do make it to the surface pass through the crust quickly and may eventually produce basalt flows such as those exposed at Sheepeater Cliffs (Fig. 5.2).

Most significantly, the basaltic magma beneath Yellowstone delivered heat to the crust. Most crustal material melts at lower temperature than does basalt and, consequently, rising basaltic magmas from the mantle can cause melting in overlying rocks. Whether the basalt made it to the surface or not, lower crustal material melted to become rhyolitic magma. These are the magmas that have historically dominated Yellowstone eruptions (Fig. 5.3).

Geophysical studies have revealed large partially melted rhyolite magma chambers under Yellowstone today, at depths as shallow as just a few kilometers. In the past, such chambers were the sources of the magmas associated with major eruptions. When the chambers emptied during Yellowstone's major eruptions, the overlying crust collapsed, producing a depression known as a

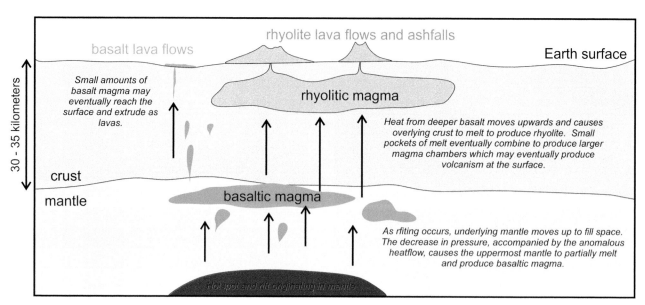

Figure 5.3   Volcanism in Yellowstone.

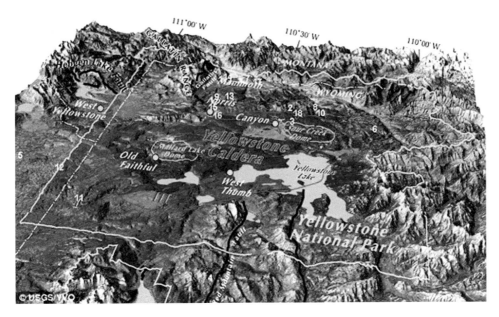

Figure 5.4 The Yellowstone Caldera, northwest Wyoming. Image from the USGS.

caldera. The present-day Yellowstone Caldera (Fig. 5.4) is a 4000-square-kilometer (1500-square-mile) shallow depression that roughly circles Yellowstone Lake and occupies about half of Yellowstone National Park. It formed after the last major Yellowstone eruption, 640,000 years ago.

5.2 Igneous petrology

5.2.1 Igneous processes

Igneous petrologists study many kinds of magmatic rocks and the processes that create them. The studies are essential as petrologists seek to understand the origin and evolution of Earth, to understand and predict phenomena such as volcanism that can be important in our daily lives, and to determine the distribution of some of Earth's most important natural resources. Igneous petrologists also investigate the distribution of igneous activity around the globe and how it relates to plate tectonics. Some petrologists even study igneous activity on other planets. But no matter what their focus, of fundamental importance to igneous petrologists is understanding the origin and nature of magma and the formation of igneous rocks.

When magma cools, it solidifies, commonly crystallizing to form individual mineral crystals. Cooling and crystallization can occur relatively quickly (for example, during eruptions at Earth's surface), by slow underground cooling over millions of years, or by mixed processes involving both slow and fast cooling. Crystal formation takes time, because atoms must migrate and bond in an organized arrangement. If cooling is too rapid for crystals to form, the melt may solidify with no visible crystals or may form a volcanic glass, termed *obsidian*. Mineral and glass fragments may be microscopic or very large (up to tens of meters in one dimension) depending on cooling rate and magma composition.

Magmas that extrude on Earth's surface as lava or pyroclastics (ash and/or related material) produce *extrusive rocks* that cool quite quickly and so have inadequate time to grow large crystals. These extrusive rocks are often associated with volcanoes (but need not be). Other magmas that crystallize at depth within Earth cool relatively slowly, producing *intrusive rocks* that may contain very large crystals. Intrusive rocks often form large rock bodies called *plutons*. So, although perhaps technically incorrect, the terms *volcanic* and *plutonic* are sometimes used to describe fine- and coarse-grained igneous rocks, respectively.

The photo in Figure 5.5 shows an outcrop of rock at the Inyo Craters, near Bishop, California. The lava that produced this rock cooled very quickly and, consequently, did not have time to form mineral crystals. Instead, the rock contains mostly black obsidian (solid volcanic glass) and thinner layers of pumice (a

Figure 5.5   Obsidian at the Inyo Craters near Bishop, California. Photo from A. Graettinger, http://inthecompanyofvolcanoes.blog spot.com.

Figure 5.6   Outcrop of granite in the Mojave Desert, California. Photo from B. Perry, CSULB.edu.

light-colored frothy mix of glass shards). Obsidian like that in Figure 5.5 is a common product of basaltic volcanism and is found all around the world.

While accounting for only about a quarter of the outcrops at Earth's surface, igneous rocks make up most of deeper parts of Earth's crust, in both continental and oceanic regions. So they are both diverse and important. They also make up much of Earth's mantle, but most igneous rocks that we study formed in the crust, simply because those that originated in the mantle rarely make it to the surface for us to see. The exceptions include (1) slivers of oceanic mantle uplifted and attached to continents through a process called *obduction*, (2) some *xenoliths* carried from the mantle to the surface by rising magmas, and (3) tectonic slices that have incorporated into mountain belts during orogenies. Although most igneous *rocks* that we see originated in the crust, the *magmas* that produced them often derived initially from the mantle. Consequently, studying igneous rocks gives us important information about the nature of the mantle and the processes that occur there.

5.2.2   *Studying igneous rocks*

Historically, igneous petrology has been based on fieldwork, often involving mapping to determine the field relationships between different kinds of rocks. Petrologists generally begin by looking at rocks in outcrops. Many rock characteristics, such as grain size, can be seen with the naked eye, or sometimes with the aid of a hand lens, while examining outcrops. Other characteristics

can only be analyzed using more sophisticated equipment, so geologists frequently break off specimens to take back to laboratories for further study. Figure 5.6 shows a geologist pointing at a large feldspar crystal in a granite outcrop in the Mojave Desert, California. The presence of both large and small mineral grains in this rock suggests it crystallized in two stages.

High magnification can reveal things not seen otherwise, so petrologists often study rocks using a microscope. The microscope shown in Figure 5.7 is a *polarizing petrographic microscope*, a fundamental tool of petrologists. It is similar in most ways to other types of microscopes, except that the light that comes from below the stage passes through a polarizing filter, and a second polarizing filter can be inserted in the column above the stage. The filters allow some rock and mineral properties to be observed that otherwise could not be seen. Petrographic microscopes reveal details that permit more accurate mineral identification and, often, reveal information about rock crystallization history. *Petrography* is the general term for the descriptive part of petrology that involves outcrops in the field, hand samples, and thin sections. Petrologists use the descriptive information to interpret rock *petrogenesis*—the origin and nature of the processes that created a rock.

To observe rock specimens with a microscope, petrologists make rock samples into *thin sections* such as the one shown in the top photo in Figure 5.8. Thin sections are very thin slices of rock that are glued to a glass slide. The rock slices are so thin (30 μm) that light can pass through most mineral grains. For example, the bottom photo in Figure 5.8 is a photograph of a thin section of granite, viewed with a petrographic microscope. Light from beneath the thin section passes easily through

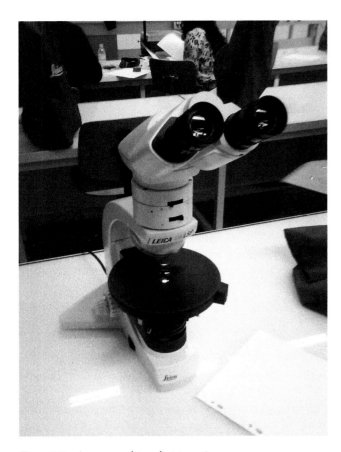

Figure 5.7   A petrographic polarizing microscope.
Photo from jd, Wikimedia Commons.

0.5 mm

Figure 5.8   A rock thin section and a view through a petrographic microscope.
Top photo from J. Schod; bottom photo from the USGS.

quartz (qtz), plagioclase (pl) and K-feldspar (kfs) grains, so they appear white—although grain boundaries are visible. The biotite (bt) and sphene (sph), however, appear darker because some light is absorbed by the very thin mineral grains. The colors of minerals, when viewed in thin section, do not always match the colors seen when examining a larger hand specimen, because the color of a hand specimen is the color of light reflected from its surface, while the color seen in thin section is the color that is transmitted by light passing through the rock slice.

5.2.3   Geochemistry

Studying igneous rocks to learn about Earth's evolution, the distribution of natural resources, the nature of volcanic hazards, and many other things requires understanding how different rocks form in different environments. Often, we gain the most important insights by examining rock and magma compositions. Thus, the study of magmas, and where and how they form, is based in large part on geochemical investigations of magma

chemistry. Although samples of (hot) lavas and magmas are occasionally collected and analyzed, most geochemical studies of igneous rocks involve analyzing rocks long after they were created by magmas.

The most widely used method for rock analysis is *X-ray fluorescence spectroscopy* (XRF), and XRF *spectrometers* are standard instruments in many geochemistry labs. Figure 5.9 shows a typical XRF spectrometer. It has a small sample port on top and a set of samples can be seen in black holders in the sample changer in front. Inside the XRF machine, an X-ray beam strikes a sample, causing atoms within the sample to fluoresce and emit X-ray radiation. Different elements emit characteristic *secondary X-rays* of different wavelengths, and the intensities at each wavelength are proportional to the amount of the element present. XRF analyses are obtained quite rapidly compared with most other analytical techniques.

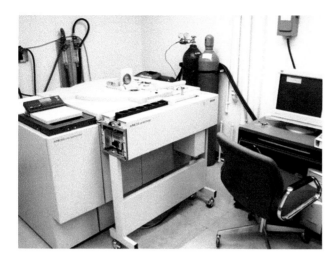

Figure 5.9   An X-ray fluorescence spectrometer (XRF).
Photo from Joel Pederson and John Shervais, Utah State University.

XRFs measure the concentration of elements ranging from major elements down to those present at the parts per million (ppm, equivalent to 0.0001 wt% level). For lower concentrations, some other analytical techniques have lower detection limits and better precision. A disadvantage of XRFs is that elements lighter than sodium cannot be easily analyzed, but the instrument can analyze most elements with atomic numbers between 11 and 41 (sodium-niobium) and some heavier ones.

A second, widely used analytical technique is *neutron activation analysis*. There are several kinds of neutron activation analysis, but the most common is *instrumental neutron activation analysis* (INAA). To obtain an INAA analysis, a sample is hit by a beam of neutrons that converts elements present to radioactive nuclides associated with the specific elements. INAA analyses cannot be done in a standard geochemistry laboratory, because the source of the neutron beam must be a (research) nuclear reactor; several different ones are available at sites in the United States. Once samples are activated, they are allowed to "cool" for a short time, to let the most radioactive elements decay, and then analyzed. As radioactive nuclides decay, they emit gamma rays with wavelengths characteristic of the elements present and intensities proportional to the amounts present. Some nuclides decay faster than others, some decay quite slowly, and there are other complications due to overlapping wavelengths. So, measuring gamma ray intensities to obtain analyses is not quickly done. Initial analyses generally take place four or five days after activation and then again a month later. Some elements require measurements a year after activation,

although many researchers do not want to wait that long and thus will not analyze elements such as gadolinium (Gd). INAA is an especially good method for analyzing rare earth, platinum group, and a few other important elements. INAA also allows measurement of the concentrations of specific isotopes.

Inductively coupled plasma mass spectrometry (ICP-MS) is a third excellent way to obtain analyses (Fig. 5.10). Sample preparation is more complicated than for XRF or INAA analyses, because samples must be dissolved, which requires time and caustic acids. Once the samples are in solution, the liquid is then passed through a plasma torch. The flame emits light having wavelengths particular to the elements present and, conventionally, light intensity is used to determine elemental concentration (ICP). More significantly, today we can send emitted ions to a mass spectrometer (MS), allowing more precise measurement and greater sensitivity when analyzing elements present in very small amounts. Although primarily used for trace elements and isotopes, some variations of ICP-MS permit analyses of major elements, too.

Petrologists also sometimes use other techniques for element and isotope analysis. All analytical techniques have advantages and disadvantages, and each technique can analyze a specific group of elements (or isotopes) over specific ranges of concentration. For the most thorough characterization, many different methods should be combined, but this is rarely possible. So, depending on purpose, one or two techniques are typically chosen, often depending on the instrumentation that is locally

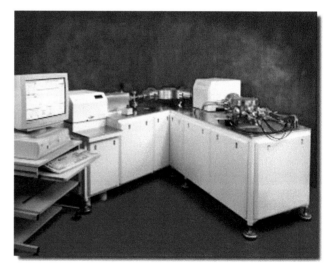

Figure 5.10   An inductively coupled plasma mass spectrometry (ICP-MS) at the United States Geological Survey.

available to the researcher. Consequently, some studies report a different number of elements than others do. Furthermore, some elements are especially problematic due to extremely low concentrations or for other reasons and so are rarely analyzed. Most commonly today, researchers use a combination of XRF and INAA to get a complete chemical characterization of an igneous rock sample.

5.3 Magma compositions

5.3.1 Similarities

Almost all magmas are dominated by the same *major elements*, and these elements are the same as those that make up most minerals. Eight of the nearly 120 known elements account for > 98% of the compositions of most magmas: oxygen, silicon, aluminum, iron, calcium, sodium, potassium, and magnesium. Consequently, these eight elements are the principle major elements in igneous rocks. Several other elements, including titanium, hydrogen, phosphorous, manganese, carbon, and sulfur, are sometimes considered major elements as well.

Petrologists generally report igneous rock compositions by listing oxide wt% values. Silicon and oxygen dominate most magmas, with silica (SiO_2) comprising 40%–80% of overall composition. Table 5.1 presents typical analyses of some common igneous rocks; most

contain six–eight oxides at levels greater than 1 wt%. The chemical oxides listed in Table 5.1 should not be confused with mineral formulas. That is, if an analysis shows that a rock contains 58.97 wt% SiO_2, it does not mean that the rock contains 58.97 wt% quartz, even though quartz is made of silica. The silica (SiO_2) in the rock may be distributed among a variety of silicate minerals, which may or may not include quartz.

Petrologists classify and group igneous rocks (and magmas) in many ways, but at a fundamental level, rocks that are relatively rich in silica, such as granite or rhyolite, are termed *felsic*. Those relatively poor in silica, such as basalt, are *mafic*. Those that fall between (andesite and syenite) are *intermediate*. Rare rocks with the lowest silica contents are termed *ultramafic*. The rocks in Table 5.1 have silica contents ranging from 40.08 (ultramafic rocks) to 77.24 (felsic rocks) wt%. Rocks with silica contents outside this range exist but are rare. Because of the high-silica content in magmas, most minerals in igneous rocks are silicates.

5.3.2 Differences

Although the major element chemistries of most magmas, and thus the chemistries of most lavas and other igneous rocks, are similar, the relative amounts of different chemical components vary quite a bit. Silica (SiO_2) content is, perhaps, the most important thing that sets magmas apart, but variations in other oxides are often

Table 5.1   Chemical analyses (oxide wt%) of some typical igneous rocks*

Oxide	Class:	Ultramafic	Mafic	Intermediate	Intermediate	Intermediate to felsic	Felsic	Felsic
	Type:	Plutonic	Volcanic	Volcanic	Plutonic	Volcanic	Plutonic	Volcanic
	Name:	Dunite	Basalt	Andesite	Syenite	Dacite	Granite	Rhyolite
SiO_2		40.08	49.80	58.97	59.54	64.50	69.22	77.24
TiO_2		0.01	2.60	1.04	0.14	0.76	0.48	0.20
Al_2O_3		0.29	14.00	17.17	18.60	14.97	15.50	10.81
Fe_2O_3		0.31	2.50	4.36	2.86	1.11	1.03	1.66
FeO		7.62	8.50	2.02	2.09	4.94	1.42	0.27
MnO		0.11	0.18	0.10	0.22	0.11	0.04	0.02
MgO		49.69	7.20	1.51	0.10	1.34	0.73	0.33
CaO		0.11	11.30	4.90	1.16	5.19	1.93	1.48
Na_2O		0.05	2.20	4.23	8.96	3.78	4.15	2.59
K_2O		0.01	0.62	2.90	4.24	3.02	4.42	4.12
P_2O_5		0.00	0.32	0.51	0.16	0.10	0.15	0.06
H_2O			0.25		1.40			0.37
other		0.58	0.10	1.55	0.40	0.10	0.30	0.65
Total		98.86	99.57	99.26	99.87	99.72	99.37	99.80

*Data are mostly from Raymond (2007). All values are wt%. Missing values indicate no analysis.

important. Some (alkalic) magmas are relatively rich in K_2O and Na_2O, for example, and the ratio FeO/MgO is variable too. So, magmas vary in composition and, accordingly, in their properties. So, too, do the lavas that form from magmas. For example, felsic lavas (lavas rich in SiO_2) tend to be very viscous. Mafic lavas (lavas relatively poor in SiO_2) are often quite fluid. Although both are liquids that may flow downhill, basaltic (mafic) lavas move at a much faster rate and sometimes move as lava "rivers."

Mafic lavas often flow long distances before solidifying, but, due to greater viscosity, felsic ones may not flow very far. Figure 5.11 (top) shows a lava river flowing from Mauna Loa Volcano in 1984. The surrounding dark and light gray material is all basalt. The darker material solidified within a few months of when this photo was taken, and the other basalt is several years old. The bottom photo in Figure 5.11 shows a mound

of felsic material—a felsic dome—in the summit crater of Colima Volcano in Mexico. This volcano is the most active volcano in Mexico and has a history of both explosive eruptions and quieter lava flows. The dome at its summit formed from gooey felsic lava magma that erupted in the crater bottom but was so viscous that it piled up instead of flowing outward. The dome disintegrated during later eruptions, after this photograph was taken, producing lavas that flowed down the sides of the volcano.

Adding a further complication, magmas are generally not entirely molten, so the distinction between magma and solid rock is not as sharp as we might expect. Different minerals crystallize at different temperatures, and consequently many magmas are mixtures of melt and solid (crystalline) components as they cool but before they are completely solid. Besides crystals, some magmas and lavas contain rock fragments incorporated when the magma was only partially crystallized.

Although all glowing lavas appear very hot, they have a wide range of temperatures that depends on composition. Most lavas at Earth's surface are between 700 °C and 1300 °C. Felsic lavas are at the low end of this range, and mafic lavas at the high end. Rare carbonate-rich lavas called *carbonatites*, which do not contain significant amounts of SiO_2, may be as cool as 600 °C, and at the other end of the temperature spectrum, ultramafic lavas may have temperatures greater than 1500 °C. Figure 5.12 shows the crater at the top of Ol Doinyo Lengai in Tanzania, an active volcano that produces carbonatite lavas. This is the only carbonatite volcano that is known to have erupted in the past 10,000 years. The cone-shaped structure in this photo is where a lava

Figure 5.11   (Top) Lava flowing from an eruption of Mauna Loa, Hawaii, in 1984. (Bottom) A felsic dome in the summit crater of Colima Volcano, Mexico, 2010. Photos from the USGS (top) and the Smithsonian Museum of Natural History (bottom).

Figure 5.12   The summit crater of Ol Doinyo Lengai Volcano in Tanzania. Photo from T. Kraft, Wikimedia Commons.

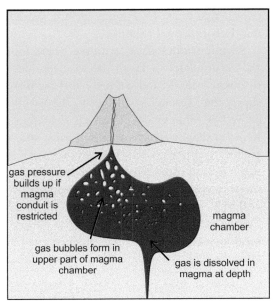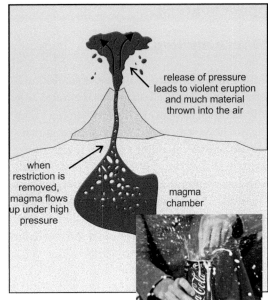

gas pressure builds up if magma conduit is restricted

gas bubbles form in upper part of magma chamber

gas is dissolved in magma at depth

magma chamber

release of pressure leads to violent eruption and much material thrown into the air

when restriction is removed, magma flows up under high pressure

magma chamber

Figure 5.13 Exploding magmas are like exploding pop bottles.

fountain occurred, and the white rock in much of the photo is solidified lava.

Most magmas contain small amounts of water vapor, carbon dioxide, or other gases such as sulfur, chlorine, or fluorine that are typically dissolved in the molten material. When magma is under pressure at depth in Earth, gases are simply minor components and dilutants. However, when magma moves toward Earth's surface, pressure decreases and the dissolved gases may form bubbles or separate from the magma completely. If the magma chamber is capped and not connected to the surface or if the connection is not large enough, pressure may build up to high levels, eventually resulting in a violent eruption when the pressure is released. As seen in Figure 5.13, The result may be explosive in much the same way that pop may explode out of a bottle if it is opened too quickly.

Figure 5.14 contrasts the eruptions of two volcanoes. The 2008 eruption of Pacaya Volcano (top) produced slow-moving lava flows. The 2009 eruption of Sarychev Volcano was much more violent, throwing ash and larger debris more than a thousand meters into the air. Relatively quiescent volcanoes, such as Pacaya are great tourist attractions—note the spectators in this photo. Explosive volcanoes such as Sarychev, are extremely violent and pose huge risks to people and infrastructure up to many 10s of kilometers away from the site of the eruptions. The contrasting types of eruptions, discussed more fully in Chapter 7, are due primarily to differences in magma chemistries.

Figure 5.14 Lava flows at the Pacaya Volcano (Guatemala, 2008), top, and the explosive Sarychev Volcano (Russia, 2009), bottom.

Photos from NASA and Greg Willis, both at Wikimedia Commons.

5.4   Magma sources

5.4.1   *Where melting occurs*

From seismic and other data, we know that the only part of Earth that is completely melted is the outer core nearly 3000 kilometers beneath Earth's surface (Fig. 5.15). The core, however, cannot be the source of magmas that reach the outer parts of Earth. For one thing, the composition of the core is wrong. It is mostly molten iron with lesser amounts of nickel and oxygen, not silica-rich material. Additionally, no known mechanism would allow magma to traverse the 3000 kilometers of mantle between the core and the crust. So, rocks in other parts of Earth must melt to produce the magmas responsible for igneous activity on or near Earth's surface.

In oceanic regions, most magmas have the composition of basalt. The oceanic crust consists of basalt too, so if the crust melted completely it could produce magmas of the right composition. However, there is no evidence of complete melting of ocean crust taking place anywhere on Earth, and if only partial melting occurs, laboratory and geochemical studies have shown that the melt would be much more SiO_2-rich than basalt is. Similar problems arise when we consider magmas in continental areas. There, we find rocks formed from both mafic and felsic magmas, and geochemical studies tell us that we cannot get mafic magmas by melting continental crust. For these and other reasons, petrologists conclude that most magmas come from the mantle beneath oceanic or continental crust. Exceptions include some felsic magmas in continental regions.

Seismic studies allow us to see where molten material accumulates in Earth, because seismic waves pass through melted, and partially melted material more slowly than through solid Earth. The seismic evidence, as well as other evidence, reveals that magmas can be found at shallow depths in both oceanic and continental regions. In many places, magmas that reach the surface, or get near to the surface, seem to come from 1–10 kilometers depth. However, some magmas exist and flow at deeper levels. Beneath Hawaii, for example, magma movement causes tremors as deep as 60 kilometers. In subduction zones, seismic evidence suggests magma movement all the way down to the upper mantle.

5.4.2   *The lithosphere and the asthenosphere*

Earth's crust and mantle have distinctly different compositions. The mantle is ultramafic overall, but the crust has a much more felsic composition. As shown in Figure 5.16, the crust and the uppermost portion of the mantle comprise the brittle *lithosphere*, the relatively rigid layer of Earth that forms the moving tectonic plates. Just about everywhere, the continental crust and lithosphere are thicker than the oceanic crust and lithosphere. The thickest crust and lithosphere occur beneath the centers of old continents and the thinnest at mid-ocean ridges. Petrologists have established that the original source of most magma that reaches the surface is in the *asthenosphere*, which is the layer that underlies the lithosphere.

The asthenosphere is mostly solid but is partially melted in some places, notably beneath mid-ocean ridges where the lithosphere is very thin and the asthenosphere is only a few kilometers below the ocean floor. In other parts of the oceans, the asthenosphere may be beneath 100 kilometers of lithosphere, and in continental regions it is generally deeper, often as deep as 200 kilometers. Earthquake waves pass through the asthenosphere more slowly than through the lithosphere, in large part due to the presence of some melt, and the upper part of the asthenosphere is often called the *low-velocity zone* (LVZ). Because the asthenosphere is partially melted in some places and at or near its melting temperature in other places, it is less rigid than the lithosphere. In fact, the apparently solid material of the asthenosphere acts in some ways like a liquid; it

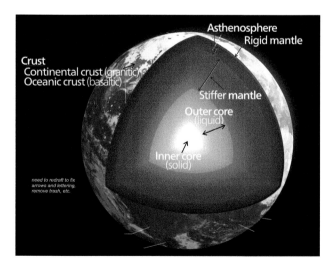

Figure 5.15   Earth's interior.
Modified from a figure by Aboluay, Wikimedia Commons.

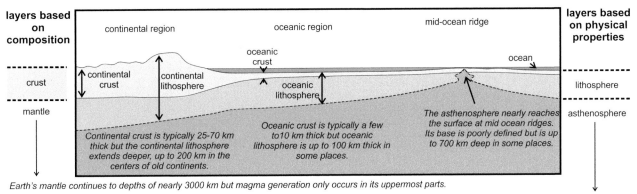

Figure 5.16 Compositional (crust and mantle) and rheological layers (layers based on physical properties) of Earth (lithosphere and asthenosphere).

convects (flows) at rates as fast as centimeters per year. As the asthenosphere moves, it carries the more rigid lithosphere above it, in part providing the mechanism for plate tectonics (described in Chapter 2).

5.4.3 Why melting occurs

Just as ice melts when the temperature goes above 0 °C (32 °F), rocks will melt if heated to temperatures above their melting temperatures. To accomplish this rock melting requires extra heat, and that poses a problem. Where is the extra heat to come from? Although radioactive decay of potassium, uranium, thorium, or other radioactive elements may create small amounts of heat, most of Earth's heat is left over from the original time of formation. This residual heat flows from Earth's interior to dissipate at the surface, and Earth has been cooling for more than 4.5 billion years. In some places, flowing magma delivers extra heat, but the origin of the heat necessary to initially create the magma is problematic.

The blue and red lines in the temperature-depth diagram on the right side of Figure 5.17 show schematically how temperature increases with depth for an average place on Earth (blue line),and a place where temperature increases faster with depth (red line). The relationships between temperature and depth, such as depicted here, are called *geotherms*. Earth's average *geothermal gradient* (the slope of the lines in Fig. 5.17) is about 25 °C/km near the surface. But, in some places, such as at mid-ocean ridges or above hot spots like Yellowstone, the gradient is greater, sometimes exceeding 50 °C/km. Other places, such as centers of old continents, have lower than average gradients.

The solid black curve in the temperature-depth diagram of Figure 5.17 is the *melting curve*; it shows the minimum temperatures at which melting can occur. If temperature-depth conditions plot in the red part of the diagram, rocks will melt. The minimum melting temperature increases with depth, but so do temperatures along the geotherms. In principle, if the geothermal gradient is high enough, the temperature may exceed the melting curve (shown where the red "high" geotherm crosses the black melting curve in Fig. 5.17). Yet, even at mid-ocean ridges or hot spots, the gradient is generally insufficient for this to happen and cause melting. Consequently, most of Earth's mantle is unmelted.

Because a normal geotherm cannot lead to melting, other mechanisms must be responsible in most cases. For instance, *tectonism* associated with mountain building can occasionally cause melting when rocks are buried by folding or faulting—because heating naturally accompanies burial. This is seen in Figure 5.18. The pressure-temperature diagram on the right side of this figure shows that if the heating is great enough, melting may occur when temperatures cross into the melting field (shown in pink). Some granites in continental regions undoubtedly form by melting of sediments and sedimentary rocks, once at or near the surface, that melted after being carried to depth.

Intruding magma, shown in Figure 5.19, is a very efficient mechanism for delivering heat that can cause melting. As shown by the black arrow in the right side of this figure, heat from magmas can cause temperature to increase without any increase in pressure. Underneath Yellowstone National Park, for example, rising magmas from the mantle are hot enough to cause overlying crustal rocks to melt. In subduction zones, magmas

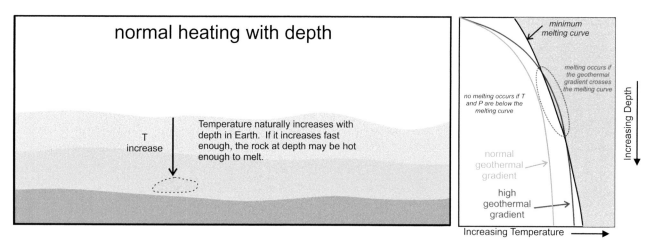

Figure 5.17   Heating with depth in Earth.

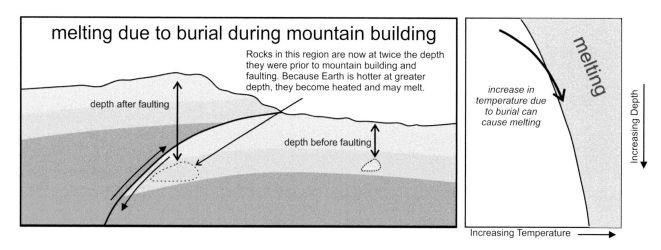

Figure 5.18   Melting during mountain building.

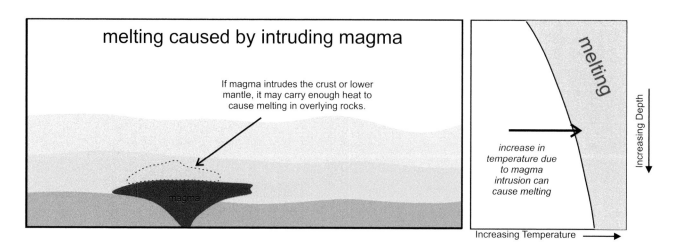

Figure 5.19   Melting caused by magmatic intrusion.

rising above a subducting plate may cause melting in the overlying continental lithosphere, creating felsic magmas that erupt in subduction zone volcanoes or crystallize underground to become plutons.

5.4.4 Decompression melting

The processes described above may all cause melting, but they do not account for the widespread melting that occurs at mid-ocean ridges. There, an additional and very significant mechanism that promotes melting is a decrease in pressure. As seen in Figure 5.20, rising mantle moves up to fill the void created by seafloor spreading. As it rises, pressure on the mantle rocks decreases. The black arrow in the pressure-temperature diagram shows how this leads to melting because rock melts at lower temperature when at low pressure, compared with high pressure. This process, called *decompression melting*, generates more magma than any other Earth process.

So, decompression melting is the key mechanism producing magmas at mid-ocean ridges, which, although we don't generally see them, are the most active volcanic settings on Earth. In these settings, partial melting of rising solid rocks produces basaltic lavas that erupt on the ocean floors, and upon cooling, are added to the spreading oceanic lithosphere. Thus, the youngest oceanic crust—shown in red in Figure 5.21—is adjacent to mid-ocean ridges, and the oldest oceanic crust (light and dark blue) is found along ocean basin margins.

Decompression melting also leads to igneous activity where continental rifting occurs, for example along the East African Rift. The East African Rift is an elongated zone that includes the Kenyan Rift and the Main Ethiopian Rift (Fig. 5.22). Along this narrow zone, the African continent has begun to split apart; the rift may eventually be the site of a new ocean basin similar to the Red Sea or the Gulf of Aden, also shown in Figure 5.22. Additionally, decompression contributes to the melting associated with more localized hot spots, like the one under Yellowstone, where warm rocks move upwards due to buoyancy.

5.4.5 Flux melting

Principles of thermodynamics tell us that two things combined will often melt at a lower temperature than they would individually. An analogy is a mix of ice and salt, which we all know melts at a lower temperature than ice alone. Similarly, rocks will begin to melt at lower temperatures when they contain water, CO_2, or another volatile compared to when dry.

The presence of water can change magma melting temperatures by hundreds of degrees. Figure 5.23 compares the conditions that cause melting (shown in red) of a basalt that contains no water and one that is saturated with water. "Dry" basalt begins to melt at temperatures in excess of 1000 °C. Depending on pressure, melting of "wet" basalt may begin at temperatures significantly lower. Because water and other volatiles lower melting temperatures in the same way that a *flux* is used to lower the melting temperature of metals, this additional mechanism for melting is called *flux melting*. If a rock is already hot, addition of only a small amount of water can promote melting, and although water is the most important geological flux, CO_2 and other gases also promote melting in some settings.

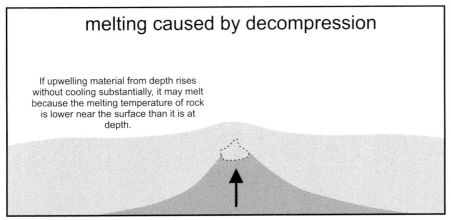
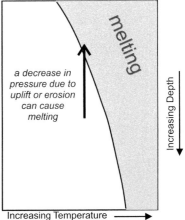

Figure 5.20 Melting caused by decompression.

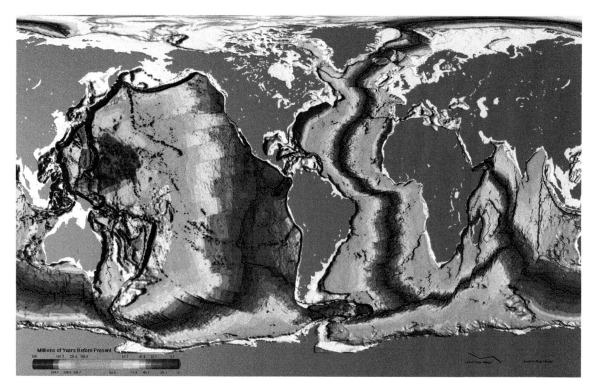

Figure 5.21   Decompression melting at a mid-ocean ridge.

Drawing from NOAA, Wikimedia Commons.

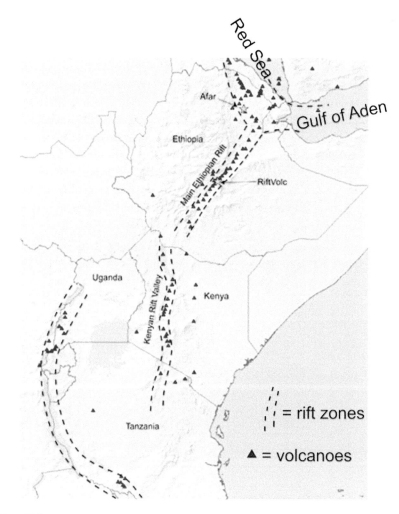

Figure 5.22   The East African Rift.

Modified from a figure by the British Geological Survey.

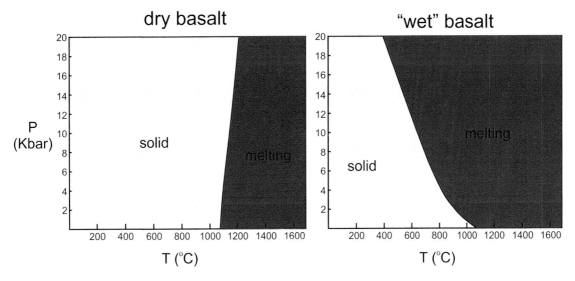

Figure 5.23   The effects of water on melting temperature of basalt.

Flux melting is especially important in subduction zones. Subducting oceanic lithosphere contains hydrous minerals that react during metamorphism to form anhydrous minerals. Consequently, as seen in Figure 5.24, water is released and migrates upwards into hot overlying mantle. This water lowers the melting temperature, causing partial melting of the ultramafic mantle to produce basaltic (mafic) magma. This magma then migrates upwards and may reach Earth's surface. More commonly, it may promote additional melting in the uppermost mantle or crust, or may become modified to produce magmas of different compositions.

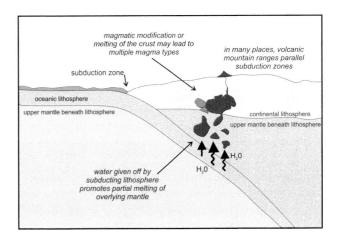

Figure 5.24   Flux melting in a subduction zone. From the USGS.

5.5   Magma movement

5.5.1   Buoyancy

Molten rock is less dense than solid rock, so molten rock is buoyant and naturally migrates upwards in much the same way that blobs of melt circulate in a lava lamp like the one in Figure 5.25. Lava lamps contain two liquids of similar density that do not mix, a little like oil and water. Heating in the base of the lamp causes one of the liquids to expand and become less dense. Buoyancy causes the blobs to migrate upwards. As they cool, they contract, regain density, and sink back to the bottom to be heated and recycled again.

Even if only a small fraction of rock melts, it wants to migrate up. Impermeability and strength of overlying rock can resist magma movement up to a point,

Figure 5.25   Lava lamp.
Image from Wollschaf, Wikimedia Commons.

but laboratory experiments suggest that if only 10% of a rock melts, the buoyancy forces are great enough to cause vertical flow. The flow may form *diapirs*, regions where mobile, less rigid material bulldozes its way through overlying rock. (Diapiric flow is different from the explosive eruptions responsible for bringing diamonds to Earth's surface. Those eruptions, called *diatremes*, are driven by gas pressure, not by magma buoyancy.)

Rising magma may create cracks that act as conduits in overlying rocks. This process, *fracture propagation*, is very important at mid-ocean ridges and produces small earthquakes, allowing seismologists to map where the fractures form and magma flows. Magma movement is easier and faster, however, if aided by fractures already in place. Sometimes these fractures reach the surface. Figure 5.26 shows lava erupting through a fracture in the eastern rift zone of Kilauea Volcano, Hawaii, in May 2018. The fracture provided a conduit for rapidly rising magma, and the rising magma caused the fracture to grow in length, leading to a classic Hawaiian *fissure eruption*. Similar fractures, at greater depth in Earth, act as conduits that carry magma toward the surface.

As a magma moves upwards, it cools and crystallizes, two processes that lead to an increase in magma density, which slows upward flow. In fact, movement will stall if the magma reaches a level in Earth where surrounding solid material has the same density as the magma does, so the magma may never reach Earth's surface. Many factors determine where magma movement will stop, including magma composition and gas content,

Figure 5.26   Volcanism in a rift zone associated with Kilauea Volcano, Hawaii, in 2018.
Photo from the USGS.

magma temperature and viscosity, and the nature and amount of crystalline material in the magma. The rate of movement is important because magmas must move relatively quickly or they will stop flowing. The temperature of the crust is also important: if the crust is warm, the magma will not cool as quickly and so will migrate more easily to upper levels. Consequently, volcanic regions, where Earth's crust is warm, tend to sustain volcanism for extended durations.

The race to the surface determines if rising magmas will erupt or if they will crystallize at depth. In general, magmas coming from the base of the lithosphere take, at most, days to years to reach the surface, provided fracture conduits are available. For example, studies of basalts in Patagonia have shown that the magma ascent rate was about 5 meters/second and that it only took the basaltic magma 2–8 hours to get from its source region to the surface. The Patagonian magma followed well-developed fracture systems. Other studies suggest that if the only upward movement is due to diapirism, the rate of ascent may be 1000 to 10,000 times slower!

5.5.2   Magma chambers and cooling

All active volcanic regions are underlain by *magma chambers*, sometimes under great pressure, where magma has pooled and collected. Geologists have drilled into magma chambers several times, including twice in Iceland and once in Hawaii. These few incidences were significant because geologists captured samples of magma in place instead of after an eruption. In Hawaii, in 2005, drillers trying to establish a well for geothermal energy production found magma at a depth of about 2.5 kilometers.

Magma chambers may exist over millions of years but are not static. Over their lifetimes, they inflate incrementally when fed by new magma and deflate during times of igneous activity. Beneath Hawaii, the chambers are at depths of 3–4 kilometers, beneath Japan 8–10 kilometers, beneath Alaska's Aleutian Islands 7–17 kilometers, and beneath Iceland 20 kilometers. In continental regions, the chambers may be deeper. Magma chambers range from very small to very large, and some can produce immense eruptions. Mt. Tambora, in Indonesia, is the site of the largest eruption of modern time. In 1815, it produced more than a trillion cubic meters (1000 cubic kilometers) of material—mostly ash and other debris thrown into the air. Although the Tambora eruption was large, prehistoric eruptions in

Yellowstone, at Long Valley (California), and at Toba (Indonesia) were much larger.

Magma chamber plumbing and geometry may be complex. The seismologists studying the Yellowstone region have found two magma chambers beneath Yellowstone National Park (Fig. 5.27). An upper rhyolitic chamber is found at 10–20 kilometers depth, and a deeper basaltic chamber is just above the base of the crust (Moho) at 30–50 kilometers depth. These chambers, like most places where magma exists, are only partly melted. Seismic wave velocities suggest the upper chamber is about 30% melted. The lower chamber may only contain a few percentage melt.

Because many magma chambers contain a crystal mush—a mix of crystals and molten material, gravity can cause them to become differentiated. Denser crystals tend to sink, so given enough time, magma chambers may become layered as crystal *cumulates* collect in their bottoms. Figure 5.28 shows dark layers of chromite that accumulated in the lower part of a magma chamber in the Bushveld Complex, South Africa. The lighter colored rock is *anorthosite*, a kind of rock composed almost entirely of the mineral plagioclase. Cumulates need not be formed by crystal settling. Some mineral crystals will

Figure 5.28   Layers of chromite surrounded by anorthosite in the Bushveld Complex, South Africa.
Photo from K. Walsh, Wikimedia Commons.

float on top of denser magmas, which also can lead to layers of different compositions. Thus, a single homogeneous chamber may become separated into zones of different compositions and produce magmas of more than one composition during subsequent eruptions. For instance, the AD 79 eruption of Mt. Vesuvius near Naples, Italy, produced bimodal volcanism thought to represent magmas from different levels in the underlying magma chamber.

5.6   Different kinds of rocks

Magma in magma chambers naturally tends to rise, but may not make it to the surface. If it crystallizes at depth, *intrusive* rock bodies including dikes and plutons of many different sizes and shapes can form. Some are shown in Figure 5.29. Magmas that form intrusive bodies are normally surrounded by very warm rocks and so cool (and crystallize) slowly, perhaps only 1 degree to several degrees in 1000 years. In contrast, magmas that reach the surface (Fig. 5.29) form *extrusive* rocks (generally called *volcanic rocks*) that cool relatively quickly. The magmas may erupt in lava flows or extrude explosively into the air, producing ash and other debris of variable size.

Cooling rate directly affects grain size, so intrusive rocks are overall coarser grained than extrusive rocks are. Many volcanic rocks contain fine grains that we can only see with the aid of a microscope, and some volcanic rocks may be partly or entirely made of volcanic glass. Igneous rocks containing crystals large enough to

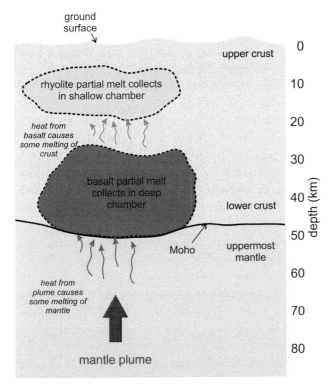

Figure 5.27   Magma chambers under Yellowstone.
Based on a figure from Huang *et al.*, 2015.

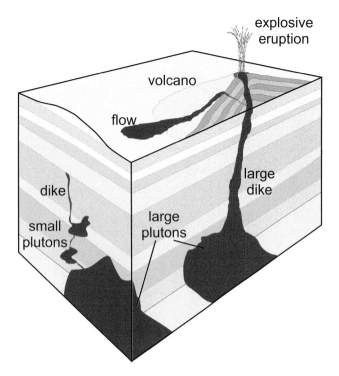

Figure 5.29　Intrusive (plutonic) and extrusive (volcanic) rocks.

Figure 5.30　A porphyritic rock from Alnö, Sweden. Photo from Mikenorton, Wikimedia Commons.

be seen with the naked eye are termed *phaneritic*; those containing crystals too small to see are *aphanitic*. Overall, due to contrasting cooling rates, most phaneritic rocks have an intrusive origin and most aphanitic rocks have an extrusive origin. There are exceptions, however. For example, fine-grained intrusive rocks can be found in narrow dikes that cool quickly after intrusion, and some coarse-grained extrusive rocks can be found in the bottom layers of thick lava flows.

Many volcanic rocks contain large crystals surrounded by very fine-grained or glassy material. Such rocks have a *porphyritic texture* and are called *porphyries*. Figure 5.30 shows such a porphyritic rock, from the Swedish island Alnö, that contains large crystals of grayish rounded quartz and blockier crystals of whitish feldspar surrounded by a darker fine-grained matrix. The two grain sizes suggest a two-stage cooling history. The large crystals, termed *phenocrysts*, grew slowly over a long time before eruption (as evidenced by their size). Subsequently, magma containing the crystals moved upward and, during rapid cooling at Earth's surface, the finer-grained material solidified. The fine-grained matrix material, termed the *groundmass*, generally contains microscopic or submicroscopic crystals (because they formed quickly and did not have time to grow large) and often glass (because cooling was so fast that no crystals could form).

Felsic magmas, because of their high viscosity, move sluggishly through the crust and may not make it to the surface. Sometimes, they may boil and lose gases and fluidity rapidly, and subsequently solidify at depth. For these reasons, felsic sills, dikes, and plutons are common. Most of the largest known plutons have felsic compositions. When many felsic plutons form in the same region, they may create very large features called *batholiths*, but individual plutons within a batholith may be difficult to distinguish if they have similar compositions.

Batholiths can be gigantic; for example, the Sierra Nevada Batholith of California, further discussed in Chapter 6, runs the entire length of the Sierra Nevada Mountains, and the Coast Range Batholith of British Columbia is even larger. Figure 5.31 shows a view of the High Sierra near Lone Pine California. The largest monolith in this photograph is Mt. Whitney, the highest peak in the Sierra Nevada Mountains. The peak left of Whitney is Keeler Needle, and left of that is Crooks Peak. All three are made of granites belonging to the Sierra Nevada Batholith.

Gooey, viscous, felsic magmas often contain much trapped gas, so if felsic magmas do make it to the surface, the result may be explosive eruptions, producing large volumes of ash or lava flows, or both. The lava may not have time to crystallize before it solidifies, so felsic eruptions may often produce flows containing *pumice* or *obsidian*. Obsidian forms when lava quenches quickly to produce a uniform glass devoid of crystals. Pumice may have the same composition as obsidian and may be largely glass, but pumice contains gas bubbles making it less dense. The black rock in Figure 5.32 is obsidian

Figure 5.31 Outcrops of granite, Sierra Nevada Batholith, California.
Photo from Cullen328, Wikimedia Commons.

Figure 5.33 Basalt flows from Kilauea Volcano, Hawaii.
Photo from D. Delso, Wikimedia Commons.

Figure 5.32 Obsidian and pumice.
Photo from D. Mayer, Wikimedia Commons.

and the gray rock below it is pumice. This photo was taken at Panum Crater, one of many volcanic craters near Mono Lake, California, about 65 kilometers north of where the rock shown in Figure 5.5 was found.

Mafic and intermediate composition magmas are hotter and less viscous than felsic magmas are, allowing them to move more quickly and have a better chance to reach the surface than their felsic counterparts do. Still, they also may form dikes or other small intrusive rock bodies and, less commonly, plutons of small to large sizes. If they reach the surface, mafic and intermediate magmas may produce lava flows associated with *volcanoes*, or *flood basalts* that spread across the land covering very large regions instead of forming discreet volcanoes. Figure 5.33 shows basalt flows from Kilauea Volcano in

Hawaii Volcanoes National Park. Kilauea is the most active of several volcanoes that form the island of Hawaii, and eruptions continue at Kilauea today. The basalt shown in the photo is just a few years old, but the oldest exposed basalts at Kilauea erupted about 2800 years ago.

5.7 Melting of minerals and rocks

5.7.1 Congruent and incongruent melting

Different minerals melt at different temperatures. Ice, for example, melts when the temperature reaches 0 °C (32 °F) at one atmosphere pressure, but gold—shown melting in Figure 5.34—melts at 1064 °C, equivalent to 1948 °F. Many minerals melt at temperatures above the melting temperatures for ice and gold. Quartz, for example melts at 1670 °C at one atmosphere pressure, and Mg-olivine (forsterite) melts at 1890 °C. When ice, gold, quartz, or forsterite melt, the composition of the melt is the same as the solid. Thus, ice melts to produce water, molten gold has the same composition as solid gold, molten quartz is SiO_2, and molten forsterite is Mg_2SiO_4. The temperatures at which melting occurs for ice, gold, quartz, and olivine—0 °C, 1064 °C, 1670 °C, and 1890 °C——are the minerals' *melting points*. Because ice, gold, quartz, and forsterite all have a single melting point, if they are subjected to heating, temperature will not increase above the melting point until all the solid has become liquid. At higher temperatures, all will be liquid; at lower temperatures, all will be solid.

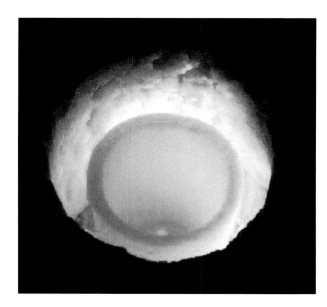

Figure 5.34 Melting gold in a crucible.
From Maksim, Wikimedia Commons.

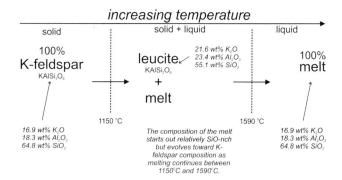

Figure 5.35 Melting of K-feldspar.

Many other materials melt in more complicated ways. For example, most of us have experience with a popsicle melting in our mouths. It melts to produce sugary, often colored, water. Eventually, all the sugar and color may melt out of the popsicle, leaving tasteless solid white ice on the popsicle stick. So, when a sugary popsicle melts, what we get in our mouths has a different composition than what is left behind, solid, on the popsicle stick. Many minerals melt in a similar way. The composition of the first melt that is produced is not the same as the original mineral.

If heated sufficiently, K-feldspar begins to melt, but the first melt to form always has a composition that is more silica-rich than the original K-feldspar. The melt and the remaining solid have different compositions but together add up to the original composition before melting began. Therefore, if the melt is more silica-rich than the original K-feldspar, the remaining solid must become more silica-poor than the original K-feldspar.

At 1 atmosphere pressure, K-feldspar ($KAlSi_3O_8$) melts to produce (solid) leucite ($KAlSi_2O_6$) and a (liquid) melt when heated to about 1,150 °C (Fig. 5.35). The first melt is more silica-rich than K-feldspar, and the leucite solid is more silica-poor than K-feldspar. If temperature continues to increase, the leucite will melt until, at about 1590 °C, it will all be gone. No solid will remain, and the composition of the melt will be the same as K-feldspar, the original starting solid. So, K-feldspar is partially melted, and solid and liquid coexist, over a range of temperature (1150 °C to 1590 °C).

Below 1150 °C, all will be solid, and above 1590 °C, all will be liquid.

Crystallization is fundamentally the opposite of melting. At Earth's surface, a cooling melt of K-feldspar composition will first crystallize leucite (at 1590 °C; Fig. 5.35). As cooling continues, more leucite will form and the amount of melt will decrease. Eventually (at 1150 °C), the leucite crystals will react with remaining melt to produce solid K-feldspar. If equilibrium between melt and crystals is maintained, for any given composition, melting and crystallization involve the same minerals and melt compositions—but they form in reverse order.

Many minerals are *solid solutions*, meaning that their chemical compositions are not fixed, and vary within limits. Solid-solution minerals melt like K-feldspar. They melt over a range of temperatures where some melt and some remaining solid coexist, and melting and crystallization temperatures are different for different compositions. Plagioclase, the most common mineral in Earth's crust, is a good example of a solid solution. Plagioclase may have any composition between anorthite ($CaAl_2Si_2O_8$) and albite ($NaAlSi_3O_8$). When a plagioclase of intermediate composition begins to melt, the first melt is more albite-rich than the original mineral was. Consequently, after partial melting occurs, the solid plagioclase that remains must be more anorthite-rich than the original mineral was. With increasing temperature, more melting takes place. The melt and solid will change compositions, but the melt will always be more albite-rich than the solid residue. Eventually, when no solid remains, the melt composition will match the original starting composition. Thus, plagioclase, like other solid solutions, begins to melt at one temperature, is completely melted at another, and between the two temperatures a melt and a solid of different compositions will coexist.

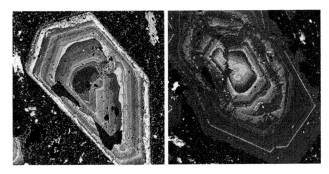

Figure 5.36   Compositional zoning in plagioclase from an igneous rock.

Images from E. Frahm, University of Minnesota Electron Microprobe Lab.

Crystallization of molten plagioclase is just the opposite of melting. The first crystals to form will be more anorthite-rich than the magma. As temperature drops, more, and larger, crystals will grow until all the melt is gone. As this occurs, the composition of the crystals will change, becoming more albite-rich. When crystallization is complete, the final crystal will have the same composition as the original melt. If equilibrium conditions are maintained, as crystallization proceeds, plagioclase crystals should be homogeneous—they should have the same composition throughout. In laboratory experiments this may happen, but in nature it sometimes does not. The two images in Figure 5.36 show compositional zoning in plagioclase feldspar from an igneous rock. They were obtained using a scanning electron microscope, and the colors show the distribution of calcium and sodium (purple zones are more sodium-rich). If the crystals had stayed in equilibrium with the melt as they grew, the compositions would be uniform and no concentric rings would be visible. But these crystals contain rims that are more Ca-rich than the cores.

Minerals like ice, halite, quartz, and forsterite melt at a specific temperature, and the composition of the liquid that forms is the same as the composition as the solid. In contrast, K-feldspar, plagioclase, and many other minerals melt over a range of temperatures, and the composition of the melt is different from the original solid until all is melted. Minerals that melt at a single temperature are said to melt *congruently*; those that melt over a range of temperatures melt *incongruently*.

5.7.2   Liquidus and solidus temperatures

Because rocks generally contain multiple minerals, most that melt incongruently, rocks melt over a range

of temperatures. When a rock or mineral is heated, the temperature at which melting begins is termed the *solidus* temperature. The solidus is the same as the minimum melting temperature previously discussed in this chapter and depicted in Figures 5.17–5.20. The temperature above which all is liquid is the *liquidus* temperature. On phase diagrams, the *solidus* and *liquidus* are names given to curves showing conditions where (partial) melting begins and ends. So, at temperatures between the solidus and liquidus, solid material (having composition different from the original solid material) will coexist with melt of a different composition. For materials that melt congruently, the solidus and liquidus temperatures are the same as the melting point. For those that melt incongruently, there is no single melting point, and solidus and liquidus temperatures may differ by hundreds of degrees.

Figure 5.37 is a schematic diagram showing the liquidus and solidus for ultramafic rock, the kind of rock that dominates Earth's mantle, but similar diagrams exist for rocks of other compositions. The top graph is a pressure-temperature graph, with pressure increasing upwards. This is the standard kind of diagram used by chemists and some geologists. Pressure increases with depth in Earth, and some geologists prefer to have pressure increasing downward, or to use a temperature-depth

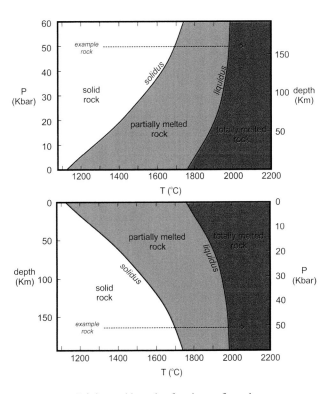

Figure 5.37   Solidus and liquidus for ultramafic rocks.

diagram (bottom) with depth increasing downward, as it does in Earth. Both diagrams show the same things, but the bottom diagram is a flipped image of the top one.

Consider, for an example, solid rock at 50 kbar pressure (160 kilometers depth). If it is heated (Fig. 5.37, dashed line), it will begin to melt at around 1700 °C. Continued heating will create increasing amounts of partial melts until temperature reaches about 1970 °C. Then, all will be melted, and the melt will have the same composition as the original rock. At temperatures below the solidus, all is solid; at temperatures between the solidus and liquidus, partial melting occurs; and at temperatures above the liquidus, all is melted.

As shown in Figure 5.37, solidus and liquidus temperatures are not the same at all pressures—melting normally occurs at higher temperature if pressure is increased. Consequently, the liquidus and solidus temperatures for rocks are greater at greater depths in Earth. As previously pointed out, temperature increases with depth in Earth, too. If it increases enough, it may intersect the mantle solidus, leading to partial melting. However, complete melting of the mantle would require temperatures to exceed the liquidus, and such conditions do not exist within Earth. Most studies have concluded that the amount of partial melting at depth in Earth rarely exceeds just a few percent.

The notion of partial melting is already complicated, but adding more complication, the products of melting may be different at different pressures. For example, at low pressure (Earth's surface or shallow depths), enstatite ($Mg_2Si_2O_6$) melts incongruently to produce forsterite (Mg_2SiO_4) and a melt that is more silica-rich than enstatite. At high pressure (deep within Earth), enstatite melts congruently. In contrast, the pressure effect on forsterite melting goes the other way. Forsterite melts congruently at 1 atmosphere but melts incongruently, at very high pressure, to produce periclase (MgO) and a liquid more silica-rich than forsterite is (Table 5.2).

Table 5.2   Melting reactions for enstatite and forsterite.

Pressure	Melting reaction
low	enstatite = forsterite + melt (incongruent)
	forsterite = melt (congruent)
high	enstatite = melt (congruent)
	forsterite = periclase + melt (incongruent)

5.7.3 Bowen's reaction series

When rocks melt or magmas crystallize, things are generally more complicated than when a single mineral melts or crystallizes. Most rocks contain more than one mineral, and different minerals melt or crystallize at different temperatures. N. L. Bowen, an early 20th-century petrologist, conducted many laboratory experiments and was the first to systematically compare melting and crystallization temperatures of common igneous minerals.

Bowen's Reaction Series (Fig. 5.38) depicts Bowen's fundamental findings. It is termed a reaction series because during melting (or crystallization) solid minerals continuously react with surrounding liquid. Bowen's Reaction Series shows the relative liquidus temperatures for common minerals from olivine (highest temperature) to quartz (lowest temperature). Plagioclase melts incongruently over a range of temperature; Bowen called this the *continuous* series. Most other minerals, depending on composition, melt sequentially over more restricted temperature ranges (*discontinuous series*). Bowen found that (mafic) minerals common in ultramafic and mafic rocks have the highest liquidus and solidus temperatures, and (felsic) minerals that are common in felsic rocks have the lowest. Consider a cooling magma: as temperature decreases, minerals higher up in the series crystallize first, followed by minerals lower down. We call minerals that melt and crystallize at high temperatures *high-temperature minerals*; those that melt and crystallize at low temperatures are *low-temperature minerals*.

Although melting and crystallization in the order depicted by Bowen's Reaction Series seems straightforward, there are many complications. No magmas follow the entire series—most crystallize only one or a few of the minerals in the series—and some magmas crystallize minerals that are not part of the series. Furthermore, some minerals melt (and crystallize) congruently and some do not. Additionally, some melting and crystallization reactions involve more than one mineral reacting together. For example, at Earth's surface, anorthite melts at about 1560 °C and diopside melts at about 1390 °C. A rock that contains both anorthite and diopside, however, will begin to melt at around 1270 °C, a much lower temperature than the melting point of either individual mineral.

For most magmas, crystallization begins at some maximum temperature and continues over a range of temperatures until everything is solid. As the process continues, different minerals form at different temperatures

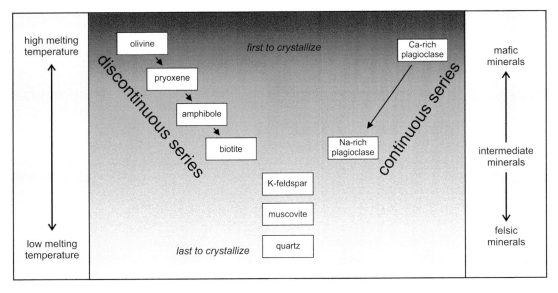

Figure 5.38 Bowen's Reaction Series.

and the composition of the magma changes. The opposite occurs during heating of a rock. Melting generally begins by melting of low-temperature minerals followed sequentially by melting of higher-temperature minerals until all is liquid. During this process, the melt continually changes composition. Bowen's Reaction Series is a model that serves to remind us that different minerals melt and crystallize at different temperatures, that mafic minerals tend to crystallize before felsic ones, and that felsic minerals melt at lower temperatures than mafic minerals do. The series does not, however, apply in detail to any known magma or rock composition.

5.8 The importance of partial melting and partial crystallization

5.8.1 *Incomplete melting*

Melting can only occur if temperature exceeds the solidus, and temperatures rarely, if ever, reach the liquidus. Because the geothermal gradient is different in different places, this means that partial melting occurs but does not occur everywhere. So, magmas generally form by melting of an originally solid *parent rock* that does not melt completely. When a rock melts only partially, producing a melt that contains melted low-temperature minerals and leaving behind solid high-temperature minerals, we call the process *anatexis*. In the mantle, for example, anatexis of ultramafic rock produces basalts.

Figure 5.39 Metasedimentary migmatite.
Photo from Chmee2, Wikimedia Commons.

Migmatites (from the Greek *migma*, meaning mixture, and *ite*, referring to rock) are rocks composed of two different components that are mixed, or swirled, together. An example is seen in Figure 5.39. Typically, migmatites contain a light-colored segregation that in many cases appears to have formed by partial melting of the darker surrounding material. In the crust, many *migmatites* are thought to have formed by anatexis associated with metamorphism of a parental sedimentary rock. The result is a mixed rock that contains both metamorphic and igneous components. The melt that develops eventually cools and crystallizes just like any other magma does, and will sometimes contain large crystals (phenocrysts) after completely solidified. If the melt migrates away from where it was produced, identifying its origin may become problematic, and the residual material left behind will not resemble the original sedimentary parent. It seems apparent, nonetheless, that large-scale anatexis of crustal rocks can produce large volumes of granitic melts that later form granitic plutons. The plutons may contain xenoliths of the original rock that melted to form the granitic magma.

5.8.2 Equilibrium or not?

In an equilibrium melting process, the melt and solid remain in contact and in chemical equilibrium as melting occurs. The system is "closed"—the overall composition does not change—so the melt and remaining solid material add up to the starting composition. Consider the equilibrium melting process that may occur when a rock is heated. Melting begins at the solidus temperature, and the first melt is formed by the melting of low-temperature (felsic) minerals. The rest of the minerals remain unmelted. Melting progresses as temperature increases, and different minerals melt at different temperatures. As the amount of melting increases, the melt composition evolves to be more like its original parent material until everything has melted. During this process, the minerals present will change, and the compositions of solid-solution mineral crystals will change as atoms migrate in and out of the solid crystals. It does not matter if the rock melts partially or completely; if melt and solids continue to react, chemical equilibrium is possible as compositions change in response to temperature. The same concept of equilibrium applies to crystallization. If equilibrium is maintained during crystallization, crystals will be homogeneous in composition and will change proportions and compositions systematically as temperature decreases. However, disequilibrium can occur if the migration of atoms through the solid crystals, or through a viscous melt, is not fast enough to keep up with cooling.

When studying rocks, petrologists often find it difficult to decide if crystals and melt stayed in equilibrium. Some volcanic rocks, however, contain zoned crystals that are evidence of disequilibrium. Figure 5.40 shows a polarizing microscope view of a large compositionally zoned grain of clinopyroxene in a basalt. The dark material surrounding the clinopyroxene is mostly volcanic glass, and the needle-shaped light-colored crystals are plagioclase. If the clinopyroxene grain were homogeneous,

Figure 5.40   A zoned clinopyroxene crystal in a basalt. Photo from R. Siddall, University College London.

the colors induced by the polarizers (called *interference colors*) would be the same in all parts of the grain. Zoning of this sort is evidence that the melt and crystals did not stay in equilibrium during crystallization. Figure 5.36 shows similar zoning in plagioclase. In both cases, the centers of the crystals grew at high temperature, and as temperature decreased, the crystals grew larger. If the minerals and melt stayed in equilibrium, the grain (no matter the size) would have a homogeneous composition, but in zoned crystals the crystal cores have compositions formed at higher temperature than the rims did. The outer zones have compositions that formed at lower temperature because atoms could not migrate into and through the crystals fast enough to maintain compositional homogeneity. Thus, only partial equilibrium was maintained. Many volcanic mineral crystals have broad homogeneous centers but are zoned near their rims, suggesting that they stayed in equilibrium with the melt until the latest stages of crystallization.

5.8.3 Fractional melting

Large-scale disequilibrium melting occurs if a melt and a solid do not continue to react together but instead become chemically isolated due to physical separation. For example, if a rock melts partially and the magma escapes upwards, the melt and remaining solid material cannot react to stay in chemical equilibrium. Figure 5.41 shows melting of an original parent rock. The first minerals to melt are low-temperature minerals

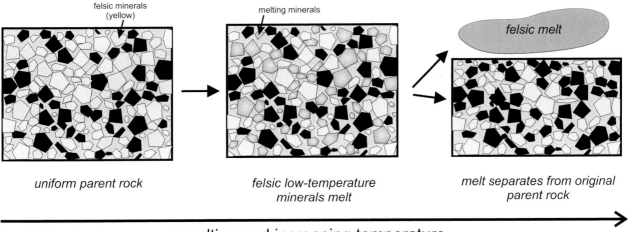

felsic minerals (yellow)

melting minerals

felsic melt

uniform parent rock

felsic low-temperature minerals melt

melt separates from original parent rock

melting and increasing temperature

Figure 5.41 Fractional melting.

(shown in yellow), so when they melt, a relatively felsic melt is produced (shown in orange). This melt may subsequently become separated from the leftovers of the original rock. Consequently, a melt of different composition from the parent has been produced and may move upwards in Earth. This process is *fractional melting*, a widespread and important process that occurs in the source region for most magmas. Low-temperature minerals always melt first, either individually or in consort with others. They are relatively silica-rich minerals compared with others in a rock, so when only fractional melting occurs, the melts are more felsic than the parent rock is. The remaining rock becomes *depleted* in felsic components and, therefore, more mafic than its parent. When this happens, felsic melts migrate upwards, leaving more mafic residue behind. So, fractional melting explains, in part, why the Earth has differentiated into a more felsic crust and more mafic mantle during its 4.6-billion-year lifetime. These generalizations always apply, but the specific products of fractional melting depend on the starting material composition and the amount of melting. Furthermore, the results will be different if the parent rock is already depleted.

Earth's mantle is ultramafic; if it melted completely it would produce ultramafic magma, but as discussed previously, complete melting cannot occur because there is no known mechanism for heating the mantle to the very high temperatures needed to melt it completely. Fractional melting, however, is widespread, and the upper mantle is the source of many magmas that move within the crust and sometimes reach the surface. Because the upper mantle has a relatively uniform composition,

fractional melting of mantle produces similar magmas worldwide, almost all mafic, equivalent to basalt compositions. More felsic magmas may also be generated in the mantle, but they are uncommon. Similarly, partial melting of subducted ocean crust, which is basaltic everywhere, generally produces magmas of intermediate composition, and fractional melting of lower continental crust produces felsic magmas (equivalent to granite).

5.8.4 *Fractional crystallization*

Fractional crystallization, the opposite of fractional melting, occurs when a magma partially crystallizes and the remaining magma becomes segregated from the crystals. In these circumstances, the new *evolved magma* will have a different composition from its *parental magma*. The evolved magma, which is more felsic than its parent was, may move upwards, leaving the high-temperature (mafic) minerals behind. Fractional crystallization, like fractional melting, has been a key process contributing to differentiation of Earth.

On a local scale, fractional crystallization explains the origins of cumulate rocks, like those shown in Figure 5.28. Such rocks form when newly formed crystals sink to the bottom of a magma chamber and no longer stay in equilibrium with the melt.

Figure 5.42, a schematic diagram, shows the principles behind fractional crystallization of a magma. While cooling, a parental magma crystallizes some high-temperature minerals. These minerals eventually sink to the bottom of the magma chamber, leaving an evolved

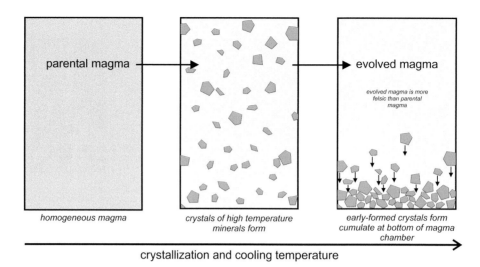

Figure 5.42   Diagram showing fractional crystallization.

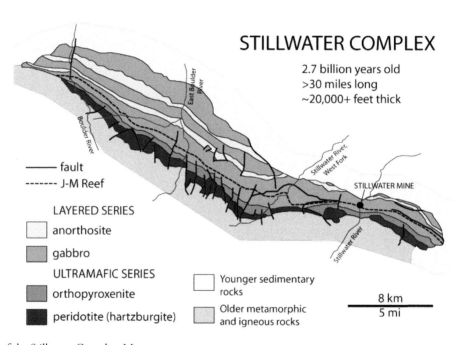

Figure 5.43   Map of the Stillwater Complex, Montana.
Map from Kurt Hollacher.

magma above. Because high-temperature minerals are mafic, the evolved melt is more felsic (less mafic) than the original parent magma. While this goes on, a cumulate rock forms at the bottom of the magma chamber, and the evolved magma may move upwards and become completely separated from the cumulate.

The Stillwater Complex near Nye, Montana, provides a spectacular and complicated example of fractional crystallization. The Stillwater Complex—a map is shown in Figure 5.43—is a large layered ultramafic and mafic series of rocks emplaced during multiple intrusive events. All intrusions involved fractional crystallization, and as each magma cooled, cumulate layers containing olivine, pyroxene, and chromite settled in the bottom of the magma chamber, leaving a less mafic magma to crystallize in the upper parts of the chambers. So, the complex is layered with more mafic rocks in the lower parts of each intrusive body. To geologists, Stillwater is famous for the spectacular rocks that are exposed and for the fascinating information revealed about fractional crystallization. Economically, Stillwater is famous because it is one of only a few platinum mines in North America.

5.8.5   *Other processes explaining variations in magma composition*

Fractional crystallization is undoubtedly the most important process that changes magma composition after a magma forms, but other mechanisms occur that can affect a magma's evolution. For example, in some settings, hot magmas may melt surrounding rocks and assimilate them into the melt. Generally, we think of this *assimilation* occurring when mafic magmas encounter more felsic rocks, because mafic magmas may be hotter than the felsic rock's melting temperature is. So, assimilation can make magma more felsic and is most likely to occur in the (felsic) crust. Some volcanic rocks contain crustal xenoliths, inclusions of rock fragments from the deep crust that were incorporated as solid pieces into the melt; often the xenoliths show evidence of partial melting. It is no stretch to assume that sometimes xenoliths melt and mix in completely. Some geochemical data, too, support the idea that crustal material has been incorporated into a mantle-derived melt.

Different magmas may also combine to produce hybrid magmas of different compositions. However, *magma mixing* is unlikely to happen if magma compositions are too different because different magmas have different melting temperatures, densities, and viscosities. Although some evidence suggests that magma mixing occurs on a small scale, most petrologists believe it is generally a minor contributor to magma diversity. A third process, *liquid immiscibility*, has also been proposed as a process that may lead to change in magma composition. Immiscible liquids unmix in much the same way that chicken soup separates into broth and fat upon cooling. Experimental evidence suggests, for instance, that sometimes a sulfide-rich melt may unmix from mafic silicate magma—a potential important process forming ore deposits, or that alkali-rich magmas may unmix from less alkaline ones. Some petrologists have invoked this last process to explain the origin of carbonatites (magma types discussed in the previous chapter) that are carbonate rich and contain little silica.

5.8.6   *Parental magmas and differentiation*

Most magmas evolve from some parental magma. They are evolved melts, not melts having the composition created during initial melting. Subsequently, as crystallization progresses, magma compositions follow what is called a *liquid line of descent*, producing a series of magmas of different compositions as fractional crystallization removes specific minerals from the melt. Only a few rare magmas may not be evolved. For example, the white veins (termed *leucosomes*) in migmatites that form by partial melting of sedimentary rocks may not have changed composition after they formed (Fig. 5.39). The leucosomes appear to have been created by partial melting of metasedimentary rock, and the melt has remained local and has not differentiated.

If solid mantle melted directly, either partially or completely, to create magma, the magma would be called a *primary magma*. Primary magmas have undergone no differentiation and have the same composition they started with. Specifically, if they come from the ultramafic mantle and were not subsequently modified, they must have a very high Mg:Fe ratio and be enriched in Cr and Ni just like mantle rocks are, and petrologists use these and other characteristics to test if magmas could be primary magmas. Most magmas fail the tests, and primary magmas are exceptionally rare, or may not exist at all. Some magmas, and rocks, however, come close to being primary, and petrologists describe them as *primitive*, meaning they have undergone only minor differentiation.

Parental magmas may be primary or primitive. The only requirement is that they lead to magmas of other compositions. If a collection of melts evolved from the same parent, they form a *magma series*. The different melts have different compositions but will share some chemical characteristics, especially trace element compositions and isotopic ratios. A challenge for petrologists is to study the compositions of an inferred magma series in order to learn the composition and source of the original parent.

Typically, petrologists begin their quest by obtaining analyses of the rocks and plotting the results on different kinds of *composition diagrams*. For example, *Harker diagrams*, first used in 1902, have SiO_2 content as the horizontal axis and other oxides plotted vertically. Figure 5.44 shows an example. SiO_2 is chosen as the abscissa because it generally shows the most variation of all oxides, and because it relates closely with magma temperature and the amount of fractional crystallization.

When looking at Harker diagrams, the principles are that (1) if derived from a common parent, rock compositions should trend smoothly and (2) the most mafic composition is closest to the parent magma composition. So, if a Harker diagram reveals smooth trends, it is possible that all the magmas derived from the same parent and that the low SiO$_2$ end of the graphs are closest to the magma's parent composition. Harker diagrams are only one kind of composition diagram; many others with different oxides on the axes are commonly used.

Figure 5.44 shows a Harker diagram for volcanic rocks from near Crater Lake, Oregon. Each point represents a different volcanic rock from the same region; the horizontal axis shows the SiO$_2$ content of the rock and the vertical axis the amount of other oxides present. The solid lines show the smoothed trends. This suggests that the different rocks may have formed from magmas that evolved sequentially from the same original parental magma. The Crater Lake magmas range from *basalt* (left) to *rhyolite* (right). On the basis of the smooth trends shown, Williams (1942) concluded that they all came from a common parent magma and that they evolved by fractional crystallization. The basalt composition is closest to that parent.

A second commonly used way to look at magma composition is to plot compositions on an AFM diagram such as the one shown in Figure 5.45. AFM diagrams eschew SiO$_2$ and instead look at alkali (Na$_2$O + K$_2$O), FeO, and MgO content. The triangle corners are A = alkali oxide wt%, F = FeO wt%, and M = MgO wt%. Many studies have found that magma series follow one of two trends, the *tholeiite* trend (data points in red in Fig. 5.45) or the *calc-alkaline* trend (data points in blue), and we easily see these on an AFM diagram. Figure 5.45 is an AFM

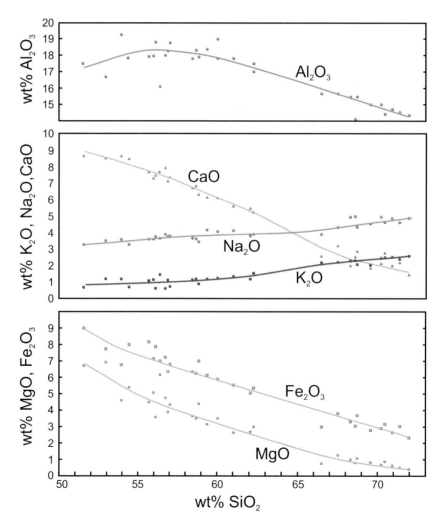

Figure 5.44 Harker diagram for volcanic rocks of the Crater Lake region, Oregon.
Data from Williams, H., 1942, *The Geology of Crater Lake National Park*, Oregon.

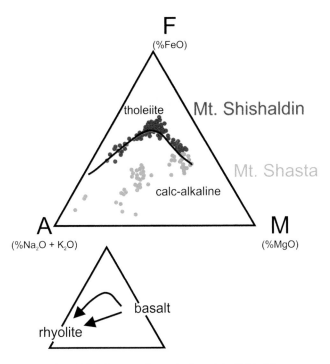

Figure 5.45 AFM diagrams for rocks from Mt. Shishaldin, Alaska, and from Mt. Shasta, California.
Diagram based on data from Zimmer *et al.* (1993).

the iron is oxidized, leading to crystallization of magnetite. Consequently, when mafic minerals crystallize, iron is removed from the magma as fast as magnesium is, and the melt Fe:Mg ratio remains about constant during differentiation. In tholeiitic magmas, olivine and pyroxene crystallize first and magnetite may not crystallize at all. Olivine and pyroxene have high Mg:Fe ratios compared with melt, and the magma becomes enriched in iron during the initial stages of crystallization. Calc-alkaline magmas are dominant in *andesitic-type* subduction zones, such as California's Cascade Mountains. Mt. Shasta is an example. Tholeiitic trends occur mostly in island arcs, such as the Aleutian Islands, and Shishaldin Volcano is an example.

5.9 The most common occurrences of melts of different compositions

5.9.1 *Ultramafic magmas*

Ultramafic magmas, which must originate in the mantle, do not reach the surface to produce ultramafic volcanic rocks today. However, they made it to the surface when the mantle melted, in part or completely, during Earth's early history. Thus, we find rare *komatiites* (volcanic rocks having ultramafic compositions) scattered around Earth in very old (Archean) terranes, and only very rarely in younger terranes. Figure 5.46 shows approximately 2.7 billion year old komatiites exposed at Pyke Hill, in western Ontario. The lack of young ultramafic volcanic

diagram comparing rocks from Shishaldin Volcano (Aleutian Islands) and Shasta Volcano (California). Each point represents an analysis of an individual rock. Shishaldin is an island arc volcano associated with an oceanic plate subducting under another oceanic plate. Shasta is a continental margin volcano where an oceanic plate is subducting under a continental plate. The Shishaldin data follow a tholeiite trend (solid line), and the Shasta data follow a calc-alkaline trend.

Whether tholeiitic or calc-alkaline, originally mafic magmas can produce rocks ranging from basalt to rhyolite, as the two examples in Figure 5.45 show. In both cases, more primitive parental magmas were basaltic and the later evolved magmas were rhyolitic. They differ, however, because tholeiitic magmas become iron-rich as they evolve, moving toward the F apex of the triangle. Calc-alkaline trends go directly from basalt to rhyolite.

The trends on an AFM diagram reveal clues about the environments in which the magmas differentiated. The difference between calc-alkaline and tholeiite trends is due to the oxidation state of iron. If iron is mostly oxidized, magnetite (Fe_3O_4), a mineral that contains oxidized iron (Fe^{3+}), crystallizes early from a melt. If the iron is mostly reduced (not oxidized, existing as Fe^{2+}), magnetite does not crystallize. In calc-alkaline magmas,

Figure 5.46 Ancient ultramafic volcanic rock at Pyke Hill in the Abitibi Belt of western Ontario.
Photo from L. Montesi, University of Maryland.

rocks is probably due to the cooling of our planet. The young Earth was much hotter than today's Earth, because of heat produced by radioactive decay and more residual heat left over from the original accretion. Because ultramafic magmas crystallize at high temperatures, a cooler Earth means that they are less likely to make it to the surface today, compared with Earth's early days. However, within Earth, where temperatures are still high, ultramafic plutonic rocks are common. They comprise most of the mantle but only reach the surface as tectonic fragments caught up in mountain building or as much smaller xenoliths carried up by magmas.

5.9.2 Mafic magmas

In contrast with ultramafic magmas, mafic magmas are common, especially at plate margins. Much basaltic (mafic) magma is generated by partial melting of the mantle beneath mid-ocean ridges, due to decompression melting. Some of the mafic melts erupt on the ocean floor as basalt, but most cool underground to form mafic plutonic rocks (gabbros). Small amounts of basalt are also occasionally produced in subduction zones when water released by subducting slabs causes flux melting of overlying mantle.

Basaltic melts also form away from plate margins. Mafic volcanoes and, often, gabbroic (mafic) dikes are commonly associated with continental rifts or with hot spots. The basalt flows shown in Figures 5.11 (top), 5.26, and 5.33 were produced by a hot spot beneath the island of Hawaii. Mafic intrusive rocks are less common, but Figure 5.47 shows a 723-million-year-old

dike, one of many similar Franklin Dikes that extend thousands of kilometers across Baffin Island and Arctic Canada. Dikes of this sort are thought to be related to a hot spot originating in the mantle.

5.9.3 Intermediate magmas

Andesite is most typically associated with subduction zones, but the exact origin of intermediate composition magmas is debated today. Perhaps the debate is because there is more than one way such magmas can form. In some places, intermediate composition magmas appear to form by partial melting of mafic rocks during subduction. In other places, it seems that andesitic magma is what is left over after a basaltic magma has partially crystallized. It also is likely that mixing of magmas, crustal melting, and crustal assimilation play important roles in the production of intermediate magmas.

Figure 5.48 shows andesite flows on Bagana Volcano. The most recent flows are dark colored compared with older flows, and both contrast with surrounding green jungle. Bagana is one of seven active volcanoes on Papua New Guinea's Bougainville Island northeast of Australia. The volcano has erupted multiple times during the last 175 years, with the last major eruptions in 1950, 1952, and 1966. More or less continuous eruptions have been occurring since the early 1970s. Bagana's andesitic lava flows are sometimes as thick as 150 meters (490 feet).

5.9.4 Felsic magmas

Like intermediate magmas, the origins of felsic magmas are somewhat debated today. If partial melting of

Figure 5.47   Olivine gabbro dike, northwestern Baffin Island.
Photo from M. Beauregard, Wikimedia Commons.

Figure 5.48   Bagana Volcano, Bougainville, Papua New Guinea.
Photograph from NASA.

ultramafic and mafic mantle and lower crust are the source of felsic melts, there must be much leftover more mafic material in the source regions. Yet, evidence for the existence of those leftovers is skimpy at best. So, although huge felsic batholiths exist, it seems inconceivable that they could be created by partial melting of highly mafic, or ultramafic, parent material. The general consensus today is that, although getting felsic magmas from anatexis of mantle or lower crust is possible, many felsic melts come from melting in the crust when sedimentary and metasedimentary rocks melt or partially melt. As occurred beneath Yellowstone, the heat needed to cause the melting may be carried into the crust by rising mafic magmas. This process has been shown to occur at hot spots, in rift zones, and in the roots of mountain chains. It may occur elsewhere on a smaller scale.

Figure 5.49 shows Mt. St. Helens, a volcano of the Cascade Range, Washington, in 1982. A small rhyolitic dome, emitting steam, is developing in the center of the summit crater. Investigations by the United States Geological Survey have concluded that St. Helens magmas originate by flux melting in the mantle, above the subducting Juan de Fuca Plate. The magma rises and collects near the base of the crust temporarily before moving up to shallower storage chambers and, ultimately, erupting. The drawing on the right shows how magma first is generated just above the subducting slab,

then accumulates near the base of the crust, and finally erupts at the surface. Temperature contour lines show that magma generation occurs at around 1200 °C.

5.10 A closer look at magma chemistry

5.10.1 Major and minor elements

Major elements, typically present at levels exceeding 1 wt%, determine most magma properties and eventually the minerals that may form. Typical major elements in igneous rocks include O, Si, Al, Fe, Ca, Na, and K. In some rocks, Ti, Mn, P, and perhaps others may be considered major elements. Rocks also contain *minor elements*. Minor elements, typically comprising 0.1 to 1 wt% of a magma or rock, substitute for major elements in minerals but are selective about which minerals they enter. For the most part, they do not affect magma properties.

Figure 5.50 shows some key elements in igneous rocks. Major and minor elements are shaded gray. The other shaded elements are *incompatible elements* in mafic minerals (discussed in detail below). The incompatible elements include large ion lithophile elements (LILE) shaded orange, heavy rare earth and related elements shaded dark green, light rare earth elements shaded light green, and high field strength elements (HFSE)

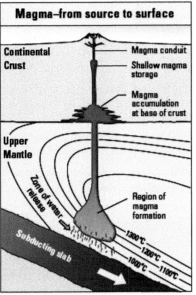

Figure 5.49   (Left) A 1982 photo of a rhyolite dome near the summit crater of Mt. St. Helens, Washington. (Right) Schematic diagram showing the origin of Mt. St. Helens magmas.

Photo and drawing from the USGS, Wikimedia Commons.

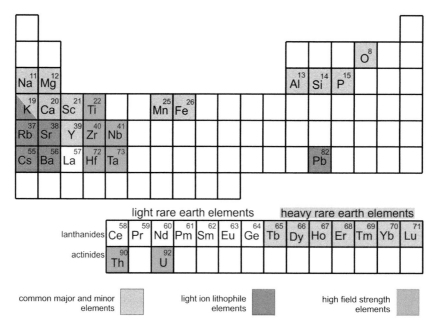

Figure 5.50   Periodic chart showing some key elements in igneous rocks.

shaded blue. Potassium (K) is considered both a major element and an LILE element.

Trace elements are elements present in very small amounts, amounts even smaller than amounts of minor elements. The amount of any trace element that can enter a growing crystal depends mostly on ionic charge and radius. Some trace elements enter growing crystals in the early stages of crystallization, but others may remain in a magma until the latest stages of crystallization. Trace elements are even more selective than minor elements about the minerals they enter and generally have insignificant effects on rock and mineral properties. Because trace elements are present in very small amounts, petrologists commonly report them in parts per million (ppm) or parts per billion (ppb) instead of weight %. 1 ppm = 0.0001 wt%. Furthermore, when plotting trace element analyses, petrologists normalize the raw data by dividing by the composition of some reference standard. Normalization means that the range of values becomes small enough so that we may plot all trace elements on a single graph.

Table 5.3 contains an analysis for a Hawaiian basalt. The analysis does not distinguish between major and minor elements (because the distinction between the two is a fuzzy one), but Mn and P together comprise only about 0.2 wt% of the rock and would be considered minor elements by most petrologists. K, too, might be considered a minor element in this rock. The analysts listed trace elements separately, reporting them in ppm

Table 5.3   Analysis of a hawaiian basalt

Major and minor elements		Trace elements	
Element	wt%	Element	ppm
O	44.66	V	292.00
Si	23.27	Cr	238.00
Fe	8.59	Zr	129.90
Ca	8.28	Sr	129.10
Al	7.20	Ni	102.60
Mg	4.27	Zn	85.40
Na	1.81	Cu	63.30
Ti	1.64	Y	40.60
K	0.46	Ba	18.40
Mn	0.13	Ce	8.30
P	0.11	Nb	4.30
		Rb	1.30
sum	100.42	sum	1113.20

Data are from Duncan, R. A., Backman, J., Peterson, L. C., *et al.*, 1990, Proceedings of the Ocean Drilling Program, Scientific Results, Vol. 115, pp. 1–11.

instead of weight %, and the total of the trace elements is about 1113 ppm, which is equivalent to 0.1113 wt%. Many other trace elements are undoubtedly present in the Hawaiian basalt but were not analyzed because they were not important to the study being conducted. Although Mn and P are minor elements in the Hawaiian basalt (and most other basalts), they may be concentrated in other kinds of rocks. The alkalis, major elements in felsic rocks, may be nearly absent in mafic rocks, although in the Hawaiian basalt they add to about 2.25 wt%. So, minor elements in some types of rocks can be major elements in others and vice versa.

We can classify and name igneous rocks based on the minerals they contain, but magmas, because they contain no minerals, must be classified in another way. Additionally, many volcanic rocks may contain glass instead of minerals or may be too fine grained for mineral identification. Thus, we often classify magmas and many igneous rocks based on their chemical composition instead of their mineralogy.

Although igneous rock chemistry varies in many ways, the silica content and alkali content of volcanic rocks form the basis of one of the most commonly used classification schemes. It is not that other compositional variations are unimportant, but many other possible variations correlate with alkali and silica content and so the classification system captures well the variation in rock compositions.

Using the *total alkalis versus silica* (TAS) system (Fig. 5.51) is straightforward, and the wt%s of silica (SiO_2) and alkali oxides ($Na_2O + K_2O$) in a rock are used to obtain a rock name. The vertical axis is the total alkali oxide content, and the horizontal axis is the silica content. In the TAS diagram, ultramafic compositions (low silica content) plot on the left and felsic compositions (high-silica content) on the right, with mafic and intermediate compositions between. The diagonal red line divides the diagram into two parts. Compositions that plot in the upper part of the diagram are *alkalic*; they are relatively rare in nature. Those plotting in the lower part of the diagram are termed *sub-alkalic* and are much more common. Sub-alkalic rocks include those that form the tholeiite and calc-alkalic series (Fig. 5.45). By far, the most common volcanic rocks are sub-alkalic: basalt, andesite, dacite, and rhyolite. Although the names in Figure 5.51 are names of volcanic rocks, they are often used to describe magma types. A dacite magma, for example, is one that could erupt to form a dacitic rock.

Although both silica and alkali content are keys when classifying magmas or volcanic rocks, silica content alone explains many variations in magma properties (Table 5.4). It is also the basis for a rock-naming scheme that is simpler than the TAS system is: simply calling rocks ultramafic, mafic, intermediate, or felsic. Variations in magma properties with silica content are profound. The more felsic a magma, the lower its eruption temperature but the greater the possibility for explosive eruptions due to high viscosity and gas content. Note that in this simpler classification scheme, alkali content correlates with silica content. This correlation is not always the case, as shown in the TAS diagram, but is the case for sub-alkalic rocks. Similar to the names in the TAS system, the rock names in Table 5.4 are often used to describe magma composition. For example, we might describe a magma as granitic if it could crystallize to form a granite.

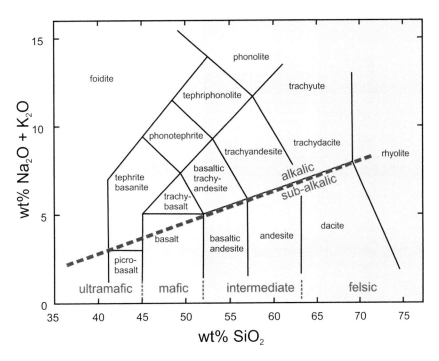

Figure 5.51 The total-alkali-silica (TAS) classification system.

Table 5.4   Properties of different kinds of magmas.

Magma type	Ultramafic	Mafic	Intermediate	Felsic
extrusive rock name	komatiite	basalt	andesite	rhyolite
intrusive rock name	peridotite	gabbro	diorite	granite
silica content (wt% SiO_2)	40–45 ⟶			71–75
mafic content (wt% FeO + MgO)	high ⟶			low
ratio alkali earth/alkali (wt% $CaO/Na_2O + K_2O$)	high ⟶			low
eruption temperature	up to 1500 °C ⟶			800 °C or less
viscosity	low ⟶			high
volatile (water, CO_2, etc.) content	low ⟶			high

5.10.2   Incompatible and compatible elements

We can divide elements in igneous rocks and magmas into two groups: those that tend to remain in a magma until the later stages of crystallization (and consequently become enriched in the magma as crystallization takes place), and those that are easily incorporated into early growing crystals (and consequently become depleted in a magma quickly). Elements that tend to remain in the magma are said to be *incompatible*, and those that enter crystals quickly are *compatible*. Petrologists use these terms, incompatible and compatible, most commonly to describe trace elements, but they apply equally well to major and minor elements. Some elements behave as compatible elements in some magma types but as incompatible in others, because the specific minerals that crystallize vary with magma composition. Trace elements, however, both compatible and incompatible, are especially useful as trackers of magma evolution. Incompatible elements include the rare earth elements, elements 57 through 71 (La–Lu), but their degree of incompatibility varies with atomic number. The rare earths and other incompatible elements are highlighted in the periodic chart of Figure 5.50.

As seen in Figure 5.52, incompatible elements fall into two main groups, a group that has large ionic radius and a group that has large ionic charge. The first group includes alkali and alkali earth elements, notably K, Rb, Cs, Sr, Ba, and several other elements. Elements that tend to concentrate in Earth's crust and mantle are called *lithophiles*. So, the alkalis and alkali earths, and elements with similar properties, are collectively termed *large ion lithophile elements*, or LILE for short (Fig. 5.50). LILEs do not fit into crystallographic sites in

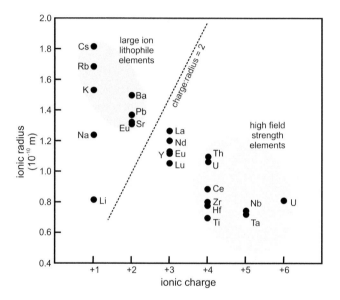

Figure 5.52   The two groups of incompatible elements.

most minerals, and the charge attracting them to growing crystals is small, so they tend to remain in a melt.

Elements that form ions with small radii and high charge are called *high field strength elements* (HFSE) (Fig. 5.50). This group of incompatible elements includes Zr, Nb, Hf, Th, U, and Ta. HFSEs have high ionic charge (+4 or greater), which means that, if they enter crystals, the charge balance is hard to obtain. Consequently, HFSEs tend to remain in a melt. For some purposes, HFSE are defined as those whose ions have a charge to radius ratio greater than 2 (to the right of the dashed line in Fig. 5.52).

Figure 5.53 compares the trace element content of an oceanic island basalt with that of a mid-ocean ridge basalt. In diagrams of this sort—sometimes called *spider diagrams*—the most incompatible elements are on

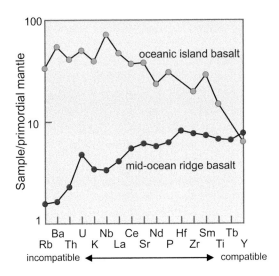

Figure 5.53 Comparing the trace element chemistry of an oceanic island basalt with a mid-ocean ridge basalt.

Data from Rollinson (1993) Using Geochemical Data: Evaluation, Presentation, Interpretation

the left, and elements are increasingly less incompatible moving to the right. The mid-ocean ridge basalt is depleted in incompatible elements compared with the oceanic island basalt. However, both have about the same concentrations of the compatible elements. This diagram, therefore, suggests that mid-ocean ridge magmas, associated with active plate spreading centers, derive from regions that have undergone significant amounts of partial melting. In contrast, the oceanic island magmas come from regions that are more primitive.

Evidence of this sort allows petrologists to conclude that the upper oceanic mantle—the source region for mid-ocean ridge basalts—is an area of active melting and recycling of material. The source of oceanic island basalts, such as those that reach the surface in Hawaii, is deeper in the mantle where melting has not removed incompatible elements. Note that the data for both basalts were normalized by dividing the analyses by an estimated composition for the primordial mantle (to keep numbers on scale). Additionally, the vertical scale is a log scale; if it were not, the trends would not be as easily seen.

Some transition elements—including nickel, cobalt, chromium, and scandium—are also important trace elements. They have small radii and (usually) +2 or +3 ionic charge, and they are incorporated into mafic minerals during the earliest stages of crystallization and tend to remain there. So, they are compatible when melting occurs in the mantle and, consequently, melts from the mantle contain them in very low amounts. They are

often used as markers that determine where in Earth a magma originated.

The concept of compatible versus incompatible elements depends on rock type, because different rocks contain different minerals that incorporate elements in different ways. Scandium may enter pyroxenes, but not enter olivine. Zirconium is easily accommodated in zircon. Phosphorus concentrates in apatite, but neither zirconium nor phosphorus go into olivine. Earth's mantle is primarily composed of olivine and pyroxene, and these minerals become enriched in scandium, nickel, titanium, chromium, and cobalt as fractional melting occurs. So, melts derived from the upper mantle are enriched in these elements, and the amount of melting that has occurred can be estimated based on trace element abundance.

5.10.3 Rare earth elements

The *lanthanide elements*, also called the *rare earth elements* (REE), having atomic numbers 57 (La) to 72 (Lu), are very important trace elements with high field strength (Fig. 5.50). Elements 57 through 64, La through Gd, are considered *light REEs*; 65–71, Tb through Lu, are *heavy REEs*. Y and Sc are sometimes grouped with the heavy REEs due to similar properties. The REEs all have similar properties: large ionic radius and, except for Eu, they are trivalent (+3). Ionic radius decreases with increasing atomic number, so La and other light REEs (largest) are more incompatible than Lu and other heavy REEs (smallest) are. If garnet crystallizes, it incorporates heavy REEs (Ho-Lu) easily, making them especially compatible.

During fractional crystallization of a magma in the mantle or the crust, incompatible elements stay preferentially in melts and so tend to move up and concentrate in Earth's outer layers. The original source region is *fertile* if there has been little removal of incompatible elements, but fertile rocks become *deplete*d rocks when melting occurs. Rocks or magmas that are rich, or only slightly depleted, in light rare earth elements are fertile, and those with strong depletions in LREE are depleted, because their chemistry indicates the degree to which they have melted. This is why trace elements are powerful indicators of magma origin. Magmas that originate in the upper levels of Earth are richer in incompatible elements than those that come from pristine mantle sources. Additionally, petrologists distinguish both fertile and depleted magma source regions within the mantle.

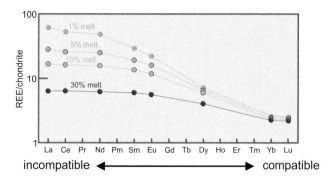

Figure 5.54   Calculated rare earth concentrations for melts derived from a garnet peridotite.

When a rock melts, the incompatible elements enter the melt quickly. Consequently, the concentration of incompatible elements will be very high after just a little bit of melting. With more melting, the concentrations decrease, because most incompatible elements are already in the melt and are diluted with further melting. This effect is especially true for elements present in rocks and magma in very small amounts, which explains why trace elements are such powerful indicators of magma origin. So, trace elements tell us how much melting has occurred to produce a magma and, although a bit more complicated, they also monitor how much crystal fraction occurs when a magma cools.

Figure 5.54 shows rare earth patterns for melts derived from a garnet peridotite. If only a small amount of melting occurs, concentrations of incompatible elements are high because they enter the melt first. With increased melting, other elements enter the melt and the concentrations of incompatible elements go down. In this diagram, the compositions were normalized by dividing by the standard composition of a chondritic meteorite, and the scale is a log scale.

Questions for thought—chapter 5

1. Why do basalt flows often have columnar jointing?
2. What causes bimodal volcanism? What are the sources of the two types of magmas?
3. Why is obsidian generally absent from intrusive igneous rocks?
4. The mantle is a long ways away. But we do have some samples to study. How do those samples make it to Earth's surface?

5. What is the difference between a magma and a lava? What determines whether molten rock will become lava?
6. Compare the chemical analyses of common igneous rocks in Table 1, which rock type has the highest concentration of silica? Which has the highest concentration of calcium? Magnesium? Sodium? Which elemental concentrations are most variable overall, and which are least variable?
7. Which are hotter: felsic or mafic magmas? So which ones do you think are more likely to make it to Earth's surface to create volcanic rocks?
8. What is the source of mafic magmas?
9. If basalt partially melts, would the resulting melt be more or less rich in silica than the unmelted basalt?
10. What is decompression melting? Why is it important in spreading ridges?
11. All other conditions being equal, will the presence of water raise or lower the melting point of a basalt? What about carbon dioxide? What effect will it have?
12. How do magmas form in subduction zones?
13. Why do magmas rise?
14. Why do some magmas stop deep in the Earth and solidify into intrusive rocks, and never erupt on the Earth's surface as lavas or other volcanic materials?
15. Igneous rocks may be aphanitic, porphyritic, or phaneritic. Are intrusive or extrusive igneous rocks most likely to be aphanitic? Explain why? Name one example of an aphanitic rock and one example of a phaneritic rock. How do porphyritic rocks form? Name one example. (This last is sort of a trick question.)
16. All other things held equal, why are batholiths more likely to be felsic than mafic?
17. Describe how K-feldspar melts. Why is it an example of incongruent melting?
18. Is a mafic igneous rock more likely to contain plagioclase that is rich in anorthite or rich in albite? Why?
19. What is the difference between the liquidus and solidus temperatures of a mineral? Would you expect a mineral that melts congruently to have different liquidus and solidus temperatures? Why or why not?
20. Which of the minerals listed in Bowen's Reaction Series are most common in mafic igneous rocks?

In intermediate igneous rocks? In felsic igneous rocks? Why is a single magma unlikely to produce the entire series?

21. Contrast and compare the composition of Earth's crust and the mantle. How was fractional melting over billions of years an important factor in producing the Earth's crust and giving it a composition different from the mantle?

22. Why are Harker diagrams useful?

23. What are AFM diagrams and how can they be useful in identifying the type of subduction zone associated with a group of volcanic rocks?

24. What are komatiites and why are they not erupting today?

25. Pegmatites are usually some of the last igneous rocks to crystallize out of a magma. Would you expect them to be relatively enriched in incompatible elements or not? Explain why.

6 Plutonic Rocks

6.1 Yosemite Valley

The first view of Yosemite Valley is a surprise. Whether you are driving from Big Oak Flats or El Portal, or descending on the Wawona Road, there comes a time when you round a corner and the valley suddenly appears before you. The view from the eastern end of the *Wawona Tunnel*, the longest highway tunnel in California, is especially remarkable. The photograph in Figure 6.1 was taken by Ansel Adams in 1938 and is one of the most famous of his career. It shows the deeply carved valley with rock walls on both sides. On the left, four or five miles away, is the prominent, near-vertical west face of *El Capitan*. Beyond the face, El Capitan's steep buttress, aptly named the *Nose*, juts out and steepens slightly at the top, and so it appears to soar upwards to its brow as it rises slightly more than 914 meters (3000 feet) from the valley floor. Beyond El Capitan, the view continues eastward to the higher Sierra Nevada Mountains. In the far distance, Half Dome, with a prominent black water streak on its north face, is silhouetted on the

Figure 6.1 Yosemite Valley, California.
Photo courtesy of Dan Anderson, Yosemite Online Library.

south side of the valley. Half Dome's summit is 1463 meters (4800 feet) above the valley floor.

Opposite El Capitan, on the south side of the valley, *Bridalveil Fall* free falls 188 meters (617 feet) between the Leaning Tower and the Cathedral Spires. The valley contains many waterfalls, and most, like Bridalveil, are fed by streams draining *hanging valleys* that enter the main valley high above its floor. Bridalveil's prominent location makes it the first waterfall seen by most valley visitors. Yet, the more famous *Yosemite Falls*, a few miles farther up the valley, is the highest waterfall in North America and one of the highest in the world. Its upper portion drops slightly more than 427 meters (1400 feet), and the entire drop, involving several falls, is 739 meters (2425 feet). Other tall Yosemite waterfalls include Ribbon Fall (491 meters; 1612 feet) and Horsetail Fall (640 meters; 2100 feet), both seasonal falls, which descend from the western and eastern ends of El Capitan, and another seasonal fall—Sentinel Fall (590 meters; 1920 feet)—which is on the opposite side of the valley a few miles farther east.

Yosemite Valley is cut into hard plutonic rocks of the Sierra Nevada Mountains. The valley floor lies at about 1200 meters (4000 feet), and cliffs rise several thousand feet on either side. Yosemite is a small valley. The main part, the part considered by many to be most beautiful, is 11–16 kilometers (7–10 miles) long and about a mile wide. Near its upper end, it splits—the main valley trends east and Little Yosemite Valley extends several miles south before it too turns east and wraps around the southern side of Half Dome.

El Capitan is mostly composed of light gray granite or rocks closely related to granite. Collectively termed *granitic*, or sometimes *granitoid*, these light-colored rocks are composed chiefly of feldspars and quartz. But, if you focus in, you can see different colors of rock exposed in El Capitan's face, suggesting that there may be significant compositional variation. In places, a much darker rock, dark gray to black, is seen in blobs and splotches. This rock becomes especially prominent as you drive farther into the Valley, around the Nose, and get a view of El Capitan's southeast face (seen in Fig. 6.2), where a large black body of rock is exposed in the *North American Wall*, so named because the body of rock is shaped like the continent. On both the southwest and southeast faces, a myriad of white to light gray dikes—small veins up to several meters thick—cut across the face in near-horizontal directions.

The plutonic rocks that make up the Yosemite region all formed between 114 and 87 million years ago, when rising magmas intruded older *country rocks*. Remnants of the older host rocks can be found both east and west of Yosemite today. In this geologic map (Fig. 6.3) of the Yosemite region, granitic rocks of the Sierra Nevada Batholith are shown in light pink. Younger sediments and volcanic rocks are in yellow and hot pink to the north and east of Yosemite National Park. Old metamorphic rocks—the rocks that the granitic magmas intruded—are in green, mostly west of the national park where the mountains give way to California's Central Valley.

The intrusions marked the beginning of the formation of the Sierra Nevada Mountains. If there were associated volcanic rocks that formed at that time, they have long since eroded away. The Yosemite plutons were initially deeply buried (several tens of kilometers) beneath

tall mountain ranges, but by about 70 million years ago, uplift and erosion had removed most of the overlying mountains. River valleys formed, and as erosion continued, they deepened and cut downward. So, plutonic rocks were exposed, initially only in valleys and later over wider areas. The region was relatively static for a long time, but about 25 million years ago, renewed uplift and tilting began the process that led to the present-day Sierras, and streams cut even deeper, exposing more into plutonic rocks.

When plutons solidify at depth, they are under great *lithostatic* (same in all directions) pressure. During uplifting, the rocks are subjected to various kinds of *directional stress* as, simultaneously, pressure decreases and rocks expand. Consequently, once solid rocks may develop *shear zones* (zones of intense strain, associated with faulting or elastic deformation) and *joints* (fractures that lack any movement parallel to the plane of the fracture). The shear zones and joints may become focuses for weathering and erosion, often leading to rock falls. These processes began the formation of the steep outcrops we see in Yosemite today.

One result of decreased pressure is *exfoliation*. As uplifted rocks expand, joints develop parallel to rock surfaces. Layers can break off in the same way that layers may be peeled from an onion. Slabs and sheets up to meters thick can separate from the main rock body and either stay in place or fall away. The photo in Figure 6.4 shows the summit of Half Dome in Yosemite Valley, California. The exfoliating slabs of rock are clear

Figure 6.2   The Nose and southeast face of El Capitan.
Photo from the National Park Service.

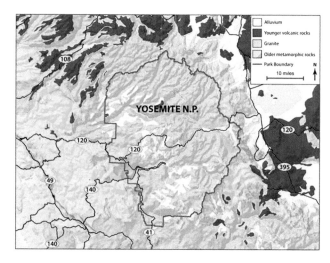

Figure 6.3   Geological map of Yosemite National Park.
Map from the National Park Service.

Figure 6.4 The summit of Half Dome in Yosemite Valley, California.

Photo from Ronnie McDonald, Wikimedia Commons.

Figure 6.5 Aerial view of Yosemite Valley and Half Dome on the right.

Photo from Tuxyso, Wikimedia Commons.

to see. For scale, look closely and you will see a parade of people going up a cabled walkway just to the right of the large pine tree. While granites and related plutonic rocks seem to provide the best examples of exfoliation, plutonic rocks of other sorts may exhibit the same features, too.

Figure 6.5 shows Yosemite Valley with peaks of the high Sierra Nevada Mountains in the background. Exfoliation created the prominent Royal Arches in the center of the photo. And, over time, exfoliation has led to the rounding of many rocky peaks, including Half Dome, the highest peak that can be seen in this photo. But, the ultimate sculptor of Yosemite National Park was ice. The region has seen extensive glaciation during the past 30 million years. The *Sherwin glaciation*, which occurred about 1 million years ago, added the final polishes. At that time, mountain icefields formed at higher elevations and flowing glaciers naturally occupied the river valleys already present. Moving ice rounded and polished mountaintops, deepened valleys, rounded and smoothed valley walls, and plucked pieces of rock that were later dropped far from their sources. Consequently, valleys were widened and obtained their present-day flat bottoms and U-shapes as shown in Figure 6.5. During this process, tributary valleys were cut off to produce hanging valleys far above the main valley floors, such as the one responsible for Sentinel Fall, shown in Figure 6.1.

6.2 Plutons of different kinds

Many names have been used to describe intrusive rock bodies but, fortunately, most see little use today. *Pluton* is a general term for any such bodies, and plutons come in many shapes and sizes. Some are depicted in Figure 6.6, Although the term *pluton* could be used to describe any of these different kinds of intrusive rocks, in practice, geologists use the word pluton mostly only to refer to blobby or irregular bodies that do not have a thin dimension like dikes and sills do.

We call large plutons *batholiths* (from the Greek *bathos* meaning depth and *lithos* meaning rock) if they are larger than 100 square kilometers in area. Figures 6.1 through 6.5 show views of small parts of the Sierra Nevada Batholith. Batholiths are the largest plutonic bodies but are generally compound (made of many individual plutons). All of the world's batholiths have an overall felsic (granitic) or, less commonly, intermediate composition. *Lopoliths*, also seen in Figure 6.6, ae dish-shaped or funnel-shaped plutons. They can be huge and are typically layered and formed, like batholiths, from multiple magmatic intrusions. Most lopoliths have an overall mafic or ultramafic composition. Figure 6.6 also includes a laccolith, dikes, stocks, and a volcanic neck.

Figure 6.7 shows examples of some different kinds of plutonic bodies. Figure 6.7a is a view of Stone

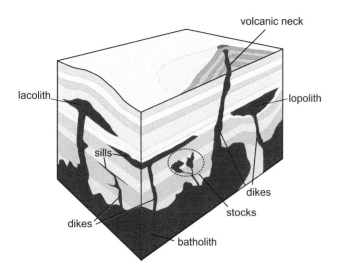

Figure 6.6   Different kinds of plutonic rock bodies.

Mountain, a small granitic pluton in western North Carolina. The pluton's overall shape is somewhat like a football but only one side can be seen. Figure 6.7b shows several individual *stocks* in the Black Hills, South Dakota. They intruded as fingers into surrounding country rock that eroded long ago and left sharp pointed spires behind.

Sills, *laccoliths*, and *dikes* range in size from small to very large and have sheet-like geometries, sometimes described as *tabula*r (Figs. 6.7c, d, e). They are relatively thin in one dimension but may be extensive in the other two. Sills (Fig. 6.7e) and laccoliths *conform* with (are bounded by) sedimentary layering, less commonly by metamorphic or igneous layering, in the country rock they intrude. They push the layers apart as they intrude. The difference between sills and laccoliths is that sills have parallel sides, and laccoliths bulge upward and so are shaped like mounds (Fig. 6.6).

Dikes are *nonconformable* sheet-like intrusions that may cut across sedimentary or other layering or through a massive rock that has no layering (Figs. 6.7c, d). Dikes generally form from magma intruding along steep or near-vertical fractures, but tectonic forces may later rotate them to different orientations. The term *vein* (Fig. 6.7f) is used by some as a synonym for dike, but most geologists reserve the word *vein* for small, irregular and branching, and sometimes discontinuous, intrusions. In contrast, dikes are larger, sometime huge, and have uniform thickness and parallel sides. Dikes and sills range from felsic to mafic compositions, but around the world mafic dikes are the most common. A special kind of dike, called a *feeder dike*, carries

magma to the surface, producing volcanic eruptions. Non-feeder dikes stall before they get there. When one observes a dike in outcrop, it is often difficult to determine whether it was a feeder dike if all overlying volcanics have been eroded. However, remnants of feeder dikes are sometimes preserved as *volcanic necks* (see Fig. 6.6), frozen magmas that once fed a volcano but now have become exposed by the removal of surrounding country rocks.

6.3   Minerals in igneous rocks

6.3.1   *Primary minerals*

We generally classify and name igneous rocks based on some combination of rock mineralogy, texture, overall composition, and occurrence. Because plutonic rocks are generally coarse grained, and because most have similar textures, the common naming systems for these rocks rely primarily on mineralogy. Although the minerals present can often be identified quickly in outcrop, perhaps with the aid of a hand lens, sometimes examining thin sections is required.

The minerals created when a rock initially forms from a crystallizing magma are the rock's *primary minerals*. Primary minerals contrast with *secondary minerals* that form later due to weathering or other forms of alteration when the rock interacts with the environment. Usually, primary minerals do not crystallize all at once, but instead form sequentially, one or more at a time, in a series such as the one described by Bowen's Reaction Series (discussed in Chapter 5). Evidence for sequential crystallization comes from the presence of phenocrysts that crystalized over much longer times than surrounding groundmass (see Fig. 5.30, Chapter 5, for an example). Additional evidence is provided by rock textures in which early euhedral or anhedral crystals are surrounded by a different mineral that filled in between the early-formed grains.

Figure 6.8 is a photograph of a thin section of gabbro. It was taken using crossed polarizing filters, so the colors seen are not true mineral colors. The gray striped euhedral to subhedral grains are plagioclase, and the mostly blue mineral that has filled in between the plagioclase crystals is anhedral augite. The texture indicates that plagioclase crystallized first and the *interstitial* (between grains) augite formed later. This is a texture common to most gabbros and to some rocks of other compositions.

a. Stone Mountain, North Carolina, is one of a number of plutons formed from magma that intruded into older metamorphic rocks of the Blue Ridge Mountains about 400 million years ago. It is composed of diorite and granodiorite and developed during several nearly contemporaneous intrusive events.

b. The Needles in the Black Hills, South Dakota, are a collection of relatively small grantitc stocks. The schist that once surrounded them has eroded away. Note the person sitting on a rock in the center of the photograph for scale.

d. This is a common sight in Precambrian shield areas. A diabase dike (about 1,100 million years old) is cutting through 1800 million year old gneisses in Kosterhavet National Park, Sweden.

c. There are three intrusions in the photo from Kosterhavet National Park in Sweden. Most of the rock is a dark colored mafic rock of gabbro composition. It has been cut by a light pinkish pegmatite dike. A later, smaller mafic intrusion cut through the petmatite.

f. Quartz veins intruding Precambrian Rock in Smith Water Canyon, Joshua Tree National Park, California. These kinds of veins are common in all igneous and metamorphic terranes.

e. The dark layer shown here is a mafic sill (probably diabase) in the Beacon Valley, near the Taylor Glacier in Antarctica.

Figure 6.7 Examples of intrusive rock bodies.

Photos from Djgazso (a), Doug Knuth (b), Thomas Eliasson (c, d), and Rob Hannawacker (f)—all at Wikimedia Commons; and from Callan Bentley (e).

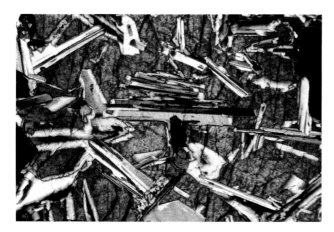

Figure 6.8   Gabbro in thin section.
Photo from Alex Strekeisen.

The number of common primary minerals in igneous rocks is limited (Table 6.1). No matter the overall rock composition, the dominant minerals are almost always some combination of quartz, feldspar, mica, amphibole, pyroxene, or olivine. However, two kinds of feldspars, two major kinds of micas, several possible amphiboles, and several possible pyroxenes may be present. Additionally, these minerals, except quartz, can have variable compositions that reflect magma composition. In all rocks, the major minerals are accompanied by some amount of less abundant accessory minerals such as the ones listed in the last row of Table 6.1.

6.3.2   Secondary minerals

Secondary minerals include many alteration products of primary minerals. In felsic and intermediate rocks, for example, feldspars and micas commonly alter partially or wholly to epidote, chlorite, or clays. Many secondary minerals are hydrous, forming by reaction of primary minerals with water flowing through a rock. They may form by *in situ* alteration of an original mineral grain or in secondary veins. The photo in Figure 6.9 shows the Ruin Granite from near Globe, Arizona. The rock contains quartz (gray, but hard to see), orthoclase (pinkish), and plagioclase (white). Individual plagioclase grains are altering to light green secondary epidote; additional epidote is seen in secondary veins. Besides epidote and clays, other common hydrous minerals that form from feldspars include minerals of the zeolite group.

Many ferromagnesian rocks contain minerals that are unstable at Earth's surface. In these rocks, mafic minerals commonly alter partly or wholly when exposed to water and air. Typical alteration products include chlorite and serpentine, both hydrous minerals, and oxides including magnetite and hematite. Figure 6.10 shows an altered rock called *serpentinite* from the Maurienne Valley in the French Alps. The primary minerals in this rock—mostly olivine and augite—have been completely replaced by secondary minerals, including dark green and gray serpentine and iron oxides. Besides serpentine and oxides, other common secondary minerals in ferromagnesian rocks include chlorite, carbonates, and hydroxides. Ultramafic rocks, in particular, alter and weather rapidly. Some of the world's largest ultramafic bodies, which may appear unaltered when viewed in outcrop, are made nearly entirely of serpentine.

If you look up the definitions of different kinds of igneous rocks, often you will find diagrams like the one shown in Figure 6.11 that shows the relative amounts (termed the *modes*) of the minerals present in different rock types. There are many variations of modal diagrams for igneous rocks, but Figure 6.11 is typical—it includes and compares the minerals in five of the most common plutonic rocks, and potential minerals in five of the most common volcanic rocks. The relative amounts of different minerals, generally expressed as wt%s, are the rocks' mineral modes.

Modal diagrams of this sort are generalizations. For example, the diagram suggests that—besides quartz and K-feldspar—granites contain plagioclase, muscovite, biotite, and amphibole, but many granites lack one or several of these minerals. So, modal diagrams are useful generalizations but incorrect in detail because mineral assemblages are variable for a given kind of rock. Additionally, minor minerals have been omitted from the diagram and—of great significance—many volcanic rocks contain substantial glass and sometimes no mineral crystals at all.

6.3.3   Essential and accessory minerals

Petrologists often distinguish two fundamental kinds of primary minerals in rocks. *Essential minerals* are those that must be present for a rock to have the name that it does. *Accessory minerals* include any other minerals that may or may not be present. The modal diagram in Figure 6.11 includes both essential and accessory minerals. For example, all granite—by definition—contains K-feldspar and quartz, but it need not contain plagioclase, muscovite, biotite, or amphibole. Gabbro always contains pyroxene and plagioclase and may also contain olivine or amphibole. The notion of essential

Table 6.1   Common minerals in igneous rocks.

Mineral	Typical composition and occurrences in igneous rocks
Quartz	Quartz is the only common primary SiO_2 polymorph. Tridymite or cristobalite—other SiO_2 minerals—may also be present but are generally secondary minerals. Quartz is restricted to felsic rocks.
K-feldspar	Most alkali feldspar is potassium-rich feldspar containing variable amounts of sodium. K-feldspar is restricted to felsic rocks.
Plagioclase	Plagioclase ranges from albite (Na-rich) in some felsic rocks to anorthite (Ca-rich) in mafic rocks.
Muscovite	Muscovite is restricted to felsic (rarely intermediate) rocks, and its composition is generally close to its end-member formula.
Biotite	Biotite occurs most commonly in felsic rocks, less commonly in some intermediate rocks, and rarely in any others. Composition varies primarily between biotite's Mg- and Fe-end members.
Amphibole	Amphibole varies widely in composition. It is found most often in intermediate rocks but may be present in rocks of other compositions. By far, the most common amphibole is hornblende, but several other varieties may be present, especially in alkalic rocks.
Pyroxene	Both clinopyroxene (mainly augite) and orthopyroxene may be present in mafic rocks, less commonly in intermediate rocks, and rarely in felsic rocks. In alkalic rocks, several different Na-rich pyroxenes are possible.
Olivine	Olivine's composition varies primarily between its Mg- and Fe-end members. It is generally only found in mafic or ultramafic rocks.
Accessory minerals	Besides the minerals listed above, these are also common as accessories: titanite, zircon, epidote, rutile, topaz, corundum, fluorite, magnetite, ilmenite, allanite, garnet, chromite, spinel, monazite, and tourmaline.

Figure 6.9   The Ruin Granite, Globe, Arizona.
Photo from Wayne Ranney, Earthly Musings Blog.

Figure 6.10   Serpentinite from the Maurienne Valley, Savoie, France.
Gabriel HM, Wikimedia Commons.

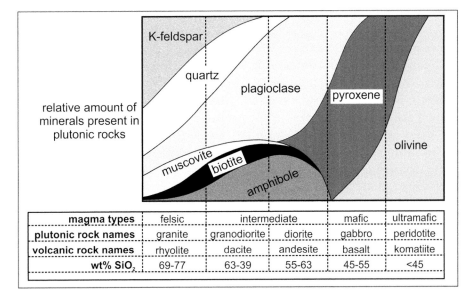

Figure 6.11   Schematic comparison of the mineral contents of different kinds of igneous rocks.

minerals, however, depends in part on the classification system used. Many systems have been proposed for naming rocks, and they do not all agree. For plutonic rocks, because they are generally phaneritic and contain no glass, essential minerals are *essential*—they must be present for rocks to have the name they do, and if present they give the rocks their names. The notion of essential minerals is less applicable to volcanic rocks because they may contain glass instead of minerals or the minerals present may be too fine grained to be readily identified.

Most plutonic rocks contain accessory minerals that may be easily overlooked if present in very small amounts. Any of the minerals listed in Table 6.1 can be accessory minerals if they are in a rock but are not essential to the rock name. If present in significant amounts, but not essential to the rock name, accessory minerals can be considered *characterizing accessory minerals*, also called *varietal minerals*. Typical characterizing accessories include biotite, muscovite, amphibole, pyroxene, olivine, and glass. Their presence or absence can be used to modify a rock name. For example, an *olivine gabbro* is a particular type of gabbro that contains olivine as an accessory mineral, and a *hornblende granite* is one that contains hornblende besides essential K-feldspar and quartz.

6.4   Different kinds of plutonic rocks

6.4.1   The IUGS system for naming plutonic rocks

Most popular websites and books describe Yosemite Valley, and the nearby Sierra Nevada Mountains, as being made of *granite*. This word is often used loosely to refer to any light-colored, coarsely crystalline igneous rock, but the term has a more restricted definition to geologists. Much of the rock in the valley walls does not fit a geologist's definition of granite, and the rocks are not the same everywhere. Many geologists duck the semantic problem and refer to Yosemite rocks as *granitic*, or as being *granitoids*. These terms are generally understood to refer to any light-colored plutonic rock composed primarily of quartz and feldspar (either alkali feldspar or plagioclase), although occasionally there is debate about whether the terms can be used to refer to rocks such as *syenite* that contain minimal quartz.

The older geological literature contains many names for igneous rocks (plutonic and volcanic), and often the

names have been used in inconsistent ways. In 1991, the number of names exceeded 1500. To resolve inconsistencies, the International Union of Geological Sciences (*IUGS*) developed a standardized classification scheme and nomenclature. The basic IUGS classification system for common kinds of plutonic rocks, shown in Figure 6.12, is based on the relative amounts of quartz (Q), alkali feldspar (A), and plagioclase (P) (top triangle) or on the relative amounts of feldspathoids (F), alkali feldspar (A), and plagioclase (P) (upside-down bottom triangle) that a rock contains. The wt%s of quartz, alkali feldspar, plagioclase, and feldspathoids can be used to plot a point on the diagram and obtain a rock name.

Petrologists debated the merits of the IUGS system for some time after the IUGS created it, and although the system has some shortfalls, today the debate has waned and the IUGS system is widely used. The entire IUGS classification system divides igneous rocks into 10 categories based mostly on rock chemistry and secondarily on origin. Most common plutonic rocks, including those in Yosemite, fall into the IUGS subsystem for *phaneritic feldspathic rocks* (shown in Figure 6.12). This subsystem applies to all plutonic rocks that contain more than 10% total quartz and feldspar (alkali feldspar or plagioclase).

The IUGS diagram for plutonic rocks has two triangular parts. The top triangle in the diagram is for rocks that contain quartz, and the bottom (upside down) triangle is for rocks that contain feldspathoids (generally leucite or nepheline). Rocks that contain quartz never contain feldspathoids and vice versa, so it is generally clear which half (top or bottom) of the diagram to use. Due to coarse grain size of plutonic rocks, it is often possible to make rough determinations of the QAPF proportions while examining outcrops or hand specimens—something generally not possible for volcanic rocks. More accurate determinations can be made by examining a rock in thin section with a petrographic microscope. As modifiers to rock names, the IUGS suggests using the prefixes *mela-* (meaning dark colored) and *leuco-* (meaning light-colored) when appropriate. A *melagabbro*, for example is a gabbro that has a darker color than most gabbros, and a *leucogranite* is a granite that has a lighter color than most granites. These prefixes, and other commonly used adjectives for igneous rocks, are listed in Table 6.2.

Figure 6.13 contains photographs of some typical plutonic rocks; many of those shown are from the Yosemite region. Petrologists often divide plutonic rocks into two basic groups: *quartzofeldspathic rocks* and *ferromagnesian*

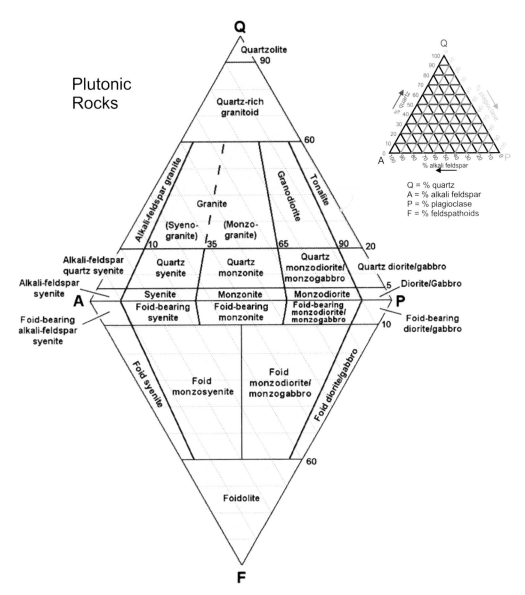

Q = % quartz
A = % alkali feldspar
P = % plagioclase
F = % feldspathoids

Figure 6.12 The basic IUGS classification system for phaneritic feldspar-bearing plutonic rocks, slightly modified from Le Bas and Streckeisen (1991).

Table 6.2   Adjectives and prefixes commonly used to modify iugs igneous rock names.

Term/prefix	Description	Examples
mela-	a rock that is darker than normal varieties of the same rock	*Melagabbro* is a gabbro that is darker than normal (generally because it contains less plagioclase than normal).
leuco-	a rock that is lighter than normal varieties of the same rock	*Leucogranite* is a granite that contains no, or very small amounts of, dark-colored minerals.
aphanitic	a rock with mineral crystals too small to see with the naked eye; this texture is uncommon for plutonic rocks except some dikes and sills	Most volcanic rocks are aphanitic, but very few plutonic rocks are.
phaneritic	a rock with mineral crystals large enough to see with the naked eye	Most plutonic rocks are phaneritic; a few, rare volcanic rocks are too.
porphyritic	a rock with one or more of the mineral species forming larger crystals (*phenocrysts*) that are surrounded by a finer-grained *groundmass* (also called a *matrix*)	Any composition igneous rock (plutonic or volcanic) can be porphyritic. Porphyritic granite, for example, typically contains large K-feldspar crystals surrounded by finer quartz, biotite, and plagioclase.

(Continued)

Table 6.2   (Continued)

Term/prefix	Description	Examples
pegmatitic	a rock that is very coarse grained	Most pegmatites are very coarse-grained granitic rocks containing conspicuous K-feldspar, quartz, and micas.
equigranular	a rock in which all mineral grains are about the same size	Gabbro typically exhibits an equigranular texture.
idiomorphic granular	a rock in which most crystals are euhedral (have flat crystal faces)	Such rocks are exceptionally rare because the crystals generally do not fit together to maintain euhedral shapes.
hypidiomorphic granular	a rock that contains a mix of euhedral, subhedral, or anhedral crystals	This is the most common texture in plutonic rocks, especially rocks such as gabbro that contain large feldspar crystals.
xenomorphic (also termed allotriomorphic) granular	a rock containing only anhedral crystals that may have a "sugary" texture	Aplites, very fine-grained rocks of granitic composition, are typical examples.
glassy (vitreous)	a rock that contains glass	Obsidian and pumice are entirely or nearly entirely glass. Other volcanic rocks may be glassy to various degrees. Few intrusive rocks contain glass; the exceptions are rapidly quenched dikes.

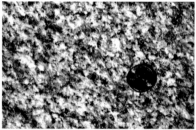

a. This is a pegmatite from Corsica, France. It contains black amhibole, white quartz, and pink K-feldspar.

b. This aplite vein (quartz and K-feldspar) is very fine grained in the center and coarser toward the outside. It is intruding a diorite (mostly plagioclase and clinopyroxene). The sample comes from the Iron Mountains in the Czech Republic.

c. This porphyritic granite comes from Rock Creek Canyon, in the Sierra Nevada Mountains, California. The large phenocrysts are K-feldspar. Surrounding minerals are fine-grained quartz (gray), plagioclase (white), and biotite (black).

d. Many granites have a pinkish hue, but not all. The El Capitan Granite (Yosemite Valley) has an hypidiomorphic granular texture. It contains quartz (light gray), K-feldspar (white), biotite (black), and plagioclase (indistinguishable from K-feldspar).

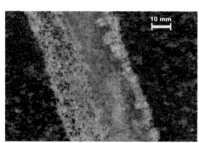

e. This photo shows a relatively fine-grained granodiorite from the Massif Centrale of France. The rock is dominated by quartz and plagioclase, with lesser amounts of K-feldspar, biotite, and hornblende.

f. The Half Dome Granodiorite (Yosemite Valley) has a pronounced porphryitic texture with large lathes of hornblende surrounded by other minerals in the rock.

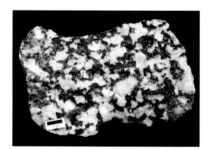

g. This is a sample of diorite that contains conspicuous light colored feldspar and quartz with black hornblende and biotite.

h. This melagabbro sample comes from a Geopark in Albertov, Prague, Czech Republic. The rock is almost entirely plagioclase (white) and clinopyroxene (black).

I. This garnet peridotite (partly serpentinized) from Alp Arami, near Bellinzona, Switzerland, is part of the collection of the Université de Neuchâtel. The coin is 23 millimeters across, for scale.

Figure 6.13   Examples of Plutonic Igneous Rocks.

Photos a, b, c, e, g, h, and i from Quentin Scouflaire, Lysippos, Wilson44691, Rudolf Pohl, Michael C. Rygel, Chmee2, and Woudloper, all at Wikimedia Commons. Photos d and f from N. King Huber (USGS).

rocks. Quartzofeldspathic rocks, commonly light-colored, are made mostly of combinations of quartz and feldspar (alkali feldspar or plagioclase), with lesser amounts of biotite and hornblende, but the proportions vary greatly. Ferromagnesian rocks, commonly dark colored, are made mostly of ferromagnesian minerals, chiefly pyroxene, olivine, and, perhaps amphibole. All the rocks shown in Figure 6.13 are quartzofeldspathic, with the exception of the gabbro (h) and the garnet peridotite (i).

The basic QAPF classification system for plutonic rocks, depicted in Figure 6.12, works well for felsic and intermediate quartzofeldspathic rocks but not well for mafic rocks that contain plagioclase and very little quartz or alkali feldspar (and so plot near the P corner). These plagioclase-rich rocks include diorite, gabbro, norite, and anorthosite. So, distinguishing diorite from gabbro, for example, must be based on one or more of the following: (1) plagioclase composition (in diorite, plagioclase is Na-rich; in gabbro, it is Ca-rich), (2) the ferromagnesian minerals present (hornblende or biotite in diorite; clinopyroxene, orthopyroxene, or olivine in gabbro), (3) the generally darker color that gabbro has compared with diorite, or (4) the speckled black-and-white appearance of common diorite. Furthermore, the QAPF system is not applicable to ferromagnesian rocks that contain no feldspar or quartz. Consequently, the IUGS created separate, but still triangular, naming systems for gabbroic rocks, including gabbro, norite, and anorthosite, and for ultramafic rocks (discussed later in the section "Naming Mafic and Ultramafic Rocks").

6.4.2 Aplite and pegmatite

As discussed in the previous chapter, magmas do not crystallize all at once, and their compositions change as crystallization proceeds. The last remaining liquid portion may crystallize separately from already formed crystals, either as segregations within a larger pluton or as intrusions into surrounding rocks. These *late stage* magmas contain leftovers and so are typically enriched in dissolved *volatiles* (commonly water and sometimes chlorine, fluorine, or carbon dioxide) and sometimes anomalous amounts of incompatible elements that did not enter earlier formed minerals. Consequently, late stage magmas often contain high concentrations of boron, beryllium, lithium, uranium, thorium, cesium, and other generally minor elements. Additionally, because the first minerals to crystallize from any magma are relatively mafic,

evolved magmas are also more felsic than their parents are and richer in silicon, aluminum, and potassium.

If late stage magmas lose their volatiles, they may cool quickly and form very fine-grained rocks called *aplites* that contain crystals that cannot be seen without a hand lens. The aplite dike shown in Figures 6.13a and 6.14, has intruded into a metamorphic rock near Sables d'Olonne in western France. Aplites are typically narrow, from a centimeter to perhaps a meter in width, white, gray, or pink dikes and veins, sometimes described as having a *sugary* texture. They intrude larger felsic or intermediate plutons or less commonly mafic plutons and are often associated with nearby larger granitic bodies. Often, aplites occur in swarms of roughly parallel veins traversing a much larger host rock.

Figure 6.14 Aplite intrusion in gneiss.
From Arlette1, Wikimedia Commons.

Figure 6.15 A pegmatite vein.
Photo from Arlette1, Wikimedia Commons.

Instead of forming aplites, many late stage magmas crystallize to form *pegmatites*, a name used for any coarse grained granite or other igneous rock containing crystals several centimeters, or longer, in length. The coarse rock in Figure 6.13a contains conspicuous white quartz, black tourmaline, and pink K-feldspar. The dike shown in Figure 6.15 is a pegmatite vein that intruded gneiss near Sables d'Olonne in western France. Note the pencil for scale. The word *pegmatite* comes from a Greek word for *bind together*, in reference to typical pegmatite textures that show intricate crystal intergrowths. The adjective *pegmatitic* is used to describe any uncommonly coarse-grained plutonic rock.

Granitic pegmatites are typically associated with granitic plutons but may be separated by short distances. Intermediate and mafic pegmatites are usually in the margins of larger intermediate and mafic intrusive bodies. However, most pegmatites have a granitic composition, and although pegmatites of other compositions exist, petrologists generally think of very coarse-grained quartz and K-feldspar-rich rocks when they hear the word *pegmatite*. Pegmatites are generally small compared to other plutonic bodies, typically no more than a few hundred meters in their longest dimension and often much smaller. They may be dikes or sills or small regions of large mineral concentrations within more typical plutonic rocks. Large pegmatites are often mineralogically zoned, with inner zones having distinct mineralogies compared to outer zones.

No matter the composition, crystal size sets pegmatites apart from other intrusive rock bodies. The minimum crystal dimension in pegmatites is, by definition, 1 centimeter, and some pegmatites contain crystals of up to 10–15 meters (up to 50 feet) in their longest dimension. Figure 6.16 shows a pegmatite in South Dakota's Black Hills—there is a marking pen for scale. The large black crystals are tourmaline, most of the lighter colored crystals are quartz, and there is also some highly reflective silver muscovite present.

Pegmatites contain most of the largest crystals in the world, with sizes comparable to the gypsum crystals from the *Cueva de los Cristales* in Mexico, described in Chapter 4. Generally, large crystals are the result of very slow crystal growth, but the large crystals in pegmatites owe their size more to the presence of water and other volatiles in the magmas. These volatiles keep the magmas fluid, allowing ions to migrate easily. The result is easily growing fewer, but larger, crystals. Although most pegmatites form from unique magmas, some similarly textured coarse rocks may also form by *metamorphic* or *metasomatic* (alteration) processes.

Granitic pegmatites contain all the minerals typically associated with granite, including essential quartz and K-feldspar, and often accessory plagioclase, micas, or amphibole. They also may contain less common minerals made of relatively rare elements. Pegmatites that

Figure 6.16   Tourmaline-containing pegmatite from the Black Hills, South Dakota.

Photo from Steven Reynolds.

Figure 6.17   Large beryl crystals.

From the Smithsonian National Museum of Natural History.

contain significant boron typically contain tourmaline (the black mineral seen in Fig. 6.16). Pegmatites rich in lithium may contain lithium minerals such as lepidolite or spodumene. Pegmatites rich in beryllium may contain beryl. Less exotic but often large and spectacular apatite (the most common phosphorus mineral) and fluorite (the most common fluorine mineral) crystals are also common. Rare cesium, tantalum, and niobium minerals may also be in pegmatites.

High concentrations of uncommon elements mean that some pegmatites are valuable ore bodies. Mining is typically short lived, due to the small sizes of most pegmatites, but may be lucrative. Some pegmatites are mined for lithium, beryllium, boron, zirconium, tin, tungsten, or rare earth elements. Additionally, because they host large euhedral crystals, pegmatites can be important sources of gemstones—including emerald, aquamarine, tourmaline, topaz, fluorite, apatite, ruby, and sapphire.

Pegmatites are also sources of spectacular mineral specimens that grace museum collections. The 11-centimeter-tall emerald crystals in Figure 6.17 are on display at the Smithsonian National Museum of Natural History in Washington, DC. Emerald is a variety of beryl, and this specimen contains some of the most beautiful and perfectly formed beryl crystals in the world. The specimen was collected in 1971 from a pegmatite near Stony Point, North Carolina.

6.5 The southeast face of El Capitan

The southeast face of El Capitan (seen in Fig. 6.2) provides a fascinating view of many of the different kinds of plutonic rocks that can be found in the Yosemite region. Putnam *et al.* (2015) created a geologic map of the 914-meter (3000-feet) precipice, identifying more than half a dozen distinctive rock types (Fig. 6.18). To compile the map, Putnam made many trips up and down the face, gathered information from rock climbers who visited places he did not, took high-resolution photographs from the valley floor, and used laser topographic scans to measure the shape of the face.

Because some kinds of rocks weather and erode more rapidly than others, rocks exposed in outcrops at Earth's surface often give a highly biased indication of the bedrock below. Additionally, due to soil and vegetation, the contacts between different rock units may be difficult to see. The map of El Capitan avoids these problems because it is a map of a vertical slice through a batholith instead of outcrops at the surface.

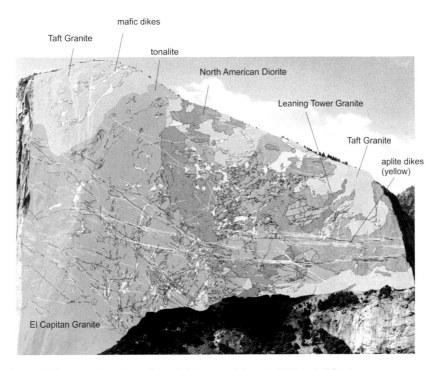

Figure 6.18 Geological map of the vertical southeast face of El Capitan, Yosemite Valley, California. Modified from Putnam *et al.* (2015).

Putnam *et al.*'s map reveals that El Capitan is a complicated mix of several major plutonic rock types that cut across and through each other. About two-thirds is granite (Table 6.3), mostly the 106 million-year-old El Capitan Granite. Fifteen percent of El Capitan is the slightly younger Taft Granite, which crops out mostly in the upper portions of the face. The Taft Granite intruded into the El Capitan Granite. These relationships allow the relative ages of the two units to be determined. The still younger Leaning Tower Granite (about

Table 6.3   Rocks making up the southeast face of el capitan, yosemite valley, california.*

Southeast face area %	Map unit	Description	Mineralogical composition
2	Aplite and pegmatite dikes	white to light gray; generally fine grained; locally coarse	quartz and alkali feldspar with rare accessory garnet and muscovite
15	North American Diorite	dark gray to black; fine to medium grained	plagioclase and augite with accessory biotite and hornblende
2	Leaning Tower Granite	medium to dark gray; medium to fine grained; commonly contains 1–5 centimeter clots of biotite or other mafic minerals	quartz and alkali feldspar with accessory biotite
< 1	Pegmatite bodies	white to light gray; coarse-grained granite	quartz and alkali feldspar
14	Tonalite of gray bands	medium to dark gray; medium to fine grained	mostly quartz and plagioclase, but biotite and minor hornblende are common
3	Dikes of Oceans	medium to dark gray; medium to fine-grained diorite to granodiorite	mineralogically variable; plagioclase rich with K-feldspar and quartz; typically biotite rich and hornblende poor
14	Taft Granite	light gray; medium to fine grained; equigranular (not porphyritic)	Quartz and K-feldspar dominate; biotite averages 3%
49	El Capitan Granite	light gray; medium to coarse grained; commonly porphyritic (alkali feldspar phenocrysts up to 2 centimeters)	Mostly quartz and K-feldspar with lesser plagioclase; biotite < 10%

*These rock units appear as map units in Figure 6.18. All are about 200 million years old. In this table, they are ordered from youngest (at the top) to oldest (at the bottom).

Figure 6.19   Sample of the Half Dome Granodiorite.
Photo from the National Park Service.

2% of the face) cuts through both of the others. Of the non-granitic rocks, 14% of the face is a gray tonalite; it is found mostly between the Taft Granite and the El Capitan Granite. Fifteen percent of the face is the North American Diorite. About 5% of the rocks exposed in the face are in dikes. Some are narrow and made of fine-grained aplite or pegmatite (both granitic); others are wider, blobbier, and composed of diorite or granodiorite. The different rocks can be distinguished by slightly different colors and textures (Table 6.3).

The rocks comprising El Capitan are typical of the entire Yosemite region and of the Sierra Nevada Mountains overall. Yosemite is a mosaic of many different intrusive bodies that formed over millions of years, including some additional rock types not seen in El Capitan's face. Figure 6.19 shows a sample of the Half Dome Granodiorite. It crops out just a few kilometers east of El Capitan on the north side of the valley in the Royal Arches and North Dome. The Royal Arches are prominent in Figure 6.5; North Dome is the rounded peak above the Arches. This granodiorite also crops out on the opposite side of the valley from the Arches, at Glacier Point. Farther east, it forms the vertical cliff face of Half Dome itself, seen in Figure 6.5. The granodiorite is medium to coarse grained and contains conspicuous plates of biotite and crystals of hornblende. It is the youngest of the major plutonic rocks in the region. The granite, granodiorite, tonalite, quartz monzonite, and quartz monzodiorite found in Yosemite are all considered granitoids, and this explains why sometimes the imprecise name *granite* is used for all of them. Other rocks—including quartz diorite, diorite, and gabbro (none of which are granitoids)—are present too, but in lesser amounts and are not called granite.

6.6 The Sierra Nevada Batholith

Felsic and intermediate magmas naturally rise toward Earth's surface, but many stall and cool slowly 5–30 kilometers beneath the surface, producing plutonic rocks and often batholiths made of many individual plutons of similar but not identical age. Although by definition a batholith must be larger than 100 square kilometers, this definition is misleading because most are significantly larger than the minimum. In western North America, the Coast Range Batholith of western Canada and Alaska is 182,500 square kilometers (73,000 square miles)(), the Idaho Batholith is 45,000 square kilometers

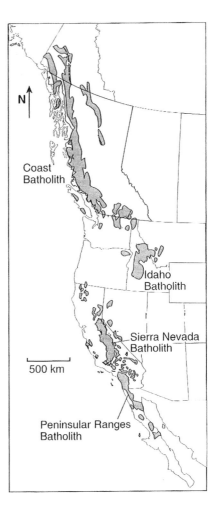

Figure 6.20 Major batholiths of western North America. D. Perkins drawing, based on Winter (2009).

(18,000 square miles), and the Sierra Nevada Batholith is 40,000 square kilometers (16,000 square miles). These batholiths, shown in Figure 6.20, along with the Peninsular Ranges Batholith, form in belts that roughly parallel the West Coast. Similar batholiths are found in the Andes Mountains of South America. Nova Scotia's South Mountain Batholith, at 3000 square miles (7300 square kilometers), is the largest in the Appalachian Mountains.

Batholiths form deep in Earth and are dominated by granitoids. They have complex histories and irregular shapes and their plutons often cut through or around earlier intrusions. Although often looking similar, individual plutons are generally differentiable by their chemistry, mineralogy, age, texture, or crosscutting relationships. The magmas creating batholiths originate near the base of the crust, and geochemical studies suggest that many plutons form from multiple magma pulses, although the contacts between them are hard to see in an outcrop. Although the rocks exposed in El Capitan are older, most of the plutonic rocks of Yosemite formed during the Cretaceous Period, about 113 to 87 million years ago. They are all part of the Sierra Nevada Batholith. Figure 5.31 (previous chapter) shows Mt. Whitney, California, the highest point in the batholith and in the lower 48 states. The entire batholith, composed of hundreds of plutons ranging from small to huge in size, took more than 100 million years to form completely.

6.6.1 The Gold Rush

The batholith rocks of California were the source of the gold that sparked the short-lived *California gold rush* beginning in 1848. Valuable metals, including gold, silver, and copper, can be dissolved from granitoids and concentrated in veins called *lodes*. The California gold rush started with *placer mining*—mining of gold from stream deposits (placers) in foothill valleys. The placers were soon exhausted, but miners successfully followed the gold upstream and discovered the *Mother Lode*— the source of the placer gold—in the Sierra Nevada Mountains. The lode is a zone of veins ranging up to 6 kilometers wide and nearly 200 kilometers long, stretching from the mountains northwest of Sacramento to the foothills southwest of Yosemite National Park. Although placer mining ceased after a few years, following the initial gold rush, the Mother Lode produced significant amounts of gold for about 100 years.

6.7   The American Cordillera

The Coast Range Batholith, Idaho Batholith, Sierra Nevada Batholith, and Peninsular Ranges Batholith are all part of the *American Cordillera*, a chain of mountain ranges, intermontane basins between the ranges, and plateaus extending from northern Alaska through North America, Central America, and South America and ending in Antarctica (Fig. 6.21). The Cordillera includes the Cascade Mountains of Washington and Oregon and other volcanic mountain chains, coast ranges such as those along the western margin of North America and the Andes, and interior ranges including the Sierra Nevada and Rocky Mountains. The entire cordillera owes its origin to the interaction of tectonic plates and subduction along the eastern margin of the *Pacific Ring of Fire*. Subducting slabs generated magmas that led to volcanic and plutonic rocks, and converging plates caused deformation and uplift, ultimately leading

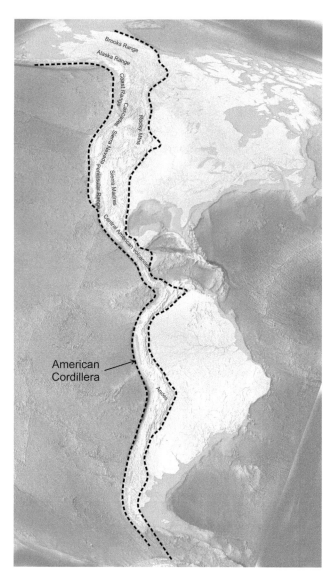

Figure 6.21   The dashed lines show the outline of the American Cordillera. It includes many mountain ranges and the lands between them. Some of the most important ranges are labeled.

to exhumation of once deep-seated rocks. Volcanism still occurs today in Alaska, in the Cascade Mountains, in parts of Central America, and in many parts of the Andes Mountains. In other places, like the high Sierra Nevada Mountains, plutonic rocks that were once deeply buried have been uplifted and exposed by tens to hundreds of millions of years of erosion.

Although some regions are dominated by volcanic rocks and some by plutonic rocks, both volcanic and plutonic rocks are found along the entire length of the Cordillera. As Figure 6.22 shows, they have similar compositions— both groups range from mafic to felsic. The vertical axis in Figure 6.22 (n) indicates the number of occurrences and, although many more volcanic rocks have been analyzed

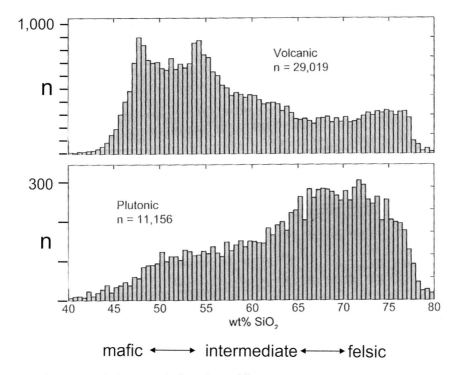

Figure 6.22 Silica content of volcanic and plutonic rocks from the cordillera.

Modified from Lundstrom, C. C., & Glazner, A. F. (2016). Silicic magmatism and the volcanic-plutonic connection. *Elements*, 12(2), 91–96. DOI:10.2113/gselements.12.2.91.

than plutonic rocks, the similarity in composition range (40%–80% SiO_2) is clear. Plutonic gabbro, diorite, tonalite, granodiorite, and granite are matched by volcanic basalt, andesite, dacite, and rhyolite. Overall, though, the plutonic rocks are more felsic than the volcanic rocks.

The close proximity of volcanic and plutonic rocks and their similar compositions support the historical view that the plutonic rocks are remnants of magma chambers that once fed active volcanoes above (see Fig. 5.29, previous chapter, and Fig. 6.6, this chapter) and that plutonic and volcanic rocks come from the same sources and processes. Some recent investigators, however, have suggested that the relationships may not be quite so simple, and research continues today to illuminate what Lundstrom and Glazner (2016) call an "enigmatic relationship" between silicic volcanic and plutonic rocks.

6.8 Mafic and ultramafic plutonic rocks

6.8.1 Naming mafic and ultramafic rocks

Mafic and ultramafic plutonic rocks are dominated by various combinations of plagioclase, pyroxene, and

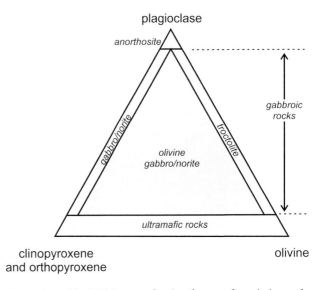

Figure 6.23 The IUGS system for classifying mafic and ultramafic rocks.

olivine. As shown in Figure 6.23, this allows rocks to be divided into general categories: those that are mostly plagioclase are called *anorthosites*; those that are mostly pyroxene and olivine are *ultramafic rocks*. Rocks that are mostly pyroxene and plagioclase, or pyroxene and plagioclase with olivine, are *gabbroic rocks*.

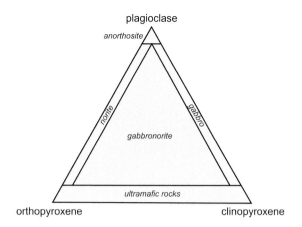

Figure 6.24   The IUGS system for naming gabbroic rocks.

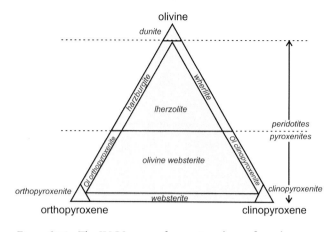

Figure 6.25   The IUGS system for naming ultramafic rocks.

Accessory amphibole, generally hornblende, and biotite may also be present in mafic rocks, but not in ultramafic rocks. Olivine, an essential mineral in many ultramafic rocks, is absent or only an accessory in mafic rocks. Especially silica-rich gabbro may contain small amounts of quartz. Other gabbro, deficient in silica, may contain feldspathoids such as nepheline, analcite, or leucite. In both mafic and ultramafic rocks, apatite, magnetite, ilmenite, and several other minerals are common in very minor amounts.

The standard IUGS system for naming gabbroic rocks, shown in Figure 6.24, is based on the proportions of plagioclase, clinopyroxene (generally augite), and orthopyroxene in a rock. *Gabbro* contains mostly plagioclase and clinopyroxene, *norite* contains mostly plagioclase and orthopyroxene, and rocks in between are termed *gabbronorite*.

For classifying ultramafic rocks, the IUGS uses the proportions of olivine, orthopyroxene, and clinopyroxene (Fig. 6.25). *Peridotites*, which all contain more than 50 wt% olivine, include *harzburgite* (orthopyroxene >> clinopyroxene), *wehrlite* (clinopyroxene >> orthopyroxene), and *lherzolites* (subequal amounts of orthopyroxene and clinopyroxene). *Pyroxenites* may be orthopyroxene-rich (*orthopyroxenite*), clinopyroxene-rich (*clinopyroxenite*), or in-between (*websterite*). The modifier olivine is put in front of a pyroxenite name if a rock contains more than 5% olivine. A different triangular diagram (not shown) is used for the rare ultramafic rocks that contain hornblende.

Mafic and ultramafic plutonic rocks can crystallize in the same way that their more felsic cousins (granitic and intermediate compositions) crystallize—as a magma body cools slowly to produce a large

Figure 6.26   Orthopyroxene cumulate from the Stillwater Complex, Montana.

Photo from Kurt Hollacher.

rock body of uniform composition and texture. Often, however, the first-formed mineral crystals—high-temperature minerals such as chromite, olivine, pyroxene, or Ca-rich plagioclase—settle to the bottom of the magma chamber, producing cumulates (see Fig. 5.42 and the discussion of fractional crystallization and the Stillwater Complex in the previous chapter). Less commonly, early-formed plagioclase may float to the top of a magma chamber and accumulate. No matter how they form, cumulate minerals eventually become surrounded by a *matrix* composed of lower-temperature finer-grained mineral crystals. Figure 6.26 shows a cumulate from the Stillwater Complex, Montana. The dominant brown crystals are cumulus orthopyroxene. White plagioclase has filled space between. The less abundant green crystals are intracumulus augite.

Very fine-grained plutonic rocks may be difficult to distinguish from volcanic rocks. For example, a hand sample of very fine-grained (aphanitic) gabbro, called *diabase*, or *dolerite*, may be indistinguishable from (volcanic) basalt. However, most mafic and ultramafic plutonic rocks are coarse grained (phaneritic), containing crystals millimeters across or larger. They are usually equigranular but may be porphyritic. Some gabbros, for example, contain very large plagioclase *phenocrysts* surrounded by groundmass. Rarely, mafic and ultramafic rocks are exceptionally coarse grained and may be described as *pegmatitic*.

Mafic and ultramafic magmas crystallize in steps over a range of temperature. So, high-temperature minerals (e.g., olivine, spinel, or magnetite) commonly are enclosed by lower-temperature minerals (augite, hornblende, or pyroxene). Plagioclase is often the last mineral to crystallize and, consequently, may fill interstices between larger Fe- and Mg-rich mineral crystals. In some rocks, however, especially some diabase and gabbro, plagioclase crystallizes first and becomes included in later-formed pyroxene. Although plagioclase and pyroxene dominate mafic rocks, lesser amounts of amphibole, olivine, oxides, and other minerals may also be present.

If smaller grains of an early-formed mineral are completely enclosed by larger later-formed mineral crystals, the rock has a *poikilitic texture*. The most common type of poikilitic texture, an *ophitic texture*, is found in some diabase and gabbro when plagioclase crystallizes first and randomly oriented plagioclase lathes are trapped in later-formed pyroxene. The cobblestone in Figure 6.27 from Lausitz, southeastern Germany, shows an exceptionally well-developed ophitic texture. The white lathes are plagioclase, and the surrounding dark material is clinopyroxene.

6.8.2 Occurrences of mafic plutonic rocks

Mafic plutonic rocks, generally called simply *gabbros*, or *gabbroic rocks*, are not as abundant as granitic rocks are, but gabbroic rocks are more widespread. Different varieties of these mafic rocks vary slightly in mineralogy and chemical composition depending on their origin. True *gabbros* contain abundant essential plagioclase and augite (clinopyroxene); other *gabbroic* rocks contain orthopyroxene instead of clinopyroxene. Characterizing accessory minerals include olivine and amphibole, and gabbroic rocks typically contain small amounts of other accessory minerals too. Gabbroic rocks, having 45–52 wt% SiO_2 and subequal amounts of FeO and MgO, have compositions equivalent to the volcanic rock *basalt*.

Figure 6.27 Cobblestone showing ophitic texture. Photo from Lysippos, Wikimedia Commons.

Yosemite Valley contains relatively few mafic rocks, but mafic rocks are common in other places, especially under the world's oceans, an area of about 362 million square kilometers (139.7 million square miles). Oceans cover 71% of our planet, everywhere underlain by oceanic crust up to 10–15 kilometers (6–9 miles) thick, and that crust is almost entirely made of *gabbro*, such as that shown in Figure 6.13h. Gabbro is arguably the most common rock type in Earth's crust, and most gabbro on Earth is created by decompression melting at mid-ocean ridges. Less commonly, gabbro is seen on continents, in plutons related to continental volcanism.

In continental regions, when mafic magmas make it to the surface, basaltic volcanism occurs. Sometimes, however, magma stalls, producing large volumes of gabbro that crystallize in deep magma chambers that later may be uplifted and exposed by erosion at Earth's surface. The connection between mafic magma chambers—perhaps later to become plutons—and mafic volcanism is well established. Figure 6.28, for example, shows gabbro outcrops on Meall Meadhonach, northernmost Scotland, that were uncovered after uplift and erosion of Earth's crust. Nearby basalt occurrences of about the same age suggest that the same mafic magma may have produced both plutonic and volcanic rocks.

Gabbros are also found in the lowermost levels of flood basalts, such as the Columbia River flood basalts of the northwestern United States and the Deccan Traps of India. There, thick sequences of basalt flows accumulated, and the flows on the bottom cooled slowly enough to become coarsely crystalline; thus, the magma became gabbro instead of basalt. Still other gabbros are

Figure 6.28   Gabbro outcrops in northern Scotland.
Photo from Anne Burgess.

Figure 6.29   Coarse crystals of clinopyroxene in peridotite.
Photo from Theklan, Wikimedia Commons.

found associated with ultramafic rock in layered igneous complexes, including the Stillwater Complex of Montana and the Bushveld Complex of South Africa.

Because they generally contain abundant mafic minerals (pyroxene and sometimes hornblende or olivine) and at most very small amounts of quartz, most gabbroic rocks are dense and moderately to darkly colored. Mafic minerals decompose easily to form chlorite or serpentine and other green or greenish minerals, so sometimes, especially when weathered, gabbroic rocks may have a greenish hue.

6.8.3   Occurrences of ultramafic plutonic rocks

Like their mafic cousins, *ultramafic plutonic rocks* vary in composition and mineralogy. Compared to gabbros, they contain less silica (less than 45 wt%), generally more MgO, and lesser amounts of FeO. Their common minerals are olivine, clinopyroxene, and orthopyroxene, allowing these rocks to be divided into two main groups distinguished by olivine content. As shown in Figure 6.25, *peridotites* contain more than 50% olivine; *pyroxenites* are more than half pyroxene (either orthopyroxene or clinopyroxene).

Ultramafic rocks, because they are mostly made of mafic minerals, are commonly dark colored (if unaltered), but some peridotites are light green because they contain abundant olivine. The peridotite in Figure 6.29 contains large conspicuous green clinopyroxene crystals surrounded by finer-grained olivine weathered to a light brown color. Figure 6.13i shows a peridotite that contains garnet. Because ultramafic rocks are quite unstable at Earth's surface and at shallow depths in Earth, they

often alter to produce green serpentinites that are mostly made of serpentine, chlorite, and iron oxides (Fig. 6.10).

Most peridotites and pyroxenites are found in layered igneous complexes that originated deep within Earth. We see them exposed in mountain belts or in the eroded roots of mountain belts exposed in Precambrian shields. Ultramafic plutonic rocks formed during the Phanerozoic Period (younger than Precambrian) are rare, and Phanerozoic ultramafic volcanic rocks are even rarer. Some ultramafic rocks, pieces of the mantle, can be found in *ophiolite complexes*—wedges of ocean crust and mantle that were thrust onto continental margins above subduction zones and can now be seen in outcrops.

6.8.4   Oceanic crust, lithosphere, and ophiolites

New oceanic crust forms at mid-ocean ridges as magma rises to fill space created when plates diverge. The spreading lithosphere, composed of crust and some underlying mantle, cools and becomes denser during spreading. So, the youngest seafloor and highest elevations are at the ridges, and the oldest seafloor and lowest elevations are at ocean margins. The ocean lithosphere may not be the same composition and structure everywhere, but all evidence suggests that there are some standard components. The evidence comes from many sources including drill core, seismic studies, laboratory experiments, grab sampling from the ocean floor, and dredging. The best information, however, may come from studies of *ophiolites*. Ophiolites are parts of oceanic crust and mantle that were uplifted and added to continental margins

Table 6.4   Some of the many well-known and most-studied ophiolite complexes

Ophiolite	Mean age	Notes
Coast Range Ophiolite (California)	155 million years (Jurassic)	scattered outcrops from Santa Barbara north to San Francisco
Semail Ophiolite (Oman and the United Arab Emirates)	95 million years (Cretaceous)	large and very well exposed
Troodos Ophiolite (Cyprus)	92 million years (Cretaceous)	especially well studied due to economical copper deposits associated with the ophiolite
Macquarie Island Ophiolite (Tasmania, Australia)	10–40 million years (Cenozoic)	named a UNESCO World Heritage Site in 1997
Bay of Islands Ophiolite (Newfoundland)	485 million years (Ordovician)	well-exposed nearly complete ophiolite sequence; named a UNESCO World Heritage Site in 1987
Zambales ophiolites (Philippines)	45 million years (Cenozoic)	several ophiolites of about the same age

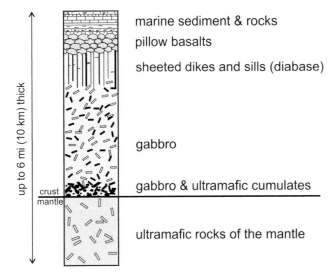

Figure 6.30   An idealized cross section of an ophiolite.

or are exposed in islands. The term *ophiolite* derives from the Greek words *ophio* (snake) and *lite* (stone), referring to the commonly green color of the rocks that make up ophiolites. There are many ophiolites around the world, but most are small or very fragmented; Table 6.4 lists some of the best-known and studied ones.

During subduction, the oceanic lithosphere generally descends beneath a continent, but sometimes pieces of oceanic lithosphere are scraped off and added to a continent. We call this process *obduction*, and obduction creates ophiolites. Obducted ocean lithosphere is most often found in mountain belts formed during continental collision, evidence that ocean basins once were present. Every ophiolite provides a partial cross section of the oceanic crust and mantle, and when information from many ophiolites is combined with the other

evidence, a standardized model of the ocean lithosphere, such as the one shown in Figure 6.30, emerges. Most ophiolites are tens or hundreds of millions of years old because the processes of seafloor spreading and obduction are slow. The Macquarie Island Ophiolite, the youngest known, is still more than 10 million years old.

A "complete" ophiolite, something that is uncommon, includes all the layers of rock shown in Figure 6.30. These layers correspond to the sediments and rocks being created by seafloor spreading at mid-oceanic ridges and make up a cross section of the oceanic lithosphere. At its top, the lithosphere is overlain by muds and other sediments that may lithify to form shale or chert. These sediments increase in thickness from mid-ocean ridges to ocean margins. Beneath the sedimentary layers, the rocks are all igneous rocks. A layer of basalt, often containing *pillow lavas* that form tube-like bodies when lavas erupt on the ocean floor, underlies the sediments. These basalts, like many ocean floor basalts, are highly altered by interaction with seawater.

The magmas that create ocean-floor lavas rise from magma chambers below, following fractures and creating vertical, parallel, mafic dikes. The many dikes create a *sheeted dike complex* beneath the basalts. At still greater depth, a thick layer of *gabbro* is the remains of once liquid basaltic magma chambers. The gabbro layer, accounting for most of the oceanic crust by volume, typically contains mafic to ultramafic cumulates in its lowest levels. Beneath the gabbro layer, peridotites (harzburgites and lherzolites) of the oceanic mantle can be found. Figure 6.31 shows gabbro layers—the browner layers contain significant amounts of orthopyroxene compared with the lighter colored layers.

Figure 6.31 Layered gabbros from the Troodos Ophiolite in
Cyprus.
Photo courtesy of W. D. Cunningham.

6.8.5 *Layered mafic intrusions*

Layered mafic intrusions are bodies of mafic and ultramafic rock formed by multiple intrusions. They range
in size from quite small to very large and can be found
in just about every geological setting (Table 6.5). Most
are *lopoliths*, funnel-shaped intrusive bodies with flat
upper surfaces, that formed during multiple intrusions
(Fig. 6.6). Most layered mafic complexes are Precambrian, but some famous ones, including the Skaergård
and Dufek intrusions, are much younger. They all
formed over long times, estimated to be as long a million years in some cases. Many, but not all, layered
mafic intrusions are associated with flood basalts of
similar age.

Table 6.5   Some of the most studied layered mafic intrusions.

Complex name	Geologic age & time of formation	Location	Area (square kilometers)
Bushveld	Precambrana, 2 billion years ago	South Africa	66,000
Dufek	Jurassic, 184 million years ago	Antarctica	50,000
Duluth	Precambrian, 1.1 billion years ago	Minnesota, USA	4700
Stillwater	Precambrian, 2.7 billion years ago	Montana, USA	4400
Muskox	Precambrian, 1.2 billion years ago	Northwest Territories, Canada	3500
Skaergård	Eocene, 56 million years ago	Greenland	100

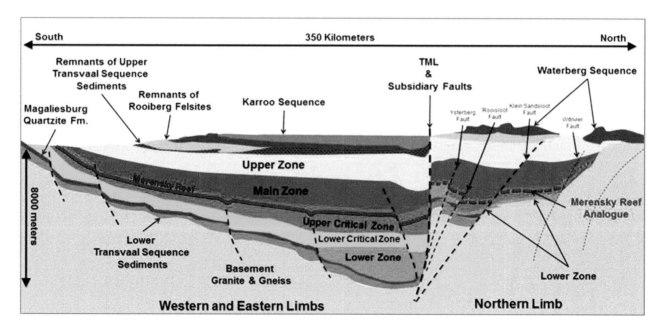

Figure 6.32   Cross-section of the Bushveld Complex in South Africa. Figure from Geologyforinvestors.com.

The largest layered mafic intrusion is the Bushveld Complex in South Africa near Johannesburg. Figure 6.32 shows a cross-section of the Bushveld. The complex has an area of more than 65,000 square kilometers, covering 15% of South Africa, and is up to 10 kilometers thick in places. The Bushveld is host to many rich ore deposits and produces most of the world's platinum-group metals (platinum, palladium, osmium, iridium, rhodium, and ruthenium) and significant amounts of other metals.

Layered mafic complexes include typical mafic and ultramafic rocks and sometimes more felsic rocks. They are all characterized by multiple intrusions and fractionation due to cumulate formation. A generic model has ultramafic rocks (both peridotites and pyroxenites) at the bottom of the complex grading into mafic rocks (norites, gabbros, and anorthosites) higher up—with perhaps intermediate and felsic rock bodies at the top. However, sometimes the sequences repeat, some rock compositions are missing, or there has been significant deformation after emplacement—making the geometry difficult to interpret. The Bushveld Complex has peridotite, harzburgite, orthopyroxenite, and chromite cumulates near its base, with norite, anorthosite, and gabbro above. A several kilometers thick layer of granitic rocks is found at the very top. Some of the many layers in the complex are shown by different colors in Figure 6.32.

Layered mafic complexes are important natural laboratories. By studying them, geologists have answered some key questions about fractionation and how Earth differentiated over time. Field investigations of the spectacular rocks have confirmed what was known from laboratory experiments—that magmas evolve as igneous minerals crystallize over a range of temperature in predictable order.

Layered complexes are also important for the ore bodies they contain. In some complexes, chromite and other oxide and sulfide mineral cumulates provide rich concentrations of chromium, platinum-group elements, titanium, and vanadium. For example, the *Merensky Reef* of the Bushveld Complex is a 30- to 90-centimeters-thick layer of norite that includes cumulate sulfide and chromite layers. It produces chromium, but more importantly contains most of the world's known reserves of platinum, palladium, and several other key industrial elements. Gold, silver, and copper are produced as ancillary products.

The Stillwater Complex (Montana) contains the *J-M Reef*, an ore body similar too, but smaller than, the

Merensky Reef. The J-M Reef is also mined for platinum and palladium. The reef is a 1- to 3-meter-thick body of altered ultramafic rock that contains pockets of very high-grade ore. Figure 6.33 shows an ore rock (3 centimeters across) from the reef. It is an altered ultramafic rock that now contains whitish tremolite, golden chalcopyrite, and bronze pyrrhotite. The chalcopyrite and pyrrhotite are the targets of mining because they contain exceptionally high amounts of platinum and palladium (Fig. 6.33).

6.8.6 *Mafic dikes and sills*

The most common and widespread kinds of mafic intrusions are dikes. Most mafic dikes and sills consist of *diabase* (a fine-grained equivalent of gabbro). Figure 6.34 shows diabase dike cutting, from left to right, through granite at Schoodic Point in Acadia National

Figure 6.33　Ore rock from the J-M Reef of the Stillwater Complex, Montana.
Photo from James St. John.

Figure 6.34　A diabase dike.
Photo from Kent Miller, National Park Service.

Park, Maine. Mafic dikes, such as the one shown, are fed by magmas of mantle origin, and some dikes are the conduits for rising magma that feeds erupting volcanoes or that erupts as flood basalt. Some mafic dikes branch, and others are directed along rock layers to create sills.

Mafic intrusions may involve a single intrusive event or multiple intrusions, or *pulses*, of the same kind of magma. Others form from multiple intrusions of magmas having different compositions. Dikes and sills range from very small to large, with thicknesses from less than a centimeter to many hundreds of meters, and they may extend laterally for long distances. Some of the largest dikes and sills became differentiated when early-formed crystals were separated from the remaining magma.

Individual mafic dikes or sills are widespread. Often, they are in sharp color-contrast with the country rocks around them. They may weather and erode differently from surrounding rock and so be especially easy to pick out. Figure 6.35 shows two peaks in the Tetons, on the Wyoming-Idaho border in Grand Teton National Park. Mt. Moran (top) and Middle Teton (bottom) both have diabase dikes (indicated with arrows) cutting through their summits. The erosion-resistant diabase is one reason the peaks are so tall. In many Precambrian terranes, dikes cut across ancient igneous and metamorphic rocks, appearing to continue forever and disappear into the distance, as shown in Figure 6.7d.

Mafic sills, like mafic dikes, are quite widespread, but because they are typically horizontal and conform to surrounding country rock, they are more difficult to notice. Often, only their top surface can be seen, and some are quite large and easily mistaken for basalt flows. Figure 6.36 shows the Palisades Sill (about 300 meters thick) on the Hudson River near New York City. Outcrops of this sill resemble a columnar basalt in some places, so it was originally mistaken for a basalt flow (Fig. 6.36).

6.8.7   Mafic dike swarms

Mafic dikes intrude along fractures, so in regions where there are many fractures, there may be many dikes in *dike swarms*. Dike swarms can cover vast regions and include hundreds of individual dikes that are of about the same age. They may be arranged in linear, parallel,

Figure 6.35   Two Teton peaks cut by diabase dikes: Mt. Moran (top) and Middle Teton (bottom).

These photos modified from photos by Paul Hermans and Acroterion, both at Wikimedia Commons.

Figure 6.36   The Palisades Sill in New Jersey.
Photo from CrankyScorpion, Wikimedia Commons.

or radiating patterns. Most are Precambrian, which many geologists interpret to be due to an overall hotter Earth at that time that allowed large volumes of mafic magma to make it more easily from the mantle into the crust.

The Mackenzie Dike Swarm, shown in Figure 6.37, is one of several dozen dike swarms in the Canadian Shield. It is the largest dike swarm in the world and covers a huge area of the shield—more than 500 kilometers (310 miles) wide and 3000 kilometers (1900 miles) long—from near the Great Lakes to the Arctic. The dikes are in a fan-like pattern and all seem to radiate from a point on southern Victoria Island in the Arctic. Most petrologists believe that a large mantle plume carried heat and magma up in that region. The rising plume caused the crust to fracture, providing conduits for dike formation. The Muskox layered mafic intrusion, the Coppermine River and Ekalulia flood basalts (also shown in Fig. 6.37), and several major plutonic complexes were all emplaced in the region at the same time—additional evidence for an active mantle plume that provide magma for both plutonic and volcanic rocks.

Shiprock, in the northwest corner of New Mexico, provides a good example of radiating dikes on

Figure 6.38 Shiprock in the distance with a dike in the foreground, northwestern New Mexico. From el ui, Wikimedia Commons.

a scale smaller than dike swarms like the Mackenzie Dikes. Shiprock's pointed peak is a volcanic neck, the erosional remnant of a volcano, made mostly of volcanic breccia and a rare highly potassium-rich mafic rock called *lamprophyre*. Half a dozen major lamprophyre dikes radiate outward several miles from Shiprock; the most prominent is easily seen in Figure 6.38, with the main peak of Shiprock in the background. Because the dikes are erosion resistant, they now stand as thin fins above the surrounding desert floor.

Questions for thought—chapter 6

1. Plutonic rock bodies come in many different sizes. Some are very small and some are huge. name and describe several small ones, and several large ones, and explain how they form.

2. Gabbro is an intrusive igneous rock and basalt is an extrusive igneous rock. Which will contain larger crystals? Often the minerals in a typical gabbro are the same as the minerals in a typical basalt. Explain why?

3. Different varieties of plutonic rock contain different minerals. For example, quartz and K-feldspar are common in granites but are never found in gabbros. Plagioclase and pyroxene dominate gabbros but are largely absent from granites. Explain the reason for these contrasts.

4. Secondary minerals are common in plutonic rocks of all kinds. How do they form and what charac-

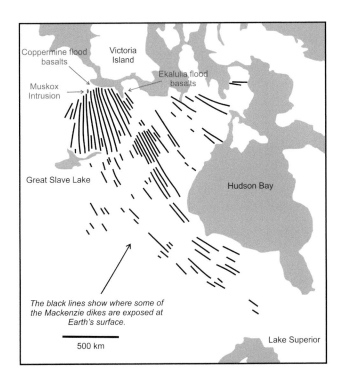

Figure 6.37 The Mackenzie Dike Swarm.

teristics might you see that help you identify them as secondary? Epidote is a common secondary mineral in granite and some other felsic rocks. Serpentine, talc, and chlorite are common secondary minerals in mafic rocks. Why the difference? Why are the most common secondary minerals in felsic and mafic rocks different?

5. Explain how the IUGS system for naming plutonic rocks works. That is, what information is used to obtain a rock name? The system does not work for all kinds of rocks – for what kinds is it problematic? Why? What about those rocks makes the system imprecise?

6. What are the differences between aplites and pegmatites? Often, they have similar compositions and contain similar minerals. So, why are they different – in other words, what causes the differences?

7. Pegmatites are often sources of spectacular museum mineral specimens. Why? What makes the specimens special and why do they form in pegmatites but not in other kinds of intrusive rocks?

8. What is the American Cordillera? Where is it found and what kinds of rocks are found there? How did it form and why is it where it is?

9. The southeast face of El Capitan reveals a typical pluton. But, here we see it in cross section, which make the few pretty unique. How did the face come to be exposed? When we look at it, we see many contrasting rocks and colors. How did it come to be this way?

10. Why do many ultramafic rocks almost entirely consist of serpentine, chlorite, and iron oxides? Are these primary minerals?

11. The most common rock rocks that make up plutons that we see on continents are granite, granodiorite, tonalite, monzonite, and quartz diorite. Gabbros and other mafic plutonic rocks are much less common. Why?

12. What are ophiolites? What are the major components the comprise them? How do they form and why do we find them where we do?

13. How do layered mafic intrusions form? What kinds of rocks do they contain? (Name the rocks.) Why are layered mafic intrusions commonly associated with economically ore deposits?

14. Consider the McKenzie Dike Swarm – it is shown on a map in Figure 6.37. Look at the map – How do you think all these dikes of the same composition formed?

15. Figure 6.38 shows a dike in the foreground with Shiprock in the background. Explain how both formed and what the relationship is between them. Be careful to think about which formed first and explain why you think that.

7 Volcanoes and Their Products

7.1 Tambora and Toba

The largest volcanic eruption in recent history occurred in 1815 when Mt. Tambora, on the island of Sumbawa, Indonesia, exploded and ejected 160 cubic kilometers of material into the air, creating the large depression seen in Figure 7.1. Depressions that form this way are called *caldera*. The Tambora caldera is 6 kilometers in diameter and 1100 meters deep. It formed when Tambora's estimated 4000-meter-high peak blew away and the ground collapsed as the magma chamber below emptied. Tambora is still an active volcano, and several minor eruptions have extruded lava on the caldera's floor since the big eruption in 1815.

The sound of Tambora's 1815 eruption was heard more than 1400 kilometers (870 miles) away in the Molucca Islands (Fig. 7.2). Thick layers of volcanic ash were deposited over a huge region, traveling thousands of kilometers from the eruption site. The ash eventually hardened to become *tuff*, a type of volcanic rock formed by consolidation of volcanic airfall deposits.

It is estimated that Tambora's eruption killed at least 70,000 people, some because of the immediate effects of the explosion and ashflows but most because of subsequent starvation and disease. Although the death toll on Sumbawa and nearby Lombok Island was huge, Tambora's effects were felt worldwide. In the months following the eruption, volcanic ash and sulfuric acid, carried around the globe by atmospheric circulation, blocked incoming sunlight and caused the *year without summer*. Average global temperatures fell by 0.5–1 °C (1–2 °F), leading to crop failure and major food shortage in the Northern Hemisphere. Famine led to food riots in northern Europe and precipitated a typhus epidemic in Ireland. The effects were not as great in North America, but still, frost killed most crops in New England and

Figure 7.1 Tambora volcano, Sumbawa Island, Indonesia.
Photo from Jialiang Gao, Wikimedia Commons.

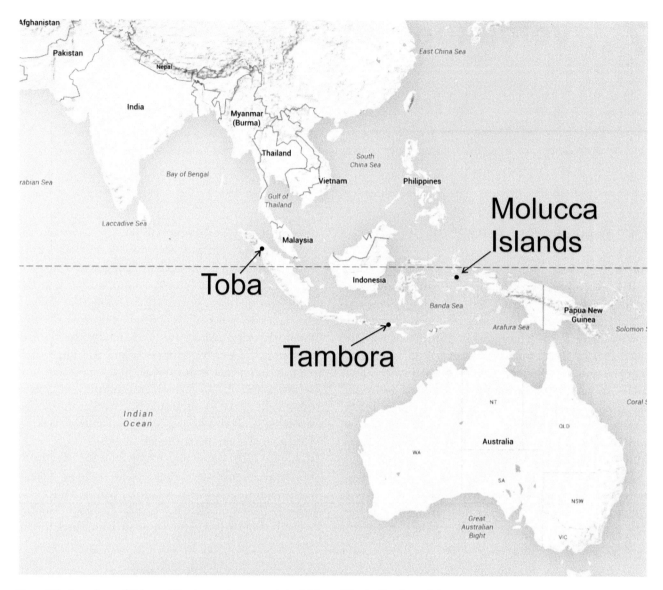

Figure 7.2   Locations of Toba and Tambora volcanoes and the Molucca Islands, Southeast Asia.

parts of Pennsylvania and the region experienced snow-fall in June and July.

We have recorded history documenting Tambora's effects, but it was a relatively small event compared to many earlier prehistoric eruptions. Scientists interpret the history of those eruptions, and their effects on humans, using incomplete geological, anthropological, and other evidence. Toba Volcano, which erupted multiple times about 74,000 years ago, is an example.

Figure 7.3 shows Lake Toba, in northern Sumatra, 2500 kilometers (1550 miles) northeast of Tambora Volcano (Fig. 7.2). It is the largest volcanic lake in the world, 100 kilometers long, 30 kilometers wide, and more than 505 meters deep. This lake occupies a large

caldera, in many ways similar to the one at Tambora, where the ground collapsed when lava spewed from a magma chamber below. The caldera was created during the eruption of a *supervolcano* named *Toba*, thousands of times larger than typical volcanoes and 100 times larger than Tambora.

Toba erupted several times, but the last eruption—the one that created the present-day caldera—was the largest, and it was the largest eruption known to have occurred on Earth in the last 25 million years. Subsequent activity, many years after the main eruption, created small cinder cones and volcanic island in the center of the lake, as well as several younger volcanoes along the lake rim.

Geologists have identified a small number of other supervolcano eruptions that were larger than Toba, such as the largest known eruption, the Wah Wah Springs eruption, that occurred in Utah 30 million years ago, and an eruption that created the Fish Canyon Tuff in southwest Colorado 28 million years ago. Figure 7.4 shows several different layers of the Fish Canyon Tuff—thick layers of consolidate volcanic ash and coarser debris. The ash layers vary in color due to slightly different compositions, and are easily eroded, as seen in this photograph. The eruption, centered at La Garita Caldera in southwest Colorado, produced an estimated 5000 cubic kilometers of volcanic ash.

Although there were larger eruptions, Toba is especially significant because it may have affected humans and human diversity. The eruption produced more than

Figure 7.3 Lake Toba, northern Sumatra.
Photo from Rusian Hidayat, Wikimedia Commons.

Figure 7.4 The Fish Canyon Tuff, southwestern Colorado.
Photo from G. Thomas, Wikimedia Commons.

2800 cubic kilometers of ash that circled the globe, depositing ash layers 15 centimeters (6 inches) deep over most of South Asia and into surrounding seas. Noxious sulfur-rich gases were spread in all directions. Ash and sulfuric acid in the air led to global temperature change, generally estimated to average 3–5 °C of cooling around the globe. (Some researchers, however, think it could have been much more.) Whatever the exact amount of cooling, the effects were a global ice age that lasted years and an overall Earth cooling that lasted, perhaps, 1000 years.

Although human ancestors have been around for more than 10 million years, scientists studying mutation rates have concluded that humans could not have been evolving independently that long—we are all too similar to each other. Even before geologists dated the Toba eruption, geneticists had concluded that humans descended from a small number of ancestors slightly more than 70,000 years ago. There is some debate, but most archaeologists and geneticists believe that Toba's eruption caused a *genetic bottleneck*.

Most humans, 74,000 years ago, lived west of Indonesia, in a downwind direction from Toba. So, the direct effects of ash and gases, the longer-term climate change, and loss of food must have been significant. So, some scientists have concluded that the once-larger human population was reduced to a small number survivors. Perhaps as few as 100 or 1000 humans existed after Toba, greatly limiting genetic diversity. Although we do not know exactly where all humans lived before the Toba eruption, ample evidence suggests that there was major migration—after the effects of the eruption waned—from Africa north to Europe from between 70,000 and 60,000 years ago. Some studies also suggest that populations of humans in southern India and on some islands upwind of Toba survived the eruption.

7.2 Volcanology

Mt. Vesuvius, in southern Italy, is the only active volcano on the mainland of Europe. It is best known for its eruption of AD 79, which destroyed nearby Pompeii and Herculaneum. It has erupted many times since then—the last time in 1944. It is considered one of the most dangerous volcanoes in the world because of the many people who live nearby and because eruptions of Vesuvius are often very violent. Because of its historical significance in a region of high population, Mt. Vesuvius has long been the focus of scientific investigation.

Volcanology, derived from the name of the Roman god of fire, *Vulcan*, is the study of volcanoes. It is a hybrid science, involving investigations by geologists, geophysicists, geochemists, geodesists, archaeologists, and others. Figure 7.5 shows the first volcanological observatory, the *Vesuvius Observatory* near Mt. Vesuvius and Naples, that was founded in 1841. Today other dedicated volcano laboratories exist, including notably the United States Geological Survey's *Hawaiian Volcano Observatory* on the Big Island of Hawaii.

Geologists classify volcanoes as *active, dormant*, or *extinct*. The distinctions are imprecise and sometimes used in different ways by different people. Active volcanoes are those that are erupting today or that have erupted recently. Tambora is an example of an active volcano, although it has not erupted in 200 years. The Global Volcanism Program defines "recently" as meaning in the past 10,000 years, but some scientists are not that specific. Dormant volcanoes are those that are presently inactive but could reasonably be expected to erupt in the future. Extinct volcanoes are those that will probably never erupt again. Scientists, however, are fallible, and some presumed extinct volcanoes have come back to life. Volcanologists study all three types, but their tools and approaches vary.

Those studying active volcanoes are concerned with predicting volcanic eruptions and protecting human lives. They use several different approaches. They may install seismographs to measure Earth tremors, because tremors often suggest magma movement and possible eruption. They may monitor uplift and deformation of Earth's surface using automated land-based surveying systems and satellite measurements, because magma flowing into a region or moving toward the surface often produces bulges in the land above. They may measure gas emissions and temperatures using direct monitoring instruments or, remotely, infrared

spectroscopy, because changing gas compositions, increased emissions, or temperature increases often precede an eruption. Furthermore, they may use other kinds of measurements that also yield valuable information, including measurement of gravity, magnetism, and electrical resistivity.

Some volcanologists focus their studies on past eruptions—on dormant or extinct volcanoes—to learn how and why volcanoes erupt, to investigate Earth's evolution, or perhaps to discover the effects that volcanoes had on civilization. Their studies include geological mapping and determining the composition and age of volcanic rocks. Those volcanologists collect rock samples in the field, make thin sections for examining rocks with microscopes, and analyze rocks using the same instruments described in the previous two chapters. Key questions include when volcanic activity first began and how it has changed over time. Additional topics involve the ways that volcanic activity has affected Earth's atmosphere, climate, soil formation, and energy and ore deposits.

7.3 Volcanic eruptions

Figure 7.6 shows volcanoes that have erupted between 1986 and fall 2018. More than 50 different volcanoes are considered very active today. More than 500 volcanoes have erupted since the beginning of recorded history, and volcanic activity dates to the early days of Earth. Volcanic rocks more than 3 billion years old are found around the globe, but we cannot estimate the total number of eruptions that have occurred on Earth because much of the evidence is gone—covered or erased by later geological events.

7.3.1 Effusive vs. explosive eruptions

7.3.1.1 Effusive eruptions

Volcanic eruptions occur when magmas reach Earth's surface, and eruptions vary in violence and size. Some are small and only produce *spatter cones* (steep-sided mounds of welded lava fragments) around small *vents*. Other eruptions may produce larger *lava fountains*, or flows from *fissures*, that yield localized lava flows. Eruptions of this sort are happening on Hawaii today. The eruptions may be single events or multiple eruptions in the same area, but usually the volume and violence of eruptions are relatively small. These sorts of eruptions are termed *effusive eruptions*.

Figure 7.5 The Vesuvius Observatory near Naples.
Public domain photo from Carlomorino, Wikimedia Commons.

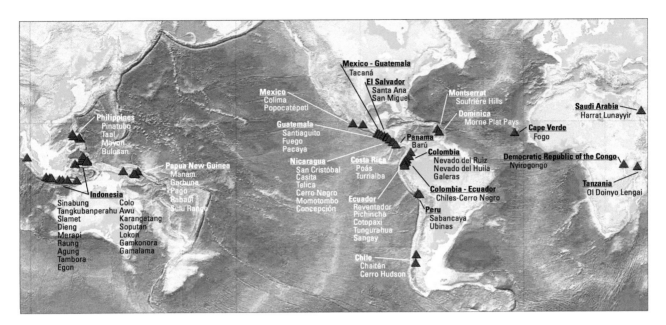

Figure 7.6   Erupting volcanoes 1986–2018.
Map from the USGS Volcanoes Disaster Assistance Program.

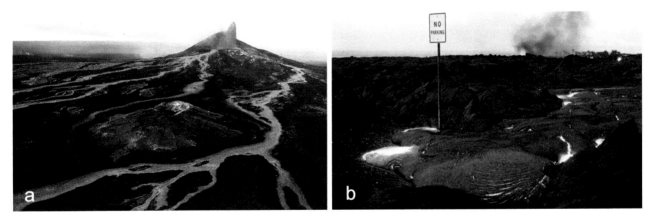

Figure 7.7   Two views of lava flowing from Kilauea Volcano, Hawaii Volcanoes National Park.
Photos from (a) USGS and (b) courtesy of Hawaii Volcano Observatory and Oscar Voss.

Figure 7.7 (left) shows a June 2, 1986 eruption producing lava flows from a lava vent on the side of Pu'u 'Ō'ō cone of Kilauea Volcano, Hawaii. The glowing lava rivers flowed for days and have flowed intermittently since then. Pu'u 'Ō'ō eruptions were nearly continuous for more than 40 years, and the lava sometimes traveled downhill several kilometers and covered roads. The photo on the right in Figure 7.7 shows partially solidified basalt and a no longer needed "No Parking" sign that restricted parking on the Chain of Craters road. Some Hawaiian flows travel greater distances than the flow shown, but never in excess of 5 or 10 kilometers.

Typical basalt flows at the surface travel quite slowly. For example, the top photo in Figure 7.8 shows flows emanating from a lava fountain on Mt. Etna, Italy, in 2001. Although only moving 2–3 meters per hour, the lava eventually reached the town of Nicolisi and caused considerable damage. However, if lava is confined underground in lava tubes, where it can maintain high temperatures, it can be significantly less viscous and thus flow much faster. The bottom photo in Figure 7.8 shows geologists looking through a *skylight* at lava flowing rapidly through a lava tube in Hawaii Volcanoes National Park. Flows of this sort may travel as fast as 35 kilometers an hour.

When lava cools and hardens, no matter its composition, it becomes a rock, also called *lava*. The geologists in Figure 7.8 (bottom photo) are standing on recently formed lava (rock) created by an eruption of Kilauea.

Figure 7.8   Basalt flows in Sicily and in Hawaii.
Images from Richard W. Williams, Wikimedia Commons (top) and
volcanoes.usgs.gov (bottom).

Figure 7.9   Pillow basalts near Hawaii.
Photo from NOAA.

Figure 7.10   Mt. Vesuvius with Pompeii in the foreground today.
Photo from Qfl247, Wikipedia Commons.

Lavas of this sort may contain small mineral crystals that are visible with the naked eye, but more commonly the crystals are much smaller. The rocks also may contain partly or wholly noncrystalline material (volcanic glass, termed *obsidian*), or xenoliths.

Although eruptions on land are the ones we generally think about, some effusive eruptions are submarine. Figure 7.9 shows *pillow basalts*, named for their shape, on the ocean floor near Hawaii. These pillows formed when basalt erupted under water and, consequently, cooled rapidly before it could flow any long distance. Sometimes pillow basalts may be obducted, as parts of ophiolites, onto continents and thus become more amenable to study. See Figure 6.30 of the previous chapter.

7.3.1.2   Explosive Eruptions

Effusive eruptions are only one of the two kinds of possible eruptions. The most violent eruptions are *explosive eruptions* that throw vast amounts of volcanic material into the air. The AD 79 eruption of Mt. Vesuvius that destroyed Pompeii was such an eruption. Pompeii and nearby Herculaneum were buried under 4 to 6 meters (13 to 20 feet) of volcanic ash, and 11,000 people died. Figure 7.10 shows ruins at Pompeii today, with Mt. Vesuvius in the background. The Vesuvius eruption is often thought of as a very large event, but the modern 1991 eruption of Pinatubo in the Philippines was three or four times larger. Still earlier, the prehistoric eruptions at Yellowstone or Toba may have been 1000 times larger than Pinatubo. Hawaiian eruptions are everyday events, but *supervolcanoes* the size of Yellowstone or Toba only occur on the order of every 100,000 years.

Explosive eruptions do not produce lava but instead eject volcanic *ash* and coarser material into the air. The ejected material, no matter its size, is termed *pyroclastic material*, or sometimes simply *ejecta*. Ejecta ranges from fine ash, with grain size of millimeters or smaller, to large bombs and blocks, commonly 30 to 60 centimeters (1 to 2 feet) in diameter but occasionally up to 3.5 meters (12 feet). After it falls to Earth, ejecta may remain as unconsolidated pyroclastic debris termed *tephra*, or if hot enough, fuse together (consolidate) to produce a *pyroclastic volcanic rock*, so named because such rocks are formed from "fire" and contain clasts of material cemented together much like a sedimentary rock. *Tuff* is the general term for fine- to medium-grained pyroclastic rocks, and *welded tuffs* are tuffs that form (weld together) immediately during cooling after material settles to the ground. Every eruption produces some combination of lava flows and pyroclastics. The flows and pyroclastics may collect in one place, producing a classic volcano-shaped mountain—a steep-sided conical mountain with a crater on top.

Figure 7.11 shows Mayon Volcano in the Philippines, a very active volcano that has erupted more than 50 times in the past four centuries. The eruptions usually involve huge volumes of pyroclastic material thrown into the air. Some of the material settles on the flanks of the volcano, producing flows of hot gas and ash that travel down the volcano's flanks, as seen here. This photo was taken during a 2001 eruption, but the most destructive eruption was in 1814 when tephra completely covered the nearby town of Cagsawa and caused several thousand fatalities. The total amount of ash was nearly 10 meters deep in some places, and some volcanologists believe that airborne ash from the eruption

Figure 7.11 Eruption of Mayon Volcano, Philippines, in 1981. Photo from Jaycee Esmeria, Wikimedia Commons.

contributed to the *year without summer* created when Tambora erupted the following year.

7.3.2 A spectrum of volcanic eruptions

Table 7.1 lists and compares some of the more famous eruptions that geologists and others have studied and

Table 7.1   A spectrum of volcanic eruptions.*

Volcano	Setting	Year of eruption	Volcanic Explosivity Index (VEI)	Volume km³
Kilauea, Hawaii	oceanic island hot spot	1983	3–4	0.1
Mauna Loa, Hawaii	oceanic island hot spot	1984	4	0.22
Mauna Loa, Hawaii	oceanic island hot spot	1976	4	0.375
Mt. Pelée, Martinique	subduction zone	1902	4	0.5
Mount St. Helens	subduction zone	1980	4	0.7
Surtsey, Iceland	mid-ocean ridge and hot spot	1963	4	1
Eyjafjallajökull, Iceland	mid-ocean ridge and hot spot	2010	4	2
Vesuvius, Italy	subduction zone	79	5	3
Pinatubo, Philippines	subduction zone	1991	5	10
Novarupta, Alaska	subduction zone	1912	6	13
Krakatoa, Indonesia	subduction zone	1883	6	18
Santorini, Greece	subduction zone	1650 ya	6	60
Mazama (Crater Lake), Oregon	subduction zone	7700 ya	7	75
Tambora, Indonesia	subduction zone	1815	7	150
Yellowstone, Wyoming	continental rift/hot spot	1.2 mya	7	280
Valles, New Mexico	continental rift/hot spot	1.4 mya	7	300
Long Valley, California	unknown	746,000 ya	7	500
Yellowstone, Wyoming	continental rift/hot spot	602,000 ya	7–8	1000
Taupo (Oruani)	subduction zone	26,500	8	1170
Yellowstone, Wyoming	continental rift/hot spot	2 mya	8	2500
Toba, Indonesia	subduction zone	74,000 ya	8	2800
La Garita (Fish Canyon)	subduction zone	28 mya	8	5000
Wah Springs, Utah	subduction zone	30 mya	8	5500

Data compiled from many different sources.

*ya = years ago; mya = million years ago; the VEI, a measure of volcano explosiveness, is based primarily on the volume of pyroclastic material ejected (see text for additional details).

most of the eruptions discussed in this chapter. The purpose of this table is to emphasize the wide variations in eruption locations and sizes.

These eruptions range from simple Hawaiian-type events to huge catastrophic explosions many thousands of times larger. Most eruptions occur at tectonic plate boundaries, either in a *subduction zone*, where one plate slides under another, or at *mid-ocean ridges*, where rising magmas produce new oceanic crust. Some volcanoes occur in intraplate settings where anomalously high heat flows from the mantle to Earth's surface, producing oceanic islands such as the Hawaiian Islands or continental hot spots including the active volcanic region at present-day Yellowstone National Park. Mid-ocean ridge eruptions, which are perhaps the most common eruptions on Earth, are rarely seen and have small direct impact on people. Consequently, they do not receive as much attention as eruptions that occur on land. The volcanic explosivity index, discussed later, is a logarithmic scale based primarily on the amount of pyroclastic material an eruption produces. The largest eruptions in this table, therefore, ejected more than 50,000 times more material than the small ones at the top of the table.

7.3.3   *Volcanic landforms*

7.3.3.1   *Shield volcanoes*

Figure 7.12 shows Mauna Loa Volcano, Hawaii, seen from halfway up Mauna Kea, 40 kilometers (25 miles) away. The slope of this volcano is gentle because the basaltic magma that was extruded had low viscosity and so spread out quickly after eruption. Mauna Loa has been erupting for at least 700,000 years and emerged above sea level about 400,000 years ago. No eruptions

have occurred at its summit since 1984, but vent eruptions away from the summit, like the one shown at Kilauea in Figure 7.7 (this chapter) and Figure 5.26 (Chapter 5), are occurring today.

Mauna Loa is an example of a *shield volcano*, the largest kind of volcanic landform. Shield volcanoes are huge with shallow slopes, typified by volcanoes in Hawaii. Eruptions commonly occur at a central vent associated with a crater at the volcano's summit, but *flank eruptions* are common on the slopes or even near the base of a volcano. Some flank eruptions, called *fissure eruptions*, are associated with long crack or rift systems. Flank eruptions, including fissure eruptions, may occur many kilometers away from the main summit. Shield volcanoes form over long periods, generally from many individual eruptions that involve multiple basalt flows. Mauna Kea and Mauna Loa, shield volcanoes on the big island of Hawaii, are so large that seeing their shapes without going to a different Hawaiian island is difficult. The parts of Mauna Kea and Mauna Loa above water are only 40–50 kilometers across, but both volcanoes extend beneath the sea a much greater distance.

Multiple flows in the same area can produce large *lava fields*, up to several square kilometers, next to volcanoes. Much larger, hundreds to thousands of square kilometer sized *lava plateaus* can form from a succession of eruptions, mostly from fissures instead of individual volcanic cones. Such eruptions produce *flood basalts*, often producing basalt flows that are many tens of meters thick and that travel hundreds of kilometers.

The Columbia River flood basalts, classic examples of flood basalts, were deposited between 6 and 17.5 million years ago. They cover most of southeastern Washington, a large part of northern Oregon, and part of Idaho, as shown in the map of Figure 7.13. The photograph in Figure 7.13 shows outcrops in the Moses Coulee near

Figure 7.12   Mauna Loa Volcano, Hawaii.
Photo from Madereugeneandrew, Wikimedia Commons.

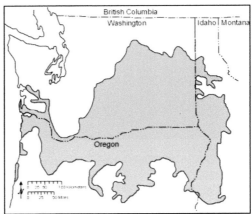

Figure 7.13 Outcrop and map of the extent of the Columbia River flood basalts.
Photo from Photo from William Borg, Wikimedia Commons; map from Paul Link, Idaho State University.

Wenatchee, Washington, a very small example of the Columbia River basalts. Overall, the basalts have a volume of about 75,000 cubic kilometers (42,000 cubic miles).

The flood basalts of Washington, Oregon, and Idaho cover a large region, but an even larger flood basalt terrain is found in India. India's ancient (66 million years old) Deccan Traps, a large igneous province on the Deccan Plateau of west-central India, is one of the largest volcanic features on Earth. The flows consist of many layers of solidified basalt that together are more than 2000 meters (6562 feet) thick. Today, the Deccan basalt covers more than 500,000 square kilometers (193,000 square miles) and has a volume of 512,000 cubic kilometers (123,000 cubic miles). Geologists have determined that the flows may have been three times larger when they originally formed, with large amounts of basalt later removed by erosion or covered by younger rocks.

Cooling flood basalts often shrink and develop columnar fractures and so become *columnar basalts*, such as the ones shown in the photograph of the Columbia River basalts in Figure 7.13. Erosion has uncovered the vertical columns to leave steep cliffs called *palisades*. Perhaps the world's most famous columnar basalts are those of the Giant's Causeway— shown in Figure 7.14. The causeway is part of a lava plateau on the north coast of Northern Ireland. The basalt is part of a large volcanic plateau called the *Thulean Plateau* that formed about 60 million years ago. In this view, the tops of columns can be seen, contrasting with the side view of the Columbia River basalts shown in Figure 7.13.

Figure 7.14 Giant's Causeway, Northern Ireland.
Photo from Code Poet, Wikimedia Commons.

7.3.3.2 Composite volcanoes

The panoramic image in Figure 7.15 was taken from the top of Paulina Peak, Oregon, looking northwest in July 2010. It shows Mount Bachelor (left), Broken Top, and the Three Sisters (right) in Oregon. These are three of the more than 80, some quite small, *composite volcanoes* in the Cascade Mountain Range of British Columbia, Washington, Oregon, and California. The volcanoes shown have been inactive for more than 1000 years, but other cascade volcanoes have been active recently. St. Helens, Lassen Peak, Hood, Rainier, Shasta, Glacier Peak, and Baker have all erupted during the past 300 years.

Compared with Hawaiian and other shield volcanoes, *composite volcanoes*, also called *stratovolcanoes*, are

Figure 7.15   Mount Bachelor, Broken Top, and the Three Sisters, Oregon.
Photo from Gabeguss, Wikimedia Commons.

generally smaller with steeper sides. Mt. St. Helens, the composite volcano near Seattle that is well known for its May 18, 1980 eruption, is about 3000 meters tall and 20 kilometers in diameter. It is a medium-sized composite volcano; others are perhaps twice as large as Mt. St. Helens. In contrast with composite volcanoes, large shield volcanoes like those in Hawaii are more than 4000 meters above sea level and up to 200 kilometers in diameter.

Both shield volcanoes and composite volcanoes form during multiple eruptions. However, in contrast with most shield volcanoes, composite volcanoes consist of combinations of lava flows and pyroclastic debris of variable composition. Basalt is usually absent or a minor component.

7.3.3.3   *Small features*

Composite, and especially shield volcanoes, may be large, but some individual volcanic eruptions are quite small. For example, eruptions may produce *cinder cones*, also called *scoria cones*, that range from being a few meters to hundreds of meters high. Figure 7.16 shows the SP cinder cone and crater near Flagstaff, Arizona. This is a basaltic cinder cone, and solidified basalt flows can be seen in the top of the photo. The cinder cone is about 250 meters (820 feet) high. It shows typical features of young cinder cones—it has steep sides and a large bowl-like crater at its summit.

Cinder cones are often found in *volcanic fields* where multiple cones are next to each other. Typical fields contain 10 to 100 separate volcanoes that are commonly associated with lava flows that fill valleys between cones. Some cones may erupt only once, but many erupt multiple times over many years. Figure 7.17 shows several cinder cones in northern Harrat Lunayyir, Saudi Arabia. Dark-colored lava flows (black) wrap around the large cone near the center of the photo.

Figure 7.16   The SP Crater in the San Francisco Volcanic Field, Arizona.
Photo from the USGS.

7.3.3.4   *Caldera*

Some volcanoes, including Toba and Tambora, have associated *caldera* that formed when ground collapses due to emptying of a magma chamber below. Calderas may be created quickly during explosive eruptions but also over longer times during effusive eruptions. In either case, when enough magma has been ejected, the empty magma chamber may be unable to support the weight of the ground above. The total area that collapses may be a few tens of square kilometers or much larger, up to thousands of square kilometers. If a caldera is very

their rims (see Fig. 5.4 in Chapter 5). Today, Yellowstone is the site of many hot springs and geysers, fueled by heat from magma below. Although there have been

large, it is generally difficult to identify the specific volcanoes that caused the caldera to form, because there may have been many.

Figure 7.18 shows picturesque Crater Lake, 20.6 square miles in size, in southern Oregon. The lake is a perfect example of a small caldera. Lake waters come only from snowmelt and rain and today occupy a caldera that formed about 7700 years ago when Mt. Mazama, a Cascade Range volcano, erupted and the ground subsided after 50 cubic kilometers of felsic magma was expelled. Mt. Mazama's eruptive history, however, extends back at least 400,000 years, and the volcano is still considered active today. Within a few hundred years after the last major eruption, smaller eruptions created Wizard Island (shown in this photograph) and a few other small cones in the caldera. Crater Lake is the deepest lake in the United States (594 meters; 1949 feet) and ranks in the top 10 worldwide.

The Yellowstone Caldera, 55 by 72 kilometers (34 by 45 miles), in northwest Wyoming, is the site of several huge eruptions between 2.1 million and 640 thousand years ago. The eruptions spread thick ash over much of the western United States. The earliest event, called the *Huckleberry Ridge eruption*, was one of the largest ever to occur on Earth. The last three major eruptions created calderas near present-day Yellowstone Park: The Henry's Fork Caldera, the Island Park Caldera, and the Yellowstone Caldera. These caldera are so large that they are difficult to see from the ground, but they can be traced in aerial views due to slightly elevated topography at

Figure 7.17 Cinder cones and lava flows in northern Harrat Lunayyir, Saudi Arabia.
Photo from USGS.

Figure 7.18 Crater Lake, Crater Lake National Park, south-central Oregon.
Photo from Dagmara Mach, Wikimedia Commons.

no major eruptions in more than half a million years, Yellowstone is still active and has the potential to erupt again in the future.

Yellowstone's volcanism stems from a *hot spot* under the North American continent—the same hot spot as the one that led to the Columbia River flood basalts, discussed earlier. As shown in Figure 7.19, the Yellowstone hot spot has caused volcanic activity for more than 15 million years. As the continent has moved southwest over the hot spot, the site of eruptions has migrated northeast to its present-day location. The numbers in Figure 7.19 show where the hot spot was in the past, from 15 million years ago to present.

7.3.3.5 Eruption columns and their deposits

Mt. St. Helens, in Washington, erupted on May 18, 1980 at about 8 a.m. The eruption, shown in Figure 7.20, was not particularly large compared to other explosive eruptions, but it was the only major eruption in the continental United States to occur in nearly 40 years. The eruption killed 57 people and destroyed many homes and much infrastructure. In contrast with *effusive eruptions*, which produce only flows, *explosive eruptions*, such as the Mt. St. Helens eruption shown here, throw magma and sometimes already solidified pyroclastic material into the air. Such eruptions produce large *eruption columns* that carry pyroclastic debris up to tens of kilometers into the air, occasionally reaching the stratosphere. See Figure 5.14, in Chapter 5, for a comparison of effusive and explosive eruptions.

Explosive volcanoes may erupt many times. For example, Mayon Volcano, shown in Figure 7.11, has exploded more than 50 times in the past 400 years. Other volcanoes may erupt for weeks or months and then become dormant. Figure 7.21 shows an eruption of Mt. Pinatubo, in the Philippines, in June 1991. The volcano has erupted many times, but the 1991 eruption was the only one in modern history. In contrast with Mt. St. Helens, Pinatubo exploded many times over an 8-day period. The large cloud of volcanic ash and gas above the volcano is impressive—this

Figure 7.20   Mt. St. Helens eruption in 1980.
Photo from the USGS.

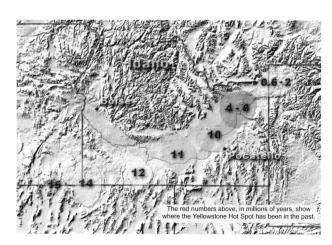

Figure 7.19   Path of the Yellowstone hot spot.
Image from Kevin Case, Wikimedia Commons.

Figure 7.21   Mt. Pinatubo erupting in 1991.
Photograph from the USGS.

photograph was taken three days before the volcano's last major explosion. In total, Pinatubo produced more than 10 times the amount of pyroclastic material produced by Mt. St. Helens (see Table 7.1), and, in the 20th century, only the 1912 eruption of Novarupta in Alaska was larger. More than 350 people died during Pinatubo's 1991 eruptions, most from collapsing roofs due to ashfall. Disease in evacuation camps and mudflows caused additional deaths, bringing the total death toll to 722 people.

Novarupta Volcano, in a remote part of Alaska's Aleutian Islands, is shown in Figure 7.22. When this volcano erupted in June 1912, it produced the largest eruption by volume in the 20th century. Three explosive episodes took place in over a little less than three days, producing 20 times more material than the 1980 eruption of Mount St. Helens and 30% more material than Pinatubo. The eruption ended when a lava *dome*, like the one shown here in Figure 7.22, formed in the summit crater and plugged the vent. The dome in this 1987 photo is nearly 90 meters (300 feet) high and 360 meters (1200 feet) wide. Novarupta still shows signs of activity today.

During an explosive eruption, lava cools quickly, and there may be insufficient time for atoms to arrange themselves in crystals. So, cooling lava may become glass before or after it reaches the ground. Pyroclastic material—including crystals, glass, and other fragmental material thrown into the air during an eruption—is quite variable in composition and contains particles of many grain sizes. Airborne material may be deposited to create *ashfalls*, like those that characterized the Tambora and Toba eruptions and created the Fish Canyon Tuff (Fig. 7.4), affecting life in profound ways. Although

pyroclastic material ranges from fine ash to quite large fragments (sometimes as large as footballs or basketballs, or even larger), the largest sizes fall to Earth quickly and collect close to a volcano. Finer-grained material, however, may spread out and create airfalls that cover larger regions. The finest material, as discussed previously, may remain suspended for days or longer and may travel around the globe, causing a volcanic winter like the *year without summer* created by Tambora or affecting climate for years.

Volcanic ash, when it settles, can be solid and dense, or very light, depending on how much air is between ash fragments. It may remain as unconsolidated ash, but often eruptions of felsic and intermediate magmas produce material that becomes welded during cooling and consequently becomes a pyroclastic rock. Although much less common, mafic magmas may produce pyroclastic rocks, too, most notably when mafic magmas interact with water, creating steam explosions that blast material into the air.

The Bandelier Tuff, in Bandelier National Monument (Fig. 7.23), New Mexico, contains the remains of homes carved into the tuff by Ancestral Puebloans more than 11,000 years ago. The soft tuff, made from glassy volcanic ash, is perfect for carving homes. It was deposited as ashfalls during eruption of the Valles Caldera Volcano 1.14 million years ago. Tuffs like the Bandelier Tuff are the most common products of explosive eruptions.

Different eruptions, and different times during a single eruption, often produce pyroclastic material of slightly different grain size or composition. When it

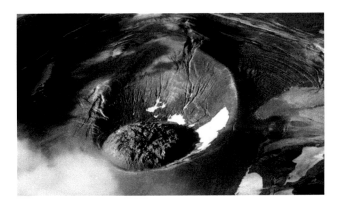

Figure 7.22 Novarupta Volcano in 1987.
Photograph from the National Park Service.

Figure 7.23 The Bandelier Tuff, New Mexico.
Photo from Douglas Perkins, Wikimedia Commons.

Figure 7.24 Outcrop of Bishop Tuff, near Mammoth Lakes, California.

Photo from Roy A. Bailey, Wikimedia Commons.

Figure 7.25 Divisadero Tuff, Sierra Madre Occidental volcanic field, Mexico.

From Cataclasite, Wikimedia Commons.

settles, we commonly see layered deposits such as that seen in the bottom part of the outcrop of Bishop Tuff, near Mammoth Lakes, California (Fig. 7.24). This tuff, however, appears to have formed in two ways. The bottom part of the outcrop shows layers that are typical for ashfall deposits

After losing its momentum, a rising eruption column may be overcome by gravity and collapse directly back to Earth. Clouds of hot gas (up to 1000 °C; 1800 °F), ash, and rock can rush down a volcano's side at up to 700 kilometers/hour (450 miles per hour) in a *pyroclastic flow* or *surge*. Figure 7.11 shows one example, and the top layer of ash in Figure 7.24, which is unstratified, probably formed from a pyroclastic flow too. Flows and surges are some of the deadliest volcanic events, but fortunately, today there is usually ample warning before they occur. They are sometimes called *nuée ardentes* (French for "glowing clouds"), a term first used to describe the deadly 1902 eruption of Mount Pelée on Martinique because, in the dark, the pyroclastic flows glowed red. The distinction between flows and surges is a hazy one based on the ratio of gaseous to solid material. Flows contain more solids and, due to their density, often follow and are confined to river valleys. Surges, however, are more gaseous than flows and, consequently, less dense. They rise to cross ridges and hills. Upon settling and cooling, pyroclastic material from both flows and surges typically fuses to produce tuffs called *ignimbrites*, testimonials to violent eruptions of the past.

The rock shown in Figure 7.25, from the Sierra Madre Occidental volcanic field in Mexico, is an ignimbrite containing consolidated volcanic ash with coarser-grained rock fragments of variable composition. Some of the fragments are pumice that has been squashed into flattened shapes called *fiamme*. The fiamme weathered and eroded, in places leaving long skinny holes.

7.4 Xenoliths and volatiles

7.4.1 Xenoliths

Magmas are composed of molten or semi-molten rock, sometimes xenoliths, and usually small amounts of volatiles (gases). Xenoliths and volcanic gases give us samples and valuable chemical data about regions deep within Earth—information that we cannot obtain in any other way. Xenoliths in magmas include individual crystals and rock fragments that are products of partial magma crystallization, or that were scavenged from surrounding solid rock that a magma passed through on its way to the surface. The general term *xenolith* describes all such inclusions, and the term *xenocryst* is used if an inclusion comprises a single mineral crystal. True xenoliths, sometimes called *exotic xenoliths*, have a distinctly different origin than the magma that incorporates them. Other xenoliths, which may be genetically related to the magma, are sometimes called *autoliths* or *cognate xenoliths/xenocrysts*. The amounts and natures of xenoliths vary greatly for different magmas and eruptions.

Figure 7.26 shows a dark-colored pyroxenite xenolith in a volcanic rock from La Palma, Canary Islands.

Figure 7.26 Pyroxenite xenolith in a volcanic rock from the Canary Islands.
Photo from Siim Sepp, sandatlas.org.

Figure 7.27 Vesicles.
Photo from Chmee2, Wikimedia Commons.

Besides the xenolith, the rock contains visible crystals of K-feldspar (tan colored) and lesser amounts of black biotite in a gray groundmass. The large (7 centimeters across) pyroxenite fragment is a piece of rock that was plucked from deep in the Earth as the magma moved toward the surface.

7.4.2 Volatiles

Magma may contain up to 7% volatiles (dissolved gases). Water vapor is the most common volatile, carbon dioxide too may be significant, and lesser amounts of other gases are typically present. In particular, besides water and carbon dioxide, magmas typically contain sulfur, either in the form of sulfur dioxide or as hydrogen sulfide. The amount of sulfur varies from volcano to volcano. When a magma reaches Earth's surface, gas pressure is released and the dissolved gases may escape into the atmosphere. Commonly, however, some gas bubbles become trapped, forming open holes called *vesicles*, which are preserved in solidified *vesicular* volcanic rock. Figure 7.27 shows a colorful lizard on top of vesicular basalt on Lanzarote Island in the Canary Islands, Spain.

In the absence of a volcanic eruption, volcanic gases may reach the surface to form *fumaroles*, openings where volcanic gases are emitted for prolonged times. Some fumaroles, such as the one in Figure 7.28 on Mauna Loa Volcano, Hawaii, are associated with significant sulfur deposits because, when gases reach the surface, water vapor, carbon dioxide, and other gases escape into the air, leaving sulfur behind. The inset in

Figure 7.28 Sulfur deposits on Mauna Loa Volcano, Hawaii.
Photo from the USGS.

Figure 7.28 shows massive sulfur at Mauna Loa, with a glove for scale.

Gas pressure is a major force—really, the only major force—behind violent volcanic eruptions. Dissolved volatiles form bubbles as magma moves upward and confining pressure decreases. The bubbles expand as pressure is released with continued upward movement (see Figure 5.13 in Chapter 5). The expanding gas not only provides gas pressure, but also dilutes the magma, making it less dense and adding buoyancy that enhances upward movement. When gas expansion and upward movement are fast enough, the result is an explosive eruption. Consequently, the main factor determining whether an eruption is effusive or explosive is the amount of volatiles that a magma contains.

Explosive eruptions often put on a spectacular pyro-technic display, such as the one shown in Figure 7.29. The photograph shows the eruption of Yasur Volcano, in Vanuatu, 1600 kilometers (1000 miles) northeast of Australia. Yasur is a stratovolcano that is about 360 meters (1185 feet) high. It was created by subduction of the Indo-Australian Plate under the Pacific Plate. Eruptions have occurred here, more or less continuously, for a few hundred years. The explosive eruptions produce amazing shows when viewed at night.

All magma contains some volatiles, but gases escape more easily from thin, fluid magmas faster than from thick gooey ones. Consequently, gas pressure may be released slowly or quickly depending on magma composition, and fluid magmas tend not to become highly pressurized and so may erupt quietly. In contrast, viscous magmas—gooey magmas, often containing large amounts of gas that cannot escape—tend to produce more violent eruptions. Thus, viscosity limits and controls volatile content that, in turn, influences the nature of eruptions. The most viscous magmas are felsic and the least viscous magmas are mafic. So, basaltic (mafic) eruptions are generally quieter and less dangerous than rhyolitic (felsic) ones.

Thus, viscosity explains why the volcanoes in Hawaii, associated with mafic magmas, have not blown their tops like Mt. St. Helens did in 1980 (Fig. 7.20) and in several more recent, smaller eruptions. Figure 7.30 shows Mt. St. Helens in 2004. A small explosive eruption, emitting ash and steam, is occurring in the summit crater—24 years after the top of the mountain blew apart in the major event that killed several dozen people. The hill of material in the center of the crater is a dome

made of very gooey felsic magma and rock. Domes of this sort are common features of felsic volcanoes.

7.4.3 *Hydrovolcanic eruptions*

Explosive eruptions are not always caused by volatiles in magma. Hot magmas may interact with ocean water, surface water, or groundwater, producing steam-powered *hydrovolcanic eruptions*. Hydrovolcanic eruptions can produce powerful explosions, especially in oceanic regions. On November 20, 2013, the Japanese island of Nishinoshima was created by a series of spectacular eruptions and explosions that lasted many weeks. Figure 7.31 shows the island during its early days of formation. This infrared image shows lava erupting at the island summit and near the shore. During the formation

Figure 7.30   Ash and steam escaping from Mt. St. Helens, 2004. Photo from the USGS.

Figure 7.29   Mt. Yasur eruption, Tanna Island, Vanuatu. Photo from Rolfcosar, Wikimedia Commons.

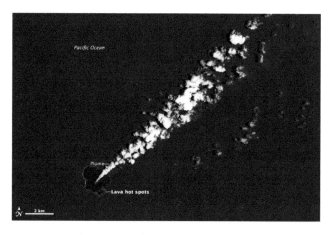

Figure 7.31   Short-wave infrared image of Nishinoshima Island, Japan. Photo from NASA.

of this island, lava, ash, and rock fragments, propelled by steam derived from ocean water, exploded into the air. The result, today, is that an island exists where none existed previously.

On continents, hydrovolcanic eruptions may produce shallow craters called *maars*, often surrounded by unconsolidated deposits called *tuff rings*. Figure 7.32 shows a maar and tuff ring at Kilbourne Hole, in the Potrillo Volcanic Field of New Mexico. Maars form when an explosive eruption creates a broad flat-bottomed hole in the ground, and tuff rings—a little like dunes—contain material ejected during the explosion. The tuff rings at Kilbourne Hole appear as narrow ridges that surround the shallow crater. The tuff lies on top of black basalt that predates the maar formation. Maars range in size from 60 to 8000 meters in diameter and 10 to 200 meters in depth. The maar shown here is a relatively small one.

Hydrovolcanic eruptions can produce eruption columns that are mixtures of pyroclastic material and water. Due to their high density, the columns often collapse on themselves to produce surges, called *base surges*, similar to, but often more dynamic than, surges produced by a collapsing eruption column associated with a normal "dry" volcano. In such a surge, a fast-moving cloud of steam and debris hugs the ground and spreads outward from the center, traveling as much as a kilometer from the eruption site. The surges have geometry and motion similar to what occurs during nuclear explosions—clouds rolling outwards in a big ring.

Figure 7.32 Kilbourne Hole, a maar, and tuff ring near El Paso, Texas.

Photo from Akanawa, Wikimedia Commons.

Explosive hydrovolcanic eruptions may cause surrounding country rock to fragment and become part or all of the ejecta, producing what is called a *phreatic eruption*. Such eruptions produce columns that are composed of rock fragments and steam, with very little or no lava. Much of the volcanic activity of Mt. St. Helens, in 1980, involved phreatic eruptions, and Kilauea Volcano, in Hawaii, known today for its typical effusive eruptions, has been more violent in the past. On May 10, 1924, a huge phreatic eruption began at Kilauea's summit; it lasted nearly three weeks. Eruption columns rose to altitudes of more than 1830 meters (6000 feet), and huge rocks—up to 8 tons—were tossed more than 800 meters (half a mile) from the crater. After the eruption subsided, a large crater 950 meters (3150 feet) across and 400 meters (1300 feet) deep was left behind.

If an eruption includes both lava and rock debris, as occurred at Nishinoshima, it is a *phreatomagmatic* eruption. The 1991 eruption of Pinatubo was also, in large part, phreatomagmatic. The term *Surtseyan* (discussed later) is sometimes used in place of *phreatomagmatic* because the Icelandic island of Surtsey was born in such an eruption.

7.5 Classifying eruptions

Some eruptions are hundreds of, or even a million, times more violent than others. Volcanologists classify eruptions as different types depending on eruption size and violence. The chart in Figure 7.33 compares different types of eruptions and some of their physical characteristics. The right-most column shows idealized cross sections depicting the various kinds of volcanic landforms that might form during different kinds of eruptions. The least violent eruptions are called *Hawaiian* eruptions. They are followed, with increasing degrees of explosiveness, by *Strombolian*, *Surtseyan*, *Vulcanian*, *Peléan*, *Plinian*, *ultra-Plinian*, and *supervolcanoes*. The different types of eruptions are named after classic volcanoes.

In 1982, Chris Newhall of the United States Geological Survey and Stephen Self at the University of Hawaii developed a quantitative measure to compare volcano explosiveness: the volcanic explosivity index (VEI). This index is based primarily on the volume of pyroclastic material ejected by the volcano, with some consideration of eruption column height and duration of eruption. The VEI scale is a logarithmic scale. A volcano

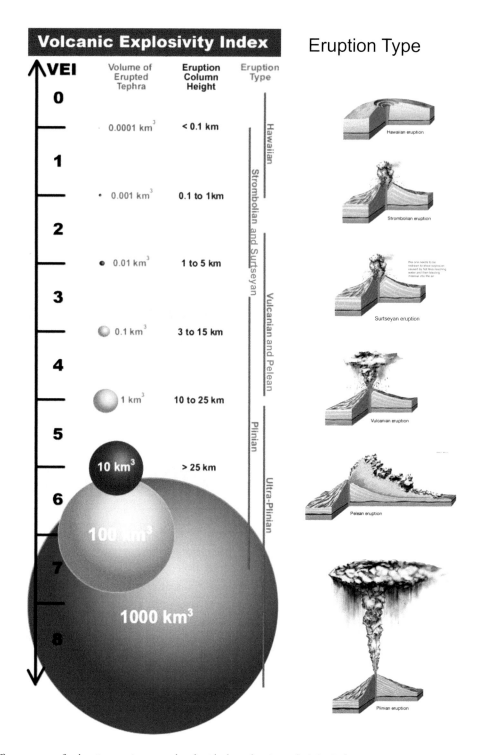

Figure 7.33   Different types of volcanic eruptions correlated with the volcanic explosivity index.
This figure was created by combining and modifying images from britannica.com, geology.com, and several other sources.

with a VEI of 6, then, is 10 times more powerful than one with a VEI of 5, and 100 times more powerful than a volcano with a VEI of 4. As shown in Figure 7.33, the VEI scale correlates well with eruption types. Volcanoes with low VEI (Hawaiian or Strombolian type) are commonplace today, but those with high index (ultra-Plinian) occur rarely. Only four eruptions with VEI of 7, and none with VEI of 8, have occurred during the past 10,000 years. Supervolcanoes are exceptionally rare. Toba, 74,000 years ago, was the last one.

7.5.1 Hawaiian eruptions

Relatively tame *Hawaiian eruptions* are effusive, characterized by fluid basaltic flows that over time often lead to the creation of shield volcanoes. Figure 7.34 shows Hawaiian basalt. After it solidifies, basaltic lava produces flows of two types: *'a'ā* and *pāhoehoe*—both can be seen in Figure 7.34. The 'a'ā appears black, and the pāhoehoe is dark gray. The two names derive from Hawaiian words that describe the flows' textures. 'A'ā flows are characterized by rough top surfaces composed of angular pieces of lava. Pāhoehoe flows are smooth, sometimes called "ropy." The differences are explained by the rate at which the lava flows. 'A'ā forms when lava flows rapidly and sort of tumbles over itself at the leading edge of a flow, but pāhoehoe flows more like a (very slow) smooth river or stream. Besides basalt flows, Hawaiian eruptions often produce cinder cones associated with lava fountains; additionally, large lava lakes may form in caldera. For other photos of Hawaiian basalt flows, see Figures 7.7, 7.8, and 7.12, earlier in this chapter.

7.5.2 Strombolian eruptions

Strombolian eruptions, named after the very active island volcano Stromboli, near Sicily, are characterized by eruptions that toss incandescent ejecta of various sizes as high as several hundred meters. Figure 7.35 shows an eruption of Pacaya Volcano in Guatemala. This volcano first erupted 23,000 years ago and has erupted many times since then. It became especially active after 1961. Pacaya is one of many active volcanoes in Central America that are characterized by Strombolian eruptions. They are the result of subduction of the Cocos Plate beneath the Caribbean Plate. Figure 7.29 is another photo of a Strombolian eruption.

Strombolian eruptions are powered by gas bubbles in mafic magma. When the bubbles reach the surface, they may pop open, leading to many brief but spectacular episodes of pyrotechnics. The lavas, which are generally basalt, typically produce cinder cones during eruptions. Strombolian eruptions are shorter lived and involve fewer flows than Hawaiian eruptions but may lead to formation of small to large composite volcanoes composed of solidified lavas, ash, pumice, and other tephra.

7.5.3 Vulcanian eruptions

Figure 7.36 shows Santiaguito Volcano in Guatemala during a 2016 eruption. This type of eruption is called a Vulcanian eruption, named after a small volcanic island about 25 kilometers (16 miles) north of Sicily. Santiaguito and Vulcano are both active volcanoes, but Vulcano has not erupted in more than 125 years. The last eruption, occurring between 1888 and 1890, deposited several meters of pyroclastic material on the island. *Vulcanian eruptions* are named after this last major eruption. Vulcanian eruptions are explosive, involving felsic or intermediate (rhyolitic or andesitic) magmas that contain gas bubbles like more mafic Strombolian magmas do. Due to greater magma viscosity, Vulcanian

Figure 7.34   Solidified 'a'ā (rougher and black) and pāhoehoe (smoother and gray) lavas.
Photo from Steven Reynolds.

Figure 7.35   Eruption of Pacaya Volcano, Guatemala.
Photo from Rolfcosar, Wikimedia Commons.

Figure 7.36   Santiaguito Volcano erupting in 2016.

Photo from La Coordinadora Nacional para la Reducción de Desastres, Guatemala.

Figure 7.37   The town of St. Pierre, Martinique, after the 1902 eruption of Mt. Pelée

Photo from DeGolyer Library, Southern Methodist University, Wikimedia Commons.

eruptions are generally more explosive than Strombolian ones and may create eruptions columns that rise as far as 10–20 kilometers (6–12 miles) into the atmosphere. These eruptions typically involve several small explosions and are sometimes partly hydrovolcanic. Because it rises to the high altitudes, fine ash can be spread over wide regions. Additionally, collapsing eruption columns may create pyroclastic flows (nuée ardentes) that travel downhill at speeds as fast as 150 kilometers (90 miles) per hour. And, melting of snow and ice at high elevations may release water that combines with ash to create fast-moving volcanic mudflows called *lahars* that rush downhill and wreak havoc.

7.5.4 Peléan eruptions

Peléan eruptions are named after Mt. Pelée on the island of Martinique in the Caribbean. The May 8, 1902 eruption of Pelée was one of the worst volcanic disasters in

history (Fig. 7.37), killing 29,000 people and destroying Martinique's capital city of St. Pierre. In the past 500 years, only the eruption of Tambora (1815) has been responsible for more direct deaths; however, several other eruptions have killed more people due to lingering climate change. Peléan eruptions are similar to Vulcanian eruptions in some ways, but in contrast with them, Peléan eruptions often occur as a single large explosion.

Pyroclastic flows are an essential part of the definition of a Peléan eruption. During Peléan eruptions, gooey viscous felsic to intermediate lavas may collect to form a bulging dome (bottom of Fig. 5.11 in Chapter 5; Fig. 7.30, this chapter) at the volcano's summit. The dome has a crust and may be stable for a while, but if the crust bursts, the dome may collapse, releasing large amounts of pyroclastic material that rush downhill. This is what happened in 1902—the destruction of St. Pierre was caused by a nuée ardentes resulting from the collapse of a magma dome on Pelée volcano's summit.

7.5.5 Plinian eruptions

Plinian eruptions, sometimes called *Vesuvian*, are named for the historical eruption of Mount Vesuvius in AD 79 that was chronicled by Pliny the Younger. Plinian eruptions vary from large to super-large. Even the smallest Plinian eruptions produce at least a cubic kilometer of ejecta and send gas and ash through the troposphere and into the stratosphere. Eruptions are often accompanied by loud sounds of explosion that can be heard far away from the eruption site. Some Plinian events are short lived, perhaps lasting less than a day, but others may continue for many months.

Redoubt Volcano, located about 160 kilometers (100 miles) southwest of Anchorage near the Cook Inlet, is a stratovolcano that has erupted four times since 1900. Redoubt's Plinian and sub-Plinian explosive eruptions, like most large explosive eruptions, involve primarily felsic to intermediate magmas. In 1989, this volcano began erupting on December 14 and eruptions continued for more than six months. The photo in Figure 7.38, taken on April 21, 1990, shows an eruption characterized by major explosions that produced ash columns rising to more than 20 kilometers (12 miles) altitude. During this eruption, pyroclastic flows and lahars flowed down the north flank of the mountain.

Ultra-Plinian eruptions produce more than 100 cubic kilometers of ejecta, and *supervolcanoes*, the largest of Plinian eruptions, produce more than 1000 cubic kilometers. Fortunately, supervolcano eruptions are rare—the remains of about 20 have been identified around the globe. The last one to erupt was the Taupo Volcano of New Zealand, about 26,500 years ago. Of the volcanoes previously discussed in this chapter, La Garita (Fish Canyon), Toba, and two of the Yellowstone eruptions fit in this category.

During the largest Plinian eruptions, gas pressure initially builds up in large, generally felsic (rhyolitic) magma chambers and then increases as magma moves upwards through narrow conduits. Upon eruption, huge eruption columns form and the ejecta may incorporate shattered pieces of solid rock picked up near the surface.

Gas continues to expand, powering upward movement, and the columns that characterize these eruptions may rise to elevations of 50 kilometers (30 miles) above Earth. The largest Plinian eruptions do not involve individual explosive events but instead columns that persist for hours or days. As gravity takes over, thick layers of tephra are deposited over large regions, and dangerous pyroclastic flows can speed down the volcano's sides and travel hundreds of kilometers outward. Additionally, fast moving lahars may cause immense damage.

7.5.6 Surtseyan eruptions

Surtseyan eruptions are named after the Island of Surtsey, shown in Figure 7.39. The island is 35 kilometers (20 miles) south of Iceland. It is in shallow water (about 130 meters deep) and formed over 3½ years during nearly continuous eruptions that began in 1963. The eruptions continued until early June 1967, when the island reached about 2.7 square kilometers (1.0 square mile) in size. Since then, erosion by wind and waves has caused the island to decrease in size. Surtsey was originally thought to represent an island forming on a mid-ocean ridge. But, more recently, it has been attributed, at least in part, to a hot spot.

Surtseyan eruptions, sometimes called *hydrovolcanic eruptions*, involve magma-water interactions that produce large volumes of steam and, consequently, violent steam explosions producing a great deal of ash. The eruptions may wax and wane as they persist over years. Although common in oceanic regions, similar eruptions can occur on continents when rising magma encounters

Figure 7.38 Redoubt Volcano, Aleutian Islands, Alaska.
Photo from R. Clucas, Wikimedia Commons.

Figure 7.39 Surtsey Island forming in 1963.
Photo from NOAA.

groundwater, and eruptions are often associated with base surges and the formation of maars and tuff rings (Fig. 7.32).

7.6   From magma to rock

Chapter 5 discussed the origin of magmas and their highly variable chemistries. Although generally dominated by molten rock, magmas may be partially solid, containing mineral crystals or solid glass fragments and sometimes rock fragments at the time of eruption. Consequently, when magmas extrude at Earth's surface, the resulting rocks have diverse compositions, textures, and properties. Magmas vary from volcano to volcano, and even a single volcano may produce different magmas, and therefore different rocks, at different times. Table 6.2, in Chapter 6, defines some common words used to describe volcanic rocks, including most importantly *aphanitic, porphyritic, equigranular,* and *glassy* (*vitreous*).

As discussed in Chapter 5, after eruption, lava will cool quickly, and there is little time for mineral crystals to grow. Consequently, crystals, if present, may not be visible with the naked eye, or even with a microscope, and most volcanic rocks are aphanitic. Figure 7.40 shows a typical felsic volcanic rock, rhyolite. The minerals it contains (K-feldspar and quartz) are so fine grained that they cannot be seen in this photo. The reddish color of the rock is because this specimen contains K-feldspar, a mineral that often has a salmon or reddish color.

When crystals begin to grow in magma chambers before eruption, the eventual result may be a porphyritic rock containing relatively large mineral crystals (*phenocrysts*), surrounded by a fine-grained *groundmass*. Groundmass crystals need not be microscopic, and if large enough to see, grains may appear euhedral or subhedral. When fine grained, they are typically anhedral. Because groundmass solidifies rapidly during later stages of cooling, it may be partly or entirely volcanic glass.

Figure 7.41 shows a porphyritic latite. Latite is a kind of volcanic rock that contains about equal amounts of plagioclase and K-feldspar, and sometimes a small amount of quartz. The largest, somewhat blocky crystals in this photo are phenocrysts of gray plagioclase. This rock also contains phenocrysts of black clinopyroxene, which are just barely visible. The dark colored groundmass, between the larger phenocrysts, is a mix of feldspars, augite, biotite, and magnetite. The combination of larger phenocrysts in a sea of fine-grained material reflects the rock's two-stage cooling history. See also Figure 5.30 in Chapter 5 for another view of a porphyry.

Phenocrysts in porphyritic volcanic rocks may be quite small. Figure 7.42 is a microscope view of a thin section of a basalt from France. This photo was taken using crossed polarizing filters, so the colors are not true mineral colors. Most of the brightly colored mineral crystals are phenocrysts of augite, a kind of pyroxene; the largest phenocrysts are about 1 cm long. Smaller and hard to see gray and white phenocrysts are crystals of plagioclase, all much less than a centimeter in length. In contrast with the matrix in Figure 7.41, the matrix in this rock is mostly volcanic glass; glass always appears black when viewed with crossed polarizing filters.

Sometimes cooling is so rapid that rocks may contain large amounts of obsidian (volcanic glass). Felsic

Figure 7.40   Sample of rhyolite from Castle Rock, Colorado.
Photo from D. Perkins, GeoDIL.

Figure 7.41   Porphyritic latite from the Bearpaw Mountains, northcentral Montana.
Photo from D. Perkins, GeoDIL.

Figure 7.42 A porphyritic basalt seen in thin section.
Photo from j. m. derochette.

Figure 7.43 Outcrop of obsidian and pumice, Inyo Domes, California.
Photo from the USGS.

Figure 7.44 Pumice, with a flake of biotite on top, and scoria.
Photos from (a) Ra'ike and (b) Avenue, both at Wikimedia Commons.

(Si-rich) magmas, which start at lower temperatures than mafic magmas do, may solidify partially or entirely to form obsidian. Shiny and dense black obsidian is often found in the margins of felsic flows where it is commonly associated with light-colored *pumice*, a porous (low-density) volcanic rock, containing open holes called vesicles created by gas bubbles created at the time of eruption as pressure was released. Although felsic rocks, due to their high silica content, are the most likely to contain obsidian and pumice, volcanic rocks of all compositions contain variable amounts of glass.

Figure 7.43 shows obsidian (black) and pumice (gray) in an outcrop near Mono Lake, California. The two have the same composition, but the pumice is full of vesicles (holes) that were once occupied by volcanic gases. The obsidian solidified under pressure, so no gas was released to produce bubbles.

Pumice is not the only possible kind of vesicular rock. Figure 7.44a shows pumice, and Figure 7.44b shows scoria—both contain vesicles. Most pumice has a rhyolitic (felsic) composition. *Scoria*, also called *cinder*, is an extremely vesicular form of basalt (mafic). It often appears to have the same texture as pumice and may contain small (< 1 mm) vesicles with fine mineral crystals and only small amounts of glass. The pumice in this figure comes from Mendig, Germany; it has a small flake of black biotite on its top. The scoria is from Mt. Tarawera in New Zealand. Scoria is found in many places, and the red variety (sometimes called *lava rock*) seen here is used for landscaping. Some pumice contains so many holes that it will float on water like a sponge. Scoria, however, is denser and always sinks.

Secondary minerals are common in volcanic rocks, because minerals that form at high temperature are typically unstable under Earth surface conditions. So, they may react to form any of the secondary minerals discussed in the previous chapter. Additionally, volcanic glass is unstable and will, over time, *devitrify* (become partially or wholly crystalline). Consequently, it is uncommon to find obsidian older than 10 or 20 million years. Figure 7.45 shows a pig carved out of *snowflake obsidian*. The white "snowflakes" are made of cristobalite (a polymorph of quartz) crystals that formed by devitrification of once homogeneous solid obsidian glass.

Secondary minerals also may form when flowing pore or surface water deposits minerals in vesicles, creating features called *amygdules*. Figure 7.46 shows a basalt with white minerals on vesicle walls. Zeolites are typical amygdule minerals, but calcite, quartz, and several others are also common. Amygdaloidal rocks, such as the one shown in Figure 7.46 from Mauritius, are common and widespread.

Figure 7.45   Pig carved from snowflake obsidian.
Photo from Adrian Pingstone, Wikimedia Commons.

Figure 7.46   Amygdules in vesicular basalt from Mauritius.
Photo from Evelyn Mervine, AGU Blog.

7.7   Naming volcanic rocks

Some volcanic rocks can be named based on a combination of texture and mineralogy. The common primary minerals in volcanic rocks are the same as those in plutonic rocks: generally, quartz, K-feldspar, plagioclase, muscovite, biotite, amphibole, pyroxene, or olivine (Table 6.1 in Chapter 6). These are also the *essential minerals* sometimes used to assign rock names. Volcanic rocks also contain the same *accessory minerals* that are found in plutonic rocks and that sometimes can be used as varietal names. However, many volcanic rocks contain small or large amounts of volcanic glass in lieu of minerals or contain mineral crystals that are too fine grained to be easily identified.

Although sometimes absent, when present, phenocrysts help distinguish different kinds of volcanic rocks. Figure 7.47 shows the common phenocrysts in rocks ranging from felsic to ultramafic. These phenocryst minerals

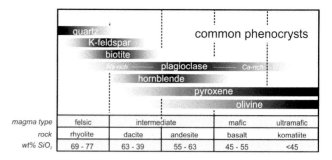

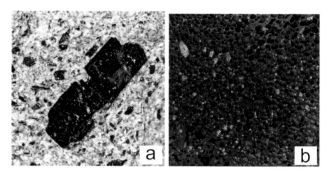

Figure 7.47   The most common phenocrysts in volcanic rocks.

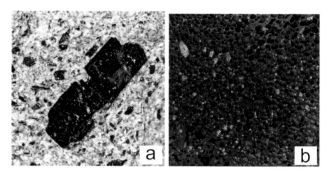

Figure 7.48   Porphyritic andesite (a) and a vesicular olivine basalt (b).
Photos from Piotr Sosnowski and Siim Sepp, both from Wikimedia Commons.

are the same minerals that may be present as microscopic crystals in the groundmass. From felsic to ultramafic, the volcanic rocks are rhyolite, dacite, andesite, basalt, and komatiite. Quartz and K-feldspar are generally restricted to relatively felsic rocks, olivine and pyroxene to relatively mafic rocks, and the other minerals to intermediate rocks. Plagioclase is a solid-solution mineral in volcanic rocks of many sorts and varies from being Na-rich for felsic rocks to being Ca-rich for mafic and ultramafic rocks. Figure 7.47 is only an approximation, because the minerals present as phenocrysts depend on many different things.

Volcanic rock names are commonly modified based on characterizing accessory minerals or texture. The thin-section view in Figure 7.48a of a *porphyritic andesite* shows an intermediate composition rock that contains small lathe-like plagioclase and a single 1-centimeter-long phenocryst of black augite (pyroxene). Smaller augite crystals are also present mixed with plagioclase. As another example, the *vesicular olivine basalt* shown in Figure 7.48b (from La Palma, Canary Islands) contains many 3–5 millimeter-sized vesicles and relatively large green olivine phenocrysts in a fine-grained matrix. A different basalt (not shown) might be described as a *vitreous basalt* if it contained a lot of obvious glass. However, many basalts contain neither olivine nor glass.

Often—adding a significant complication to assigning rock names—the typical fine grain size in volcanic rocks can make mineral identification difficult or impossible (see, for example, Figure 7.40), even with aid of thin sections and a petrographic microscope. And, many volcanic rocks are largely or entirely glass. For example, all the black material in Figure 7.42 is glass. For porphyritic volcanic rocks, we can often identify the phenocryst minerals—which give a hint to overall rock composition. Yet, in some porphyries, where the number of phenocrysts is small compared to the amount of groundmass, phenocryst mineralogy may not be representative of the entire rock. So, sometimes classifying and naming volcanic rocks based on mineralogy is problematic.

7.7.1 Identifying rocks in the field

Because identifying different kinds of fine-grained rocks in outcrops, or in hand specimens, is problematic, geologists sometimes use approximate classification systems. Often, the best we can do when naming volcanic rocks without the aid of analytical instruments is to note rock color and texture. Rock color depends on the minerals present and their grain sizes. Usually, rocks that contain abundant feldspars and quartz have a light color; those that contain abundant mafic minerals (Fe- and Mg-rich minerals such as amphibole, pyroxene, and olivine) have a dark color. The percentage of dark-colored minerals is the rock's *color index*, and in general the higher the index, the more mafic a rock is.

Figure 7.49 illustrates a simple naming system that can be used when examining volcanic rocks in the field. The top part of the diagram provides a rock name. The bottom part of the diagram lists the (likely) minerals present, although they may not be large enough to be identified. Alkali feldspar, the kind of feldspar that dominates felsic rocks, is often pinkish or tan. Because felsic rocks generally contain alkali feldspar, felsic rocks are generally light-colored (low color index) and may have a pink or tan hue, and mafic rocks are dark colored (high color index). Intermediate composition rocks fall somewhere between. Unfortunately, some fine-grained rocks may appear to be darkly colored, even if they are felsic. Often, the best that can be done when looking at rocks in the field is to call them *basalt* if dark colored, *andesite* if white or gray, and *rhyolite* if pinkish or tan. Upon thin section examination, or after obtaining a chemical analysis, the names may have to be corrected.

	Felsic (light colored; often pink)	Intermediate (light to dark gray)	Mafic (dark colored)
Fine	Rhyolite	Andesite	Basalt
Vesicular	Pumice		Scoria
Glassy	Obsidian		
	Common Minerals Present		
	Quartz K-feldspar Na-plagioclase	Na-Ca Plagioclase Amphibole	Ca-Plagioclase Pyroxene

Figure 7.49 A simple classification scheme for naming volcanic rocks in the field.

7.7.2 IUGS classification

The IUGS classification system, introduced in Chapter 6, divides volcanic rocks into 6–10 fundamental groups, each of which contain several rock types (Table 7.2). Unfortunately, identifying and naming rocks cannot be done the same way for all groups. Some names are assigned based on the minerals a rock contains, some based on the overall rock chemistry, and some by rock texture. So, rocks belonging to the different groups are classified in different ways. The most important groups of volcanic rocks are *pyroclastic rocks* and *feldspathic (feldspar containing) non-pyroclastic rocks*. These groups include the most common volcanic rocks at Earth's surface. Rocks belonging to other groups are quite rare and will not be discussed further here.

7.7.2.1 IUGS system for naming feldspathic effusive rocks

The most common effusive rocks contain quartz, alkali feldspar, and plagioclase, singly or in combination. If these minerals can be identified and their relative proportions determined, the proportions yield an IUGS rock name. The modes (volume %) of quartz (Q), alkali feldspar (A), plagioclase (P), and feldspathoids (F) are determined—most commonly by examining a thin section with a petrographic microscope—and the percent values are normalized so that they add up to 100%.

Once mineral modes have been determined, the results are plotted on a QAPF diagram, shown in Figure 7.50, to determine a rock name. A similar

Table 7.2   Six rock categories from the iugs classification system for volcanic rocks.

Group	Definition	Basis for classification	Most common rock names
pyroclastic rocks	fragmental rocks composed of airfall (minerals, crystals, or glass, ejected from a volcanic vent)	rock texture and size of clasts	agglomerate, lapilli tuff, coarse tuff, fine tuff
feldspathic effusive rocks	rocks in which feldspars (plagioclase, K-feldspar) and sometimes quartz are essential components	minerals present or overall rock chemical composition	basalt, andesite, rhyolite, dacite, trachyte, phonolite, tephrite, foidolite
ultramafic rocks	rocks with relatively low SiO_2 content that generally contain essential olivine, orthopyroxene, or clinopyroxene	relative amounts of the different kinds of pyroxenes and olivine present	peridotite (e.g., dunite, harzburgite, lherzolite, and wehrlite), pyroxenite (e.g., orthopyroxenite, websterite, and clinopyroxenite)
carbonatites	volcanic rocks that contain more than 50% carbonate minerals	the kind of carbonate present (e.g., calcite, dolomite, siderite)	calcite-carbonatite, dolomite-carbonatite, ferrocarbonatite
melilitic and related rocks	rocks that contain essential melilite, kalsilite, leucite, or other quite rare low-SiO_2 minerals	somewhat variable for different kinds of rocks, but generally based on mineral proportions	there are many different names; none are common
high-Mg rocks	ultramafic lavas and high-Mg basalts	rock chemical composition	komatiite, picrite, and other less common rock names

diagram, with different names for plutonic rocks, was presented in Chapter 6. The sizes and shapes of the different named fields for volcanic rocks are similar, but not identical, to the fields in the diagram for plutonic rocks. Despite some lingering debate and some shortcomings, the IUGS system shown here gives useful and consistent results for any rock that contains at least 10% by volume of identifiable quartz, alkali feldspar, plagioclase, and feldspathoids combined. It is the most widely used classification scheme today.

Two different triangles (top and bottom of Figure 7.50) are used for rocks that contain quartz (the top triangle) and the (less common) rocks that do not contain any quartz due to low SiO_2 content (upside-down bottom triangle). The two share the AP line. If rocks do not contain quartz, they often contain a mineral belonging to the *feldspathoid group*. So, for low-SiO_2 rocks, feldspathoids replace quartz, and naming is based on the modes of alkali feldspar, plagioclase, and a feldspathoid. Because quartz and feldspathoids are never found together, it is generally clear whether to use the top or bottom half of the diagram. Some of the rock names in this system include the term *foid*, which refers to any feldspathoid mineral, the most common being nepheline and leucite.

Most igneous rocks contain minerals other than quartz, alkali feldspar, plagioclase, and feldspathoids. Some may contain abundant mafic minerals such as olivine or pyroxene, but the IUGS system does not

consider this. Consequently, two rocks with dissimilar appearance may have the same name using the IUGS system. Other effusive rocks may be extremely mafic, containing very small amounts of, or no, quartz and feldspars. For such rocks, the IUGS classification system for ultramafic rocks must be used. Furthermore, when looking at a hand specimen, it can sometimes be difficult to distinguish plagioclase from alkali feldspar, making the standard IUGS system unworkable.

7.7.2.2   *The TAS alternative*

Determining modes can be challenging. For *plutonic* (phaneritic) rocks, a rough determination may be possible while examining outcrops or hand specimens in the field. More accurate analysis, however, generally requires painstaking examination of a thin section, but using the IUGS system for plutonic rocks is generally practicable. In contrast, determining mineral modes for *volcanic* rocks can be difficult or impossible due to small grain size. Although some volcanic rocks contain larger phenocrysts, the fine grain size in general may mean that minerals cannot be identified in a hand specimen or in a thin section. Moreover, if phenocrysts are present, they may not be the same minerals as minerals in the groundmass, so the modes of phenocrysts may not reflect the true mineral modes. Adding more difficulty, many volcanic rocks contain significant amounts of glass.

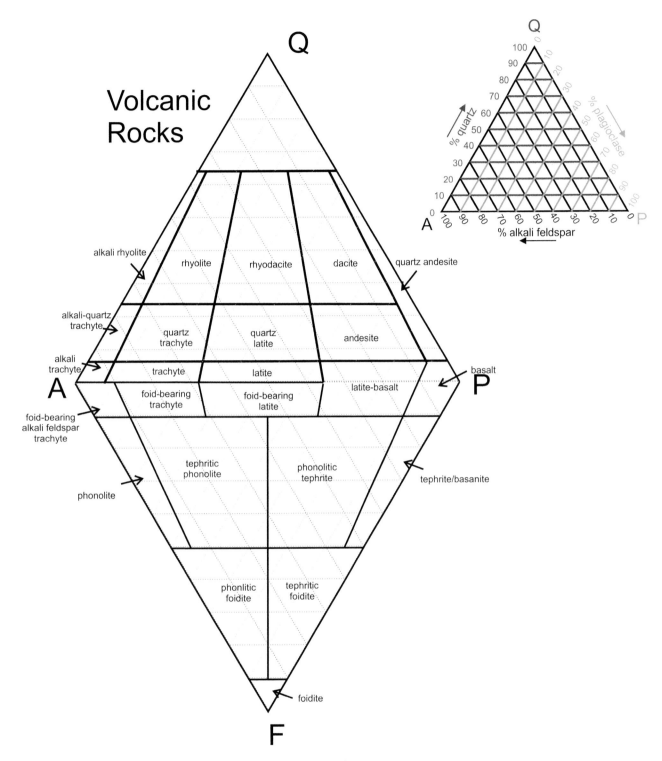

Figure 7.50 The IUGS classification system for feldspathic effusive rocks.

For the reasons described above, the IUGS recommends using the TAS (total-alkali-silica) diagram (shown here in Figure 7.51) to name rocks if mineral modes cannot be determined. The TAS system is, today, more often used than the QAPF system to name volcanic rocks. This figure is the same diagram presented in Chapter 5 to name magmas, so this system gives a rock and its parent magma the same name. A second advantage is that even rocks that contain no visible minerals, or that contain large amounts of glass, can be unambiguously named.

In the TAS diagram, the vertical axis is the total alkali oxide ($Na_2O + K_2O$) content of a rock and the horizontal axis is the silica (SiO_2) content of the rock. The diagonal red line divides rocks into alkalic (top) and sub-alkalic (bottom) categories. To use the TAS system, we must know the chemical composition of a rock, so we can plot rock composition appropriately. Acquiring whole rock chemical analyses is not difficult but is considerably more complicated than making thin sections. Doing analyses can be time consuming and requires specialized instruments. The most common analysis methods used today are X-ray fluorescence spectroscopy (XRF) and inductively coupled plasma mass spectrometry (ICP-MS). Chapter 5 contains descriptions and photographs of these instruments.

The TAS and QAPF classification systems are closely related, and the two systems yield the same, or nearly the same, rock names if we can apply them both. The horizontal axis in the TAS diagram (wt% SiO_2) correlates with the vertical axis in QAPF diagram (the amount of quartz or feldspathoids present in a rock) because the modal amounts of either quartz or feldspathoids are functions of SiO_2 concentrations. Rocks that contain abundant quartz are relatively rich in SiO_2; those that contain less quartz contain less SiO_2. Rocks that contain feldspathoids (foidites) are even poorer in SiO_2. The vertical axis in the TAS diagram (wt% $Na_2O + K_2O$) correlates with the horizontal axis in the QAPF diagram, because feldspathic rocks high in $Na_2O + K_2O$ generally contain abundant K-feldspar; those that contain less $Na_2O + K_2O$ contain more plagioclase.

7.7.2.3 IUGS system for naming pyroclastic rocks

Pyroclastic material, material ejected into the air during an eruption, is quite variable, being composed of clasts of different compositions that range from fine ash to large bombs and blocks. So, once it reaches the ground, the material becomes tephra that can contain clasts of different sizes (Table 7.3). The clasts themselves may be crystals, crystal fragments, glass fragments, or rock fragments. Furthermore, although some tephra and, consequently, some pyroclastic rocks are composed entirely of pyroclastic material, others are not. Tephra or rocks may contain pieces of local country rock or sediments that mixed with the pyroclastic material or may contain mineral or rock fragments formed after the initial volcanic event. Additionally, weathering or erosion may alter or modify pyroclastic material, producing *reworked* tephra, or pyroclastic rocks formed from reworked tephra. So, pyroclastic rocks are diverse in both origin and properties.

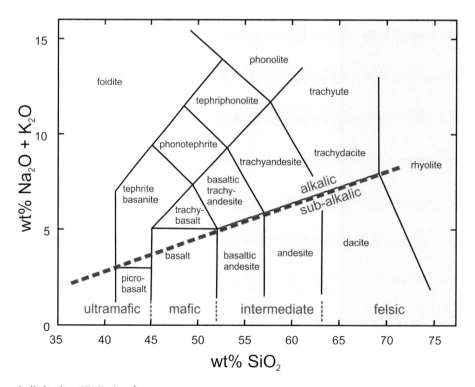

Figure 7.51   The total-alkali-silica (TAS) classification system.

Table 7.3 Different types of pyroclastic materials and rocks.*

Clast size (mm)	Pyroclast	Pyroclastic rock
> 64	bomb (molten or semi-molten at time of eruptions) block (solid at time of eruption)	agglomerate (mostly rounded clasts) pyroclastic breccia (mostly angular clasts)
64–2	lapillus	lapilli tuff, lapillistone
2–1/16	coarse ash	coarse ash tuff
< 1/16	fine ash	fine ash tuff

*Clast size refers to diameter or to longest dimension.

The geologist shown in Figure 7.52a is sampling tephra horizons near Hekla Volcano in southern Iceland. The material in this outcrop is all volcanic ash. The thick and light-colored layer at the center of the photo has the composition of rhyolite; other layers are more mafic. The geologist in Figure 7.52b is examining pumice blocks at the edge of a pyroclastic flow from Mount St. Helens. The pyroclastic material in this flow is significantly coarser than the material from Hekla.

The principal types of pyroclasts (see Table 7.3) are bombs, blocks, lapilli, and ash, distinguished by their sizes. For the largest size, the term *bomb* is reserved for clasts formed from material ejected from a volcanic vent while molten or semi-molten. Bombs develop "bread-crust" or other surface textures, suggesting that they were cooled quickly, shaped, and solidified as they traveled through air. In contrast, pyroclastic *blocks* have an angular or blocky texture, suggesting that they were already solid at the time of eruption (Fig. 7.52b). The term *lapilli* (*lapillus*, singular) refers to any pyroclast

with a mean diameter of 2 to 64 millimeters, and *ash* (sometimes divided into fine ash and coarse ash) refers to finer material.

If a pyroclastic rock contains clasts of uniform size, we easily name it using the names listed in Table 7.3. However, many pyroclastic rocks contain a mix of clast sizes, so we often invoke a more complicated naming scheme (shown in Fig. 7.53). To use this system, the amounts of grains of different sizes are determined and normalized to 100% to provide modes. The result can then be plotted on the triangle

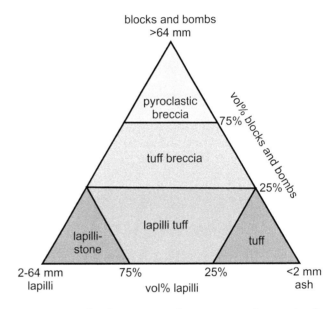

Figure 7.53 Classification system for naming pyroclastic rocks of mixed grains size.

Figure 7.52 Volcanic ash (a) and blocks of pumice (b).
Photo from Dentren and Donald A. Swanson, both at Wikimedia Commons.

Table 7.4   Some terms used to describe pyroclastic rocks.

Terms describing nature of clasts	Terms describing composition	Terms describing genesis
vitric: tephra or rock composed partially or entirely of glassy clasts	rhyolitic: having the composition of rhyolite	airfall: uniform deposit of material which has been ejected during a volcanic eruption
crystalline: composed of crystalline clasts	dacitic: having the composition of dacite	flow: deposit from a fast-moving current of tephra with some gas
lithic: composed of clasts of rock fragments	andesitic: having the composition of andesite	surge: deposit from a fast-moving current of tephra
	basaltic: having the composition of basalt	lahar: a mudflow composed of pyroclastic material, rocky debris, and water

to obtain a rock name. The names in Table 7.3 and Figure 7.53 can be made more specific by using modifiers that describe the nature of the clasts, the composition of the rock, or its origin. Table 7.4 lists some of the more commonly used modifiers. A *vitric tuff*, for example, is a tuff dominated by glassy clasts. A *basaltic tuff-breccia* is one having the overall composition of a basalt.

Questions for thought—chapter 7

1. How do volcanologists monitor a volcano to determine if an eruption is imminent? List and describe several different kinds of information they collect.
2. The "Ring of Fire" encircles the Pacific Ocean. What characterized this ring? Why is it there, and what causes it? And, why is there not a similar ring around the Atlantic Ocean?
3. What are pyroclastic materials? Why do very violent volcanic eruptions usually produce more pyroclastic materials than lava flows?
4. Why do mafic lava flows produce shield volcanoes and flood basalts, but not felsic volcanic eruptions?
5. What are composite volcanoes? How do they form? How do they differ from shield volcanoes?
6. How do ignimbrites differ from tephra? Both are typically made of felsic material, not mafic. Why? Why not mafic?
7. What are vesicles and what produces them in volcanic rocks? What was present in the magma that led to the production of vesicles?

8. Why are some volcanic eruptions far more explosive than others? (There are at least two key parts to the answer.) And, why are felsic and intermediate composite volcanic eruptions more likely to be explosive than eruptions from shield volcanoes?
9. What is the VEI? How is it determined? What kinds of eruptions rank lowest on the VEI scale, and what kinds rank highest? Consider volcanic activity on Hawaii, the eruption of Mt. St. Helens in 1980, and the eruption of Toba 74,000 years ago. How did they rank on the VEI scale and how life threatening were each of them?
10. What are aa and pahoehoe lava flows? In what kind of lava do they most commonly occur? How do they form and why are there differences?
11. How do vulcanian eruptions differ from strombolian eruptions? Which eruptions are more violent?
12. What are the major differences between an aphanitic rhyolite and a rhyolite porphyry? How do each form? And, what about the processes of formation explains their differences?
13. Obsidian, volcanic glass, is common in many basalts that we see today. But, only in relatively young basalts. What is obsidian and how does it form? Why is it absent from Precambrian rocks on Earth?
14. Why are volcanic rocks often more difficult to properly identify and name in the field than identifying and naming plutonic rocks?
15. Why don't geologists use the IUGS system for classifying igneous rocks in the field? And, what are some advantages of the TAS system over the IUGS system?

8 Sediments and Sedimentary Rocks

8.1 Sedimentation during the formation of the Appalachian Mountains

The timeline seen in Figure 8.1 shows the part of the geologic time scale covering the period when the Appalachian Mountains started forming, 500 to 350 million years ago. The mountain range originated during three pulses of mountain building—called *orogenies*; the black bars in this figure show the two earliest orogenies. Orogenies do not start and stop abruptly, but the Taconic Orogeny occurred around 480 to 445 million years ago, and the Acadian Orogeny occurred about 390 to 360 million years ago. The third major Appalachian orogeny, the Alleghanian Orogeny (not shown in Fig. 8.1), occurred around 325 to 260 million years ago. Each of the three orogenies created a tall mountain range near where the present-day Appalachians are found. But, between orogenies, and since the last one, a great deal of erosion occurred. Thus, most of the mountains that were once there are now gone.

The Taconic Orogeny happened in phases when small microcontinents and volcanic islands—similar to present day Japan, Philippines, and Indonesia—merged westward and collided with ancient North America. This produced a lengthy mountain chain that once extended from eastern Canada south to Georgia. Remnants of the chain are found in the Taconic Mountains (part of the Appalachians along the eastern border of New York State) and in a narrow belt that extends intermittently south to Georgia.

Figure 8.2, a cross section, shows merging volcanoes and microcontinents that caused the Taconic Mountains to be pushed upward. Volcanism and plutonism also added mass to the mountains. As the mountains grew higher, the steep topography naturally led to erosion. Typical orogenies deliver vast amounts of eroded sediment to nearby basins and the Taconic was no exception. Rivers transported huge amounts of sediments to the west, and the sediment collected in a shallow inland sea between the mountains and the main part of what is now North America. The deposits created a huge delta complex—more than 483 kilometers (300 miles) across in the east-west direction—larger than any deltas in the world today. The delta contained sediments of many sorts, including gravel, sand, silt, and mud. These sediments, in layers up to 300 meters (1000 feet) thick, are called the *Queenston Delta* (labeled in Fig. 8.2). The largest accumulations were in regions that are now New York and Quebec. Lesser amounts were deposited in eastern Ontario, Pennsylvania, and some in Ohio.

Most of the Queenston sediments accumulated between 451 and 446 million years ago—shown by a yellow bar on the timeline in Figure 8.1. The sediments later became sedimentary rocks, including various sorts of conglomerate, sandstone, siltstone, and shale. And many of these Queenston rocks have a reddish color due to iron oxidation,

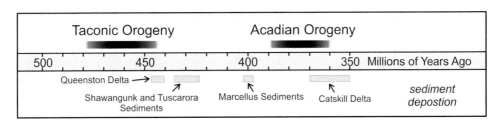

Figure 8.1 Timeline for the first two orogenies that contributed to formation of the Appalachian Mountains.

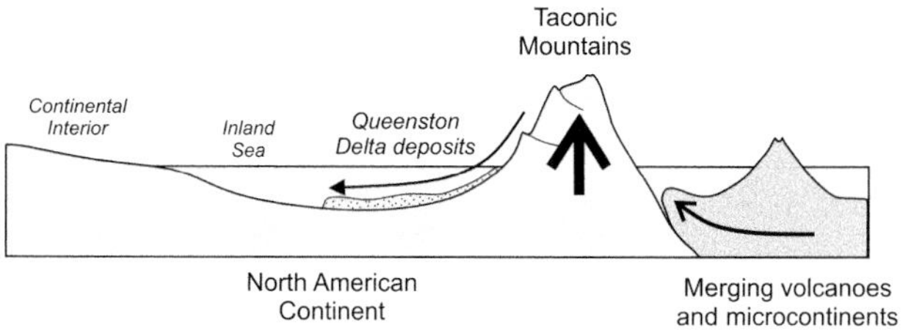

Figure 8.2   The formation of the Taconic Mountains, 480 to 455 million years ago.

suggesting sediment deposition on land or in oxygen-rich waters. The coarsest and sandiest sediments collected closest to the Taconic Mountains; most of the sediments farther west were fine muds or silts, because coarse material settled before being carried that far.

Erosion of the Taconic highlands continued long after the formation of the Queenston Delta had slowed. For example, coarse conglomerates of the Shawangunk Formation in New York, New Jersey, and Pennsylvania, and of the equivalent Tuscarora Formation farther south, are evidence of major rivers that continued to deliver eroded material to an ocean basin. Figure 8.3 shows a major outcrop of the Tuscarora Formation at Seneca Rocks, West Virginia. There, the formation consists of coarse sandstone and conglomerate, but in other places it contains lesser amounts of graywacke, siltstone, and shale. The sediments that led to these rocks were deposited in continent, delta, or shallow marine environments around 430 million years ago, shown by a yellow bar in Figure 8.1.

Around 400 million years ago, the ocean environment changed. Highlands in both the east and west caused the inland sea to become isolated and somewhat stagnant, and thus lacking in oxygen. Lack of oxygen meant that organic material did not decompose, so organic-rich clays and muds collected in a region that is now western New York, Pennsylvania, and parts of Ohio. These sediments (shown by a yellow bar in Fig. 8.1) later became the Marcellus Shale, named after a town in central New York. Typical sedimentary rocks contain no more than a few percent organic material, but the Marcellus exceeds 10% in some places. High organic content and low-oxygen concentrations led to large accumulations of petroleum and, today, the Marcellus Shale is a very productive petroleum source rock.

As shown in Figure 8.1, the Acadian Orogeny, the second major Appalachian Orogeny, occurred 20 million years after deposition of the Marcellus sediments. It occurred as small continents merged on the eastern shore of North America, affecting a region from Nova Scotia to Alabama, with greatest impacts between the mid-Atlantic states and Canada. The highest mountains were in eastern Pennsylvania and New Jersey and, like the mountains formed during the Taconic Orogeny, they soon began to erode and shed sediments into the inland sea to the west. Between 380 and 360 million years ago, rivers delivered thick sequences of highly varied material, up to 3000 meters (10,000 feet) thick, across much of New York and Pennsylvania and into Ohio as the Acadian highlands eroded away.

Figure 8.3   Outcrop of the Tuscarora Formation at Seneca Rocks, West Virginia.
Photo from Jarek Tuszynski, Wikimedia Commons.

Figure 8.4 The Catskill Formation (Devonian) near North Bend, Pennsylvania.

Photo from M. C. Rygel, Wikimedia Commons.

The sediments produced by the Acadian Orogeny (shown by a yellow bar in Fig. 8.1) collected to create the *Catskill Delta*. This more recent delta was, in many ways, similar to the Queenston Delta that formed nearly 100 million years earlier. In contrast with the Marcellus Shale, which was deposited in a low-oxygen marine environment, most of the Catskill sediments were deposited in an oxidizing environment on land, in river and stream beds, on floodplains, or in ephemeral lakes. Rocks formed from these sediments crop out extensively in the Pocono and Catskill Mountains of Pennsylvania and New York. Figure 8.4 shows an outcrop of the Catskill Formation in north-central Pennsylvania. This formation is made of sediments that accumulated over more than 20 million years as the Acadian Mountains eroded away. The Catskill Formation, which is up to 3000 meters (10,000 feet) thick in some places, includes mostly red sandstone derived from continental sediments (such as the sandstone shown in Fig. 8.4), although minor marine rocks are found in some places.

The final pulse of Appalachian mountain building, the Alleghanian Orogeny, occurred 260–325 million years ago when the Atlantic Ocean closed as the African continent collided with North America. This orogeny, longer lasting than the Taconic and Acadian orogenies, also led to thick sequences of sediments being deposited. Some sedimentation occurred to the west but, in contrast with the earlier two orogenies, huge accumulations of Alleghanian sediments are today found to the east of the main mountain belt where they are now parts of the continental shelf, the coastal plain of North America, and the adjacent Piedmont Mountains. If the earlier orogenies—the Taconic and Acadian—deposited sediments east of the mountains, they were buried or removed by subsequent events.

8.2 Sediment and sedimentary environments

Weathering and *erosion* produce sediment that later may become sedimentary rock. The two terms are closely related—weathering refers to the breakdown or disintegration of a rock in place, and erosion refers to processes that carry rock or soil away from their sources. *Terrigenous sediments*, such as the sediments that dominated the Queenston and Catskill deltas, account for most sediments. They are sediments derived from weathering and erosion of rocks on land and may include individual mineral grains or pieces of rock, sometimes biological components, chemical precipitates from water, material produced by volcanism, and (rarely) extraterrestrial contributions, including meteorites. Once created, these sediments can collect in many different environments, including most significantly the ones depicted in Figure 8.5. Some sediments accumulate in continental environments, and others are deposited in beaches, deltas, or other *transitional environments* between continental and marine. Eventually, however, blowing wind, flowing water, or moving ice transport almost all terrigenous sediment to an ocean where it becomes *marine sediment* deposited on the ocean floor.

When flowing water reaches an ocean basin, it slows and, consequently, can no longer carry large amounts of material. So, gravity deposits sediment near shorelines on *continental shelves* (Fig. 8.5). Thus, marine sediments cover about a quarter of the ocean floor—generally the portions near continents—and these shelf deposits may be several kilometers thick. In many places, undersea landslides carry shelf sediment down *continental slopes* to the *continental rise* and deeper parts of the oceans (Fig. 8.5). The coarsest material will settle first, and finer sediment may stay in suspension and move across *abyssal plains*, perhaps even to an ocean's center. Besides terrigenous sediments, the deepest parts of the oceans also contain sediments derived from marine sources, for example from the remains of marine organisms, reef deposits, volcanic debris, or chemical precipitates from seawater.

Overall, marine sediments account for roughly 90% of all sediments deposited on Earth. These sediments are highly variable, but they usually retain characteristics that provide clues to their origins. For example, terrigenous

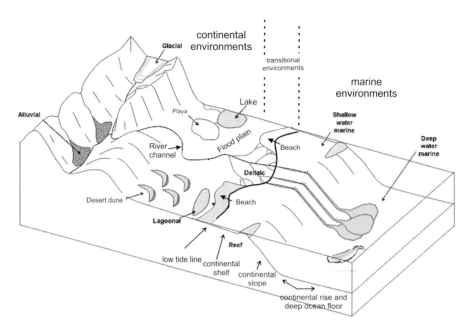

Figure 8.5　Some of the many environments where sediments may be deposited.

sediments contain minerals and rock fragments derived from erosion of mountainous regions, sediments associated with reefs commonly contain carbonate minerals and fossils, and sediment derived from volcanic terranes contains glass and volcanic rock fragments.

Once sediment settles in layers, it may harden to produce a sedimentary rock. This hardening process adds other characteristics that are preserved in the rock and that record characteristics of the deposition and rock-forming processes. So, sediments and sedimentary rocks provide us with a record—generally incomplete, however—of past geological histories, including both sediment origins and rock-forming processes.

Because we live on land, people are familiar with the ways that sediments collect on land. River and lake deposits, swamps and marshes, deserts and glaciers—we can examine sediments being deposited in all of these places. It is important to remember, however, that most of the sedimentary rocks that we see in outcrops originated as undersea deposits. Some formed from sediment deposited near shore—in deltas, on beaches, or on barrier islands. Other rocks formed from continental shelf deposits or are associated with reefs, and still others derive from sediment deposited in deeper parts of the oceans. So, Figure 8.5, which will be discussed more in Chapter 9, shows many different depositional environments. This figure, however, gives a distorted view of their relative importance, because just a few of the

environments—the marine environments—account for most of the sedimentary record.

8.3　Products of weathering and erosion

8.3.1　*Different kinds of sediments*

Figure 8.6 shows weathered debris associated with sandstone (blocky remnants in the photo) and sloping mudstone outcrops. The weathering products created here are typical, including solid residual materials and dissolved materials. The solid material, called *clastic material*, or *detritus*, consists of loose (*unconsolidated*) rock and mineral debris resting on slopes and at slope bottoms. Individual clasts—some quite large and some quite small—comprise both single mineral grains and *lithic fragments* (pieces of rock). Because they are heavy, boulders and other coarse clasts tend to stay near where they form, unless they break apart into smaller pieces. Thus, they will remain as part of the local *regolith*, a general term for any loose unconsolidated material that sits on top of bedrock. In contrast, finer material will be carried away by erosion. Coarser material near its source and finer material that is transported and deposited, perhaps by flowing water or by blowing wind, are both types of *clastic sediment*.

Figure 8.6 Weathering of sandstone and mudstone in the Grand Staircase-Escalante National Monument, Utah.
Photo by W. C. Poon, Wikimedia Commons.

Figure 8.7 Weathering of limestone, Burren National Park, County Clare, Ireland.
Photo by A. King, Wikimedia Commons.

Weathering is of two fundamental kinds—*physical weathering* and *chemical weathering*—which often work together. Figure 8.6 illustrates physical weathering—the decomposition of rock into smaller pieces. Physical weathering dominates in arid regions, such as the one shown in Figure 8.6. In wetter places, however, the dominant form of weathering may be chemical weathering, weathering that produces dissolved material by reaction with water, instead of fractured pieces of rock. Figure 8.7 shows weathered limestone in Ireland. Although some clastic sediment has collected next to the outcrops, much of the weathering occurring here is due to chemical dissolution that erodes the limestone, leaving the pitted and etched surfaces that we can see. This type of weathering is typical for limestone, and often allows geologists to identify an outcrop as limestone with a quick glance.

Weathering of limestone releases dissolved ions (Ca^{2+} and CO_3^{2-}) that flowing water may carry short or long distances before the ions are precipitated (deposited) as *chemical sediment*. Chemical weathering of other kinds of rocks may yield other ions or dissolved compounds, and thus other kinds of chemical sediment. No matter its composition, much chemical material ends up making its way to the oceans, where it becomes part of the dissolved material in seawater.

Two additional mechanisms—biochemical and organic sedimentation—both involving organic processes, also produce sediments. Plants and animals make *biochemical sediments* if they secrete hard parts such as shells and bones that remain and collect after the rest of the organism has decomposed. For example, Figure 8.8

Figure 8.8 Cockle shells on a beach, Shark Bay Marine Park, Western Australia.
Photo by B. W. Schaller, Wikimedia Commons.

shows an Australian beach covered with cockle shells. Dead plants and animals also make *organic sediments*—sometimes considered as a different category than biochemical sediments—when their dead remains, mostly intact, collect, perhaps at the bottom of a lake or in a swamp.

Thus, sediments have many diverse origins and are highly variable. Their compositions and physical characteristics reflect many things, including the nature of the parent material, the climates where they came from, how they were deposited, the kind of weathering and erosion involved, and the distances they traveled before deposition.

8.3.2 Clastic sediments

Most sediments are clastic sediments—Figure 8.9 shows some examples that contain different grain sizes. From fine to coarse, these are silt, sand, pebbles, and cobbles. Clastic sediments can be of many different compositions and grain sizes; thus, if they eventually become rocks, the resulting *clastic sedimentary rocks* will be of many kinds too. When we study these sedimentary rocks, we find that they have distinct characteristics reflecting their origins. Geologists use those characteristics to determine how and where ancient rocks formed.

Products of weathering, including both single mineral grains and lithic fragments, make up most clastic sediments and thus most clastic sedimentary rocks. Sometimes a single kind of clast makes up an entire rock, but this need not be the case. Sandstones, for example, may contain only quartz grains of uniform size. In contrast, Figure 8.10 shows a type of sedimentary rock closely related to sandstone, called *graywacke*, that contains several kinds of relatively coarse lithic fragments (that are up to 1 cm across) in a fine-grained matrix of sand-sized individual mineral grains.

The minerals present at the time a clastic sediment first collects are the sediment's *primary minerals*. Primary minerals include both individual mineral grains and minerals in lithic fragments. Most primary minerals are residual products of weathering; such materials are typically quartz and clays. Additional primary mineralogical components may include volcanic debris, such as ash, or chemical minerals deposited from water. Sediments may also contain roots, leaves, shells, pollen, or other biological material, but generally in subordinate amounts. And, after deposition, chemical reactions in sediment sometimes add small amounts of *secondary minerals*, such as clays or zeolites, that form when primary minerals react with water or air.

Erosion moves all kinds of weathering products downhill. Normal mechanical weathering processes, including abrasion or frost action, may start the erosion process, and larger-scale erosion processes—such as slides, slumps, and rock falls (collectively called *mass wasting* or *mass movement*)—may be locally important. Some clastic material remains near where it forms, but flowing water, blowing wind, and sometimes gravity can move it long distances. Eventually, some sediment may be deposited in valley bottoms, river beds, or lake bottoms. Most sediment, however, ends up making it to an ocean, although the paths followed are highly variable. No matter where it settles or what route it follows, once deposited, sediment may sit unconsolidated for a long time. Alternatively, it may be buried and become a sedimentary rock, or it may be eroded and carried away to continue through the rock cycle.

silt

sand

pebbles

cobbles

Figure 8.9   Clastic sediments of different kinds.
Silt from R. Faridi, sand from Bill Cunningham, USGS, gravel from I. Valencia, acapulcorock.com, cobbles from M. A. Wilson, Wikimedia Commons.

Figure 8.10 Sedimentary rock (graywacke) from Scotland containing mineral grains and lithic fragments.

Photo from Siim Sepp, Sandatlas.org.

Figure 8.11 Bowling Ball Beach in Greater Farallones National Marine Sanctuary 160 kilometers (100 miles) northwest of San Francisco.

Photo from NOAA.

8.3.2.1 *Mineralogical composition and grain size*

The term *siliciclastic* refers to sediments composed mostly of silicate minerals. The most common sedimentary rocks—including shale, sandstone, and conglomerate—form from siliciclastic sediments. Other, less common kinds of rocks consist of carbonates, iron oxides, and hydroxides, such as hematite, goethite, or other minerals.

Geologists classify siliciclastic sediments based on grain size and composition. Clast sizes vary from fine clay and silt to huge boulders. Figure 8.9 shows some of the finer sizes and Figure 8.11 shows some coarser small- to medium-sized rounded boulders. Small clasts are usually composed of a single mineral, generally quartz or clay. Larger clasts, such as the boulders on Bowling Ball Beach, near San Francisco (Fig. 8.11), are almost always lithic fragments composed of multiple minerals.

To classify clasts of different sizes, geologists in the United States use the *Udden-Wentworth scale*, commonly just called the *Wentworth scale*. From fine to coarse, the main classes are *clay, silt, sand, pebble* (also called *granule*), *cobble*, and *boulder* (Table 8.1). The Wentworth scale further subdivides classes by adding modifiers such as *fine, medium*, or *coarse*. Each class and subclass is associated with grains of different mean diameters, and the limiting sizes for each class and subclass vary by powers of two; each class has double the particle size from the previous smaller class. For example, small cobbles have diameters between 64 and 128 millimeters; large cobbles have diameters between 128 and 256 millimeters. An equivalent scale to the Wentworth scale, the *phi*

Table 8.1 The udden-wentworth and phi scales for classifying clastic particles.

Classification	Particle size (diameter in mm)	Phi (φ) scale
boulder	> 256	< –8
cobble	64 to 256	–6 to –8
pebble	4 to 64	–2 to –6
gravel or granule	2 to 4	–1 to –2
very coarse sand	1 to 2	0 to –1
coarse sand	1/2 to 1	1 to 0
medium sand	1/4 to 1/2	2 to 1
fine sand	1/8 to 1/4	3 to 2
very fine sand	1/16 to 1/8	4 to 3
coarse silt	1/32 to 1/16	5 to 4
medium silt	1/64 to 1/32	6 to 5
fine silt	1/128 to 1/64	7 to 6
very fine silt	1/256 to 1/128	8 to 7
clay	< 1/256	> 8

scale (φ scale), also listed in Table 8.1, is sometimes used instead of giving sizes in millimeters. The phi scale is based on log values of the grain sizes:

$$\varphi = -\log_2 d \text{ where d = mean grain diameter}$$

For example, cobbles (d = 64 to 256 millimeters), have φ values of –6 to –8.

The names of the Wentworth classes, unfortunately, have general meanings that can lead to some confusion if they are used in different ways. Geologists, for example, may talk about a *boulder* and be referring to a large rock that may or may not meet the size criteria in Table 8.1. The term *clay* is particularly problematic, because it refers to a specific grain size in the Wentworth scale, to a specific kind of mineral, and to a kind of soil horizon. And, there are some inherent contradictions created by using mineral names to refer to different clast sizes because more than one kind of mineral may typify a size class. Due to their ability to survive weathering, for example, most, but not all, clay-sized clasts consist of clay minerals. Silt-sized clasts are generally quartz or feldspar. Sand grains are generally quartz, but other minerals are possible. Larger sized clasts are even more variable.

Many sediments contain grains of mixed sizes. The term *sorting* describes the variation in grain sizes that a sediment (or a rock) contains. Sedimentologists and engineers have developed numerical scales to describe sorting, but for most purposes general descriptors such as *poorly sorted*, *moderately sorted*, and *well sorted* suffice. Well-sorted sediments, such as the sediment shown in the top photo of Figure 8.12, contain grains of relatively uniform size. The grains may, however, be made of more than one kind of mineral, like the sediment shown. Rocks derived from well-sorted sediments tend to be quite porous because there is lots of space between grains. Other sediments, which contain grains of many different sizes (Fig. 8.12, bottom), are *poorly sorted*. The graywacke in Figure 8.10 is an example of a very poorly sorted sedimentary rock. Rocks formed from poorly sorted sediments may have low porosity because space between larger grains is filled by finer material. Note that the sediment grains in the well-sorted sediment in Figure 8.12 have been quite well rounded by abrasion—they have no sharp edges or corners. In contrast, the grains in the poorly sorted sediment (bottom photo) have some angular edges and corners.

Sorting occurs during sediment transportation because the sizes of clasts that can be moved depend on wind or water velocity. Relatively coarse debris is left behind during transportation because smaller grains are more easily transported and because water (or wind) velocities slow over time. So, coarse material is more common near a sediment's source. For example, a small stream can only transport particles of small size (silt or, less commonly, sand). So, sediment becomes sorted as it moves downstream, and the coarsest stream bottom deposits are near stream headwaters. The sand on a beach is generally all of one grain size because finer material is washed or blown away and coarser material never makes it to the beach in the first place. Wind does an exceptional job of sorting sediment because, depending on

well-sorted sediment

poorly sorted sediment

Figure 8.12 Well-sorted and poorly sorted beach sediment. Photos from S. Tvelia, thisoldearth.net.

wind speed, only fine sand or smaller particles can be carried. A lightly blowing wind, for example, can only move fine clay or dust, leaving everything else behind—thus explaining why sand dunes with grains of uniform size form in desert regions. Glacial ice, in contrast with wind and water, does a very poor job of sorting because the ice can carry material of almost any size.

8.3.2.2 Clast shapes

While being transported, clasts commonly bump and scrape other clasts and, consequently, become abraded. As shown in Figure 8.12, this abrasion process removes sharp corners and edges, eventually leading to rounding of the grains. So, the shape of grains tells something about the grains' history. *Roundness* and *sphericity* are the two key characteristics most often considered when describing clast shape. Figure 8.13 shows drawings depicting variations in both. Roundness refers to the general smoothness of a grain's surface. *Well-rounded* grains have smooth surfaces with no sharp corners. In contrast, *angular* grains contain many sharp edges or corners.

Sphericity refers to whether a grain is equidimensional—roughly the same length in all directions. Highly spherical clasts approach the shape of a sphere; less spherical clasts may be quite elongated. The boulders in Figure 8.11, for example, appear spherical. Some of the grains shown in the top photo of Figure 8.12 are quite spherical, but most are somewhat elongated.

Figure 8.14 shows two samples of sand. The sand on the left is from Pismo Beach, California. It is made of several different minerals and some shell fragments. The grains are all quite angular and are not spherical. The sample on the right is from Pink Coral Sand Dunes State Park in Utah. It is entirely quartz, and the grains are well rounded and most are quite spherical.

8.3.2.3 Sediment texture and maturity

The sizes, shapes, and orientations of clasts give a sediment its *texture*, and sediment textures evolve over time. For example, newly formed sediment typically contains angular clasts that, during transportation, will become more rounded. That same sediment may contain mineral grains or lithic fragments that will eventually decompose and disappear. *Sediment maturity* refers to how evolved a sediment has become since its original deposition. Time is a key to maturity, but sediments mature at different rates in different environments. Sediments that are transported long distances mature faster than those that stay in one place.

Figure 8.15 shows a 2-meter-tall outcrop in Point Lobos State Reserve, near Carmel, California. In this outcrop, clasts of different sizes have been sorted (fine at the bottom and coarse at the top), and many sizes are present. This outcrop is a good example of immature sediment: it is poorly sorted and contains angular clasts. The clasts are of several different compositions and contain a variety of minerals. In contrast with immature sediments, the most mature sediments are well sorted and contain well-rounded grains because they have had a long history, involving transportation that included many opportunities for sorting and abrasion. Mature sediments also are composed only of very stable Earth surface minerals, such as quartz, because other minerals have had ample time to decompose.

Many beach sands are very mature sediments that contain only well-rounded quartz grains of uniform size. The top photo in Figure 8.16 shows an example. The mature sand had a lengthy history before collecting in the dunes seen here. Some beach sands, however, are

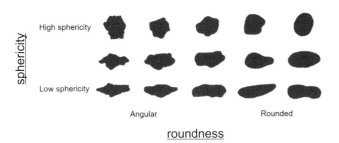

Figure 8.13 Sphericity and roundness of clasts.
Drawing from Woudloper, Wikimedia Commons.

Figure 8.14 (a) Beach sand from Pismo Beach, California, and (b) sand from Pink Coral Pink Sand Dunes State Park, Utah.
Photos from Wilson44691, Wikimedia Commons.

Figure 8.15 Outcrop of sedimentary rock at Point Lobos, California. Red pocketknife for scale.

Figure 8.16   Beach sand in Hoge Veluwe National Park, the Netherlands (top photo), and at South Point in Hawaii (bottom photo).

Photos by Ellwya and Brocken Inaglory, both at Wikimedia Commons.

not so mature. For example, sand near Hawaii's South Point (Fig. 8.16, bottom) is continuously eroded from the nearby volcanic rocks shown in the photo. This Hawaiian sand contains many angular clasts, including rock fragments, and many minerals, such as olivine, that are unstable at Earth's surface. The olivine is what gives the sand a greenish color. These contrasting photos remind us that sediment texture and composition can tell us about a sediment's origin. Additionally, textures and maturity are sometimes very important because they determine key properties such as porosity, permeability, and density, which have implications for engineering properties and petroleum resource extraction.

8.4   Transportation and deposition of clastic sediment

Clastic material is transported in several different ways. Typical transportation occurs when detritus, or

previously deposited sediment, is picked up and carried, or pushed along the ground, by flowing water, blowing wind, or glacial ice. Gravity often contributes to the effects of water, wind, and ice. And, gravity, by itself, can cause sediment to slide downhill slowly on hillsides or slopes. More dramatically, gravity can rapidly transport material when cliffs fall apart or when landslides occur on continents or submarine slopes.

8.4.1   Transportation by flowing water

When it rains, some water evaporates, some infiltrates soils or rocks, and the remainder becomes runoff. The runoff flows downhill until it evaporates, infiltrates, or reaches a stream or river. While flowing, it erodes and transports unconsolidated sediment and soil, differentially removing small and light particles of clay and silt and leaving coarser material behind. The amounts of runoff and erosion depend on many things, including soil structure, rainfall intensity, slope steepness, and slope length. Although rain, because it hits the ground with enough energy to loosen soil and sediment, can be an especially powerful erosion agent, snowmelt too leads to runoff and erosion.

A general term used for material carried by running water or wind is *load*. Thus, material moved by runoff is called the *runoff load*. Runoff and its load may only travel a short distance, but if the runoff reaches a stream, the water adds to stream flow, and the runoff load is deposited on the stream's bottom or is added to the *stream load* and carried downstream.

Streams come in all sizes, from small *rills* (equivalent to very small gullies cut into a hillside) to large rivers. Geologists use the term *fluvial* to refer to processes and materials associated with streams or rivers, no matter the size of the water body. So, materials deposited by flowing waters are called *fluvial deposits*. Rills have small loads but, in contrast, rivers move more sediment than any other Earth process does. Most sediment will settle to the bottom of slow-moving, low-energy rivers, but will be eroded and transported downstream in fast-moving, high-energy rivers. At intermediate velocities, some material will be eroded and transported and some will be left behind, leading to sorting. Because river velocities are not constant, rivers erode, transport, and deposit material at different times. Such cycles of erosion, transportation, and deposition are often seasonal due to different amounts of runoff at different times of the year.

If water velocity is too low, no clastic material is eroded or transported, and all material suspended in the

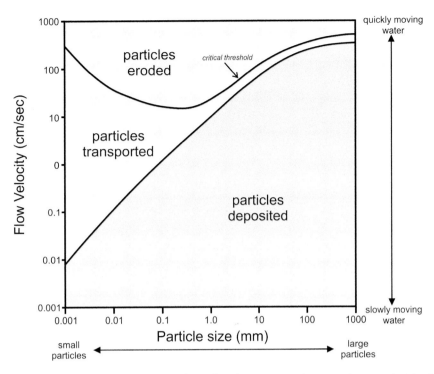

Figure 8.17 A Hjulström-Sundborg diagram depicting the relationships between particle sizes and water velocities that cause erosion, transportation, or deposition.

water will be deposited. If velocity increases, the flowing water may transport some sediment, and if velocity increases further, it will eventually reach the *critical threshold* at which erosion begins. These relationships are shown in Figure 8.17, called a *Hjulström-Sundborg diagram*, which has a logarithmic scale for both flow velocity and particle size.

The velocities at which deposition, transportation, and erosion occur depend on clast size, shape, and density. So, the numbers in Figure 8.17 are only approximations. High velocities are required to erode and transport large particles; slower velocities may be adequate for some finer material. But, somewhat surprisingly, high velocities may also be required to erode fine material, because many fine-grained sediments contain clay or other mineral material that is cohesive and sticks together, resisting erosion. This explains why the critical velocity curve shown in Figure 8.17 curves upwards on the left side of the diagram. Once picked up, however, fine material is transported easily. Besides water velocity, a stream's ability to erode and transport clasts also depends on water *turbulence*. Turbulence is generally absent in slow-moving waters and present in fast-moving waters, but also is affected by the roughness of the bed material over which the water is flowing.

Once erosion begins, clastic particles can roll along a streambed and be pushed downstream in a process called *traction* (Fig. 8.18). If a particle bumps into others, causing the others to move, the process is termed *creep*. Traction and creep are common transportation mechanisms for large clasts, such as gravel, because large clasts are too heavy to be lifted from stream bottoms. Intermediate-sized clasts may be lifted and become suspended in flowing water for brief or long times before settling back to the bottom. Sometimes clasts will bounce downstream, never rising far from the bottom, moving by a process called *saltation* (Fig. 8.18). The finest material can be lifted farther above the streambed, remain suspended for longer times, and be transported long distances.

Streams get their loads from several sources. Loads may derive from erosion and runoff in an upstream source area, from erosion of stream banks, or from scouring of a stream bottom, or they may be contributions from tributaries that join the main stream. No matter what their origin, coarse or dense particles tend to settle to a stream bottom quickly if they are lifted—or they may never leave the bottom at all. The relatively coarse material that travels along, or near, a stream bottom is called the *bed load*. Bed loads include moving

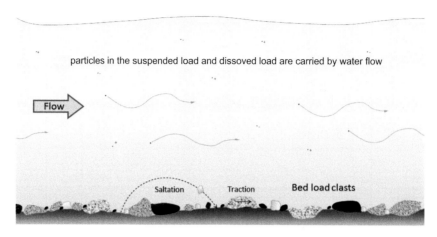

particles in the suspended load and dissoved load are carried by water flow

Flow

Saltation   Traction   Bed load clasts

Figure 8.18   Modes of transportation of sediments and dissolved ions in a stream. Modified from a figure at https://opentextbc.ca/.

material that is never out of contact with the bed, material that moves by saltation, and material that remains suspended close to the stream floor. Above the bed load, material that remains suspended intermittently or continually is called the *suspended load*; it is finer than the bed load. Suspended loads are generally more voluminous, and move much faster, than bed loads. The finest material in the suspended load, composed of clay-sized grains, may remain suspended indefinitely. Additionally, flowing water also carries a *dissolved load* consisting of dissolved ions (represented by red dots with + and – signs in Fig. 8.18), downstream.

8.4.2   Deposition by flowing water

8.4.2.1   Alluvium

Loose, unconsolidated, and sometimes poorly sorted sedimentary material deposited by water in a non-marine environment is generally called *alluvium*. In other words, alluvium is the material that fills stream or river valleys. Some geologists define *alluvial* deposits as sediment associated with floodplains and *fluvial* deposits as sediment associated directly with a flowing stream, but commonly the two adjectives are used interchangeably. Flowing water deposits alluvium when water velocity decreases. The decrease may be due to a change in stream *gradient* (steepness), perhaps when a stream flows into a valley, a lake, or an ocean. Velocity also decreases if a stream flows over a rough section of the stream bottom or if boulders or other obstructions block water flow, leaving a low-velocity shadow downstream of the obstruction. Additionally, stream velocity varies laterally across a stream and with water depth.

Flow is generally greatest in moderately deep water near the center of a channel because near banks, at the water surface, and near the streambed, friction causes velocity to slow.

8.4.2.2   Features of river deposits

Straight streams may have relatively uniform flows across their widths, but most streams are not completely straight. Some curve gently, and others, like Bolivia's Mamoré River, seen in Figure 8.19, *meander* significantly. Stream flow is always faster on the outer

Figure 8.19   1984 Landsat satellite image, 42 kilometers (70 miles) across, of the Mamoré River, Bolivia. Photo from the USGS.

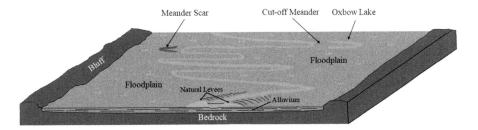

Figure 8.20 Features of alluvial deposits.

sides of curves and slower on the inner sides. Because stream velocity is not uniform, erosion can occur in some places while fluvial deposits accumulate in others. Where deposition occurs, *bars* form that may or may not be exposed above the water surface. Bars on the inside of curves are called *point bars*—many examples of point bars can be seen in Figure 8.19 where light-colored sediment has accumulated on the inner sides of meanders.

Point bars, and bars of other sorts, are commonly covered during high-flow times and exposed during low-flow times. While deposition is occurring on bars and in other places, erosion may be occurring if water scours the stream bottom. Erosion also occurs if high water velocities on the outside of meanders remove material from banks, creating *cut banks*. The outsides of the curves in Figure 8.19 are all characterized by cut banks, although they are difficult to see.

Cycles of erosion and deposition cause meandering streams to migrate across floodplains over time. As depicted in Figure 8.20, old channels can become abandoned, and *cut-off meanders* may be left behind as *oxbow lakes*, or as relict oxbow lakes that no longer hold water, sometimes called *meander scars*. Many examples of cutoff meanders and oxbows are clearly seen in Figure 8.19, and some are shown here in the drawing of Figure 8.20.

Some streams and rivers change their courses and migrate long distances across broad flat *floodplains*, leaving abandoned channels and other fluvial features over broad areas. In Figure 8.19, the floodplain is the darker green region that includes the meandering river and its cutoffs. A USGS study of streams in Indiana found that, on average, most streams migrate across their floodplains at rates of feet/year; a few, however, moved 10 times faster.

When stream flow increases during a seasonal flood, or perhaps because of a storm, a stream may top its banks. In such cases, initially, water velocity is high.

When it eventually slows, sediment is deposited adjacent to the stream banks. Often this creates *natural levees* of sand and silt that can grow higher with every flood (Fig. 8.20). If flood waters spread farther, they can cover adjacent floodplains, where they deposit fine-grained silt and clay. Because of the silt, clay, and moisture, floodplain soils are often excellent for agriculture.

If shallow streams have an ample supply of sediment, and if velocities vary over time, the streams may develop *braided channels*. Figure 8.21 shows a good example of a braided river in New Zealand. Braiding of this sort

Figure 8.21 The Rakaia River, a braided river in New Zealand.
Photo from Andrew Cooper, Wikimedia Commons.

develops when average erosion rates and average deposition rates are about equal. Braided streams are characterized by many channels and bars containing coarse material; relatively high water velocities ensure that finer material is carried away. The channels and bars migrate because erosion dominates where water moves most rapidly, and deposition dominates in other places. Water is commonly restricted to one or a few channels during times of very low flow, but water may cover the entire stream valley during high flow. So, stream sediments cycle through times of erosion, transportation, and deposition. These relationships may be seasonal or result from thunderstorms or other weather events.

Over geological time, tectonism may uplift land, causing new floodplains to form at elevations below older floodplains. In such cases, *stream terraces*—remnants of older floodplains perched above present-day floodplains—may remain. The terraces run parallel to river channels and are generally composed of poorly sorted fluvial sediment of variable thickness. Figure 8.22 shows a small river with several terrace levels next to it.

Besides tectonic uplift, terraces may develop if climate change—perhaps associated with ice ages—causes a lowering of sea level so that streams erode more deeply into Earth. Climate change may also lead to changes in river discharge that further promote terrace formation.

8.4.2.3 Alluvial fans

At the mouth of a stream, where it empties into a valley, plain, or a larger water body, stream velocity typically decreases. Consequently, gravity deposits sediments, creating *alluvial fans* or *deltas*. Alluvial fans, like the one near Badwater in Death Valley National Park, California, shown in Figure 8.23, typically form when mountain streams enter a larger valley. The Badwater fan, and similar fans, develop because steep mountain streams, carrying a great deal of sediment, slow and deposit material on shallower slopes and valley bottoms.

Fan deposits are combinations of alluvial material (floodplain deposits) and fluvial material (stream deposits), characterized by poorly sorted sediment containing pebbles and cobbles, possibly sand, and sometimes boulders. Finer sands, silts, and clays are generally transported farther downhill and deposited in valley bottoms. Fans represent dynamic environments, generally forming jumbled fan-shaped wedges that constantly move downhill. They are dry sometimes and wet other times, often containing many small ephemeral stream channels that change seasonally and so may present engineering challenges. Note, for example, how the road has been routed to circle around the most chaotic part of the Badwater fan in Death Valley (Fig. 8.23).

In many ways, deltas are wetter equivalents of alluvial fans. Deltas form when moving water enters nonmoving water or much slower moving flowing water. Small deltas commonly develop when streams enter larger rivers, but the largest deltas occur where rivers enter lakes or oceans. Figure 8.24 shows a Landsat image from the United States Geological Survey of the Selenga River delta in Russia. Deltas often have fan/wedge shapes like alluvial fans, but contain a different mix of materials. The sediments in deltas are generally well sorted and finer than those in alluvial fans and are typically silt and clay. Coarser material is left farther upstream.

Figure 8.22   Stream terraces in a Karakoram Mountain valley, Pakistan.

Photo from Hewitt *et al.* (2011), with permission of J. Hammann.

Figure 8.23   Alluvial fan in Death Valley National Park.

Photo from Marli Miller.

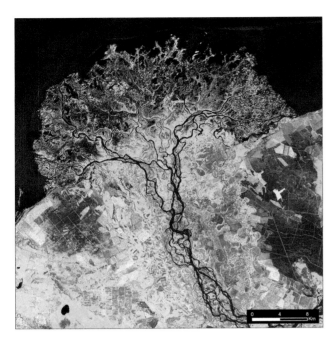

Figure 8.25 A haboob—a dust storm—engulfing Lubbock, Texas, in 2011.

Photo from the U.S. Weather Service.

Figure 8.24 Selenga River delta on the southeast shore of Lake Baikal, Russia.

"Earth as Art" satellite image courtesy of the USGS National Center for Earth Resources Observation and Science (EROS) and the National Aeronautics and Space Administration (NASA) Landsat Project Science Office.

Because water flow is continuous in deltas, deposition is an ongoing process. Delta environments are dynamic, often involving many channels, called *distributaries*, connected in a *dendritic pattern*. Such a pattern is clearly visible in Figure 8.24. Distributaries migrate and deposit sediment in different places at different times. The leading edge of a delta, the part that pushes farthest into a lake or ocean, is characterized by a steep slope, similar to the face of a sand dune. Sediments on these underwater slopes slide and roll down the slopes just like sand grains do on the faces of dunes. Deltas may grow over time or be eroded away as fast as they are created, depending on the energy of the lake or river where they form.

8.4.3 Transportation by wind

Figure 8.25 shows a kind of dust storm, called a *haboob*, that is about to cover part of Lubbock, Texas. Haboobs (from the Arabic word for blasting or drifting) are intense clouds of dust carried by a weather front. They are common in arid regions throughout the world. Haboobs are one kind of *eolian transportation*—one way that sediment can be transported by wind.

Eolian erosion and transportation are especially effective in dry regions with little vegetation or with poor soils lacking humus. Moisture, vegetation, and humus would otherwise hold the soils or sediments together, reducing erosion. Threshold values for wind erosion are greater than for water erosion, but wind-blown grains can strike other grains, causing them to move even if wind velocity is below threshold levels.

Transportation by wind is, in many ways, like transportation by water. Water, however, is denser and more powerful than wind is and thus can move particles of greater size. Winds can only move particles of sand size or smaller. And, for the most part, wind-blown sand-sized grains cannot remain in the air indefinitely. They mostly move by saltation, bouncing across the ground, but may also move by creep or traction. So, sand makes up the bed load while fine silt and dust particles (< 0.2 mm in average radius) move as a suspended load, commonly called the *dustload*, above the sand.

8.4.4 Deposition by wind

Wind-borne sediments will be deposited when wind speed slows. Deposition may occur because of a change in weather, when flowing wind currents spread out, or sometimes because of a wind shadow caused by rocks, vegetation, houses, or snow fences. When large amounts of sand-sized material are deposited, *sand dunes* may form, and when fine material is deposited, dust and silt may collect.

8.4.4.1   Sand Dunes

Figure 8.26 shows sand dunes in Coral Pink Sand Dunes State Park, Utah. Dunes like these can only develop if there is an original source of sand, and generally the source cannot be too far away, because sand is not normally carried long distances by wind. Utah's Coral Pink Sand Dunes, for example, contain loose sand eroded during the past 10–15 thousand years from nearby Navajo Sandstone. The Navajo formed about 200 million years ago when preexisting dunes became hardened and became rocks. The dunes that exist today may eventually continue through the rock cycle and become new sandstone that is eroded at some time in the future.

Large dunes form if large amounts of sand are deposited—often because of a topographic low where wind velocity slows. So, in the Coral Pink Sand Dunes, strong winds carry sand east from source regions. The winds flow through a gap between the Moquith and Moccasin Mountains, and when they emerge into a valley east of the mountains, the winds spread out and slow, depositing the pink dunes.

Most dunes are asymmetrical piles of unconsolidated sand, usually steeper on the downwind side. Dunes migrate as sand erodes on the upwind side, is carried over the top, and is deposited and slides down the *slip face* on the downwind side. Although dunes migrate horizontally across land, the sand on the slip face is deposited in inclined layers. This deposition often produces *crossbeds*—layers of sand at an angle to the horizontal. Thus, Utah's Navajo Sandstone contains huge petrified crossbeds that contribute to the spectacular scenery of Zion National Park; Figure 8.27 shows an example from the park.

Sand dunes can extend over large areas and reach heights up to several hundred meters. For example, the

Figure 8.27   Crossbedded Navajo Sandstone in Zion National Park. From R. Clausen, Wikimedia Commons.

Simpson Dunes of Australia cover more than 170,000 square kilometers (65,000 square miles). The tallest known dunes are in the Namib Desert of Namibia and exceed 365 meters (1200 feet). The most commonly occurring dunes have crescent shapes, but dunes are quite variable. Some are localized with crescent or elliptical shapes, others are star-shaped, and still others form long linear ridges.

8.4.4.2   Dust and loess

Wind-blown dust consists of silt and clay-sized material. Sediments of fine grain size are normally cohesive and hard to erode, but when erosion occurs, the result is hazy skies or sometimes large dust storms such as the one shown in Figure 8.25. Dustload materials can be carried to great heights, sometimes thousands of meters above Earth's surface, and material can remain in the air for long times. Consequently, wind-blown dust and silt can travel far and spread over large areas. Dust from the Sahara Desert, for example, is carried across the Atlantic Ocean between late spring and fall, at altitudes of 2 kilometers to 4.5 kilometers (1.2 to 4.8 miles). This dust can make it all the way to the western Gulf of Mexico, where it is responsible for hazy skies in Texas and other Gulf Coast states.

Wind-blown dust is a key component of many soils, including soils far away from where the dust originated. Wind-blown dust is also deposited in oceans and is a significant component of some deep ocean sediments. If large amounts of dust are deposited on land in thick layers, the deposits produce highly erodible—but otherwise excellent for agriculture—soils called *loess*.

Figure 8.26   One of the largest dunes in Coral Pink Sand Dunes State Park, Utah.
Photo from F. Kovalchek, Wikimedia Commons.

Figure 8.28   Eroded loess deposits in central Israel.
Photo from Eman, Wikimedia Commons.

Figure 8.28 shows a canyon eroded in loess of the Negev Desert in Central Israel.

8.4.5 Sediment transportation by glaciers and glacial sediments

8.4.5.1 Continental ice sheets and alpine glaciers

Glaciers are of two fundamental types: *continental ice sheets* and *alpine glaciers*. Figure 8.29 shows examples of both: the Greenland Ice Sheet, a continental ice sheet, and Switzerland's Aletsch Glacier, an alpine glacier. These, and most glaciers around the world, are shrinking today and thus adding water to oceans, causing a rise in sea level.

Continental ice sheets are huge masses of ice that flow from regions of accumulation near Earth's poles to regions of warmer climate. Such ice sheets, which may cover regions greater than 50,000 square kilometers (20,000 square miles) today, only exist in Antarctica and Greenland. They are up to 3500 meters (12,000 feet) thick. During the last period of worldwide glaciation, which ended about 10,000 years ago, the Laurentide Ice Sheet covered much of North America, and other ice sheets covered Europe and South America. The photo on the left in Figure 8.29 shows the part of Greenland's continental ice sheet where it enters the ocean on Greenland's east coast. This ice sheet is the remnant of a much larger sheet that once extended from Greenland across Canada and into much of the northern United States.

Alpine glaciers, sometimes called *valley glaciers*, are much smaller than continental ice sheets. There are many alpine glaciers around the world, but counting is problematic because many are disappearing and others are very small. They are found in most major mountain ranges worldwide. Alpine glaciers flow down mountain valleys, beginning at high elevations in bowl-shaped valleys called *cirques*. Some alpine glaciers exist only in a small cirque in upland areas, but many glaciers flow long distances downhill. The lengthiest glaciers in the world, such as the Fedchenko Glacier, Tajikistan, are up to 100 kilometers (60 miles) long. The longest glacier in North America, the Nabesna Glacier in the Wrangell Mountains, Alaska, is more than 80 kilometers (40 miles) long. The glacier shown on the right in Figure 8.29 (Switzerland's Aletsch Glacier) is the largest glacier in Europe's Alps. It is shrinking today but had a length of about 23 kilometers (14 miles) in 2014.

Figure 8.30 shows a cross section of an alpine glacier. For alpine glaciers, snow accumulation is most significant at high elevations near the glacier's head, in a region called the *accumulation zone*. For continental glaciers, snow accumulation is generally greatest in cold climate regions closest to the north or south poles. As snow accumulates, fresh snow compacts the snow beneath it. In both continental and alpine glaciers compaction, crushing, recrystallization, some melting, and refreezing eventually transform snow into *firn*, an intermediate product, and eventually into glacial ice. Layers of snow, firn, and ice are highlighted in Figure 8.30. As snow becomes ice, porosity decreases from more than 80% to nothing over one to several years, and consequently glacial ice is much denser than the snow from which it forms. Glacial ice, the snow, and firn riding on it, and suspended material within it flow downhill under the influence of gravity or outward from polar origins. Typically, flow rates are on the order of meters/day for alpine glaciers but range from less than a meter/year to as much as 30 meters/day. Continental ice sheets, by comparison, flow much slower.

While glaciers are flowing, they lose ice through melting, sublimation, and evaporation (Fig. 8.30). We call these processes, collectively, glacial *ablation*. At high elevations or in polar regions, in a glacier's accumulation zone, snowfall is greater than losses due to ablation. At lower elevations or distant from poles, in a glacier's *ablation zone*, ablation rates are greater than snow accumulation is. As seen in Figure 8.30, an *equilibrium line*, which may move seasonally, divides the zone of accumulation from the ablation zone. Glacial ice always flows downhill, but at the leading edge of a glacier, called the

Figure 8.29   Two kinds of glaciers: Greenland's ice sheet, where it enters the North Atlantic in southeastern Greenland, and the Aletsch Glacier, the longest alpine glacier in the Alps.

Left photo from Hannes Grobbe, right photo from J. Simon, both at Wikimedia Commons.

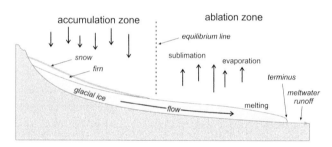

Figure 8.30   Regions of ice, firn, and snow in an alpine glacier.

Modified from http://glaciers.pdx.edu/Projects/LearnAboutGlaciers/Skagit/Basics00.html.

Figure 8.31   Rock material in the Viedma Glacier, Patagonia, Argentina.

Photo from L. Quinn, Wikimedia Commons.

terminus, or the *snout*, ablation is great enough to stop farther advancement of ice.

Although glacier ice movement is much slower than the movement of wind and water is, glaciers can pick up and carry large amounts of clastic material. The finest material is silt-sized *rock flour* created by grinding that abrades material from bedrock. Much of the material in loess derives from this fine rock flour. However, glaciers pluck and scrape material of all sizes from the ground beneath them, or from valley sides and bottoms as they move forward. Alpine glaciers also gain material from rock falls at their margins. Consequently, clasts of all sizes, including large boulders, can make up a glacier's load. Figure 8.31 shows much medium-sized to very coarse material in the glacial ice of Patagonia. This clastic material is carried at all levels within glacial ice, including on the surface, although larger and heavier clasts slowly settle and concentrate in the lower levels.

Glacial ice melts when advancing glaciers reach warmer regions or when climate changes. When melting occurs, all entrained material is deposited, sometimes producing large voluminous deposits that cover large regions. Occasionally, however, only individual rocks are left as evidence of a glacier's past presence. Such rocks, termed *glacial erratics*, generally consist of different material from the bedrock on which they sit. Erratics range in size from cobbles or pebbles to very large boulders. The world's largest erratic (Fig. 8.32), the *Big Rock* erratic near Okotoks, 48 kilometers (30 miles) south of Calgary, Alberta, is a quartzite boulder that weighs more than 16,500 tons and is 18 by 40 meters in area (60 by 135 feet) and 9 meters (30 feet) high.

Figure 8.32 Big Rock erratic, near Calgary, Alberta.
Photo from Coaxial, Wikimedia Commons.

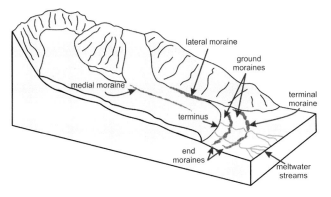

Figure 8.33 Moraines and other glacial features.

8.4.5.2 Glacial drift

Glacial drift is a general term for any clastic material deposited by glacier. Drift is typically a mix of coarse material, including sand, gravel, and boulders, in a matrix of fine silt and clay. Glacial drift is significantly different from stream sediments because glacial loads contain much coarse material, and stream deposits are mostly well sorted, sand-sized or smaller. Drift is conveniently divided into two kinds of deposits: *till* is material deposited directly from ice, and *stratified drift* is material deposited by glacial meltwater downstream of a glacier's terminus. Till is generally unsorted, while stratified drift may become moderately sorted during transport by water and may be deposited in layers that have different compositions. Till and stratified drift can develop soil horizons, but the soil is generally poor for agriculture due to poor sorting and an abundance of large rocks.

8.4.5.3 Till

Moraine is a general term for landforms made of till; several different kinds of moraines are labeled in Figure 8.33. *End moraines*, which form at a glacial terminus as ice melts, leave ridges and piles of glacial debris oriented generally perpendicular to the direction of ice flow. Because ice advances or retreats depending on seasonal and longer-term climate changes, glaciers commonly leave more than one end moraine behind; two are shown in Figure 8.33. The end moraine formed at the farthest point of glacier advance is called the *terminal moraine. Ground moraines* consist of material deposited from glacial ice beneath a glacier. After

Figure 8.34 Lateral moraine deposits in a glacial valley near Lake Louise, Alberta, Canada.
Photo from Wilson44691, Wikimedia Commons.

glacial ice is gone, ground moraines are characterized by uneven topography, typically with many small ridges and basins. Ground moraines may be small or extensive and may be closely associated with terminal moraines.

Not all moraines form at the ends of glaciers. *Lateral moraines*, like the one labeled in Figure 8.33 and the moraine shown in Figure 8.34, are deposited along the sides of mountain glaciers where melting ice leaves debris behind. The ice melted long ago and disappeared from the glacial valley shown in Figure 8.34, but the lateral moraine is testimony to the former presence of an alpine glacier. *Medial moraines*, found in the centers of glacial valleys, typically form when two glaciers merge, creating trains of till in the center of what is, or was, glacial ice. Medial moraines contain material that was formerly part of lateral moraines belonging to each of the joining glaciers.

Till deposits are not stratified (they are unlayered) and contain clasts of many sizes, from rock flour to large angular rock fragments. Large pebbles or boulders may have facets (smooth faces) or striations created during transportation, but angular rock fragments are commonly present because significant rounding may not occur during the short distances of transport in ice. Most tills are deposited as lateral moraines or end moraines along the edges or at the terminus of a glacier. Ground moraines and medial moraines are less common. No matter their origin, if tills harden to become rocks, we call them *tillites*. Figure 8.35 shows the Moelv Tillite, 50 kilometers (30 miles) north of Oslo, Norway, with a hand lens for scale. This tillite is typical, consisting of angular blocks of many sizes in a fine-grained matrix that formed from rock flour and coarser clastic material. Tillites are found around the globe; some are associated with modern-day glaciers or ice ages of the recent past. Others provide evidence of former times of glaciation in regions that, most recently, have been too warm to sustain glacial ice.

Sometimes, moving glacial ice reworks (modifies) previously deposited glacial material. When glaciers advance over older till, for example, they can produce *drumlins* such as the ones shown in Figure 8.36. Drumlins are elongated, asymmetrical, oval hills of glacial debris that formed under moving ice. These streamlined mounds have a steep slope on their front face (down and to the left in the photo) that records the direction of glacial flow and may be up to 2 kilometers (1.25 miles) long. Drumlins are often found in groups called *drumlin fields*. Some spectacular drumlin fields are found in the Great Lakes region of North America, including the field in Wisconsin shown in Figure 8.36.

Figure 8.36   Multiple drumlins near Madison, Wisconsin, in the winter.

Photo from Guell Collection, University of Wisconsin-Oshkosh and Drumlin Area Land Trust.

The largest Great Lakes drumlin field is in western New York, where more than 10,000 drumlins are associated with the glacially scoured Finger Lakes. These drumlins formed during an ice age about 18,000 years ago, near the terminus of a continental ice sheet that covered Canada and the northernmost United States.

8.4.5.4   Stratified drift

8.4.5.4.1   Outwash

When glaciers melt during warm times of the year, streams flow from their tops, bottoms, and sides. These streams carry large sediment loads that include rock flour and lots of coarser material. Eventually this material is deposited as *outwash* composed of moderately sorted layers of stratified drift, in large *outwash plains* in front of the glaciers. Seasonal meltwater streams, often braided, are commonly present in outwash plains. The streams deposit coarser materials near their headwaters at the glacier terminus and clays and silts farther away. The largest outwash plains form after some glacial retreat has occurred, so they lie between the present-day glacier terminus and the former terminal moraine. Eventually, a glacier that deposits outwash may disappear, but the distinctive material in the outwash plain is evidence of a glacier's former presence.

8.4.5.4.2   Kettle lakes and meltwater lakes

Kettle lakes, also called *kettle holes*, are common in outwash plains. Figure 8.37 shows a kettle lake in the Isunngua region of central-western Greenland. Such lakes

Figure 8.35   The Moelv Tillite in Norway.

Photo from Mahlum, Wikimedia Commons.

Figure 8.37   A kettle lake in central-western Greenland.
Photo from Algkalv, Wikimedia Commons.

Figure 8.38   Pleistocene varved clay near Baumkirchen, Tyrol, Austria.
From H. Hammer, Wikimedia Commons.

form when glacial drift buries a block of ice. When the ice melts, it leaves a depression that collects fine sediments and often water, producing a small lake with a silty bottom. Kettle lakes can remain long after a glacier is gone. The Prairie Pothole Region, an area of the Great Plains extending from Iowa and Minnesota through the Dakotas and into Saskatchewan and Alberta, once contained thousands of *potholes*. The small, round pothole lakes originated as kettle lakes at the end of the last major period of continental glaciation, about 10,000 years ago. But today, in many parts of the Great Plains, the vast majority of potholes have been drained and filled for agriculture.

Glacial scouring often leaves topographic lows that can be the sites of *meltwater lakes*. As a glacier melts, water flows into these depressions and deposits sediment, often on an annual cycle. The greatest sediment accumulation occurs in spring, when discharge and sediment load are high. Coarser materials collect that, in the fall, are overlain by finer layers of slightly different color. So, the alternating layers record years like tree rings do. A good example of such a deposit is shown in Figure 8.38 (which includes a centimeter tape for scale). Layers, such as those shown in Figure 8.38, are called *varves*. The different layers trap samples of Earth's atmosphere, pollen, and other artifacts of interest to geologists trying to reconstruct the past. Scientists can collect samples from meltwater lakes and use them to interpret past climates and glacial histories. Many studies focus on active meltwater lakes near present-day glaciers, but ancient meltwater lakes, now no longer near glaciers, are studied, too.

8.4.5.4.3   Kames and Eskers

Depressions in ice, caused by streams flowing on the tops, or at the sides of glaciers, commonly collect sediments.

Subsequently, when the glacial ice melts, irregularly shaped, often steep-sided hills of glacial debris can be left behind. These hills or mounds, called *kames*, consist of sand, gravel, and till. Kames are highly variable but are, typically, meters to tens of meters in longest dimension. Water flowing on the ground beneath glacial ice, too, can create river channels that collect sediment. After the glacial ice melts, the channel sediment may be left behind as an *esker*, a ridge that resembles an upside-down stream. Eskers contain stratified drift, mostly gravel and coarse sands, deposited by water that once flowed through tunnels beneath glacial ice.

Figure 8.39 shows good examples of kames and related landforms in western Sweden. The kames are the lumpy hills and piles of debris in the foreground. Behind the kames, running diagonally from the lower right to the upper left, is a prominent esker. Also present are many kettle lakes. Eskers, kames, and kettle lakes are commonly associated because they all form near glacier margins where meltwater carries a large amount of sediment.

8.4.6   *Mass movements by gravity*

Flowing water, blowing wind, and moving ice—the most common agents that move sediment and rock debris—act as fluids that entrain and carry solid

Figure 8.39   Kames, kettles, and an esker at Fulufjället, western Sweden.
Photo from H. Lokrantz, Wikimedia Commons.

Table 8.2   Different kinds of mass movement.

material involved	Velocity			
	< 1 cm/yr	> 1 km/hr		> 5 km/hr
rock		rock slide		rock avalanche or rock fall
unconsolidated sediment	creep	Earthflow, debris flow or slump	Mudflow or debris slide	debris **fall or avalanche**

material. Gravity, too, can move detritus but may do so without the assistance of any fluid. Such mass movements are generally termed *rock falls* or *landslides* by the layperson but *mass movements* or *mass wasting* by geologists. Mass movements occur on just about any slope; sometimes the movement is fast and sometimes it is slow. When the movement is fast and in an inhabited area, disaster may result. For example, at 10:35 a.m., on March 22, 2014, a major landslide occurred near Oso, Washington. The slide crossed the North Fork of the Stillaguamish River and covered a residential neighborhood, killing 43 people and destroying 50 homes.

Mass movements are highly variable and may involve solid rock, looser unconsolidated material, or both.

Some mass movements, called *slides*, are episodal events that occur relatively quickly. Other mass movements, called *flows*, can be short-lived or relatively long lasting. Table 8.2 contains the most common names used to describe the various types of mass movements. There is a great deal of overlap, and many events fall between two categories. Figure 8.40 shows a few examples of the many kinds of failures and flows.

Falls, slides, and slumps are all kinds of rock or debris failures that may occur without warning. Because they are sudden and unexpected, they can be deadly. *Falls* occur when material falls off a hillslope or falls from a cliff. *Rock falls* involve only pieces of rock—cobbles, boulders, etc. *Debris falls* involve soil

Rockfall on the road to the east entrance, Zion National Park, in 2016. Photo from the National Park Service.

Loch Beoraid, northwest Scotland. Photo from David Jarman, Wikimedia Commons.

Super Sauze earthflow near Barcelonnette, France, June 2009, from HylgeriaK Wikimedia Commons.

Mudflow after the 1980 eruption of Mt. St. Helens near the Cowlitz River at Castle Rock, Washington. Photo from Lyn Topinka, Wikimedia Commons.

A debris slide in Guerrero, Mexico. Photo taken in 1989. From Not home, Wikimedia Commons.

Figure 8.40　Examples of different kinds of mass movements.

and *regolith*, too. *Slides* occur when material slides down a hillside along a plane of weakness. Often these planes are bedding planes between layers of sedimentary rocks, but slides can occur on other surfaces too. Slides generally produce piles of *talus*—coarse material, including boulders, that collects on and at the bottom of slopes.

Slumping is related to sliding but is a slightly different form of slope failure. When slumping occurs, an arc-shaped break, called the *failure surface*, or *slump trace*, develops between sediments (and sometimes rocks) at Earth's surface and material at depth. The failure surface is the bottom of a *slump block*. The top of the block, typically vegetated, remains undisturbed as the block rotates on the failure surface and moves downhill.

Some slumps are natural—for example, many occur on river banks. Others occur because people remove material that was stabilizing a hillside, perhaps by building a road. Both natural and unnatural slumps can be triggered by earthquakes, rainstorms, or even a clap of thunder. Slides differ from slumps in that there is no rotation of the sliding rock mass along a curved surface.

Flows—including *creep*, *earthflows*, and *debris flows*—move more slowly than slides and falls. *Creep*, the slowest kind of mass movement, may continue for extended times on any slope. Some results of creep are bulging hillsides, bent trees that have curved to grow vertically, and inclined telephone poles. *Earthflows* (involving mostly soil) and *debris flows* (involving soil, rock, and organic debris) move faster than creep, travelling on the order of kilometers/hour. Most earthflows and debris flows remain localized on a single slope and, in any kind of flow, if the flowing material is supersaturated with water, it moves faster than if dry. *Mudflows*, involving water-saturated fine muds, for example, move at velocities up to several kilometers/hour. Mudflows usually result from heavy rains, but melting snow on erupting volcanoes may produce them as well. Once mudflows start to move, they may engulf valleys. They can travel long distances and so are potentially very dangerous to people.

8.5 Diagenesis and lithification

After sediment is deposited and buried, it doesn't remain unchanged. It is altered—sometimes quickly and sometimes slowly—by a combination of chemical, physical, and biological processes, collectively termed *diagenesis*. Diagenesis begins as soon as sediment is deposited, but

it occurs more rapidly in some sediments than in others. The process involves physical, mineralogical, and chemical changes. For example, sediment texture can change and new (secondary) minerals can form. Some iron minerals may oxidize, and thus sediment can take on a red or brown color. At the same time, organic material may completely disappear or, alternatively, if oxygen content is low, organic matter may be converted to *coal* or to *kerogen* (the first step toward petroleum). Eventually, continued diagenesis may eventually cause loose unconsolidated material to harden and become a rock. This overall process that turns loose material into a hardened rock, generally termed *lithification*, is gradational and has many manifestations. Consequently, distinguishing between other diagenetic processes and lithification is often arbitrary or impossible. Adding more complications, diagenesis commonly continues after lithification has occurred.

Figure 8.41 summarizes the key processes involved in diagenesis. These processes may operate in sequence or simultaneously. Additionally, they may operate to different degrees. For example, due to compaction, the clasts in sedimentary rocks may all be in contact, much like the way marbles in a bucket are in contact, with cement between. In other, less common sedimentary rocks, the clasts may all be separated by significant amounts of cement or other secondary minerals, giving the rock a texture equivalent to chocolate chip ice cream.

Diagenesis begins as sediment piles up because the weight of overlying material causes *compaction*. All sediments contain substantial intergranular water that is slowly squeezed out during compaction. As water is expelled, sediment clasts are pushed closer together. If the clasts have a thin dimension, they may line up as compaction occurs, giving the sediment distinctive planar texture called a *foliation*. Clay mineral grains, in particular, are generally sheet-like and thin in one dimension. So, for sediments that include clays, compaction may lead to parallel alignment of clay mineral grains and a foliation. For other sediments, in which the mineral grains or clasts are more equidimensional, compaction pushes grains closer together without developing an oriented rock texture. Although compaction by itself can cause clay-rich sediments to lithify and become soft rocks called *claystones*, most sediments have too much space between grains to become rocks by compaction only.

During diagenesis, pore water flowing between mineral grains may deposit a cement that glues individual clasts together. This process, termed *cementation*, is

Diagenesis

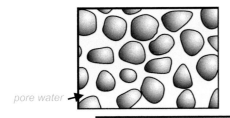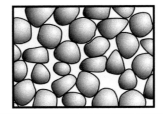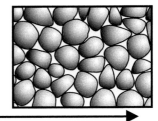

pore water

Compaction

During compaction, water is squeezed out of loose sediment and grains may eventually all be in contact.

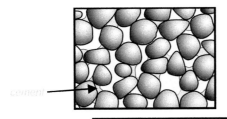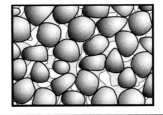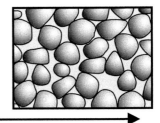

cement

Cementation

During cementation, chemical cements are deposited from the pore water. They glue individual mineral grains together. Eventually all the water may be gone and the sediment may be entirely lithified.

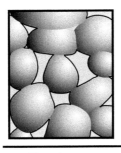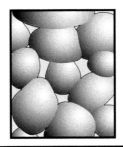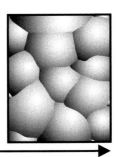

Recrystallization

During recrystallization, small mineral grains may grow into each other and join to become larger crystals of the same mineral. In extreme cases, all the original material may become joined in a mosaic pattern.

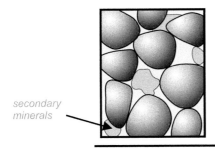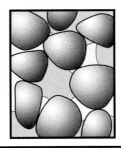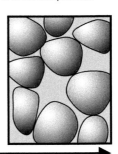

secondary minerals

Recrystallization

During recrystallization, some of the original minerals may dissolve and then precipitate to form new secondary minerals in spaces between grains.

Figure 8.41 The key processes that occur during diagenesis and lead to lithification.

often the key process that causes lithification and the creation of hardened rock. The process mostly occurs below the water table, involving precipitation of minerals from groundwater, producing most commonly calcite, quartz, iron oxides, and clay minerals as cement. Cementation is generally not a fast process, often requiring millions of years, because flow of groundwater is slow and large amounts of water are required.

During diagenesis and lithification, compaction, loss of water, and sometimes heating due to burial can cause chemical reactions leading to *recrystallization. Recrystallization* can change already-existing minerals into new, more stable ones or can cause small mineral grains to combine to form larger ones. Rocks of any sort may recrystallize, but limestones and other chemical sediments recrystallize especially quickly because their relatively high solubility in water means that their ions are more mobile.

8.6   Sedimentary layers and structures

8.6.1   The nature of sediments

Sediments and sedimentary rocks vary greatly and on all scales. On the largest scale, they vary from place to place. On a small scale, they vary within single rock outcrops or even within a single hand specimen. Most outcrops contain layers of different compositions and thicknesses that reflect different sediment sources, different types of deposition, or both. Sediments and sedimentary rocks also contain layering and many other kinds of *sedimentary structures* (including graded bedding and crossbedding, discussed below). Most, but not all, of these structures reflect the conditions at the time of sediment deposition or shortly afterwards, and thus they reveal clues about the environment of deposition. We call such structures *primary sedimentary structures.* Other sedimentary structures, *secondary sedimentary structures*, form during diagenesis, often long after deposition has occurred, and will not be discussed further in this chapter. Sedimentary structures are of many kinds. Some are large and easily seen in outcrop, others require closer examination, and still others require, occasionally, a microscope to be seen.

8.6.2   Layering in sedimentary rocks

When sediment settles from water and is deposited on an ocean floor, it typically forms extensive horizontal sheets and layers. Sediment deposited in lake bottoms,

too, may produce extensive layers. Sediment deposited by a river forms horizontal layers too, but the layers may be confined to a river channel. In fact, with just a few exceptions, such as the slopes of sand dunes, gravity ensures that all sedimentary layers are horizontal. The layers may fill depressions or become thinner in places where they cross high points, but the overall layering does not slope in any direction. So, if sedimentary layers in an outcrop are sloping, it generally means that the layers were tilted after they formed. This observation is called *the principle of original horizontality.*

Strata are horizontal layers, or formerly horizontal layers, that can be seen in sediments or sedimentary rocks, and a rock that contains visible strata is termed *stratified.* (A single layer is properly called a *stratum*, but the term is generally used in the plural form.) Colors, grain sizes, and other textural contrasts commonly differentiate strata. These differences may be original, dating from the time of sedimentation and diagenesis, or produced by weathering, or both. For example, the 5–10 cm thick alternating layers of shale and sandstone shown in Figure 8.42 show both color and texture

Figure 8.42   Strata in the Pt. Loma Formation, near San Diego, California.

From Zimbres, Wikicommons.

contrasts, and the sandstone has weathered away more than the shale has. Strata need not be parallel or horizontal as seen in this photo. Another example of strata, in the sandstone shown in Figure 8.27, is clearly neither horizontal nor parallel.

Layered sediments with contrasting strata form in several ways. Changes in water velocity, for example, can cause different sized or textured materials to be deposited. Alternatively, strata can develop if the composition of sediment being deposited changes—perhaps because one source of sediment becomes exhausted and replaced by another. Stratification may also develop if erosion removes sediments from the upper surface of a layer and other sediment, of different composition, collects on top. Erosional breaks of this sort are common in sediments and sedimentary rocks, but are easily overlooked.

Some sediments and sedimentary rocks are jumbled mixes of grain sizes. Such sediments and rocks, for example, may originate as glacial deposits or may form from landslides, rock falls, or debris flows. Some river deposits produce conglomerates composed of cobbles and pebbles in a sea of fine-grained matrix. However, most sedimentary rocks, and most layers within sedimentary rocks, have experienced some degree of sorting. Coarse grains tend to be with coarse grains, and fine grains with fine grains. Yet, although rocks may contain both fine-grained and coarser-grained layers, layers are generally present.

The boundaries between layers may be distinct or hazy. If a stratum is clearly compositionally, texturally, or in another way distinguishable from layers above and below, we call it a *bed*. Geologists, for example, may talk about a *limestone bed* when they refer to an obvious layer of limestone surrounded by rocks of different compositions. Identifiable planar boundaries that separate one bed from others above or below it are called *bedding planes*. Although bed tops and bottoms must be well defined, beds are generally not entirely uniform and often contain observable but less sharply defined strata within them.

8.6.3　Variable thickness of layering

Distinct sedimentary rock layers may be paper-thin to many meters thick. In general, homogeneous layers are thick if the source of sediment remained constant and if the environment was stable and unchanging at the time of deposition. These factors allow ample time for thick uniform sediment to accumulate. In dynamic environments, however, distinct layers are generally quite thin. For example, the many contrasting layers shown in Figure 8.42 were deposited one layer at a time during submarine landslides onto a continental shelf. The different slides carried sediments of different compositions, and the entire stack of sediments accumulated over a long time.

Geologists use some general terms to describe bed thickness. Very thick layers that show little stratification (layering) within them, sometimes defined as being greater than 3 meters thick, are termed *massive*. The uppermost layers of sandstone in Figure 8.43 are massive. Layers, whether massive or not, on the order of meters thick are termed *thick*. Still thinner layers, such as those in the bottom half of the photo, have medium or thin thicknesses.

We call rocks that have very thin layers—less than 1 centimeter thick—*laminated*. The sandstone shown in Figure 8.44 is a good example. Note that the layers in this photo are mostly, but not entirely, parallel. Ripples, created by the wind that deposited the sediment, disrupt the parallel layers. Some laminated sediments and rocks exhibit *rhythmic layering*—meaning that they contain alternating parallel layers that have distinctly different characteristics. *Varves* (see the example in Fig. 8.38) that form in meltwater lakes, for example, typically show rhythmic layering caused by seasonal changes in deposition.

8.6.4　Crossbedding

Horizontal, or near-horizontal, strata are the most commonly observed type of sedimentary layering. However, some beds—like the ones shown in Figure 8.45—have parallel top and bottom bedding planes but contain inclined layers between. The inclined layers are called *crossbeds*. The dip of the beds records the direction that wind or water was flowing at the time of deposition; think of sand grains sliding down the face of a sand dune. The sandstone in Figure 8.45 is a petrified sand dune, and the wind that deposited the sand was mostly flowing from left to right. If rocks contain more than one layer of crossbeds, the boundaries between the layers generally represent erosional surfaces. Crossbeds are extremely common in beach deposits, sand dunes, and river sediments. Figure 8.27 shows an outcrop in Zion National Park that displays huge *crossbeds* preserved in the Navajo Sandstone since its formation from desert sand dunes nearly 200 million years ago. And Figure 8.4, an outcrop of the Catskill Formation, shows crossbeds created by deposition in river channels.

Figure 8.43   Sandstone outcrops in Arches National Park, Utah.

Photo by Mav, Wikimedia Commons.

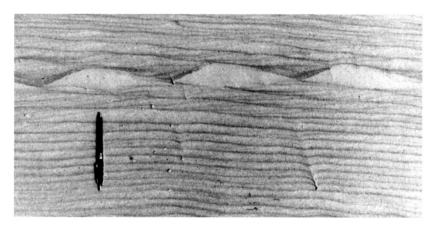

Figure 8.44   Laminations in a sandstone.

Photo from Ralph Hunter, USGS.

Figure 8.45   Crossbedding in sandstone in Coyote Gulch, southeast of Escalante, Utah.

Photo from G. Thomas, Wikimedia Commons.

8.6.5 *Graded bedding*

Many clastic rocks display another common texture—called *graded bedding*—characterized by a systematic change in grain size from coarse at the base of the bed to fine at the top. Figure 8.46 shows a good example. Clasts in the conglomerate in this figure vary (grade) from coarse at the bottom to fine at the top. Sometimes graded bedding involves very fine grains or is so subtle that we can only see it with a hand lens. Graded bedding generally forms because coarse, heavy grains are deposited from suspension before lighter grains are, so sediments pile up with coarse grains on the bottom and fine grains on the top. Grading forms, for example when river velocity slows and grains of different sizes, starting with large followed by small, settle to the river bed. This kind of texture may develop when submarine sediments slide down a continental slope and slowly settle in deeper water. The coarser material settles first and finer sediment is deposited on top. Repetitive marine deposits of this sort are termed *turbidites*. For example, the strata shown in Figure 8.42 are turbidite deposits and, if examined with a hand lens, exhibit graded bedding. (At the scale of the photos shown, however, the grading cannot be seen.) Some rare rocks exhibit inverse graded bedding, with coarse material on top of fine. This type of bedding is most commonly associated with debris flows related to kinds of mass wasting. Figure 8.15 shows one example of inverse grading created when submarine landslides deposited poorly sorted material in undersea canyons.

8.6.6 *Bedforms*

The structures and textures of sediments within a layer can be distinctive, but other key characteristics, called *bedforms*, form on the tops and bottoms of beds. These features can be petrified and preserved in sedimentary rocks. For example, flowing water can create *ripple marks* (Figs. 8.47 and 8.48) like the ripples on modern-day beaches. Ripple marks form most commonly in shallow-water deposits but also can be created by blowing wind. Symmetrical ripples form when water is oscillating, as it does on a beach (Fig. 8.47). Asymmetric ripples record flow directions of rivers or streams, with the steep side downstream.

Figure 8.47 Cross-sectional view of ripple cross-laminated sandstone, showing symmetry indicative of a wave origin. Entrada Formation, Jurassic, San Rafael Swell, Utah.
Photo from T. Cope.

Figure 8.46 Graded bedding in the Rockfish Conglomerate, central Virginia.
Photo from C. Bentley.

Figure 8.48 A pile of eroded, rippled beds that all contain well-formed ripple marks. Carmel Formation, Utah.
Photo from T. Cope.

8.6.6.1 Mudcracks

Mudcracks, also called *desiccation cracks*, are common in mud puddles today and were common in the past. The cracks form when mud shrinks during drying. Subsequently, the cracks may be preserved when the sediment is lithified. So, muddy sedimentary rocks (e.g., shales) can preserve cracks from millions of years ago. Figure 8.49 shows mudcracks in a mudstone in Glacier National Park, Montana, that formed this way. If present, mudcracks suggest that the original sediments dried rapidly.

8.6.6.2 Raindrop marks

Raindrop marks, caused by impaction of raindrops, too, are sometimes preserved in sedimentary rocks—suggesting that sediment was exposed at the surface just before rapid burial. Figure 8.50 shows raindrop marks in a rippled sandstone formed more than 300 million years ago.

Mudcracks and raindrop impressions form on the tops of sedimentary layers when the sediment is exposed to the atmosphere or to weather. Other structures form when a layer is covered by another layer of sediment. Figure 8.51 shows *sole marks*, one example of a structure that forms this way. Sole marks may be created and preserved if coarse sedimentary material is deposited on mud. A rapid deposit of coarse material can scour the finer mud that it settles on, leaving grooves that record flow direction. Sole marks can also form if the weight of overlying material pushes it into the underlying mud. These grooves and depressions may become filled with

Figure 8.50   Fossil raindrop impressions on the top of a wave-rippled sandstone from the Horton Bluff Formation (Mississippian), near Avonport, Nova Scotia.
Photo from M. C. Rygel, Wikimedia Commons.

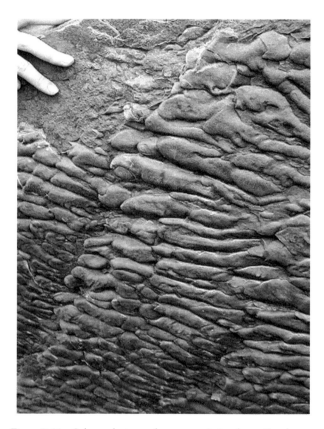

Figure 8.51   Sole marks in sandstone near Laingsburg, South Africa.
Photo from D. L. Reid, https://serc.carleton.edu/NAGTWorkshops/sedimentary/images/sole_marks.html.

Figure 8.49   Mudstone that has preserved mudcracks.
Photo from Wenatchee Valley College Commons.

material that, when lithified, preserves the sole marks. The sole marks in Figure 8.51 were created by water flow (from upper left to lower right) that scoured grooves in underlying sediment. Sand filled the grooves that, when lithified, preserved the structures shown.

Figure 8.52 Flame structures in an outcrop, Inyo County, California.

Photo from T. Cope.

8.6.6.3 Flame structures

Flame structures, such as the ones shown in Figure 8.52, are another kind of sole mark. They form between two sedimentary layers when the weight of an overlying bed forces underlying material to push up and into the overlying layer. The result is bed tops with flame-shaped protrusions into the overlying layers—seen several centimeters above the finger in Figure 8.52. In this photo, the "flames" are made of sandy material that pushed into somewhat coarser material above. Graded bedding—present but hard to see in the photo—commonly accompanies flame structures. In this photo, the bases of the layers contain coarser grains than the "flames" on their tops.

8.7 Different kinds of sedimentary rocks

8.7.1 Siliciclastic sedimentary rocks

Different kinds of sedimentary rocks have different characteristics. Large differences may be visible in outcrops, in hand samples, or in drill cores. Smaller differences may require a hand lens or a microscope. Properties such as grain composition, grain size, color, and the thickness of strata allow different sedimentary *lithologies*—kinds of sedimentary rocks—to be distinguished and named. The most common lithologies are siliciclastic.

Geologists name siliciclastic sedimentary rocks based primarily on grain size (Table 8.3). Confusion can arise,

Table 8.3 Classifying siliciclastic rocks.

Clast size	Sedimentary rock		Major component of unconsolidated sediment	Clast sizes
very fine grained	mudrock — mudstones	claystone	clay	more than 50% clay-sized
		siltstone	silt, generally quartz	more than 50% silt-sized
		shale	clay	clay- and silt-sized
small to medium	sandstone		sand, most commonly quartz	sand
very coarse	conglomerate (rounded grains) or breccia (angular grains)		gravel or coarser sediment	pebble
			gravel or coarser sediment	cobble
			gravel or coarser sediment	boulder

though, because some names used for siliciclastic rocks are the same words used in the Udden-Wentworth scale (Table 8.1) to classify sediments by mean grain diameters. But rocks, in contrast with well-sorted sediments, may contain grains that include a range of sizes. Additionally, the term *clay* refers to clasts of the finest grain sizes and also to a specific kind of mineral. Yet, clay-sized clasts are not always composed of clay minerals. And finally, the names mudrock and mudstone sound similar but do not mean the same things. Mudrock is a general term for very fine-grained rocks that may be either mudstone or shale.

Table 8.3 presents a general classification scheme for siliciclastic rocks. The finest-grained rocks, claystones, consist primarily of clay-sized grains (< 1/256 mm), and the minerals present are mostly clays. Siltstones comprise silt-sized grains (1/256 to 1/16 mm), mostly quartz. Sandstones comprise sand-sized grains (1/16 to 2 mm), generally quartz. Coarser rocks, including conglomerate and breccia, generally include gravel, pebbles, cobbles, or boulders of many sorts in a matrix of finer material.

8.7.1.1 Mudrocks

The finest-grained siliciclastic rocks, generally termed *mudrocks*, consist of silt- and clay-sized clasts. Mudrocks include two kinds of *mudstones (siltstones and claystones)* and *shales*. All contain combinations of very small clay- and silt-sized clasts, but the ratios vary. Claystones comprise mostly clay-sized grains, and siltstones are slightly coarser (mostly silt-sized grains). Shales, too, contain combinations of clay- and silt-sized grains. The difference between mudstones (siltstones and claystones) and shales is that mudstones break into blocky pieces, while shales have *fissility*, a property that means the rocks *cleave* (break) into thin sheets because mineral grains aligned like a stack of very small pieces of paper during compaction.

Figure 8.53 shows outcrop and closer views of siltstone (top photos) and shale (bottom photos). Both rock types often weather and erode to produce steep debris-covered slopes such as the one shown in the bottom left photo. These rocks can be difficult to distinguish with just a glance, but siltstones fracture to produce curved surfaces and shales break into flaky pieces. Additionally, an experienced geologist may scratch them, or even chew on small pieces of them, to find out if the dominant minerals present are clay (soft) or quartz (hard)—which distinguishes one rock type from the other.

Mudrocks form in low-energy environments, because in high-energy environments the fine grains they contain would not be deposited. Thus, mudrocks most often form in lagoons and other protected near-shore environments, in deep offshore waters, or in lakes. Organic-rich shales, one kind of mudrock, are the source of most petroleum, but getting the petroleum out of the ground is often problematic because shales are typically neither porous nor permeable due to their parallel alignment of grains.

8.7.1.2 Sandstones

Sandstones are clastic rocks that contain mainly sand-sized grains (1/16 to 2 millimeters diameter). Grain size, however, can be variable; sandstones that contain significant amounts of fine-grained clasts grade into mudstones, and sandstones that contain significant

Outcrop (left) and closeup (right) Balls Bluff Siltstone, Luck Stone Quarry, Manasses, Virginia

Rochester Shale, Niagara Gorge, New York

Roadcut with exposed shale, southeastern Kentucky

Figure 8.53   Siltstone (top) and shale (bottom).

Top photos from Greg Willis, bottom left photo from Wilson44691, and bottom right photo from Pollinator, all at Wikimedia Commons.

amounts of large clasts grade into conglomerates. Graded bedding is commonly present in such poorly sorted rocks.

Sandstones are common and form in many different settings. In terrestrial (continental) environments, sand, which later becomes sandstone, can accumulate in river or lake bottoms, glacial outwash or alluvial fans, and deserts. In marine environments, sand typically accumulates in beaches, deltas, or offshore bars. Common sandstones are dominated by medium-sized grains of quartz, because quartz is the most common and stable product of weathering. But the definition of sandstone is based only on grain size, and there are different kinds of sandstones distinguished by their mineralogical compositions.

Arenites are sandstones that contain less than 15% matrix between grains, and even more specifically, *quartz arenite* is the name given to a variety of arenite, which is generally very well sorted, containing more than 90% quartz. Figure 8.54 shows an example of quartz arenite. There are many kinds of arenites; the most significant are listed in Table 8.4. *Quartz arenites* are quartz-rich, *arkoses* contain significant amounts of feldspar, *lithic sandstones* contain rock fragments, and *graywackes* contain more than 15% clay minerals.

Table 8.4   Common kinds of arenites.

Variety of arenite	Clast characteristics
quartz arenite	more than 90% quartz
arkose	more than 25% feldspar; the rest mostly quartz
lithic sandstone	more than 5% rock fragments; the rest mostly quartz
graywacke	more than 15% clay; the rest mostly quartz, feldspar, and rock fragments

Figure 8.55   Arkose from Mt. Tom, Massachusetts. Photo from R. Weller.

Arkoses, such as the one shown in Figure 8.55, are sandstones that are generally reddish, orangish, or gray and contain more than 25% feldspar. Typical arkoses contain a mix of grain sizes, but coarse sand and some larger clasts are usually present in significant amounts. The arkose in Figure 8.55 contains conspicuous amounts of pebble-sized clasts besides the dominant sand. The reddish color of this sample, and most arkose, is due to the color of the K-feldspar clasts that it contains. Arkose most commonly forms from sediment produced by weathering of granitic rocks. The sediment must be buried rapidly or else chemical decomposition will replace the feldspar with clay minerals.

Lithic sandstones are sandstones with clasts that are more than 5% of rock fragments. The presence of feldspar or lithic fragments in sandstone implies that the parent sediment did not travel far from its source before deposition, or else the feldspar or fragments would have decomposed or broken apart.

Graywackes, sometimes just called *wackes* or *dirty sandstones*, are sandstones that contain more than 15% clay. They have a dark color and contain poorly sorted and angular sand-sized grains in a finer-grained matrix. Figure 8.56 shows a typical graywacke. The layers of

Figure 8.54   Sandstone hand specimen about 12 cm across. Photo from D. Perkins, GeoDIL.

Figure 8.56   Tilted graywacke blocks near Port Logan, western Scotland.
From R. and T. Clough, Wikimedia Commons.

Figure 8.57   The Navajo Sandstone in Zion National Park, Utah.
Photo from R. Luck, Wikimedia Commons.

sediment were originally deposited horizontally before being tilted, breaking up, and falling into the sea.

Graywackes typically contain less quartz, and more feldspar and rock fragments, than other kinds of sandstone. The matrix material is generally clay, which, along with volcanic rock fragments, is what gives the rocks a dark color. Crossbedding, graded bedding, and bedforms are common in graywackes, and the mix of grain sizes and minerals in graywackes suggests that the parent sediments were deposited near their source, before sorting or chemical decomposition could occur. Many graywackes form from sediments deposited by submarine landslides near continental margins.

Quartz-rich sandstones resist weathering much more than mudrocks do. (See the contrasts shown in Figure 8.6.) So, in contrast with mudrocks, sandstone outcrops often form cliffs like the Tuscarora Sandstone in Figure 8.3 and the Navajo Sandstone cliff shown here in Figure 8.57. Note that prominent crossbeds remain from the original deposition of the sand dunes that later became this rock. Similar crossbeds are shown in Figure 8.27. The black staining on the outcrop in Figure 8.57 is *desert varnish*, made of iron and manganese oxides, which is typical on exposed rocks in arid environments.

8.7.1.3   Conglomerates

Conglomerates consist of rounded to subangular (partially rounded) pebbles, cobbles, and boulders separated by a finer matrix. Because strong currents are required to transport and round coarse material, conglomerates typically form in high-energy environments, including beaches or stream channels. Thus, they often have small aerial extent. The pebbles, cobbles, and boulders are typically deposited first, and subsequently finer-grained matrix material fills gaps between the clasts.

In conglomerates, the large clasts may not be of uniform composition and size. The top photo shown in Figure 8.58 is a conglomerate from Cyprus that contains clasts of many different compositions. Most are lithic fragments. The bottom photo shows a conglomerate from the Carmelo Formation at Point Lobos, California. Besides conglomerate, this outcrop also contains sandstone. The matrix of the conglomerates seen in Figure 8.58, like most conglomerates, is composed of fine to sand-sized grains held together by a cement such as silica.

8.7.1.4   Breccias

Breccias are similar to conglomerates and contain clasts and matrix of similar compositions, but in breccias, the clasts are angular instead of rounded. Breccias form where rock and coarse mineral material collect with little time for sorting or rounding. The angular nature of clasts suggests that they were not transported far from their places of origin. Figure 8.59 shows a classic example from Marble Canyon in Death Valley National Park. The light-colored angular blocks in this outcrop are made of limestone that has been slightly metamorphosed to become marble. Less common clasts of other compositions are also present, including a gray piece of shale just above and to the right of the hand.

Figure 8.59 Breccia in Mosaic Canyon, Death Valley National Park.

Photo from M. C. Rygel, Wikimedia Commons.

The key characteristic of breccias is that they consist of material that has not been transported far. Conglomerates such as the one shown in the bottom photo of Figure 8.58 may include some angular clasts and, although uncommon, some rocks have characteristics partway between breccia and conglomerate.

8.7.2 Biochemical and related sedimentary rocks

When marine animals or plants die, their constituent soft tissues rapidly decompose and disappear. Waves, flowing water, or scavenging animals then break hard parts such as shells or bones into small pieces, producing clastic material composed of hard remnants of once-living organisms. These sediments are *biochemical sediments* that may become *biochemical rocks*. Biochemical limestone, for example, which forms in shallow marine environments, is produced from skeletal parts and shells made of calcite (calcium carbonate). Such limestones may contain easily seen fossils, but sometimes mechanical weathering breaks shells and other organic debris into such fine material that fossils are unrecognizable. Additionally, recrystallization during lithification may cause fossil material to combine and become part of larger mineral grains, thus removing the fossils altogether.

The top photo in Figure 8.60 shows a petrified oyster bed near Del Mar, California. The shells have been little altered or decomposed since they collected in a shell bed before lithification produced a limestone. Although calcite dominates biochemical sediments, other minerals are possible too. The bottom photo in Figure 8.60

Figure 8.58 Two examples of conglomerates, from Cyprus (top) and Pt. Lobos, California (bottom).

Top photo from Siim Sepp, sandatlas.org; bottom photo from Brocken Inaglory, Wikimedia Commons.

Some breccias form from material that falls from outcrops and collects on talus slopes; others form from sediments deposited in canyons or cracks or that collect on an alluvial fan. Still others are from debris flow deposits.

Figure 8.60   Oyster limestone on the beach at Del Mar, California (top) and an 8 cm across sample of diatomite from near Lompoc, California (bottom).

Top photo from B. Perry, bottom photo from J. St. John, Wikimedia Commons.

shows a *diatomite*, a very porous rock made from skeletal remains. It, like all diatomites, is composed of quartz that was once skeletal parts of marine diatoms (algae-like organisms that are neither plants nor animals). After diatoms die, their remains settle to the seafloor, where they may be lithified to produce a rock such as the one shown. Diatoms and radiolaria (animal-like microorganisms) also deposit silica that, when recrystallized, can become *chert*, an exceptionally fine-grained and hard rock composed of a variety of silica called *chalcedony*. Dark-colored varieties of chert are called *flint*.

If plant matter collects in swamps or other wet environments, it often becomes buried and compressed. As it is compressed, it also is heated, and eventually can be converted to coal, a type of *organic sedimentary rock*. In the marine environment, a different kind of organic sedimentary rock may be associated with coral reefs if precipitation

of calcite by coral and other marine organisms leads to significant rock formation. Organic sedimentary rocks, such as coal and reef rock, are closely related to biochemical sedimentary rocks. But they are not made of detritus and instead consist of the primary material that was deposited in place (not transported), and these rocks are commonly assigned to their own category.

8.7.3   Chemical sediments and rocks

Chemical components dissolved in water may derive from chemical weathering or impure rainfall or can be picked up from river banks or bottoms. These dissolved materials make up a stream's *dissolved load*. The loads are generally much smaller than suspended loads or bed loads (although such may not be the case for rivers that receive large chemical contributions from agriculture, industry, or other anthropogenic sources). Dissolved loads are eventually delivered to lakes or to the oceans where they accumulate until concentrations reach the saturation limit. Except under unusual circumstances, if that limit is passed, water becomes *oversaturated*, precipitation will occur, and mineral deposits will form. Such deposits generally contain interlocking crystals that grew together as they formed.

8.7.3.1   Evaporite deposits

The Dead Sea, a salt lake between Jordan and Israel, is one of the saltiest water bodies on Earth, much saltier than common ocean water. So, in the Dead Sea region, precipitation often leaves salt in thin to thick layers and crusts such as the salt deposits shown in Figure 8.61.

The Dead Sea deposits and similar ones elsewhere form when evaporation leads to oversaturation, and

Figure 8.61   Dead Sea salt deposits.
Photo from T. Monto, Wikimedia Commons.

minerals that crystallize during evaporation are called *evaporite minerals*. Table 8.5 lists the more common evaporite minerals and includes all that are mentioned in this chapter. Most of these minerals are chemical salts that have high solubility in water. This group includes many, some quite rare, alkali (Na or K) or alkali earth (Ca or Mg) chlorides, sulfates, carbonates, and borates. Some evaporite deposits are ephemeral. For example, groundwater reaching the surface commonly deposits gypsum crystals on shales in arid regions, but when it rains, the gypsum dissolves and disappears.

Waters of both the Mediterranean Sea and the Gulf of Mexico have deposited thick layers of evaporite minerals in the past when they became isolated from the larger oceans and evaporation led to oversaturation. In both places, when ocean water evaporated, precipitation started with minerals that were least soluble in water and progressed to those that were most soluble. So, initial precipitates were calcite ($CaCO_3$) and other carbonates. Subsequently, gypsum (a sulfate) and halite (a chloride) were deposited—and then more sulfates and chlorides, including anhydrite, sylvite, carnallite, langbeinite, polyhalite,

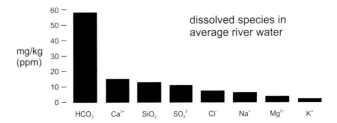

Figure 8.62 The average composition of river water.

and kainite. Massive layers of all these minerals underlie both the Mediterranean and the Gulf of Mexico today.

Evaporation of continental water may also deposit minerals, but the deposits are generally thinner than marine deposits are, and different minerals form because fresh continental water does not have the same composition as seawater does. Continental evaporites typically contain minerals such as borax (a borate), epsomite (a sulfate), and trona (a carbonate) and sometimes other halides or sulfates.

Figure 8.62 shows the average composition of river waters, thus giving an idea of the chemical species that are generally present in continental waters. However, the dissolved loads of streams and lakes vary greatly depending on the nature of underlying bedrock, the source of runoff, seasonality, discharge, and many other factors. The dissolved load of most fresh water consists primarily of HCO_3^- (bicarbonate) and Ca^{2+} ions, as well as sometimes dissolved SiO_2. Other common cations include Na^+, K^+, and Mg^{2+} and other common anions include Cl^- and SO_4^{2-} (sulfate). Most of these dissolved materials will make it to the ocean, thus explaining why oceans are salty. However, dissolved silica (SiO_2), a common dissolved compound in lakes and rivers, is largely absent from ocean water.

8.7.3.2 Saline lake deposits

Precipitation of chemical sediments from fresh surface waters—including streams and lakes—is generally minor. But, in closed basins, such as the one occupied by Utah's Great Salt Lake, evaporation causes precipitation of many different chlorides and sulfates, including halite, sylvite, barite, gypsum, bischofite, thenardite, mirabilite, sylvite, and lithium chlorides. Figure 8.63 shows salt deposits on the shore of the Great Salt Lake.

Lakes whose waters contain large amounts of dissolved salts, like the Great Salt Lake, are called *saline lakes*. Some saline lakes are *perennial* (forming every year after spring runoff) and some are *playas* (that flood sporadically). Parts of Utah's Great Salt Lake are playas that are sources of many valuable minerals. In California, 600 miles southwest of the Great Salt Lake, Searles

Table 8.5   Some of the most common evaporite minerals.

Mineral class	Mineral name	Chemical composition
chlorides	halite	NaCl
	sylvite	KCl
	bischofite	$MgCl_2$
	kainite	$KMg(SO_4)Cl\cdot3H_2O$
	carnallite	$KMgCl_3\cdot6H_2O$
sulfates	anhydrite	$CaSO_4$
	gypsum	$CaSO_4\cdot2H_2O$
	barite	$BaSO_4$
	thenardite	Na_2SO_4
	mirabilite	$Na_2SO_4\cdot10H_2O$
	kieserite	$MgSO_4\cdot H_2O$
	langbeinite	$K_2Mg_2(SO_4)_3$
	polyhalite	$K_2Ca_2Mg(SO_4)_6\cdot H_2O$
	kainite	$KMg(SO_4)Cl\cdot3H_2O$
	epsomite	$MgSO_4\cdot7H_2O$
carbonates	dolomite	$CaMg(CO_3)_2$
	calcite	$CaCO_3$
	magnesite	$MgCO_3$
	trona	$Na_3(HCO_3)(CO_3)\cdot2H_2O$
borate	borax	$Na_2B_4O_7\cdot10H_2O$

Figure 8.63　View of Utah's Bonneville Salt Flats from a rest area
　　　　　　on Interstate 80.
Photo from Famartin, Wikimedia Commons.

Lake is a dry playa that has flooded often in the last
150,000 years. The lake is 19 kilometers (12 miles) long
and 13 kilometers (8 miles) at its widest point. Every
time the lake floods, more mineral deposits are left
behind. Searles Lake yields 1.7 million tons of borax,
epsomite, trona, and other industrial minerals each year.

8.7.3.3　Other chemical mineral deposits

Evaporation is not the only process that causes chemical sediments to precipitate. Deposition may also be
caused, for example, by temperature change. Chemical
limestone, less abundant than limestone formed from
clastic material, may precipitate when cold ocean water
heats up, if the water is saturated, or nearly saturated,
in calcium carbonate. This process happens on ocean
floors when cold water masses move into warmer areas.

In continental settings, *tufa* (a variety of limestone)
and *travertine* (a less porous variety of tufa) are commonly deposited at hot springs when cold, saturated
groundwater is warmed. Figure 8.64 shows examples
of both. The Mono Lake towers (top photo) formed
when cold calcium-saturated groundwater and warm
CO_2-saturated lake waters combined to precipitate tufa
(top). The travertine in Yellowstone (bottom photo)
formed when hot mineral-rich waters evaporated at
the surface. Chemical deposits can produce spectacular
outcrops. For example, the *Trona Pinnacles* (misnamed,
because they are not made of the mineral trona), a destination tourist location at Searles Lake, rise to 140 feet
(40 meters) above the dry lake floor.

Figure 8.64　Tufa in Mono Lake, California (top), and at
　　　　　　Mammoth Hot Springs, Yellowstone National Park
　　　　　　(bottom).
Photos from Vezoy and I. Sagdejev, Wikimedia Commons.

Precipitation of chemical sediments can also occur
due to pH (acidity) change. For example, *marl* forms on
some lake or stream bottoms when water saturated in
calcium carbonate becomes less acidic. In some streams,
the formation and dissolution of marl are diurnal,
because plants produce oxygen from CO_2 (decreasing
acidity) during the day and animals produce CO_2 at
night (increasing acidity).

Chemical precipitation can also create secondary
minerals in rocks and sediments. Figure 8.65 shows
jasper (a variety of inorganic chert), made of SiO_2
stained by red iron oxide, formed when groundwater
deposited silica in preexisting rock nearly 3.5 billion
years ago. Chert, which can also form through organic
processes, includes many distinct varieties, including
flint (black or gray because of included organic matter), jasper (red or yellow because of included iron

Figure 8.65 Banded marble bar chert in Western Australia.
Photo from G. Churchard, Wikimedia Commons.

oxides), petrified wood (preserved by silica), and agate (silica with concentrically layered rings of distinctive colors).

8.7.4 Carbonate rocks

Figure 8.66 shows a limestone outcrop in southern England. Carbonate rocks, such as the one shown, make up 10%–20% of all sedimentary rocks, and many form cliffs. Most carbonate rocks are of marine origin, forming in warm, clear, shallow waters. They are either *limestones* (primarily composed of calcite or aragonite, both calcium carbonate minerals) or *dolostones* (mostly composed of dolomite, a calcium magnesium carbonate mineral). The vast majority of limestones are detrital rocks formed by accumulation of organic debris that may include shell, coral, algal, or fecal material. Other (chemical) limestones form by direct precipitation of calcite from ocean or, less commonly, continental waters.

Figure 8.67 shows an outcrop of a dolostone from western Canada (pocketknife for scale). The many holes and pockets in this outcrop are typical of many carbonate rock outcrops because of the ease with which calcite and dolomite dissolve in some surface waters. Fossils are common in carbonate rocks of all sorts, and this dolostone contains many *stromatoporoid* (a marine organism related to sponges) fossils that did not dissolve as easily as surrounding rock did, giving the outcrop a conspicuous blobby appearance.

Some dolostones form directly through precipitation from water, like many limestones do, but most form by diagenesis of originally calcite-rich sediment when

Figure 8.66 Crumbling limestone cliffs of Emmetts Hill, Isle of Purbeck, on the south coast of England.
Photo from J. Champion, Wikimedia Commons.

Mg-rich water flows through and changes calcite into dolomite. Both limestone and dolostone dissolve in acidic waters, so they are commonly associated with cave systems created when slightly acidic groundwater flows through carbonate rock. Additionally, because of their solubilities, these carbonate rocks may be quite porous and permeable, making them potential reservoirs for petroleum.

We can distinguish limestone from dolostone using the *acid test*. If we place a drop of dilute hydrochloric acid on limestone, it reacts (effervesces) and gives off CO_2 bubbles immediately. In contrast, dolostone reacts slowly, or not at all, unless the rock is ground to a powder. Color may also sometimes distinguish limestone from dolostone. Dolomite often contains some iron replacing magnesium, so when dolostones weather they may gain a tan or brownish color, in contrast with typical limestones that are white or light gray.

Some carbonate rocks have textures equivalent to that of siliciclastic rocks—containing grains of variable sizes and shapes. Others are more massive, reflecting

Figure 8.67   The Cairn Formation, a dolostone near Canmore,
              Alberta.

Photo from Georgialh, Wikimedia Commons.

formation by chemical precipitation or recrystallization
after formation. Still others may have textures created by
living organisms. The standard *Folk Classification System*
for carbonate rocks divides limestones into two main
groups: those that are detrital, called *allochemical carbonates*, and those that are chemical precipitates, called
orthochemical carbonates. Many carbonate rocks contain
both allochemical and orthochemical components.

8.7.4.1 Allochemical carbonate rocks

Allochemical carbonates contain grains, called *allochems*,
cemented together by fine-grained mud-like carbonate
material called *micrite* or coarser-grained carbonate material called *sparite*. Because carbonate minerals are quite
soluble in most natural water, allochems generally cannot have been transported far before deposition. Typical
allochems include *fossils*, *ooids*, *peloids*, or carbonate clasts.

Fossils may be intact and up to centimeters in size, but
many are only fragments of once larger fossils, and many
are very fine grained. The size depends on the organism
that created the fossils and how much the fossils were broken before lithification. *Ooids* are spheres of calcite that
have a concentric, snowball-like structure created when
carbonate layers grow around a small seed particle. Ooids
generally form when particles roll around in shallow
water, adding additional material to their outsides in the
same way that children make large snowballs by rolling
them. Figure 8.68 shows two views of ooids. The top is
a standard photo, and the bottom is a photo of a thin
section (a very thin slice of rock) taken with a microscope.

Peloids, the most common of which are fecal pellets
from marine organisms, are small aggregates of calcite
grains of variable size. *Carbonate clasts*, also called *lime-
clasts*, are detrital fragments of limestone or partially
cemented muds that formed *in situ*, or they are detritus
from nearby eroded bedrock.

2.0 mm

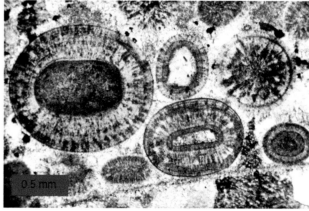

0.5 mm

Figure 8.68   Two views of ooids in the Carmel Limestone, central
              Utah.

Photos from M. A. Wilson, Wikimedia Commons.

8.7.4.2 *Orthochemical carbonate rocks*

Orthochemical carbonates are rocks that do not contain any allochems. Many orthochemical limestones are made only of micrite, others only of sparite. The micrite may form during diagenesis or may precipitate from water. Sparite can precipitate directly during the early stages of diagenesis, but much of it forms later, as micrite recrystallizes over time. Some orthochemical limestones contain chert (silica), mostly of a biological origin (from diatoms, radiolarians, or sponges), that may be in small grains or quite large inclusions up to meters in long dimensions. Detrital clay minerals and quartz grains, too, may be present in orthochemical carbonate rocks.

Questions for thought—chapter 8

1. What were the three major orogenies that produced the Appalachian Mountains? What caused each orogeny? And how long ago did those orogenies take place?

2. What is the difference between weathering and erosion?

3. How can a rock be weathered but not eroded? I suppose we should ask the opposite question – can a rock be eroded but not weathered? Explain.

4. What are clastic sediments and how does grains size influence their rates of erosion?

5. What is the difference between chemical and physical weathering? Describe how the two may work together to cause an outcrop of rock to decompose.

6. What are the typical products of chemical weathering? What kinds of rocks (chemical sedimentary rocks) may eventually form?

7. How are arkose and graywacke different from a normal quartz-rich sandstone? Be specific about what makes up each. Speculate about why one or the other may form from sediments in a particular environment.

8. What is sorting? Why will well-sorted sediments likely be more porous than poorly sorted sediments?

9. Why are stream and desert sediments generally better sorted than a glacial till?

10. Desert sediments are generally made of sand, glacial till can be made of all sorts of different kinds of material. Explain these differences. And, be clear about why desert sediments are sand and not some other kind of sediment.

11. Describe some contrasting characteristics of an immature and a mature sediment. What do the shape, size, and sphericity of clast tell us about its likely transportation history?

12. How may sediment particles be transported in nature? Describe several processes.

13. Is wind or water more effective in moving larger-sized particles? Why? What factors determine whether a particle will be transported by flowing water or not? Why may a stream be able to pickup and transport sand but not clay-sized particles?

14. Why are some streams braided? Describe the processes of erosion and deposition that characterize a braided stream.

15. What may cause stream terraces to form? Hint - There are two general ways this can happen.

16. How do alluvial fans form and how well-sorted are their deposits? Where are alluvial fans most likely to occur?

17. What is the difference between continental ice sheets and alpine glaciers? In the past, continental ice sheets covered large parts of our planet, but not today. Where are continental ice sheets currently found? Anywhere, or all the entirely gone?

18. What is glacial drift? What are the two major types of glacial drift?

19. How does glacial drift compare with alluvium? What are any similarities, and what are the differences?

20. What is the difference between diagenesis and lithification? And, what kinds of processes/steps are involved in diagenesis?

21. What minerals make up the most common cements in sedimentary rocks? Name 3 or 4. Why these materials, why not others?

22. What is the principle of original horizontality? How can this principle be important in determining if sedimentary rocks have been tectonically disturbed after deposition?

23. What causes cross-bedding to form? Is it formed by wind, by water, or by both? How is cross-bedding an exception to the principle of original horizontality?

24. What is the difference between a graded bed and a massive one? What causes graded bedding to form? Make a drawing of graded bedding showing how grain size varies.

25. Porous rocks contain lots of holes. How can a porous rock be impermeable? And, is it possible for permeable rock NOT to be porous?

26. What kinds of depositional environments produce abundant sands? Name three (two should be rather obvious). Speculate about how you could distinguish between sands of the different environments.

27. What is the difference between a breccia and a conglomerate?

28. What types of depositional environments are responsible for the origin of conglomerates? What types of depositional environments are responsible for the origin of breccias? Explain the differences – why do breccias and conglomerates form from sediments deposited in different environments.

29. How do evaporites form? Name three minerals most commonly found in evaporites. Why these minerals? Why are they the most common? Why aren't other more common minerals such as quartz or clays found in evaporites?

9 Stratigraphy

9.1 Rocks of the Grand Canyon

Figure 9.1 is an aerial view that includes the Grand Canyon in Grand Canyon National Park. It is a nearly 480-kilometer (300-mile) long sinuous canyon carved by the Colorado River in northern Arizona's Colorado Plateau. The colorful canyon walls include layers that are whitish or gray, and other layers that are red, green, and purple. Some of the colors can be seen in this view.

The Colorado River originates in northern Colorado and flows through portions of Utah and northern Arizona before entering the Grand Canyon. Several major tributaries, including the Little Colorado River and the Havasu River, contribute additional water to the river within the park. Erosion by these and other smaller tributaries has produced a dendritic (branching like a tree) drainage pattern that is easily seen in Figure 9.1.

The Grand Canyon was carved by simultaneous uplifting of land and downcutting by erosion. In places, it is more than 16 kilometers (10 miles) wide and more than 1850 meters (6000 feet) deep. Uplift and erosion of the Colorado Plateau began 75 million years ago, at the time that North America's Rocky Mountains began to form. Some recent studies suggest that erosion by river systems that preceded the present-day Colorado River may have started 60 or 70 million years ago, but the general consensus is that most of the erosion that created the present canyon occurred during the last 5–6 million years.

This chapter is about the relationships of sedimentary rocks of different types. The field of study that deals with such relationships is called *stratigraphy*. Most of the examples in this chapter are from the Colorado Plateau Region, especially the area near the border between Arizona and Utah, because the region contains much well-exposed rock. The area has three National Parks that are known for their spectacular scenery: Grand Canyon, Zion Canyon, and Bryce Canyon National Parks. Farther north in Utah, Capitol Reef National Park, Canyonlands National

Figure 9.1   Google Earth view of the Grand Canyon and surrounding region. Created from a Google Earth image by D. Perkins.

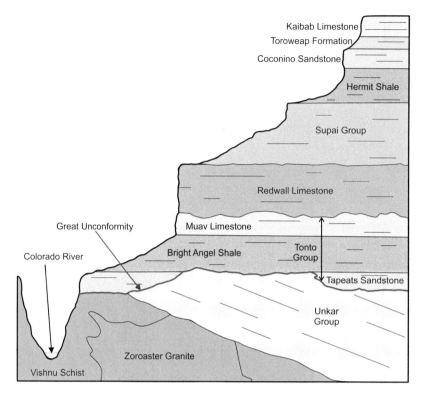

Figure 9.2   Rock units of the Grand Canyon.
Drawing modified from several that are common on the internet.

Park, and Arches National Park include much of the same geology. Although the examples in this chapter are from a relatively small area, the principles of stratigraphy exhibited in the Colorado Plateau apply equally anywhere. Additionally, this chapter focuses on sedimentary rocks younger than Precambrian age, because many Precambrian rocks are igneous and metamorphic, which adds complications that we are not going to consider here.

Figure 9.2 shows the most prominent formations exposed in the Grand Canyon walls. The oldest rocks in the region crop out in the canyon bottom near the Colorado River. They make up what geologists call the *Precambrian basement* (because Precambrian is their geological age and they form the basement for the overlying sediments) The basement rocks include the Vishnu Schist, the Zoroaster Granite, and other metamorphic and plutonic rocks that formed almost 1.75 billion years ago (bya). In some places the ancient metamorphic and plutonic rocks are covered by slightly younger, but still ancient (1.1 bya), sloping group of sedimentary rocks belonging to the Unkar Group.

Above a break called the Great Unconformity, much younger horizontal sedimentary rocks make up all of the canyon walls. Outcrops of these sedimentary rocks vary greatly—with steep slopes in some places and vertical

cliffs in others. Erosion-resistant formations dominated by limestones and sandstones are responsible for the cliffs; softer shales and mudstones have eroded to produce debris-covered slopes. Fossil and other evidence suggests that most of the formations derive from beach or near-shore marine sediments. Other formations— less common—contain material that was deposited in swamps, lakes, or rivers, and some sandstone formed from desert sands. Thus, the rocks exposed here represent many different sedimentary environments.

Figure 9.3 shows the youngest rock exposed at the Grand Canyon, the Kaibab Limestone that forms a cliff at the canyon rim. The 280-million-year-old Kaibab forms both the South Rim and the North Rim of the canyon and provides a horizontal surface for many overlooks and other features of Grand Canyon National Park. The Kaibab is erosion resistant and is the bedrock found near the surface that is responsible for the gently sloping Coconino and Kaibab Plateaus to the south and north of the canyon. In some parts of northern Arizona, younger rocks lie above the Kaibab, but near the Grand Canyon, the younger rocks have been removed by erosion.

Overall, the sedimentary rocks of the approximately 1-mile-deep Grand Canyon represent 260 million years

Figure 9.3 View of the Kaibab Limestone at Zuni Point on the Grand Canyon rim.
Photo from A. F. Borchert, Wikimedia Commons.

of Earth history, averaging out to about 0.01 millimeter of rock per year. But, the rocks did not all form at the same rate, and many rocks that were once present have since eroded away. Thus, the record of Earth's past recorded in the Grand Canyon walls is skimpy and incomplete.

9.2 Formations, groups, and members

Some sedimentary deposits are local, perhaps restricted to a narrow river channel or valley, but many, especially marine sediments, span large regions. So, when lithified, the sediments may become rocks with small or large areal extent. A series of beds, or strata, that are recognizable on a regional scale are often grouped and called a *formation*. Formations may not be uniform over large distances, but they have recognizable characteristics, including compositions and textures, that allow outcrops in one place to be matched with those in other places.

Figure 9.4 shows two examples of limestone outcrops. The Redwall Limestone (top photo) is a prominent cliff-former in the Grand Canyon. Its red color comes from iron oxides eroded from rocks that lie above it. This limestone layer is a good example of a formation that is found over a wide region. Equivalent limestone, but usually gray, exists in eight states in the Colorado Plateau, Rocky Mountains, and the Great Plains regions. Although it is the same formation, the Redwall Limestone has different names in different places. In the Black Hills (South Dakota), it is called

the Pahasapa Limestone; in Wyoming and Montana, it is called the Madison Limestone (photo on the bottom) or the Guernsey Limestone; and in Colorado, it is called the Leadville Limestone. The thickness of the formation is nearly 610 meters (2000 feet) in some places.

Figure 9.5 shows where outcrops of the Navajo Sandstone can be found in the western United States. Formations, like the Navajo, are named primarily based on their lithology (rock type), not when they formed. Consequently, some formations may have slightly different ages in different places. For example, the Navajo is a thick—up to 700 meters in some places—petrified desert sand formation that formed 178 to 192 million years ago from an extensive *erg* (accumulation of desert sand) in the western United States. Navajo outcrops are found in seven western states, but the desert migrated from east to west over tens of millions of years. Thus, the Navajo is oldest at its eastern limits and youngest in the west. The formation is also thickest in the west.

Formations are the fundamental units of most geological maps but are sometimes combined into *groups* of formations that share features or significance. Figure 9.6 shows an example. The *Wingate Sandstone* is a prominent cliff-forming formation composed of 200-million-year-old petrified sand dunes. The Wingate—found in Arizona, Utah, Colorado, and Nevada—is a member of the *Glen Canyon Group*, which in most places contains four formations. The Wingate Sandstone and the Navajo Sandstone (mentioned earlier) are responsible for much of the spectacular scenery in Canyonlands National

Figure 9.4   Redwall Limestone in the Grand Canyon (top) and a Madison Limestone outcrop near Helena, Montana (bottom). Photos from Tobi 87 and Montanabw, both at Wikicommons.

Figure 9.5   Area extent of the Navajo Sandstone.

Figure 9.6   The formations of the Glen Canyon Group.

Figure 9.7 Some outcrops of formations of the Glen Canyon Group in Capital Reef National Park, Utah. From M. Chan, University of Utah.

Park and other national parks of Utah. The formations of the Glen Canyon Group are all about the same age and reflect widespread and generally arid conditions over much of the western United States. The Wingate and Navajo sandstones formed from desert sands during arid times. The other two formations of the Glen Canyon Group (the Moenave and the Kayenta formations) contain mixed beds of red to brown siltstone and sandstone. These two formations formed from river, lake, and floodplain deposits during slightly wetter times.

Figure 9.7 shows outcrops of the Glen Canyon Group in Capital Reef National Park, Utah. The two major sandstones, the Navajo and the Wingate, are easily seen. At the top, the Navajo appears with its typical frosty white and rounded outcrops. Lower down, the Wingate Sandstone is a darker red and has vertical joints forming columns. The Moenave Formation is absent here, but the thinner bedded rocks below the whitish Navajo Sandstone make up the Kayenta Formation. The absence of one member of a group is not unusual. Where this photo was taken, the appropriate sediment was never deposited in the first place, so the Moenave Formation never formed. In some parts of Utah, the Wingate is absent but the other three members are present. Although it did not happen with the Glen Canyon Group, a formation may be missing because, although sediment that could make up the formation was deposited, erosion removed the sediment prior to lithification.

Some formations consist of only one kind of rock. The Tropic Shale in southern Utah is a good example. It is a thick and rather uniform 85- to 90-million-year-old mudrock—at least 60 million years younger than the Glen Canyon Group but found in the same general area. Other formations may contain strata of different rock types created from different kinds of sediments deposited in the same area. Thus, the Moenave and the Kayenta formations contain both siltstone and sandstone, and many formations formed from ocean shore and near-shore sediments contain mixed layers of sandstone and shale. Consequently, geologists sometimes divide formations into *members*—recognizable layers of strata within a formation that may have only limited areal extent. Additionally, individual members may be composed of smaller *beds* that can be distinguished by color, grain size, or composition.

9.3 Stratigraphy

The field of stratigraphy involves the study of rock layers (strata). Some principles of stratigraphy can be applied to volcanic or metamorphic rocks, but usually when geologists mention stratigraphy they are thinking of

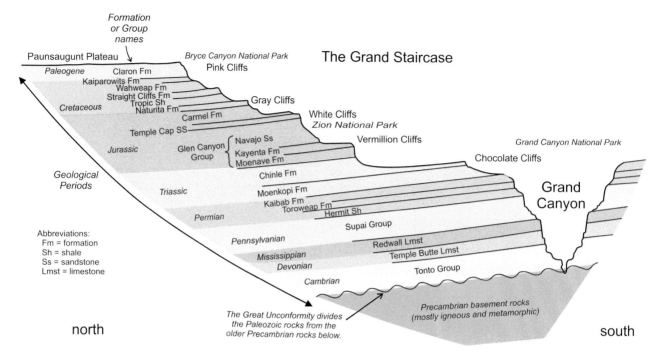

Figure 9.8 Stratigraphy of the Grand Staircase and the Grand Canyon.

sedimentary rocks only. The key principles that make stratigraphy possible were presented in the previous chapter:

- Sedimentary rocks have distinctive characteristics that allow different kinds to be told apart.
- Different kinds of sediments, and subsequently sedimentary rocks, represent different depositional environments.

Figure 9.8 shows an approximately 100-mile north-south cross section of the geology between Paunsaugunt Plateau near Bryce Canyon National Park in Utah and Grand Canyon National Park in Arizona. At the lowest levels, igneous and metamorphic basement rocks are exposed near the Colorado River in the bottom the Grand Canyon. Above the basement, the rocks are all sedimentary up to the Paunsaugunt Plateau, where some volcanic rocks are found. The region from the north rim of the Grand Canyon to the Paunsaugunt Plateau is called the *Grand Staircase*, because it is marked by several steep cliff outcrops formed of erosion-resistant rocks. The rocks exposed in the staircase include many kinds and represent almost 500 million years of Earth history.

From lowest elevation to highest (south to north), the cliffs of the Grand Staircase are the Chocolate Cliffs, the Vermillion Cliffs, the White Cliffs, the Gray Cliffs, and the Pink Cliffs. The different cliffs are labeled in Figure 9.8. Sandstone makes up most of the cliffs, and two of the cliff-forming sandstones are sandstones from the Glen Canyon Group. The Vermillion Cliffs are outcrops

of the Moenave and Kayenta formations, and the White Cliffs consist of Navajo Sandstone. The Pink Cliffs consist of the Claron Formation, which is younger than the Glen Canyon Group. The slopes between cliffs were formed by erosion of shale and other soft sedimentary rocks. Figure 9.9 shows the gray cliffs (bottom) with the pink cliffs above. The other steps in the staircase are older than these and so are buried and not exposed at the surface where this photo was taken.

Figure 9.2 showed the principal formations and groups exposed in the walls of the Grand Canyon. The Glen Canyon Group and the Tropic Shale, discussed earlier, are not found in the canyon because they are younger than the youngest formation present in the canyon, the Kaibab Limestone. As shown in Figure 9.8, if present, these younger formations would lie above the Kaibab Limestone, but there are no rocks above the Kaibab in the Grand Canyon region. Glen Canyon Group rocks and the Tropic Shale are, however, seen 100 miles north of the Grand Canyon in Utah, notably in the area between Zion National Park and Bryce Canyon National Park (Fig. 9.8). In these parks, the Kaibab Limestone and other older formations are not exposed at the surface. They exist in the *subsurface*, with thousands of feet of younger rocks covering them.

The stratigraphy and formations of the Grand Staircase and Grand Canyon are typical of continental regions where relatively flat-lying sedimentary rocks are

Figure 9.9   Two of the cliffs in the Grand Staircase.
Photo from Tucker Kirby, Wikimedia Commons.

found. If a sedimentary layer or formation extends over a large area, outcrops made of identical kinds of rocks can be *correlated* (matched with rock units of the same age but in different places), even if a valley or other erosional feature separates them. The areal extent, thickness, and compositions of different kinds of beds, and their relationships with beds above and below them, allow geologists to reconstruct geological histories. So, studying stratigraphy is important for the wealth of information it provides about Earth's evolution, the evolution of life, and the history of civilization. It is also important because it can guide coal, oil, gas, and mineral exploration.

Preservation of older rocks, of course, has not been as complete as preservation of younger rocks, because older rocks have had more time to be eroded. Additionally, the rocks exposed at the surface are usually the youngest rocks in a region, so even if older rocks are present underground, we may have no outcrops of them to study. Consequently, stratigraphic investigations are most easily carried out on relatively young rocks. Yet, stratigraphic principles can be applied to older rocks, just not with as much certainty.

9.3.1   *Nicholas Steno's contributions*

Nicholas Steno was a Danish scientist who made landmark discoveries in anatomy and various branches of geology, during the mid- and late 17th century. In his

later years, he became a scholar of religion and eventually a bishop. His investigations of sedimentary rocks and fossils provided much of the foundation for modern stratigraphic studies today.

In 1669, Steno published his *Dissertationis Prodromus*. In that scholarly work, he described the nature of rock strata and argued that a single, uniform layer of strata must have formed from sediments deposited over time in a single depositional environment. According to Steno, physical, biological, and other environmental factors determine the nature of sediment deposited. Consequently, if the environment changed, different kinds of sediments, leading to different kinds of rocks, were deposited, producing adjacent contrasting strata separated by what Steno called *bedding planes*. To geologists, individual beds, separated by bedding plains, are the smallest division of any formation.

Figure 9.10 shows two examples of contacts between beds. The photograph on the left shows a sandstone bed above a conglomerate; the contact between the two is quite sharp. These beds were deposited horizontally and tilted later. The outcrop photo on the right shows sedimentary rocks in New South Wales, Australia. Several distinct rock layers are present, all separated by near-horizontal bedding plains. The different layers may all belong to the same formation (or the same member of a formation), but there is great vertical variability. In both photos, the contacts are sharp, and the contrasting strata reflect significant variations in the environments when the sediments were deposited.

Figure 9.10   Two examples of beds and contacts.
Photos from M. Miller, University of Oregon, and D. Vernon, Wikimedia Commons.

Steno presented what we now recognize as four fundamental tenets, or principles, of modern sedimentology: (1) the principle of original horizontality, (2) the law of superposition, (3) the principle of lateral continuity, and (4) the principle of crosscutting relationships. These tenets are at the heart of stratigraphy and are taken for granted today. But at Steno's time, they were not so obvious, in large part because the rate at which geological processes occur, and the magnitude of geological time, had yet to be understood.

9.3.1.1   The principle of original horizontality

Figure 9.11 shows layers of sediment that were deposited in an ocean basin. They are horizontal, extending a long distance toward the center of the basin but they are truncated where they run into the continent. This figure is a depiction of Steno's *principle of original horizontality*—the principle that sediments are deposited in horizontal layers, and the layers are continuous unless some other rock or other obstruction gets in the way. The principle is based in part on the observations that suspended sediment, when it settles from muddy water, forms horizontal layers and that salt and other minerals deposited by evaporating water form horizontal layers. Thus, if we see layers of rock that are tilted, as in Figure 9.10, the tilting must have occurred after the rock formed. Today we know that this principle is not completely correct. Non-horizontal layers of sand, for example, may be deposited on the faces of sand dunes, and some sediments drape over topographic highs or fill lows. Nonetheless, most layers of sediment are deposited horizontally, and the rare exceptions are generally recognized.

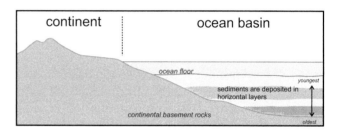

Figure 9.11   Diagram showing horizontal sedimentary layers truncated by continental rocks.

9.3.1.2   The law of superposition

Figure 9.11 also depicts Steno's *law of superposition*—the principle that new sediment, whether deposited by water, wind, ice, or some other agent, collects on top of older preexisting material. So, unless rocks have been overturned during mountain building, in a stack of horizontal or nearly horizontal strata, the oldest beds are at the bottom and the youngest at the top.

9.3.1.3   The principle of lateral continuity

Figure 9.12 shows horizontal strata exposed in buttes of Monument Valley, Arizona, just south of the Utah border. Four formations account for the spectacular scenery. Most prominent is the De Chelly Sandstone, because it comprises the vertical cliffs. Below it, the Organ Rock Shale forms slopes. Two other formations, part of the Chinle Formation and the Moenkopi Formation, cap the buttes. The four formations are discontinuous—they crop out in the buttes but are absent between the buttes. Steno's *principle of lateral continuity* states that, when deposited, layers of sediment spread out in all directions. So, the layers were originally laterally continuous. Thus, as in Monument Valley, if stratigraphic units are similar but are separated by a valley or other erosional feature, we can assume that they were originally continuous. The Moenkopi Formation—seen in this photo—is also found in the Grand Staircase (where it makes up the Chocolate Cliffs) but is mostly absent between the two locations. The principle of lateral continuity tells us that the formation was once continuous across the entire region.

9.3.1.4 The Principle of Crosscutting Relationships

Figure 9.13 shows the Great Unconformity—discussed later in this chapter—in the Grand Canyon. Flat-lying layers of strata lie above, and truncate, sloping lower layers below. (This unconformity is also shown in the stratigraphic column of the Grand Canyon in Fig. 9.2.) The *principle of crosscutting relationships* is the observation that if a geological unit (or other feature) cuts

Figure 9.12 Monument Valley.
Modified photo from Tobi 87, Wikimedia Commons.

Figure 9.13 The Great Unconformity in the Grand Canyon.
Modified photo from C. M. Morris, Wikimedia Commons.

across or through another, it must be the younger of the two units (or features). So, the upper layers of rock in this photo must be younger. The sloping formations formed first and were tilted by tectonism before being eroded. Subsequently, horizontal sediments were deposited on top and became the horizontal sedimentary rocks seen here.

Figure 9.14 shows another example of crosscutting relationships. This outcrop near Winkelman, Arizona, contains an igneous intrusion, a near-vertical dark-colored dike, cutting through limestone. If a dike or other intrusion cuts across strata, it must be younger than the strata. So, if we can determine how old the dike is, we know the strata are older.

9.3.1.5   *The principle of inclusions*

About 160 years after Steno's time, geologist Charles Lyell articulated one other principle that we use to determine relative ages. Lyell's *principle of inclusions* is the observation that any clasts in a sedimentary rock must be older than the rock that contains them. For example, the conglomerate shown in Figure 9.15 contains cobbles of older rocks. Similarly, sandstones and other finer-grained sedimentary rocks contain individual mineral grains eroded from older rocks. The principle of inclusions also applies to some igneous rocks that contain pieces of rock (xenoliths) older than the igneous rock itself. So, if we could find the source of the clasts shown in Figure 9.15, we

would know that the source rock was older than the conglomerate.

9.3.2   *William Smith's contributions*

For 20 years, beginning around 1790, William Smith was a professional geologist in southern England. Smith worked mostly for coal companies but also studied bedrock exposed in canal excavations. In 1801, he drew a rough sketch of what would become "The Map That Changed the World" (which inspired

Figure 9.15   Example of a conglomerate. (See also the photos of conglomerates in Chapter 8, Figure 8.58.)
Photo from D. Mayer, Wikimedia Commons.

Figure 9.14   Diabase dike cutting through horizontal limestone beds near Winkelman, Arizona.
Photo from Jstuby, Wikimedia Commons.

a book of that name) shown in Figure 9.16. This first geological map of England represented the first large-scale applications of Steno's Laws. Smith was able to construct his map because he recognized the uniqueness and significance of distinct rock layers, noting that the same sequence of layers was exposed in different places across a large region. He also discovered that fossils could be used to match layers in different places. Several years after Smith, Georges Cuvier and Alexandre Brongniart did similar studies and constructed the first geological map of the Paris Basin in France. Thus, Nicholas Steno, William Smith, George Cuvier, and Alexandre Brongniart ushered in the science of stratigraphy.

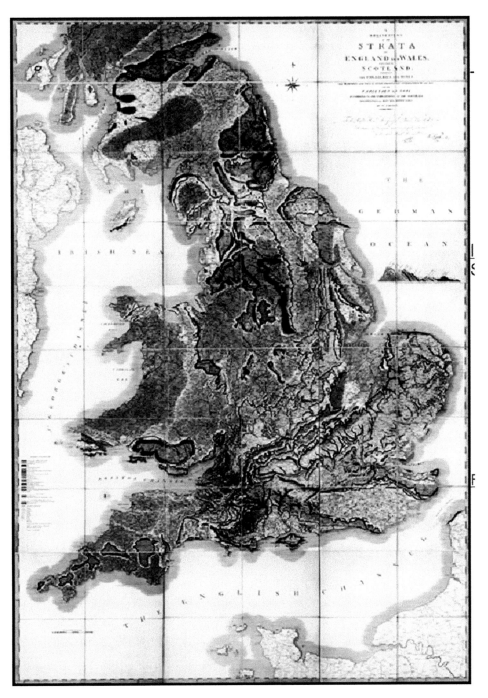

Figure 9.16 The first geological map of Britain, published by William Smith in 1815.
Image from Life Science Image Gallery, Wikimedia Commons.

9.4 Interpreting the environment of deposition

Although early stratigraphers focused on describing rocks, it was not long before they realized that sediments, and thus sedimentary rocks, contained characteristics that reflected conditions when the sediments were deposited. For example, different kinds of sediments are deposited in river channels, glacial outwash plains, and ocean basins (Fig. 9.17). Additionally, within ocean basins, different kinds of sediments may be deposited in shallow (near-shore), moderate depth, and the deepest parts of an ocean. These differences occur because the processes that occur in different environments are different, due to variations in physical conditions and sometimes because of different life forms that may be present. Thus, sedimentary rocks often have distinctive lithologies, sedimentary structures, fossils, and other characteristics that give clues about their origins. Geologists can often interpret *depositional environments*, also called *sedimentary environments*, of old rocks by comparing the rocks with sediments being deposited today.

Interpretations, however, are not always easy, because sedimentary rocks are highly variable and often quite complex. Their natures not only reflect sedimentary settings but also climate, topography, source of the sediment, and many other things. Additionally, many sediments undergo significant changes during lithification or subsequent diagenesis, and sometimes during weathering or metamorphism, that may make interpretations problematic. Furthermore, some of the oldest rocks—for example, banded iron formations—have no modern equivalents to use for comparison.

On a large scale, geologists divide depositional environments into three principle types reflecting three different kinds of depositional settings: *continental* (also called *terrestrial*), *transitional* (shore or near-shore regions), and *marine* (ocean). The main characteristics of each are listed in Table 9.1. Each of these major environment types is further divided into subtypes. For example, marine deposits may collect on continental shelves, in reef settings, or in deep marine basins (Table 9.1). Each setting yields different rocks. The different rocks and settings are distinguished based on the lithologies, distinctive primary and secondary sedimentary structures in the rocks (e.g., specific kinds of crossbeds, ripple marks, or mudcracks), and fossils that may be present.

Continental environments are quite variable, but all are characterized by terrigenous sediments (sediments derived by erosion of rocks on land). These environments include river and stream channels, floodplains, alluvial fans, deserts, glaciated regions, swamps, and

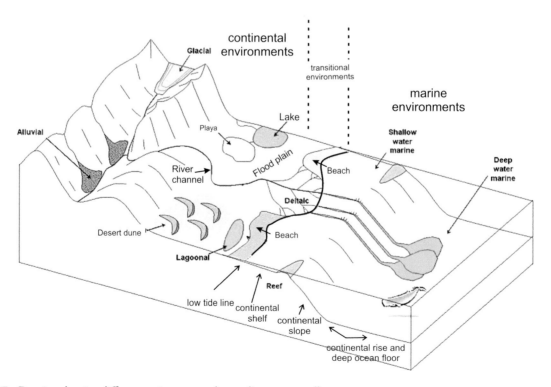

Figure 9.17   Drawing showing different environments where sediments can collect.
This figure also appeared in a previous chapter.

Table 9.1 The major characteristics of different depositional environments.

Continental (on land)

Depositional environment	Common sedimentary lithologies	Common sedimentary structures	Common fossils
Desert	Sandstone	Well-sorted, large scale crossbedding	Footprints, terrestrial fossils
Alluvial fan	Conglomerate, sandstone	Poor sorting, crossbedding	Few
Fluvial (streambed)	Conglomerate, sandstone	Crossbedding, ripple marks	Few
Fluvial (floodplain)	Sandstone, shale	Mudcracks, crossbedding	Terrestrial animals and plants
Soils (paleosols)	Shale, siltstones, freshwater limestones	Paleosol horizons	Root casts, animal burrows
Glacial (till)	Tillite	Very poor sorting, massive	Few
Glacial (outwash)	Conglomerate, sandstone	Crossbedding, ripple marks	Few
Swamp	Coal, shale	Mudcracks, crossbedding	Plants
Lacustrine (lake)	Sandstone, shale, freshwater limestone	Mudcracks, possible varves	Fish, other freshwater aquatic organisms

Brackish and Other Transitional Environments (Where Land Meets Ocean)

Depositional Environment	Common Sedimentary Lithologies	Common Sedimentary Structures	Common Fossils
Delta	Shale, siltstone, coal, conglomerate	Crossbedding, ripple marks, mudcracks	Plants, invertebrates
Beach	Sandstone	Well-sorted, crossbedding	Invertebrates, trace fossils
Tidal flat	Shale, siltstone, sandstone	Crossbedding, ripple marks, mudcracks	Invertebrates, trace fossils

Marine (Ocean)

Depositional Environment	Common Sedimentary Lithologies	Common Sedimentary Structures	Common Fossils
Restricted basin	Evaporites, limestone	Thin beds, mudcracks, salt casts	Few
Shelf	Limestone, dolostone, shale, sandstone	Crossbedding, ripple marks	Invertebrates, fish
Reef	Limestone, dolostone	Massive	Corals
Slope	Shale, graywacke	Graded beds, turbidites	Microfossils
Deep marine	Shale, chert, limestone (including chalk)	Thin bedding	Microfossils

Based in part on a table from R. Dawes, Wenatchee Valley College.

lakes. Clastic sediments from fine muds to coarse conglomerates are dominant. Less commonly, coal or freshwater limestone may be present. The four formations of the Glen Canyon Group (the Wingate, Moenave, Kayenta, and Navajo formations discussed earlier in this chapter) are examples of continental sedimentary rocks.

Figure 9.18 shows a transitional environment, a tidal flat in Oman. If the muddy sediments eventually become lithified, they will turn into mudstone or shale. Tidal flats are only one of many kinds of transitional environments. Others include shorelines, beaches, deltas, and lagoons. These are often dynamic environments characterized by both erosion and deposition. If present, fossils are generally shallow-water marine organisms, but some continental plant and animal fossils may be present as well. Mudstone, siltstone, and sandstone

Figure 9.18 Playing soccer on tidal flat, Oman.
Photo taken by R. Coveney, Wikimedia Commons.

Figure 9.19   Rock with a brachiopod fossil from northeastern Ohio.

From James St. John, Wikimedia Commons.

are the most common rocks that form in transitional environments, but other rock types are possible.

Continental and transitional environments produce many kinds of rocks, but most sedimentary rocks derive from sediments deposited in marine environments. Figure 9.19 shows one example. It is a photo of the Lowell Formation—a limestone in northeastern Ohio. This formation, about 320 million years old, contains abundant brachiopod (a type of shelled organism) fossils, such as the one shown.

Marine sedimentary rocks may be limestone, like the Lowell Formation, but are more commonly fine-grained clastic rocks and sometimes evaporites. Shallow-water environments produce sandstone and shale and, perhaps, reef limestones. Deeper deposits may be fine-grained muds that become shales or graywackes. Chert (sedimentary rock composed of microcrystalline quartz) is commonly deposited in the centers of deep oceans. In all marine rocks, fossils reflecting the environment of deposition may be present. These fossils include fish, corals, shelled organisms, or plankton and other microscopic organisms.

9.5   Rock lithology

9.5.1   *Lithologic variation*

Geologists use the term *lithology* in two ways. On a small scale, *lithology* refers to the overall physical and chemical nature, including grain composition, grain size, color,

thickness of strata, and composition, of a particular stratum, rock, or rock formation. On a larger scale, it is also used as a synonym for rock type. For example, the lithology of a particular formation might be described as an *eolian sandstone*. The field of *stratigraphy* involves both kinds of lithology—the study of the nature and variation of rock characteristics, as well as the different kinds of rocks and formations that may be found in one place or region.

9.5.2   *Different kinds of stratigraphy*

Geologists investigate rock lithologies and stratigraphy using several approaches. The main kinds of stratigraphy fall into three areas: *lithostratigraphy*, *biostratigraphy*, and *chronostratigraphy*. Lithostratigraphy involves identification and classification of strata based on their lithologic (physical) characteristics. Biostratigraphy involves identifying and correlating rocks of similar ages based on the fossils they contain. Chronostratigraphy—closely related to biostratigraphy—involves rock age too, but the goal is to assign absolute ages to lithographic units. Thus, biostratigraphers may determine that one formation is older than another, and chronostratigraphers may be more concerned with how many million years ago a particular formation formed. Other kinds of stratigraphy, all closely related to the three main kinds, include *chemostratigraphy* (study of the variation in rock chemistry), *cyclostratigraphy* (study of variations in sedimentary rocks due to long-term climate cycles), *magnetostratigraphy* (study of variations in the magnetic fields recorded by rocks), and *archaeological stratigraphy* (study of the stratigraphy associated with archaeological studies).

9.5.3   *Lithostratigraphy*

On a large scale, lithologies vary from place to place, reflecting different environments of deposition, and although sedimentary layers may cover large areas, they do not continue forever. Strata typically become thinner and eventually disappear altogether as the distance from sediment source increases. The composition of the sediment, too, may change gradually as the distance from the source increases. Commonly, for example, the densest and the coarsest mineral grains are deposited first, near the source, and larger amounts of lighter and finer sediment are deposited farther away. Horizontal variations within a single stratigraphic unit may be profound or subtle. If gradational, they may be most easily seen by comparing samples collected from outcrops that are separated by long distances.

On a smaller scale, individual lithological units also show variation. When viewed from a distance, a single rock layer may appear quite uniform. Closer examination, however, commonly reveals heterogeneity in both vertical and horizontal directions. Figure 9.20 shows vertical variation, including variations in grain size and grain composition in a beach sand at Chennai, India. Dense minerals, including magnetite, ilmenite, and garnet, are concentrated in the darker layers that alternate with layers that contain mostly quartz. If this sand becomes lithified, the resulting rock will preserve these variations. Other kinds of vertical variations are possible. For example, in some beds, small amounts of heavy mineral grains, composed of the densest minerals, may be present in the bottom of the bed but absent from the top.

As shown in Figure 9.21, the nature of the contacts that separate lithological units is also variable. In the vertical direction, contacts may be sharp, like the lowest contact in the photo on the left, reflecting significant changes in lithology. Such contacts are commonly planar, but occasionally, mostly due to erosion of a sedimentary layer before other sediment is deposited on top, they may be irregular surfaces. Sometimes *gradational* contacts form instead of sharp contacts. For example, the center of the rock in the left photo shows one kind of gradational contact, called a *progressive contact*, where there is a gradual change from fine to coarse material moving upward. Some gradational contacts are *intercalated contacts*, distinguished by mixed layering of two different lithologies over some vertical distance before one lithology entirely replaces the other. The photo on the right in Figure 9.21 is a good example. It shows Utah's Castlegate Formation, a formation that includes continental sandstone intercalated with abundant coal intervals.

Vertical variations in lithology reflect differences in depositional environment or in source of sediment. Thus, a stream may deposit different material during

Figure 9.20   Example of vertical variation in grain size (graded bedding).

Photo from M. A. Wilson, Wikimedia Commons.

sharp contacts and gradational contacts

intercalated contacts

Figure 9.21   Different kinds of lithologic contacts.

Photos from Callan Bentley (left) and L. Moscardelli (right).

wet years and dry years, producing a variety of rock types. In some places, desert sands may replace stream deposits during times of drought. Marine deposits may change if sea level rises or falls. And a landslide may cut off a river's sediment source.

Some vertical variations in stratigraphy are cyclic. River deposits, for example, may vary with season in response to changing water flow—greatest in spring and least in winter—depositing coarse material during times of high flow and finer material at other times. The sediments thus may contain sedimentary packages that repeat every year. In ocean settings, many marine deposits form from *turbidites*, sedimentary gravity flows that slide down continental slopes to flat ocean plains during times of high sedimentation. Multiple turbidites often occur in the same place and so produce sedimentary rocks containing individual strata reflecting many different submarine flows. And, many glacial sediments are seasonal, such as the varves shown in Figure 8.38 of Chapter 8.

9.5.3.1 *Sedimentary facies*

Marked vertical variations in lithology are common, but horizontal variations in lithology may be just as pronounced. These variations reflect different environments of deposition. For example, sand may be deposited in a beach setting and mud just a short distance away offshore. And rivers deposit much coarse material at the shallow deltas at their mouths but also transport finer sediment that settles in deeper water. In either case—beach or river mouth—sandstone and shale may be juxtaposed. Some lithologic units are uniform over long horizontal distances; others may *pinch out* as the distance from sediment source gets greater. If units pinch out, sharp contacts may be present. Alternatively, one rock type may grade horizontally into another, or two different lithologies may locally *interfinger* or on a larger scale *intertongue*, with wedges of each penetrating into the other.

Small basins may be characterized by somewhat uniform depositional environments, but large ones generally contain more than one setting having different physical, chemical, or biological characteristics. Geologists use the term *facies* to refer to any sediment or sedimentary sequence that is distinctly different from others of the same age due to variations in depositional environments. The change in the nature of sedimentation thus represents a change in *sedimentary facies*.

Consider the nature of sediment deposited in and near an ocean basin at any given time (Fig. 9.22). Onshore deposits may be any kind of continental sediments, perhaps collecting in swamps or on floodplains. Shoreline deposits will include beach sands and maybe mudflats. Shallow marine deposits are typically mostly mud, and deeper-water marine deposits may be mostly carbonates. Thus, different places have different sedimentary environments, and sediment may collect in all of them contemporaneously. The sedimentary rocks that eventually form—examples are listed in Figure 9.22—will vary laterally even if they are all of the same age.

Near the north rim of the Grand Canyon, the Moenkopi and Chinle formations overlie the Kaibab Limestone (Fig. 9.8). These two younger formations also exist in Monument Valley (Fig. 9.12). But the Kaibab Limestone is absent in Monument Valley, and the De Chelly limestone, of the same age as the Kaibab, is present in its place. The two locations are only 240 kilometers (150 miles) apart, but the lithologies are different because the environments of deposition were different. This classic example of a facies change occurs because the rocks at the Grand Canyon formed from deeper-water marine sediments than the rocks in Monument Valley did.

9.5.3.2 *Sea level change*

Sea level is rising today in response to global warming, and geological evidence chronicles many rises and falls in the past. Global changes in sea level, called *eustatic* changes, occur when the volume of water in the world's ocean changes or when the sizes of ocean basins changes. Eustatic sea level changes can be very significant. During ice ages, for example, much water is frozen at the North and South Poles. During warm times, polar ice melts and sea level rises by hundreds of feet. (Rising sea level is one reason why scientists are concerned about global warming today.) Besides

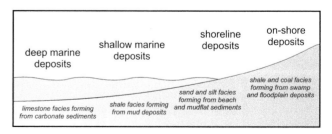

Figure 9.22   Examples of different sedimentary facies and some rocks that may form in each.

eustatic sea level variations, tectonics has caused land to rise or fall throughout Earth history, thus moving local shorelines. This movement causes changes in *relative sea level*, the difference between sea level and nearby continent elevations, that vary from place to place. So, the position of shorelines and the size and shape of marine basins have not remained constant over geological time. Because most sedimentation is associated with marine environments, and because marine environments are highly variable, changes in sea level are the most significant cause of lithologic variation over time.

Figure 9.23 shows the effects of rising or lowering relative sea level. When sea level rises or land drops, ocean waters encroach onto continents and shorelines move toward continental interiors. Geologists call this process a *transgression*. When sea level drops or land rises, the opposite—a *regression*—occurs. Seas retreat, continents grow larger, and shorelines move toward the centers of oceans. During both transgressions and regressions, the nature of sediments being deposited in a particular place will change because the environment changes. For example, deep marine sedimentation will migrate toward the continental interior during a transgression and toward the deep ocean during a regression.

So, during transgressions, beach sediments migrate inland and cover continental sediments, ocean sediments move toward land and cover beaches, and deep-water marine sediments cover shallow-water marine sediments. Thus, sedimentary facies migrate toward continental interiors. During regressions, the opposite occurs. Continental sediments will be deposited on top of former beaches, beach sands will cover shallow ocean deposits, and shallow-water marine sediments will cover deep-water sediments. Thus, sedimentary facies migrate seaward.

A classic transgression occurred when sea level rose and flooded the interior of North America 525–550 million years ago, depositing sediments that later became the Tonto Group (Fig. 9.24). This figure, a modified version of Figure 9.13, shows the formations of the Tonto Group near the bottom of the Grand Canyon. The three formations that make up the group total more than 380 meters (1200 feet) thick in some places. This group of sedimentary formations is separated from underlying Precambrian basement rocks by the Great Unconformity.

Figure 9.25 shows how the Tonto Group evolved as sea level rose and the ocean moved over land. The bottom formation of the group is the Tapeats Sandstone, formed mostly from shallow marine sands. As sea level increased, deeper-water muds were deposited, which became the Bright Angel Shale. Sitting above the Bright Angel Shale, the Muav Limestone formed from even deeper-water marine sediments. Thus, as sea level rose and ocean water moved from the south (left) to the north (right), the formations of the Tonto Group migrated toward dry land.

Figure 9.25 is a good depiction of what stratigraphers call *Walther's Law* (although it is really an observation, not a law), named after the German sedimentologist

Figure 9.24 The Tonto Group formations.
Photo from James St. John, Wikimedia Commons.

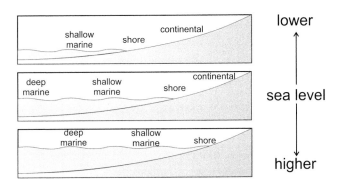

Figure 9.23 Migrating environments due to transgression or regression.

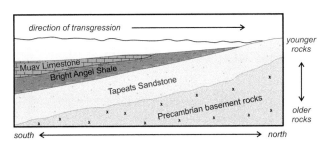

Figure 9.25 Effects of sea level change on stratigraphic facies.

Johannes Walther (1860–1937). Walther pointed out that during a transgression or regression, the vertical change in facies is the same as the horizontal variation. From the bottom of the Tonto Group up, shale replaces sandstone and then limestone replaces shale. This same sequence occurs from right to left in Figure 9.25.

9.5.3.3 Stratigraphic columns

Figure 9.26 shows two examples of *stratigraphic columns* for Arches National Park, near Moab, Utah. Geologists use stratigraphic columns like these to summarize the sequence of strata that occur in a given place. The column on the left is a simple one, just showing the formations that are present; it is typical of many stratigraphic columns that geologists produce. The column on the right is much more informative. It includes names of formations and members, absolute ages (in millions of year), geologic period names, and a scale. It also includes an outcrop profile that shows which formations weather most easily and which form cliffs. Some stratigraphic columns are even more informative, perhaps including detailed descriptions of each rock formation and member.

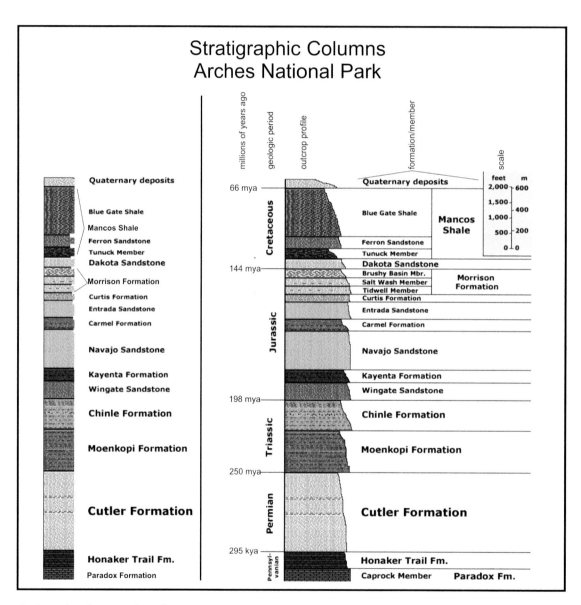

Figure 9.26   Examples of stratigraphic columns.
Modified from stratigraphic columns at www.oldearth.org/images/strat4.jpg.

9.5.3.4 Correlations

When geologists attempt to interpret and reconstruct geological histories, knowing the areal extent of different formations is important. This involves examining rocks in many different places and comparing them to see if the same formations are present. We term this process of rock matching *correlation*, and much of stratigraphy is about naming and correlating stratigraphy based on lithology, physical characteristics, fossils, or other attributes. Some strata contain very distinct layers, called *marker beds*, that make correlations easier. The best marker beds are easily distinguished and can be traced over large areas. Coal beds, for example, often make good markers, as do thin layers of volcanic ash between layers of sedimentary rock. Other strata contain distinctive fossils that make correlations relatively straightforward.

Sometimes correlations can be ambiguous, especially if considering only one kind of rock. Many shales, for example, are similar, no matter their age, and some fossils in shales persisted over many millions of years. Additionally, minor facies changes may mean that a particular lithology, perhaps a shoreline sandstone, does not look identical in different places. So, the most reliable correlations are based on sequences of rocks containing multiple strata of varying lithologies and fossil assemblages.

Sometimes complications arise because the same lithologic unit may have different names in different places. For example, stratigraphers have correlated the Tropic Shale (southern Utah), the Pierre Shale (Dakotas), the Bearpaw Shale (Montana), the Lewis Shale (Colorado), and the Mancos Shale (central Utah). All are marine shales formed at about the same time (70–100 million years ago) by deposition from the same sea. But, because separate investigators described them initially, they have their own names. Furthermore, they did not form at exactly the same time and do not have exactly the same characteristics. Similarly, the Redwall, Madison, Pahasapa, Guernsey, and Leadville limestones discussed earlier in this chapter are correlative; that is, they are the same formation but called different names in different places.

Additionally, lithostratigraphic correlation may not match rocks of identical ages over long distances because facies migrate (Walther's Law). For example, Utah's Navajo Sandstone, shown in the top photo of Figure 9.27 and in some earlier figures in this chapter, correlates lithologically with the Nugget Sandstone

(middle photo) in northern Utah and Wyoming and with the Aztec Sandstone (bottom photo) in Nevada. The three, all shown in Figure 9.27, are really the same

Navajo Sandstone in south-central Utah

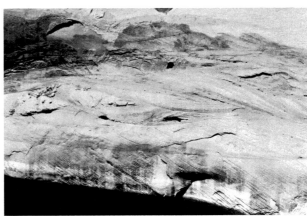
Nugget Sandstone in northern Utah

Aztec Sandstone in Red Rock Canyon, Nevada

Figure 9.27 The Navajo Sandstone, the Nugget Sandstone, and the Aztec Sandstone.

Photos from G. Willis, Wikimedia Commons (top); The Utah Geological Survey (middle); Superfish at Wikimedia Commons (bottom).

formation in different places. They all are sandstones that contain typical features associated with dune sands, but because the dunes migrated westward, the sands were deposited tens of millions of years earlier in northern Utah and Wyoming than they were in Nevada. So, the lithostratigraphic correlation is valid but the formations did not form at the same times.

Figure 9.28 shows stratigraphic columns for, and correlations between, five national parklands: Zion Canyon, Capitol Reef, Canyonlands, and Arches National Parks in Utah, and the Colorado National Monument near Grand Junction, Colorado. Different geological periods are labeled on the left, and formations belonging to each period are shown in different colors. As shown in the inset map, the columns are ordered, with the westernmost parkland on the left and easternmost on the right; the total distance from west to east is about 400 kilometers (250 miles). Dashed lines between the columns show how formations correlate between parks. For the most part, the formations continue from one park to the next. For example, the Navajo Sandstone and Carmel Formations are present in all the Utah parks (but not in Colorado). The Entrada Sandstone is absent in Zion National Park but present in the four easternmost parklands.

Note that these columns are all drafted at the same vertical scale. Thus, formation thicknesses can be compared visually, and this drawing shows that there was overall less deposition, or maybe more erosion, of Permian, Triassic, and Jurassic sediments near Capitol Reef National Park than in the other Utah parks. In some places, formations present in one park are missing from an adjacent park; this absence is shown in Figure 9.28 by converging dashed lines connecting columns. Colorado National Monument seems to be a special case—several major formations are absent there, for example the Navajo Sandstone, which can be found just 120 kilometers (75 miles) away in Arches National Park. But, overall, the correlations shown in this figure suggest that the environments across all of Utah were about the same.

9.5.4   *Biostratigraphy*

Biostratigraphy originated in the late 18th and early 19th centuries when William Smith, Georges Cuvier, Alexandre Brongniart, and others recognized that rocks that contained similar fossils probably had similar ages, and that different fossils found in rocks of similar types

suggested the rocks had different geologic ages. Since then, fossils have been one of the most important tools used by stratigraphers.

Figure 9.29 shows how geologists use fossils to correlate rocks in different places. Studies of fossil shells and other preserved animal parts and plant remains tell us that species have changed greatly over time. In a normal stack of sedimentary rocks, like those shown in this figure, the oldest rocks are on the bottom and the youngest on top (law of superposition). So, fossil plants and animals vary vertically, with the oldest fossils normally in the lowest strata and the youngest in upper strata. The sequence from bottom to top reflects the order in which new species evolved and old species became extinct. Thus, geologists can match fossils to correlate rocks of similar ages even if they are found in different places. The lines from the column on the left to the column on the right show how fossils match. Note that there is one biostratigraphic unit that exists only in the column on the right (shown by converging dashed lines). Either that unit was not deposited in the location on the left, or after deposition, it was removed by erosion.

Although some ancient fossils are similar to organisms that exist today, many are not. In fact, many plants and animals only existed for very short times—at least on geological time scales. Therefore, rocks from different places containing the same fossil fauna and flora must have formed at about the same time. For example, if one outcrop contains siltstone, and another contains shale, and if both contain the same fossils, the two outcrops are most likely of the same geological age.

In contrast with lithostratigraphy, fossil correlations do a better job of relating rocks of similar ages because they represent one time in the evolutionary record. Biostratigraphic correlation is not, however, always feasible. For example, the previously mentioned Nugget, Navajo, and Aztec sandstones contain few fossils that can be used for this purpose, largely because few organisms live in deserts and because desert environments are harsh, so few plant and animal remains are preserved.

In general, for biostratigraphic correlations to be possible and reliable, all of the following conditions must be met:
- Animals or plants must have been present at the time of deposition, which means the environment of deposition must have been amenable to life.
- Key plants and animals that were widespread need to be present, not just locally occurring organisms.

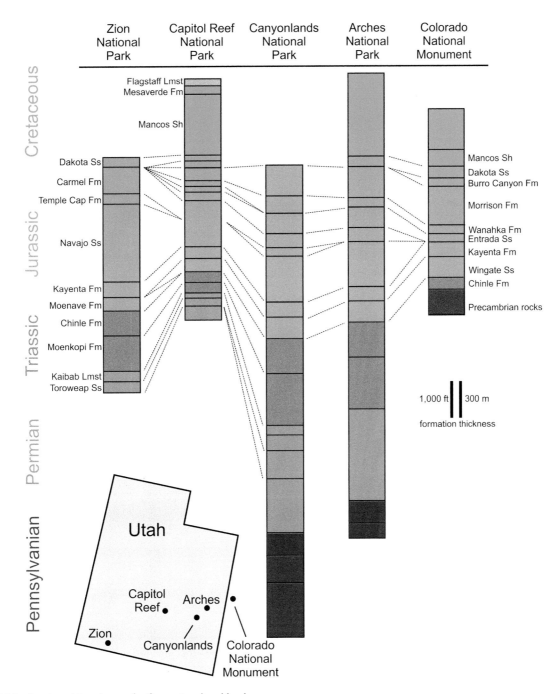

Figure 9.28 Stratigraphic columns for five national parklands.

- Fossils must have been preserved, which may not be the case for some sedimentary environments and may not be the case if diagenesis has occurred. To make precise correlations, appropriate and sufficient fossils must be present. If a sedimentary unit is many thousands of feet thick, it was deposited over a long time, and contains only a single fossil that provides little information. More is better, and the most reliable biostratigraphic correlations are based on the presence of multiple kinds of fossils.

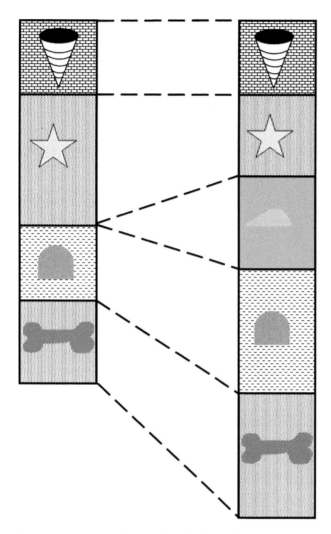

Figure 9.29 Diagram showing how fossils may be used to correlate strata.

Figure 9.30 Two commonly used index fossils: a trilobite (top) and a brachiopod (bottom).

Photos from Wikimedia Commons. Trilobite photo taken by Daiju Azuma. Brachiopod photo taken by Mark A. Wilson.

9.5.4.1 Index fossils

If a particular kind of fossil existed for long periods of geological time with little distinctive evolutionary change, it is not useful for determining rock ages. *Lingula*, for example, is a genus of brachiopods that exists today; the modern versions are very similar to fossils from 500 million years ago. Additionally, horseshoe crabs today are nearly identical to those that existed 450 million years ago. The fossils most useful for correlations are those that are easily identified, are common and widespread, could live in diverse environments, only existed for short times, and are easily preserved. Fossils with these, or most of these, characteristics are termed *index fossils*.

Figure 9.30 shows two examples of index fossils. Fossilized trilobites (top), a distinctive kind of organism related to crabs, are index fossils common in sedimentary rocks from the early Cambrian to Late Permian periods, about 521 to 252 million years ago. *Mucrospirifer*, a brachiopod (bottom), is a common index fossil in sedimentary rocks from the Middle Devonian Period, about 383 to 393 million years ago. The trilobite shown is from Japan and the brachiopod is from Ohio, but similar fossils are found in rocks of the same ages in many places around the world.

9.6 Geological time

Although early stratigraphers did not realize it, what they were really investigating was geological time, because layers of strata are ordered sequentially with the oldest strata on the bottom and the youngest on the top. The stratigraphers, however, were working under a handicap. Geological time is a little like the size of the universe or the number of atoms in our bodies—the numbers are so large that they are mysterious and almost incomprehensible. In the 17th century, Archbishop James Ussher, using the genealogies chronicled in the Bible, concluded that Earth was created in 4004 BC. Many scientists eventually took exception, and by the end of the 19th century, scientists generally accepted that the Earth was 100 million years old. This estimate, too, was in error, but it was not until the middle of the 20th century that reliable radioactive dating established Earth to be around 4.5 billion years old. So, the age of Earth or of particular rock formations or how long ago different geological events occurred are things that the earlier investigators could not be expected to appreciate.

Geochronology is a general term for investigations of the age of rocks, fossils, and other geological specimens and of key events in Earth history. These investigations rely heavily on *radiometric dating*—a method of determining how long ago a rock formed or an event occurred based on the discovery that radioactive (*parent*) atoms decay at constant rates to different atoms, called *daughter atoms*. Early interpretations of geologic time were based mostly on fossils, which led to limitations because many of the oldest rocks on Earth typically contain no fossils. In contrast, radiometric dating can be applied to many igneous and metamorphic rocks of all ages (although not as successfully to sedimentary rocks).

Geologists think about geologic time in two main ways. *Absolute time* refers to time measured in days, months, years, millions of years, etc. However, so far in this chapter, we have mostly focused on *relative time*. Relative time—largely based on lithostratigraphy and biostratigraphy—deals with the order in which events occurred or artifacts were created. Studies of relative time allow geologists to determine the sequence in which events occurred without assigning absolute ages to them. In contrast, *absolute time* refers to how long before present a specific event occurred or a rock formed. Radiometric dating is at the heart of absolute time. So, stratigraphers studying relative time might seek to determine whether the Kaibab Limestone is older or younger than the nearby Redwall Limestone, while geologists studying absolute time may wish to know how many years ago the Kaibab Limestone formed. These two kinds of geologic time are closely related but remain distinct today, in large part because

different tools and techniques are used for studying each and, in part, because the magnitude of geological time is so hard to fathom.

Geologists named formations and established relative ages long before they determined absolute ages. They established a geological time scale, giving names to different portions of geologic time, and established a chronology—the order in which the different segments of time occurred. Biostratigraphy was the original basis for most of the time scale, and most divisions were based on the presence or absence of specific fossils or fossil assemblages. So, the original time scales, developed in the 19th century, depicted relative time but contained no absolute ages. The first absolute ages were added to the scale in 1913, and geochronologists have modified them many times as more precise and accurate information has become available.

Figure 9.31 shows the major divisions of the geologic time scale used today. Some names in the time scale refer to particular places where rocks of a specific age are found, other names have lithologic origins, and others derive from Greek words dealing with the origins and nature of life.

From largest to smallest, the scale is divided into *eons*, *eras*, and *periods*. Although not shown here, geologists divide the periods into even smaller durations called *epochs*. The time scale is generally depicted as a relative scale (left side of Fig. 9.31), but the different divisions do not have the same durations. The right side of the figure shows that, in absolute time, the Precambrian Eon accounts for most of the Earth's history since Earth's origin 4.6 billion years ago. The rest of geologic time only accounts for about 11% of Earth's history. The geologic time scale has many subdivisions that are not shown in this figure. However, we have relatively few samples of ancient rocks, and many of young rocks, so the most recent part of the time scale, the Phanerozoic Eon, is conveniently divided into many more subdivisions than the Precambrian Eon is.

Figure 9.8, which shows the stratigraphy of the Grand Canyon and the Grand Staircase, has formations named and correlated with geological periods. Figure 9.28, showing the stratigraphy in southern Utah, also has geological periods listed. This is commonplace in the geological literature. Most geologists prefer to use the names in the geological time scale rather than give absolute ages, because absolute ages are often imprecisely known, especially for sedimentary rocks, and because the periods of the geologic time scale allow geologists to group rocks into bins representing different parts of geological history.

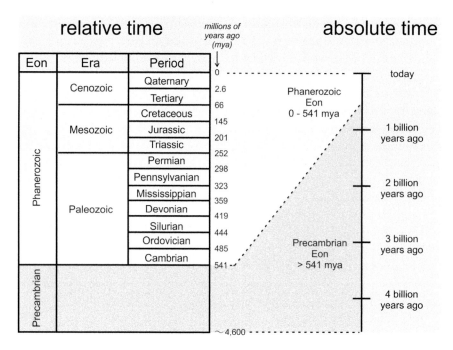

Figure 9.31   The major divisions of the geological time scale. Relative time is on the left and absolute time is on the right.

9.6.1   Chronostratigraphy

Chronostratigraphy is the branch of stratigraphy focused on determining the absolute ages of different strata. An additional aim of chronostratigraphic studies is to assign numerical age dates to specific fossils and fossil assemblages. Unfortunately, determining radiometric age dates for sedimentary rocks and fossils is often problematic or impossible because the appropriate minerals may not be present. However, layers of volcanic rocks and, sometimes, igneous intrusions are commonly associated with sedimentary rocks. These volcanic and intrusive rocks are more easily dated than sedimentary rocks are and are often the basis of chronostratigraphic correlations. Often, the ages of sedimentary strata are estimated from the ages of nearby igneous rocks by applying the law of superposition and the principle of crosscutting relationships. Additionally, geologists can sometimes date and use metamorphic events to put age limits on sedimentary rocks.

9.6.2   Completeness of the stratigraphic record

9.6.2.1   Unconformities

Gravity deposits sediments during depositional events of variable duration. Sometimes the duration is relatively short—during spring runoff, during flood years when sediment-choked rivers feed their deltas, due to undersea landslides, or for other reasons. Other times, sedimentation may take millions of years, perhaps when sediment is slowly accumulating in a deep ocean bottom. Yet even if sedimentation is continuous over a long time, the rate of deposition may not remain constant; 10 meters of sediment may be deposited in 1 year or in 1 million years. And, even in deep ocean basins, where we can expect depositional processes to be somewhat uniform, sedimentation may be faster at some times than at other times, and sometimes no sedimentation occurs.

When present, bedding planes separating different lithologies are the results of times of no sediment deposition. Geologists use the term *unconformity* to describe surfaces, like bedding plains, that separate two strata of different ages, and they call the gap in deposition a *hiatus*. The layers of sediments above (younger) and below (older) an unconformity are products of distinct depositional events. So, a hiatus is a time when deposition has stopped for a while and, commonly, erosion has removed some of the underlying older material.

Geologists distinguish between three main kinds of unconformities (Fig. 9.32). A *disconformity* (drawing a) is a gap in sedimentation involving older and younger strata that are parallel. An *angular unconformity* (drawing b) is one where the underlying beds are inclined with respect to those on top. The Great Unconformity in the Grand Canyon, labeled in Figures 9.2 and 9.8 and shown in the photo of Figures 9.13 and 9.24, is an angular unconformity. Figure 9.4 shows another example of an angular unconformity—a contact at the base of the Redwall Limestone. The third kind of unconformity, a *nonconformity* (drawing c), is one involving sedimentary rocks above igneous or metamorphic rocks (shown with a "v" symbol in the bottom of Fig. 9.32c). The Grand Canyon's Great Unconformity is both an angular unconformity and a nonconformity. The creation of a disconformity may be due only to a hiatus but angular unconformities and nonconformities always involve significant erosion.

Unconformities may represent very long times or shorter times. Figure 9.33, a close-up view of the Grand Canyon's Great Unconformity, shows the contact between Precambrian basement rocks below and the Cambrian-aged Tapeats Sandstone above. We clearly see the angular relationship between the underlying and overlying beds. The hiatus associated with the Great Unconformity lasted at least 175 million years. Higher up in the canyon, there is another major unconformity where there are no rocks between 420 and 485 million years old. This major (65-million-year-long) hiatus encompasses the Ordovician and Silurian periods. If sedimentation occurred during those times, the sediments were eroded and thus not preserved.

In addition to the two big unconformities, the canyon walls display many smaller, but distinct, unconformities. Furthermore, as previously discussed in this chapter, the canyon contains many formations with different lithologies. Most of the boundaries between the different formations must be due to unconformities. If not, the contacts between lithological units would be gradational. Some contacts show evidence that erosion occurred during a hiatus, but others do not. The amount of erosion varies across the region, and some

formations, such as the Surprise Canyon Formation, have been completely eroded away in most places.

Unconformities are good time markers; all rocks above an unconformity are younger than those below, but the duration of the hiatus that created an unconformity may be hard to determine. In some cases, the duration is thought to be short (tens or hundreds of thousands of years) and in other cases long (millions of

Figure 9.33 An unconformity (nonconformity) in the Grand Canyon, Arizona, USA.
Photo from Chris M. Morris, Wikimedia Commons.

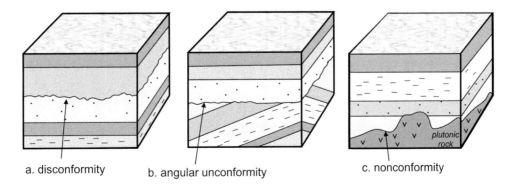

a. disconformity   b. angular unconformity   c. nonconformity

Figure 9.32 The three main types of unconformities.

years). If the duration was short, geologists often dismiss it as insignificant. Identifying unconformities can be difficult, and geologists undoubtedly overlook some unconformities, especially if a lithological boundary is parallel to the layers of strata and there is no evidence of erosion. Other unconformities are easily identified, especially if significant erosion occurred during the time of non-deposition or if the unconformity juxtaposes layers of rocks that are not parallel.

9.6.2.2 Sequences

The rocks present in the Grand Canyon and the Grand Staircase today are highly variable and formed in many different environments. Some are marine, some are transitional, and some are continental. The Tonto Group shows evidence that sedimentologists interpret as a marine transgression, but other parts of the stratigraphic column are not so easily interpreted. It appears that marine conditions and continental conditions alternated in the same geographical area. How can this be? The answer is that transgressions have flooded the central portion of North America multiple times when global sea level was high. Invading seas, called *epeiric seas*, deposited marine sediments on land that today is far from any ocean. At other times, the seas retreated and erosion occurred because the land was exposed to weathering, or, sometimes, continental sediments such as desert sands accumulated on top of marine sediments.

Figure 9.34 shows a reconstruction of what North America looked like about 100 million years ago. At that time, mountain ranges that later evolved to become the present-day Rocky Mountains and Appalachian Mountains existed in the west and east, respectively. A shallow sea, called the *Western Interior Seaway* or the *Tejas Sea*, up to 760 meters (2500 feet) deep, flooded space between the mountain ranges. The seaway extended 3200 kilometers (2000 miles) from the Gulf of Mexico to the Arctic Ocean. Ocean waters deposited marine sediments on what is now dry land in the center of the continent. During this time, the regions labeled Larimidia and Appalachia were above sea level, but continental sedimentation may have been occurring there. The oceans eventually retreated, and by about 65 million years ago, ocean levels had dropped and all of North America was high and dry.

Today, eustatic sea level is low and continents are large. We have relatively few large inland water bodies, commonly called *inland seas*, scattered around the

world. Hudson Bay, the Baltic Sea, the Black Sea, and possibly a few others might qualify. In the past, however, when sea level was higher, the world's oceans flooded continents, creating large shallow seas and depositing marine sediments. Hudson Bay, the Baltic Sea, and the Black Sea are the leftovers from the times of flooding. And the marine sedimentary rocks found in the Grand Canyon region are also relicts.

Most continents contain *cratons*, stable parts of the continental interior where little tectonic activity has occurred for a long time. As sea level rises and falls, cratons may be periodically flooded and sediments deposited. Thus, the sequence of sediments deposited on a continental interior during transgressions and subsequent regressions of an epeiric sea form a *cratonic sequence*. When seas are absent, sediments and sedimentary rocks are, most of the time, exposed to weathering and erosion. So, the tops and bottoms of cratonic sequences are unconformities, like the unconformities above and below the Tonto Group in the Grand Canyon, which, as described earlier in this chapter, formed during a marine transgression.

The relationship of epeiric seas to the deposition of a continuous series of lithologies is called *sequence stratigraphy*. At the heart of sequence stratigraphy is the notion that if there are no unconformities in a sequence, a complete record (a complete sequence) of environments of deposition has been preserved. So, any variations in

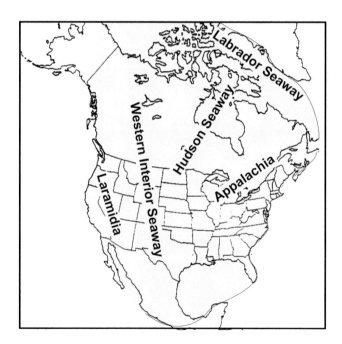

Figure 9.34 North American geography 80 million years ago. Map from W. A. Cobban and K. C. McKinney, USGS.

lithologies depend on variations in sediment sources, in sea level, in the size of the depositional basin, and in the relative rates of sea level change and sediment supply.

Sequences may reflect variations in global sea levels; sometimes they are clearly related to times of glaciation and non-glaciation. Additionally, tectonics is important because it causes continents to move, continental elevations to change, and the spreading rate and sizes of oceans to vary. So, sequences may occur with different timing on different continents. In North America, stratigraphers have identified six major cratonic sequences since the end of the Precambrian Eon, about 540 million years ago. Their relationships to stratigraphy are well established. On other continents, however, the sequence stratigraphy is not as well established—perhaps because local tectonism has played a significant role.

Figure 9.35 shows the time spanned by the Phanerozoic Eon. Six major epeiric seas invaded North America during this nearly 300-million-year-long time. From oldest to most recent, they are the Sauk, Tippecanoe, Kaskaskia, Absaroka, Zuni, and Tejas seas. Because continental topography changed over time, different seas reached different parts of the continent. The blue color in the right column of Figure 9.35 represents the relative areal extent of each of the seas. The tan color shows times and locations of no marine sedimentation. Most seas made it to the center of the continent but were substantially deeper near the margins.

The marine sedimentary rocks found in the walls of the Grand Canyon today are products of the four oldest seas shown in Figure 9.35. The continental sedimentary rocks in the canyon region formed from sediments that accumulated when seas were absent. As discussed earlier in this chapter, the walls of the Grand Canyon contain nearly a vertical mile of sedimentary rock outcrop. The sedimentary rocks represent more than 260 million years of Earth history. But much erosion occurred when the land was high and dry, removing an unknown amount of sediments and sedimentary rocks. Thus, rocks from many of those 260 million years have not been preserved and the record seen in the canyon walls is incomplete.

9.7 Subsurface stratigraphy

Most of the fundamental understandings about stratigraphy derive from studies of outcrops. Today, however, subsurface stratigraphy may be more important because many of our important energy and mineral resources come from the subsurface. Although geologists use several methods to investigate the subsurface, the most straightforward are based on drilling holes in the ground.

Petroleum resources have been used for thousands of years, but the "modern" petroleum age did not begin developing until the middle and late 19th century, when drilling techniques improved and the internal combustion engine was invented. Demand and production jumped significantly at the beginning of the 20th century as gasoline engines became more widespread and technology improved.

Figure 9.36 shows the Spindletop well in eastern Texas gushing oil in 1901. In January 1901, the completion of this 347 meter (1139 foot) deep well set new industry standards, and drilling activity expanded quickly. Early investors soon developed the largest easy-to-find-and-produce oil fields. Consequently, oil companies have had to drill deeper wells to extract oil as time has passed. In 1950, the average well was 900–1200 meters (3000–4000 feet) deep. By 2010, average

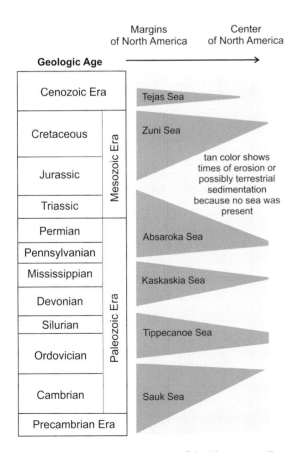

Figure 9.35 The six major epeiric seas of the Phanerozoic Eon in North America.

Figure 9.36   Photograph of Spindletop.
Photo from the Book of Texas, Wikimedia Commons.

depth had increased to perhaps 1830 meters (6000 feet). And today the deepest oil wells, although rare, are more than 10,700 meters (35,000 feet) deep. Coal and mineral companies also drill holes that provide information about geology at depth, but their drilling activities are restricted to much shallower depths.

9.7.1   Samples from deep in Earth

When traditional holes are drilled, fluid circulates to lubricate the bit at the end of the drill stem. This circulating fluid carries *borehole cuttings* to the surface. These cuttings contain chips of rock that geologists examine to learn about the rock lithologies at depth. This process, called *mud logging*, is common today, and it has traditionally been a fundamental tool used by stratigraphers to investigate the subsurface and to direct exploration activities.

Drillers can also use a kind of drill bit that collects *core samples*, such as those shown in Figure 9.37. Cylinders of solid rock are brought to the surface where they can

Figure 9.37   Geological drill core from the Waubakee Formation in Wisconsin.
Photo from Gorthian, Wikimedia Commons.

be examined using microscopes or other instruments. The core shown in this photo contains samples of the Waubakee Formation taken from a well near Milwaukee, Wisconsin. Collecting core samples is both time consuming and expensive, but because the rock in cores has not been pulverized, cores provide information that otherwise cannot be obtained.

9.7.2   Geophysical logging

Besides providing samples to study, boreholes yield a lot of geophysical data that provide information about rock types and rock properties beneath the surface. Geophysicists have long known how to measure gravity, magnetism, and other properties at Earth's surface. Wells provide access to measure these same, and other, properties at depth beneath the surface.

We call the process of collecting geophysical data from boreholes *logging*, and it is accomplished using instruments of many different types. Figure 9.38 shows well logs for a well at Walakpa Bay in the National Petroleum Reserve of Alaska. The logs, collected by instruments lowered through a drill hole, show density, magnetism, electrical conductivity, and other properties that correlate with different kinds of rocks. Today, many oil companies put great reliance on well logging,

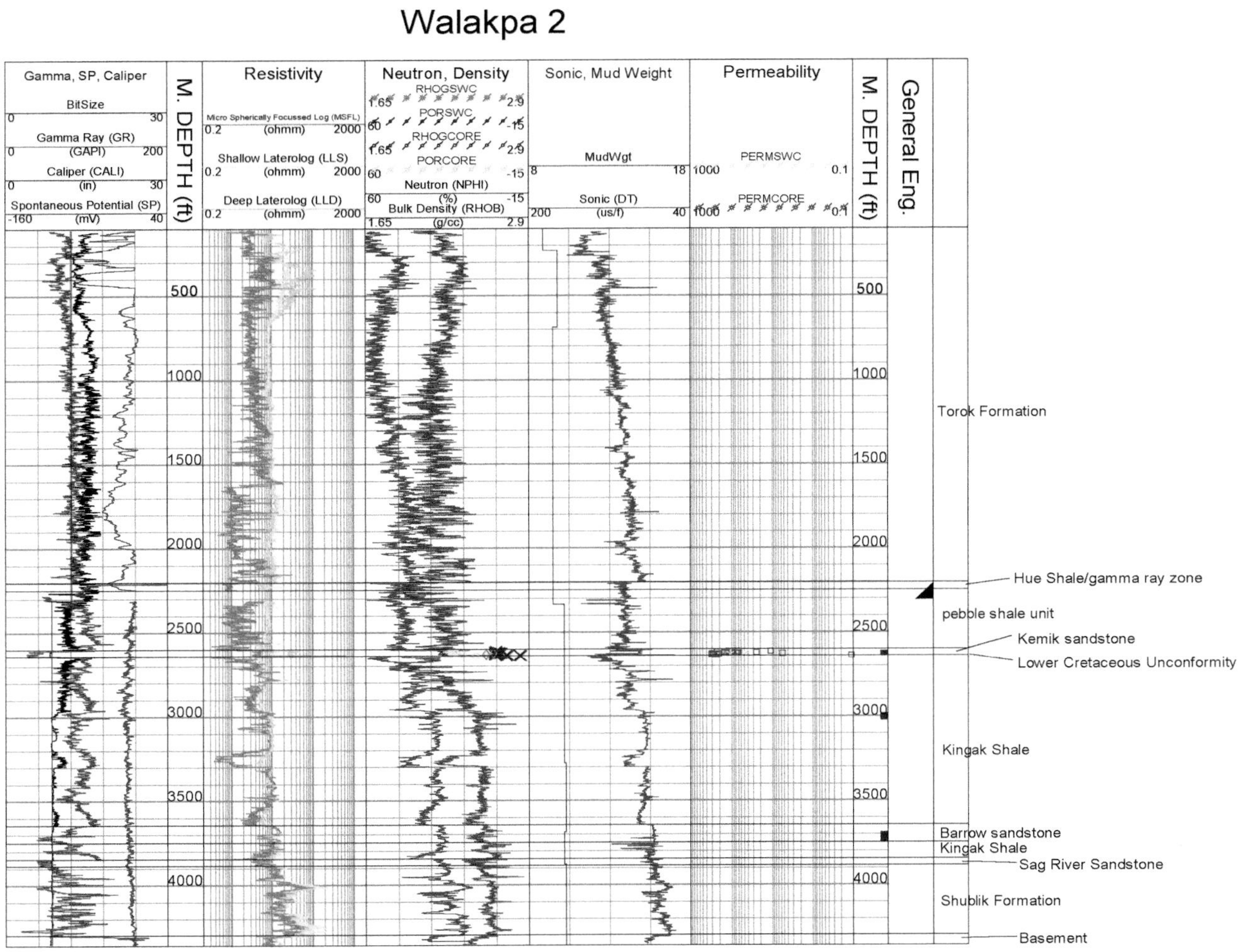

Walakpa 2

Figure 9.38 Logs from the Walakpa 2 Well, National Petroleum Reserve, Alaska.
From Wikicommons. USGS source.

because geophysical data can be efficiently gathered and yield a wide variety of information.

9.7.3 Seismic reflection surveys

When investigating a single well or a small area, samples and geophysical logs may be most practical. But, when exploring a large region, oil companies and others rely on *seismic reflection* data. Figure 9.39 shows how such data are collected at sea. A similar process is used on land, with trucks instead of ships towing hydrophones. This kind of data, first collected in the 1920s, is created by generating low-frequency (mostly sound) waves (at a *source*) that pass to some depth in Earth where they are reflected from contacts. In the example shown in this figure, an air gun generates waves that pass through the water and into the subsurface.

The waves reflect off rock and sediment contacts and are detected by hydrophones towed behind the ship. The hydrophones record the intensity of waves and the time the waves take to return to the surface. The data provide detailed information about variations in stratigraphy with depth, allowing geologists to map the subsurface. Seismic measurements can be made quite efficiently, and many such measurements can be used to map geology over large regions. Consequently, seismic reflection surveys are the most important tool used today for petroleum exploration.

Figure 9.40 shows vibroseis trucks in Austria. In the past, geophysicists used explosions as sources to generate sound waves, but today large vibrating trucks, such as the ones in the photo, are used on land and airguns are used at sea. The waves penetrate deep into Earth and reflect from bedding planes and other heterogeneities in rock strata. Receivers, *geophones* on land and

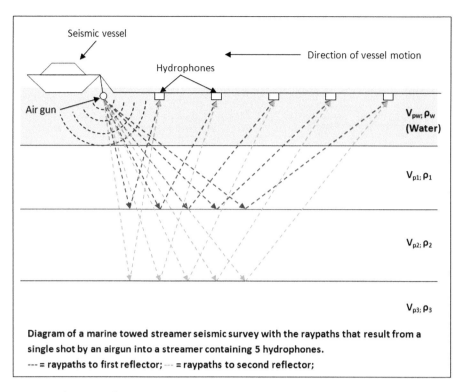

Diagram of a marine towed streamer seismic survey with the raypaths that result from a single shot by an airgun into a streamer containing 5 hydrophones.
--- = raypaths to first reflector; --- = raypaths to second reflector;

Figure 9.39 The basic process of seismic reflection surveying.
Drawing from Nwhit, Wikimedia Commons.

Figure 9.40 Vibroseis trucks in Austria.
Photo from Techcollector, Wikimedia Commons.

hydrophones at sea, record the reflected waves. The most accurate surveys use many sources, and waves reflect from many strata before being recorded by geophones or hydrophones. So, seismic studies yield huge amounts of data that must be analyzed by sophisticated computer programs.

Traditionally, seismic surveys involved a line of receivers—often on a road—that were tens of meters apart. Such studies produce a seismic profile such as the one shown in Figure 9.41—equivalent to a vertical slice—of the stratigraphy at depth. The top of the figure shows the raw survey data, and the bottom drawing shows a geological interpretation based on the data.

Improvements in technology, including faster computers and better programs, mean that more data can be handled today. So today, larger numbers of sources and receivers are used, and three-dimensional surveys covering areas as large as 130–260 square kilometers (50–100

square miles) are common results. Energy companies, and oil geologists, are most concerned about the uppermost portion of Earth's crust because producing energy from greater depths is impractical. But, some other kinds of geologists use seismic data to investigate much greater depths, including Earth's mantle and core.

Questions for thought—chapter 9

1. The oldest rocks in the Grand Canyon are metamorphic and igneous, and about 1.75 billion years old. If the Earth is 4.55 billion years old, where are the missing rocks from 4.55 to 1.75 billion years ago? Or, are there no rocks missing? Explain.
2. Why do some formations, like the Redwall Limestone, have different names in different places?
3. What kind of rock is the Navajo Formation; what did it form from? Did all parts of the Navajo form

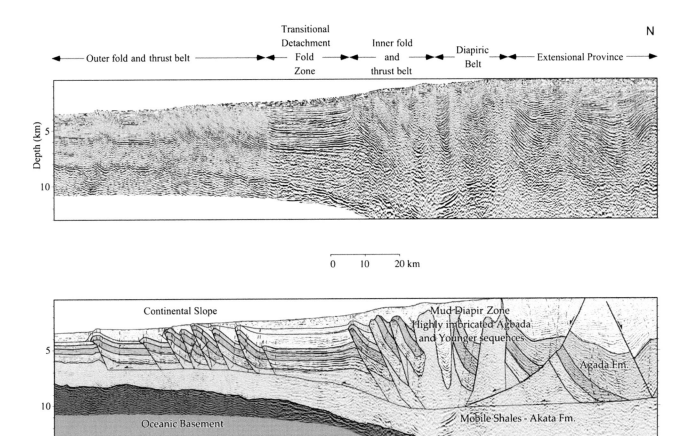

Figure 9.41 Tectonic structures drawn over a seismic profile of the Niger Delta Basin.
From the American Association of Petroleum Geologists, Wikimedia Commons.

at the same time? If not, what (geographical) part of the formation formed first? Explain.

4. What is the *Law of Superposition*? How would overturned rocks disrupt the application of the Law of Superposition?

5. What is a *xenolith*? Explain how the *Principle of Inclusions* applies to xenoliths.

6. What properties of a continental sedimentary rock allow it to be distinguished from a rock with a marine origin? Be thoughtful and speculate.

7. Describe a possible facies change from on- to off-shore of a marine coastline. What kinds of rocks may form from sediments in the different places?

8. A location has beach deposits overlying continental deposits and deep water deposits overlying the beach deposits. Does this situation indicate a marine transgression or regression? Why?

9. Explain why a complete stratigraphic column may be a compilation and not actually occur in any single location.

10. Organisms may be fossilized and preserved in rocks. Sometimes rocks contain fossils that formed long before the time at which the rock formed. How can this happen?

11. Why do stratigraphers prefer to date a rock with an assemblage of different fossil species rather than just one species?

12. Why are index fossils important in stratigraphy? What are the characteristics of a good index fossil?

13. What defines the boundary between the Precambrian and the Cambrian? That is, what is distinctive about that time in Earth history? Why is the Precambrian so long and why is it not divided into epochs?

14. How does an unconformity differ from a hiatus?

15. Why are most sharp contacts between sedimentary rocks unconformities? (Perhaps they all are!)

16. What are *epeiric seas* and how did they create cycles in sedimentary rocks? Hudson Bay and the Baltic Sea are two examples of modern epeiric seas? What characteristics do they have in common that are similar to ancient epeiric seas?

17. What is a *craton* and where is one in North America? How about other places? In what other parts of the world do we find cratons? Name several.

10 Metamorphic Rocks

10.1 Wollastonite in the Adirondack Mountains

Figure 10.1 shows a rock from the Willsboro Mine in New York. The large white crystals are lathes of the mineral *wollastonite* up to 1.5 centimeters in length. The wine-red crystals are *garnet*, in some places several millimeters in longest dimension. And the small dark green crystals are millimeter-scale *diopside*. Society uses wollastonite in ceramics, plastics, paints, and for other purposes once served by asbestos. The Willsboro Mine has produced a great deal of wollastonite in the past, but operations shut down in the early 1980s. However, the nearby Lewis Mine is producing more than 60,000 tons of wollastonite, equivalent to 10% of world production, every year.

The white region in Figure 10.2 shows where Precambrian metamorphic and igneous rocks crop out in the New York's Adirondack Mountains; much younger, relatively flat-lying unmetamorphosed sedimentary rocks

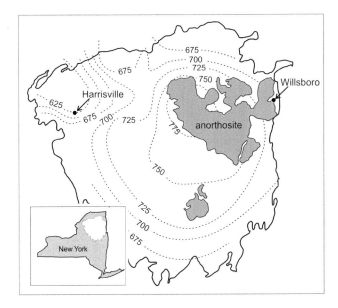

Figure 10.2 Map of the Adirondacks Region in New York. DP drawing.

(shown in yellow) surround this region. The Adirondacks contain two significant wollastonite districts, indicated on this map. An eastern district near Willsboro, east of the Adirondack High Peaks region and near the New York–Vermont border, includes the Willsboro and Lewis mines. A western district, near Harrisville on the western flanks of the Adirondack Mountains, includes the Valentine Mine. The Lewis Mine in the east and the Valentine Mine at Harrisville are the only active wollastonite mines in the United States today. In both places, wollastonite formed by metamorphism of Precambrian limestone.

Near Willsboro, major wollastonite deposits occur in a 25-kilometer (16-mile) long sinuous belt, only a few kilometers wide at its widest point. This belt, composed of metamorphosed sedimentary and igneous rocks, contains significant amounts of ore in the Willsboro,

Figure 10.1 Wollastonite ore from the Willsboro Mine, New York. Photo from Jamie Schod.

Lewis, Oak Hill, and Deerhead deposits. The ore rocks are a kind of *marble*, a name used for rock created when carbonate rocks (limestone or dolostone) are metamorphosed. The Willsboro marble formed during the Precambrian, 1155 million years ago, due to heating and metamorphism caused by intrusion of the Mt. Marcy anorthosite (pink in Fig. 10.2). Several different studies have concluded that this intrusion took place no deeper than 10 kilometers in Earth.

Figure 10.3 shows simplified geology and geometry of the wollastonite deposit at Willsboro. Three zones, distinguished by different mineral concentration, wrap around and are parallel to a contact with the Mt. Marcy anorthosite. The several meters thick inner zone, next to the anorthosite intrusion, contains high concentrations of pyroxene (diopside) and garnet with lesser amounts of wollastonite. The main ore zone, up to 50 meters thick, is mostly wollastonite but contains scattered pyroxene and garnet. The outermost zone, up to a few meters thick, is composed of a rock called *garnetite*, which is almost entirely garnet. Outside this zone, the surrounding rock is all mafic gneiss. Figure 10.3 is a generalized view of rock relationships that might have existed shortly after the wollastonite deposits formed; subsequently, much deformation occurred, and in most places, today, the zones are mixed and not as uniform as shown. The marble at Willsboro formed from limestone containing calcite, $CaCO_3$, and dolomite, $CaMg(CO_3)_2$. Yet, after metamorphism, the marble contains no calcite

or dolomite and instead is almost entirely made of silicate minerals: wollastonite ($CaSiO_3$), diopside ($CaMgSi_2O_6$), and garnet ($Ca_3Al_2Si_3O_{12}$). Although the parent limestone may have contained minor quartz (SiO_2), significant compositional change must have occurred during metamorphism—most important, silica was added to the rock while CO_2 was removed—to produce the silica-rich marble we find today. This change in composition was due to the flow of metamorphic fluids that could dissolve elements and thus move elements in and out of the limestone during metamorphism.

Although the Willsboro wollastonite deposits formed at the time of a relatively shallow anorthosite intrusion, that was not the end of metamorphism in that area. During the 100 million years of continental collision and tectonism after the intrusion, the Adirondacks and much of what is now the Appalachian Mountains and parts of eastern Canada were deformed and buried to great depth. The burial caused heating and further metamorphism. Figure 10.2 contains temperature contours that show that metamorphic temperatures were highest (close to 800 °C) near the anorthosite body, around 700 °C at Willsboro, and lower in the western Adirondacks. Although not shown in Figure 10.2, the pressures of metamorphism were also greatest near the anorthosite, roughly 8 kbar, representing a depth in Earth of about 25 kilometers. In the western Adirondacks, metamorphic pressures were about 6 kbar, equivalent to about 20 kilometers depth. Thus, the rocks at Willsboro went through (at least) two stages of metamorphism. The first was local due to an intrusion, but the second was a regional event affecting a huge area. Subsequently, uplift and erosion, which were greatest in the central Adirondacks, exposed the rocks that we find at the surface today.

10.2 Metamorphic rocks

As discussed in Chapter 2, most metamorphic rocks form when heat, pressure, or fluids cause changes in preexisting rocks. The preexisting, or parent, rocks are called *protoliths*. Protoliths can be igneous, sedimentary (as at Willsboro), or metamorphic rock of all sorts. The changes that occur during metamorphism may involve changes in rock texture, in the minerals present, and in overall rock composition. So, metamorphic rocks, and the processes that create them, are key parts of the rock cycle that also includes igneous and sedimentary processes. Thus, metamorphic rocks record Earth history.

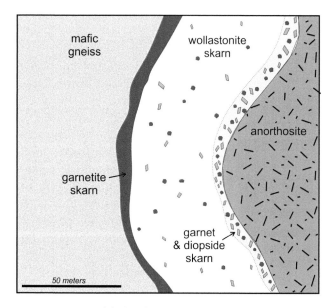

Figure 10.3   Simplified geology (map view) of metamorphic zones around anorthosite at Willsboro. Scale only approximate.

And metamorphic petrologists try to interpret the rocks to interpret the history.

The most significant causes of metamorphism are mountain-building processes that bury or exhume rocks. This kind of metamorphism, called *regional metamorphism*, creates large metamorphic *terranes*, regions characterized by distinctive metamorphic rocks and intensity of metamorphism that may vary laterally. Regional metamorphism occurs because both pressure and temperature increase with depth of burial. Although regional metamorphism, which accounts for most metamorphism, occurs at relatively deep levels within Earth, metamorphism can also occur at shallow levels or even at Earth's surface. For example, as happened at Willsboro, magma that intrudes the crust may rise close to or all the way to the surface. In such cases, heat from the magma can cause *contact metamorphism* that affects shallow or surface rocks. The effects of such metamorphism may be profound, because of the high temperature contrast between magmas and upper crustal rocks.

Occasionally metamorphism occurs without significant tectonism or magmatism. For example, metamorphism may occur because of hot water flowing through rock in areas next to hot springs or other geothermal areas. And, sometimes, water flowing through vast regions of the crust alters rocks far from mountain belts. These changes, mostly chemical in nature, can occur without significant increases in temperature and pressure. Metamorphism can also occur when rocks grind together in fault zones or when meteorites impact Earth.

10.3 Agents of metamorphism

10.3.1 Heat and temperature

Heat is thermal energy that can move (flow) from one place to another or from one substance—such as rock, magma, or water—to another. Thermal energy is high for substances at high temperature and low for substances at low temperature. Three processes can transfer heat: conduction, convection, and radiation; but within Earth, heat transfer by radiation is insignificant. *Conductive heat transfer* occurs when heat flows naturally from a place of high temperature to one of low temperature with no associated movement of matter. Thus, for example, heat is always flowing from Earth's hot interior to the cooler surface by conduction. And if a (hot) magma intrudes the (cooler) crust, the magma will cool as heat is conducted grain by grain into the surrounding

rock, causing the rock to warm. This warming initially occurs only next to the pluton, but, over time, heat is conducted farther away and warming can affect a large area.

Convective heat transfer, which is more efficient than conductive heat transfer, is the transfer of heat due to the flow of material, such as the flow of hot water or hot air. Within Earth, convection occurs mostly because of flowing water and flowing magmas. Heat transfer by water can have a significant, although generally quite local, effect. Heat transfer by convecting magmas can be much more significant and can warm huge regions of the crust. And in Earth's mantle, the slow creep of solid rock due to plate tectonics also moves heat by convection.

Different places on Earth get their heat by different combinations of conduction and convection. As described in Chapter 5, Earth's geothermal gradient, the rate at which temperature increases with depth, averages about 25 –35° °C/km near the surface in most places. This gradient, also called a *geotherm*, is due solely to conductive heat flow. The left column in Figure 10.4 shows temperature-depth relationships for a normal geotherm typical of regions where all heat transfer is by conduction. In mountain belts and other places where volcanic activity occurs, convective heat flow due to rising magmas contributes much more heat than normal conduction. Consequently, temperature increases faster with

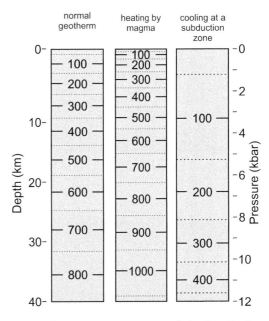

Figure 10.4 The temperature increase with depth in Earth in various settings.

depth than is normal (middle column, Figure 10.4). In some places, next to large igneous intrusions, contact metamorphism occurs and extremely high temperatures may persist for short times before the intrusions cool. In subduction zones (right column, Figure 10.4), the opposite occurs. Descending slabs of wet cool ocean lithosphere, which have been continuously carried to depth for millions of years, cool off the crust and upper mantle below. So, the rate of temperature increase with depth is less than normal.

Metamorphism occurs over a wide range of temperatures. At low temperatures (150–200 °C), metamorphism overlaps with diagenesis. At high temperatures, metamorphism gives way to igneous processes when rock begins to melt. For some composition rocks, partial melting may begin at temperatures as low as 700 °C, but other kinds of rocks may remain completely solid to temperatures as great as 1100 °C. Metamorphic temperatures are important, not just because different kinds of rocks and minerals form at different temperatures, but because temperature affects chemical reaction rates. Rocks metamorphosed at low temperature may change only very slowly, and some changes may not go to completion. Rocks that form at high temperatures generally do not have the same problems. However, there are many kinds of metamorphic rocks, and some of them are more chemically reactive than others.

10.3.2   Pressure and depth

If an external force is applied to a rock, it may cause the rock to change texture or perhaps cause new minerals to replace old ones. But it is the *pressure* that is applied that is important, not the *force*. Pressure, defined as the force applied divided by the area where it is exerted, determines if changes will occur or not. For example, if a 20-kilogram weight is applied to a very small area, it may have an effect, but if the same weight is applied to a very large area, it may not. Scientists use different units to describe pressure; some of the more commonly used are *psi* (pounds per square inch), *pascals* (equivalent to one newton per square meter, or 0.000145 psi), *bars* (equivalent to 14.5 psi), or *atmospheres* (equivalent to 14.7 psi).

Burial causes rocks to experience *lithostatic pressure*, also called *confining pressure*. Lithostatic pressure is the same in all directions and thus can cause an object to become smaller without altering its overall shape. This kind of pressure is equivalent to the pressure that

swimmers feel on their ears when they go to the bottom of the deep end of a swimming pool. The pressure on a swimmer's ears accrues because of the weight of water pushing down from above. Within Earth, the weight of rock, which is commonly three times denser than water, causes lithostatic pressure to build up quickly with depth. As shown in Figure 10.4, pressures of around 12 kbars (1.2 GPa) are reached at 40 kilometers depth, although pressure depends, in part, on the density of overlying rocks. Typical metamorphic rocks form at pressures of 0 to 10 kbars, but we find higher pressure rocks in some places. They are rare because to get to very high pressure requires that rocks are buried to great depth—an uncommon occurrence. Subsequently, getting the rocks back to the surface so we can see them is even more problematic.

10.3.3   Directed stress

Directed stress, sometimes called *differential pressure*, is also a force, which is not the same in all directions, applied to an area. For example, when we squeeze a lemon, we are applying directed stress. When we stretch a rubber band, we are also applying directed stress. Within Earth, directed stress is common due to plate tectonic processes that push large pieces of lithosphere together or pulls them apart. Unlike lithostatic pressure, high levels of directed stresses are not sustained for long because rocks deform to reduce the stress. Directed stress, thus, is commonly associated with rock folding or faulting.

Directed stress can cause new minerals to form within a rock, but much more commonly it produces deformation, fracturing, or textural changes only. Mineral grains may rotate, align, become distorted, or disintegrate. Directed stress may also cause recrystallization as grains dissolve and regrow in other places or combine to produce larger crystals. Sometimes, directed stress causes *shearing*, which means that different parts of a rock slide past each other.

During shearing, mineral grains can become elongated in one direction, and fractures can develop that give a rock a planar texture. Figure 10.5 shows a rock called *mylonite*, a highly deformed kind of rock created when fine sheared material recrystallizes, from Norway's Western Gneiss Region. Directed stress, parallel to the layering in this rock, caused feldspar (white) and biotite (black) grains to become elongated as shearing took place. While this was occurring, metamorphism

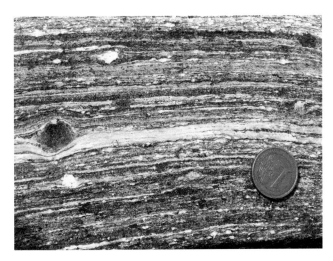

Figure 10.5 A mylonite, a highly deformed rock from Otrøy in Western Gneiss Region, Norway.

Photo from Woudloper, Wikimedia Commons.

produced wine-red garnet crystals—a single large one can be seen near the left side of the photo and many small ones are scattered throughout. The 1 euro coin is 2.3 cm across, for scale.

10.3.4 *Metamorphic fluids*

Metamorphism often involves fluids, most commonly water-rich but sometimes dominated by CO_2, sulfur, or other components. The fluids may be *magmatic* (expelled by magmas as they crystallize), *meteoric* (derived from precipitation that infiltrates the ground), released during subduction of wet lithosphere, or products of reactions that release H_2O or CO_2 from minerals. *Hydrothermal metamorphism* occurs when fluids significantly alter protolith rocks. This kind of metamorphism can affect large areas and be part of regional metamorphism, or it can be localized and part of contact metamorphism. In either case, the metamorphism involves hot, generally water-rich fluids that flow through cracks and along grain boundaries. The fluids act as catalysts and fluxes that promote reactions and large crystal growth. More important, the fluids may change the composition of the protolith by adding or removing specific elements, and the resulting rock may be of a much different composition than its parent. This composition-changing process, called *metasomatism*, creates many kinds of products. Metasomatism can, for example, create ore deposits by concentrating minerals (most commonly copper, iron, or lead sulfides) in host rocks where they did not exist previously. And

metasomatism was responsible for the wollastonite ores in the Adirondacks.

10.4 Metamorphic textures

Natural systems strive for conditions of lowest energy, and metamorphic rocks are no exceptions. For example, the mineral grains within a rock that is under pressure may move or readjust to reduce stress on the rock, thus reducing *physical energy*. Simultaneously, minerals within a rock may react together to produce new minerals of lesser *chemical energy*. And many small grains of some minerals, perhaps garnet, may combine to produce larger crystals, and thus reduce surface area and *surface energy*. All these changes can give a rock *metamorphic texture*, sometimes called a *metamorphic fabric*. Different textures are characterized by different combinations of mineral grain sizes and by the way that mineral grains align, or do not align, within a rock.

10.4.1 *Grain size and porphyroblasts*

Some metamorphic rocks are so fine-grained that seeing individual mineral crystals requires a microscope. These rocks may contain no visible layering or fractures and appear as a homogeneous mass. Geologists use the general term *hornfels* for any dark-colored fine-grained metamorphic rock that does not show visible layering. Most hornfels are quite hard and durable, because constituent grains are tightly bound together. *Biotite hornfels*, the most common kind of hornfels, are dark brown and sometimes have a slight sheen due to microscopic grains of biotite. Other hornfels may have different colors; the color depends on the rock composition. Some hornfels contain grains that become visible if a rock is weathered (because different minerals weather in different ways) but, because of the generally uniform rock color, are invisible otherwise.

Large mineral crystals have less energy than small crystals because mineral molecules on crystal surfaces are less stable than those entirely within a crystal, and large crystals have lower surface area to volume ratios. Natural systems move toward conditions of lowest energy and, consequently, with any significant amount of metamorphism, microscopic grains react to produce larger, visible (macroscopic) grains, and already visible grains tend to grow larger. These observations, together, are termed Ostwald ripening, first described by chemist Wilhelm Ostwald in 1896.

Many metamorphic rocks are bimodal, meaning they contain mineral crystals of two sizes. Such rocks, termed *porphyroblastic rocks*, contain large crystals that have grown within a sea of finer-grained mineral material. The large crystals, called *porphyroblasts*, may be anhedral but are often subhedral or euhedral. The finer material is the *groundmass*. Some of the most spectacular porphyroblastic rocks are mica-rich schists that contain chloritoid, kyanite, garnet, or staurolite as porphyroblasts. Figure 10.6 shows a schist from southern Switzerland, near Italy, that contains porphyroblasts of kyanite (blue) and staurolite (brown). The largest of the blue kyanite blades are about 4 cm long. In this rock, a silvery mica called *paragonite* constitutes most of the groundmass. Figure 10.7 shows garnet in an outcrop on Syros, one of the western Greek islands. The largest of the porphyroblasts are about 2 cm across. The groundmass in this rock is mostly quartz and feldspar with lesser amounts of biotite. Garnet is a very common porphyroblast in rocks that metamorphose at medium to high temperature, and garnet crystals are commonly euhedral to subhedral, like the crystals shown here.

Figure 10.7   Garnet-mica schist on the island of Syros, Greece. Photo from Graeme Churchard, Wikimedia Commons.

10.4.2   *Foliated metamorphic rocks*

Metamorphic rocks may be foliated or nonfoliated. *Foliated* metamorphic rocks have visible repetitive, and generally parallel, layers that may be as thin as a flake of mica or meters thick. Figure 10.8 shows an example of a foliated rock. The foliation, a series of near-parallel cracks and offsets, runs nearly vertically in this view, shown by layering in the photo and by dashed lines in the drawing. But, this specimen also has a roughly horizontal bedding plane that separates older rock (below) from younger rock (above). As shown in this figure, foliation produced during metamorphism need not be parallel to bedding layers (which are produced during deposition). Foliations form in response to directed stress and are commonly perpendicular to the stress. However, if foliation develops when a rock shears, it will be parallel to the direction of shearing. Foliated rocks are common in mountain belts where tectonic movements compress or stretch geological terranes.

Protoliths vary greatly in composition, and some compositions develop foliations more than others do because of their mineral constituents. If clays or micas, for example, are in a rock, they can align to produce a foliated fabric. But if only quartz or feldspar are present, such as in granite, the mineral grains are too equant and alignment will not occur. More than any other kind of rock, clay-rich rocks, generally called *pelites* by petrologists, tend to develop foliations of several kinds. During metamorphism, a shale protolith, for example, may sequentially evolve to become rocks called *slate*, *phyllite*,

Figure 10.6   Porphyroblasts of kyanite (blue) and staurolite (brown) in paragonite. Sample from Ticino Canton, Switzerland.
Photo from R. Lavinsky, Wikimedia Commons.

Figure 10.8 Example of a foliated metamorphic rock with visible bedding.

Original photo from Peter Davis.

Figure 10.9 Green slate from Pawlet, Vermont; 10 centimeters across.

Photo from D. Perkins, GeoDIL.

schist, or *gneiss* while being metamorphosed at increasing temperature and pressure.

10.4.2.1 Slate

Figure 10.9 shows a sample of slate. *Slate*, characterized by fine-grained clay minerals or micas that become aligned, forms from shale derived by lithification of clay-rich sediments or volcanic ash. Metamorphism may obliterate the original bedding as foliation develops perpendicular to the direction of maximum stress. This foliation, *slatey cleavage*, gives slates a property called *fissility*—an ability to break into thin sheets of rock with flat smooth surfaces. The thin sheets have historically been used for paving or roofing stone. Slates come in many colors, but various shades of gray are most common. The minerals in this rock cannot be identified in hand specimens, but in thin section quartz, feldspar, chlorite, muscovite, and biotite can be seen.

10.4.2.2 Phyllite

Figure 10.10 shows a sample of *phyllite*, a shiny foliated rock created by further metamorphism of slates. The foliation is due to parallel alignment of very small—mostly microscopic—muscovite, chlorite, or other micas, sometimes with graphite. Thus, foliation of phyllites is different from the foliation in slates that stems from clay mineral alignment and different from foliation in schists, because schists always contain

Figure 10.10 Example of phyllite.

Photo from Kurt Hollocher.

macroscopic (visible) mica mineral grains. Like slates, phyllites exhibit fissility. The rocks are typically black, gray, or green, and the fine-grained micas and graphite, which are too small to see without a microscope, give phyllites a silky/shiny appearance, or sheen, called a *phyllitic luster*. It is this luster—which is absent from slate and schist—that really defines a phyllite. Additionally, although not seen in Figure 10.10, the layering in some phyllites is deformed, giving the rocks a sort of wavy or crinkly appearance.

10.4.2.3 Schist

Figure 10.11 shows a typical schist. Schists are higher-temperature rocks than phyllites, and most form when

Figure 10.11 Muscovite schist.
Photo from Michael C. Rygel, Wikimedia Commons.

Figure 10.12 Deformed gneiss.
Photo from Chmee2, Wikimedia Commons.

phyllites are further metamorphosed. Thus, the precursors of schists are shale, slate, and phyllite. Less commonly, however, schist may form by metamorphism of fine-grained igneous rocks, such as tuff or basalt. Large and aligned flaky minerals, easily seen with the naked eye, define schists. These minerals are most commonly muscovite (such as can be seen in Fig. 10.11) or biotite in parallel or near-parallel orientations that give the rocks *schistosity*—the ability to be broken easily in one direction but not in other directions. Although most schists are mica schists, graphite, talc, chlorite, and hornblende schists are common. Quartz and feldspar are present in mica schists, often deformed or elongated parallel to the micas, and many other minerals are possible. If schists contain prominent porphyroblasts, we name them accordingly. So the schist in Figure 10.6 is a *kyanite-staurolite schist*, the schist in Figure 10.7 is a *garnet schist*, and the one in Figure 10.11 would be called a *muscovite schist*, or simply a *mica schist*.

10.4.2.4 Gneiss

Gneisses, the highest temperature-pressure kinds of foliated metamorphic rock, typify many regions that have undergone high-temperature metamorphism. The defining characteristics of most gneisses, such as the one in Figure 10.12, are that the rocks are medium- to coarse-grained and contain alternating layers of light- and dark-colored minerals that give the rock foliation called *gneissic banding*. Sometimes the banding is deformed, as is seen in this photo, but often it is in parallel layers. Gneissic banding most commonly forms

in response to directed stress, although sometimes layering may form solely due to chemical processes that concentrate different minerals in different layers. The felsic light-colored layers typically contain quartz and feldspars, and the more mafic darker layers typically contain biotite, hornblende, or pyroxene. Other accessory minerals, such as garnet are common. The gneiss in Figure 10.12, for example, a deformed granitic gneiss from the Czech Republic, contains pink K-feldspar rich layers alternating with darker layers that contain biotite.

Gneisses are often named based on their protoliths, and petrologists use the general terms *orthogneiss* for gneisses derived from igneous rocks and *paragneiss* for gneisses derived from sedimentary rocks. More specific names abound—for example, *pelitic gneisses* form by metamorphism of originally clay-rich sedimentary rocks, *granitic gneisses* (such as the one shown in Fig. 10.12) form by metamorphism of granites, and *mafic gneisses* form by metamorphism of mafic igneous rocks. Sometimes key minerals are included in rock names. For example, a garnet gneiss is a gneiss that contains conspicuous garnet crystals.

Some gneisses do not display well-defined dark- and light-colored banding but still maintain less distinct foliation. For example, the foliation in kyanite gneiss may come from alignment of light-colored kyanite crystals in an otherwise quartz- and muscovite-rich rock. And *augen gneiss*, such as the rock shown in Figure 10.13, contains large feldspar crystals-"eyes" (*augen* is German for eyes)—stretched in one direction. In Figure 10.13, the sample is oriented so the stretch direction (and thus the foliation) is horizontal.

Figure 10.13   Augen gneiss.

Photo from Eurico Zimbres, Wikimedia Commons.

Figure 10.15   Blue calcite marble.

Photo from James St. John, Wikimedia Commons.

Figure 10.14   Flat-pebble conglomerate from Death Valley, California.

Photo from Nicolas C. Barth.

A somewhat different kind of foliated metamorphic rock forms when a conglomerate experiences directed stress that flattens individual clasts in the rock, creating a *flat-pebble conglomerate* like the one shown in Figure 10.14. If a sample of "unstretched" rock is available, the clast shapes and dimensions provide a record of how much the rock was stretched, and the orientation of the clasts records the direction of maximum stress.

10.4.3   Nonfoliated metamorphic rocks

10.4.3.1   Marbles

Many nonfoliated metamorphic rocks are dominated by a single kind of mineral. In these rocks, individual mineral grains or crystals, which may start small,

recrystallize (grow together) during metamorphism to produce larger crystals. Figure 10.15, for example, shows an 8 cm wide rock consisting almost entirely of coarse blue calcite. This rock, which comes from the Valentine Mine in the western Adirondack Mountains, New York, derived from a limestone protolith. Petrologists use the term *marble* for all metamorphic carbonate rocks—rocks that form from limestone or dolostone—dominated by calcite or dolomite. (This sometimes leads to confusion because builders and others use the same word to describe any polished slab of rock.)

Marble protoliths have relatively simple chemistry; they are mostly $CaCO_3$ (calcite) or $CaMg(CO_3)_2$ (dolomite). When metamorphosed, however, the rocks often gain Ca-Mg silicate minerals such as those listed in Table 10.1. Sometimes, grossular (a kind of garnet), vesuvianite, or other Al-bearing minerals grow as well. The order of minerals in Table 10.1 is from those that generally form at lowest temperature (talc and then tremolite) to those that form at highest temperature (vesuvianite and grossular). The silica and aluminum, necessary for some of these minerals to form, come from small amounts of quartz or clay present in the parent rocks or are introduced by metasomatism.

Figure 10.16 shows a marble from the western Adirondacks, from another site near the Valentine Mine. It contains dark green diopside besides white calcite. Calcite in most marbles is white, as in this sample. Some marbles contain no carbonate minerals at all, because the carbonates that were once present have all reacted to form other, non-carbonate minerals. The wollastonite ore from the Willsboro deposit (Fig. 10.1) is an example of such a marble.

Table 10.1   Common accessory minerals in marbles.

	Mineral	Formula
temperature	Talc	$Mg_3Si_4O_{10}(OH)_2$
	tremolite	$Ca_2Mg_5Si_8O_{22}(OH)_2$
	forsterite	Mg_2SiO_4
	diopside	$CaMgSi_2O_6$
	wollastonite	$CaSiO_3$
	monticellite	$CaMgSiO_4$
	vesuvianite	$Ca_{10}Mg_2Al_4Si_9O_{34}(OH)_4$
	grossular	$Ca_3Al_2Si_3O_{12}$

Figure 10.17   An example of quartzite.
Photo from Kurt Hollocher.

10.4.3.3 *Other nonfoliated metamorphic rocks*

Figure 10.18 shows an example of a garnet amphibolite. *Amphibolites* form by metamorphism and recrystallization of basalt or other mafic igneous rocks. All amphibolites contain mostly hornblende and plagioclase, but other minerals including garnet and minor quartz are typically present. In Figure 10.18, the garnet porphyroblasts, 1–2 cm wide, are large and conspicuous; sometimes when garnet is present, it is less easily seen. Some amphibolites are strongly foliated but most are not.

Hornfels, mentioned earlier in this chapter, and greenstones are also kinds of nonfoliated metamorphic rocks. Geologists use the name *hornfels* for any dark-colored nonfoliated fine-grained metamorphic rock; *greenstones*, which are a specific kind of hornfels, form by metamorphism of basalts. Figure 10.19a shows a 10 cm wide sample of hornfels derived from a sedimentary protolith, and Figure 10.19b shows a 9 cm wide sample of greenstone from Ely, Minnesota. The greenish color is due to chlorite or epidote that grew during metamorphism. Figure 10.19c shows an outcrop of greenstone in Italy. The rock originated as an ocean-floor basalt and contains *pillows* indicative of submarine eruption.

Anthracite (Fig. 10.19d) provides another example of a nonfoliated metamorphic rock. As discussed in Chapter 14, anthracite is a type of very high-ranked coal. While most coals have sedimentary origins, anthracite forms in the same way as metamorphic rocks in mountain belts when metamorphic temperatures exceed 200 °C. Under such conditions, organic materials break down and coal loses water and other volatile

Figure 10.16   Diopside marble from the Adirondacks.
Photo from GeoDIL.

10.4.3.2 *Quartzite*

Quartzites form by metamorphism of sandstone, and most sandstones comprise mainly quartz. Figure 10.17 shows a typical quartzite. It is composed entirely of millimeter-sized quartz crystals that have grown together so that no grain boundaries are visible without a microscope. Recrystallization has produced a typical hard and shiny quartzite, and during metamorphism any original sedimentary textures were erased. Common quartzites are white or gray, but minor components may add color. The pink color in this sample comes from hematite that may have been the cement that held the sandstone together. If the protolith sandstone contained minerals besides quartz, so too will the quartzite. Thus, feldspar, titanite, rutile, magnetite, or zircon may be present in small amounts. And, if the protolith contained some clay, aluminous minerals such as kyanite may also be present.

Figure 10.18 An example of garnet amphibolite.
Photo from Kurt Hollocher.

Figure 10.20 Hornblende schist, with more or less parallel crystals of black hornblende.
Photo from Kurt Hollocher.

a. Hornfels from Riverside County, California.

c. Metamorphosed pillow basalt in Italy. Wikicommons.

b. Metamorphosed basalt from near Ely, Minnesota. DP photo.

d. Anthracite from unknown place. Photo modified from Wikicommons.

Figure 10.19 Examples of nonfoliated metamorphic rocks.
Photos from GeoDIL.com, and from Matt Affolter and Amcyrus2012, both at Wikimedia Commons.

compounds. The coal becomes hard and compact as it turns into anthracite, typically containing graphite as its main mineral.

10.4.4 Lineations

Metamorphic foliations develop when micas or other minerals become aligned. Metamorphic *lineations* develop in about the same way but involve alignment of long, skinny mineral grains, like what occurs if you

hold a bunch of sticks or pencils together in your hand. Figure 10.20 shows an example of a lineated rock, a hornblende schist; the black crystals are needle-like hornblende crystals in subparallel alignment. Lineation and foliation occur together in some rocks, but not in all. When they occur together, the direction of lineation commonly lies within foliation plains. Both fabrics—foliation and lineation—form because of directed stress and thus may not be parallel to original rock bedding.

10.5 Metamorphic reactions

Just as rock fabrics develop during metamorphism to minimize physical and surface energy, natural chemical systems generally adjust (by chemical reactions) to achieve equilibrium conditions of lowest chemical energy. The *Laws of Thermodynamics* describe reactions and equilibrium, and define minerals or *mineral assemblages* (combinations of coexisting minerals) as *stable, unstable,* or *metastable.* The term *stable* refers to minerals, rocks, or other chemical systems that are at their lowest state of chemical energy. Metamorphic petrologists, when interpreting metamorphic rocks, are often concerned about the conditions that make individual minerals or mineral assemblages stable, because this knowledge allows interpretation of a rock's history based on the minerals it contains. Quartz, feldspar, and a few other minerals are stable over very wide ranges of metamorphic conditions. But most minerals and mineral assemblages are not, and their occurrences are restricted to rocks that formed under specific conditions of pressure and temperature. Thus, they have a restricted *stability range.*

Unstable minerals are those that are in the process of breaking down by chemical reaction (to produce reaction products that are stable). For example, feldspars that are weathering in an outcrop to produce clay are unstable. Weathering at Earth's surface is a relatively quick process, but within Earth, metamorphic reaction rates are typically slow, metamorphism may take millions of years, and sometimes reactions do not achieve stability. Thus we have, at Earth's surface today, samples of minerals and rocks that are, in principle, only stable at conditions deep within Earth. The term *metastable* refers to such minerals and rocks that should, in principle, be reacting to form more stable minerals and rocks but are not. For example, diamond forms and is only stable at pressures that exist at 150–250 kilometers (93–155 miles) depth or even deeper within Earth. Diamond is made of carbon, just like the common mineral graphite that makes up pencil lead. The Laws of Thermodynamics tell us that all the diamonds that we see should react to form graphite, yet we have many examples of (metastable) diamonds to enjoy.

The important thermodynamic variables that determine the nature of chemical reactions and whether reactions will occur are temperature, pressure, volume, and entropy. A rock within Earth experiences a particular temperature and pressure, and the minerals within that rock have a specific volume and entropy. (*Entropy*, a somewhat arcane concept, is a thermodynamic quantity that describes how much energy is tied up in a mineral's structure due to atomic disorder and kinetic energy.) The Laws of Thermodynamics tell us that, at high pressure, minerals of greatest density are most stable and at low pressure, minerals of lower density are stable. Thus, for example, diamond, which is denser than graphite, is stable at high pressures deep within Earth, while graphite is stable at Earth's surface. Additionally, at high temperature, minerals of greatest entropy are stable and at low temperature, minerals of least entropy are stable.

Consider, for example, the two aluminosilicate minerals kyanite and andalusite. These minerals are polymorphs, meaning both have the same composition (Al_2SiO_5). We can write a chemical reaction relating the two:

kyanite = andalusite   [Reaction 1]
$Al_2SiO_5 = Al_2SiO_5$

Andalusite has greater entropy than kyanite, and kyanite has greater density than andalusite. Thus, if kyanite is heated sufficiently, it reacts to form andalusite as Reac-

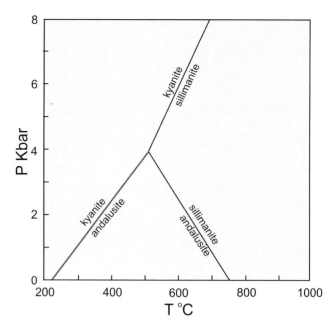

Figure 10.21   A metamorphic phase diagram showing reactions that limit the stabilities of andalusite, kyanite, and sillimanite.

tion 1 (above) proceeds to the right. And, if andalusite is squeezed sufficiently, it reacts to form denser kyanite as Reaction 1 goes to the left. The red line on the phase diagram in Figure 10.21 shows these relationships. The figure also shows that a third Al_2SiO_5 polymorph, sillimanite, is stable at high temperature. Sillimanite has greater entropy than kyanite and andalusite and so is more stable than the other two minerals at high temperature. Sillimanite is denser than andalusite, but less dense than kyanite. Thus, kyanite is stable at high pressure, andalusite is stable at low pressure, and sillimanite at intermediate pressure. The lines in Figure 10.2 show three reactions that separate *stability fields* (regions of pressure-temperature space where minerals are stable) for the three minerals.

Reaction 1, above, is a simple reaction involving only two minerals. Metamorphic reactions, however, can be much more complicated, sometimes involving four or more minerals, and reactions can be of many kinds. Table 10.2 gives some examples. Reactions 1 and 2 are *solid-solid reactions* that involve, as their name implies, only solid mineral phases. But other typical reactions involve H_2O or CO_2. When dehydration reactions, such as reactions 3–5 (Table 10.2) occur, hydrous minerals (minerals that contain H_2O) decompose to other (anhydrous) minerals as H_2O is given off as vapor. When *decarbonation reactions* such as reactions 6 and 7 occur, carbonate minerals (minerals that contain

Table 10.2 Examples of different kinds of metamorphic reactions.

Reaction type	#	Example
solid-solid reaction	1	kyanite = sillimanite $Al_2SiO_5 = Al_2SiO_5$
	2	grossular + quartz = anorthite + 2 wollastonite $Ca_3Al_2Si_3O_{12} + SiO_2 = CaAl_2Si_2O_8 + 2\ CaSiO_3$
dehydration reaction	3	muscovite + quartz = K-feldspar + sillimanite + H_2O $KAl_2(AlSi_3)O_{10}(OH)_2 + SiO_2 = KAlSi_3O_8 + Al_2SiO_5 + H_2O$
	4	gibbsite = diaspore + H_2O $Al(OH)_3 = AlO(OH) + H_2O$
	5	2 diaspore = corundum + H_2O $AlO(OH) = Al_2O_3 + H_2O$
decarbonation reaction	6	calcite + quartz = wollastonite + CO_2 $CaCO_3 + SiO_2 = CaSiO_3 + CO_2$
	7	dolomite + 2 quartz = diopside + 2 CO_2 $CaMg(CO_3)_2 + 2\ SiO_2 = CaMg\ Si_2O_6 + 2\ CO_2$
carbonation reaction	8	forsterite + 2 CO_2 = 2 magnesite + quartz $Mg_2SiO_4 + 2\ CO_2 = 2\ MgCO_3 + SiO_2$
hydration reaction	9	2 forsterite + 3 H_2O = brucite + serpentine $2\ Mg_2SiO_4 + 3\ H_2O = Mg(OH)_2 + Mg_3Si_2O_5(OH)_4$
	10	6 forsterite + talc + 9 H_2O = 5 serpentine $6\ Mg_2SiO_4 + Mg_3Si_4O_{10}(OH)_2 + 9\ H_2O = 5\ Mg_3Si_2O_5(OH)_4$

CO_2) turn into other minerals while CO_2 is released. Note that reactions 6 and 7 are key reactions that were responsible for producing the wollastonite ore rock at Willsboro.

Dehydration and decarbonation reactions are examples of *prograde reactions*, reactions that occur in response to increasing temperature. Solid-solid reactions may be prograde as well. These prograde reactions create higher-temperature rocks from lower-temperature rocks, a process called *prograde metamorphism*. Dehydration and decarbonation reactions, which essentially involve gaseous components being kicked out of minerals, take place with heating because H_2O and CO_2 in vapor form have much greater entropy than they do as components in minerals. So, at high temperature, H_2O- and CO_2-bearing minerals become unstable.

In contrast with prograde reactions, *retrograde reactions* occur in response to temperature decreases. *Carbonation* and *hydration reactions*, such as reactions 8, 9, and 10 in Table 10.2, are common examples. Retrograde reactions, and thus *retrograde metamorphism*, can be especially significant for rocks, such as some igneous rocks, that originally formed at very high temperatures but are later exposed to lower-temperature hydrothermal solutions. Figure 10.22 shows a *serpentinite*, a serpentine-rich rock created by retrograde metamorphism of an ocean-floor basalt. Reactions 9 and 10 (in Table 10.2) were largely responsible for creating this rock.

Figure 10.22 Sample of serpentinite from the Golden Gate National Recreation Area in San Francisco, an example of retrograde metamorphism.

Photo from Zimbres, Wikimedia Commons.

10.5.1 Metamorphic phase diagrams

Figure 10.21 is an example of a *metamorphic phase diagram*, and Figure 10.23 shows two more examples. Petrologists use such diagrams to show the minerals

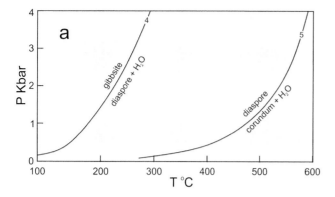

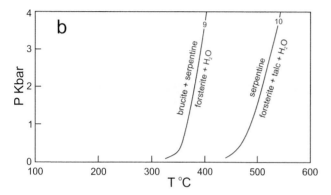

Figure 10.23   Two examples of metamorphic phase diagrams.

contains brucite and serpentine is heated. In such a case, the brucite and serpentine will react and disappear; at elevated temperatures, forsterite and talc will remain.

Phase diagrams such as the ones shown in Figures 10.21 and 10.23 are very important for two reasons. First, if petrologists find a rock that contains a specific mineral assemblage, they can use phase diagrams to estimate the conditions at which the rock formed. For example, if a metamorphosed bauxite contains gibbsite, the bauxite could not have been metamorphosed to very high temperatures. Alternatively, we can use phase diagrams to predict what will happen during metamorphism. Such predictions have important implications for engineering. For example, if clay-rich rocks containing bauxite are heated, eventually they will lose hydrous minerals and become mostly corundum. If this occurs, the properties of the rocks will change significantly because corundum is much harder than most other minerals.

10.6   Burial metamorphism

Burial metamorphism, the precursor to regional metamorphism and the lowest-energy kind of metamorphism, occurs when rocks undergoing diagenesis are buried to significant depth. Eventually, as the weight of overlying sediments increases and temperatures increase due to burial, diagenetic processes give way to metamorphic processes. This change happens most commonly in sedimentary basins where thick layers of sediment accumulate. The transition between diagenesis and burial metamorphism occurs at temperatures of 200 °C to 300 °C and depths of several kilometers, but the distinction between diagenesis and metamorphism is hazy. During burial metamorphism, heat and pressure cause minerals to recrystallize, clay minerals to turn into different clay minerals, and new minerals—most typically zeolites—to grow. Sandstone may become quartzite and shale may become slate, but commonly some bedding and other sedimentary structures persist and rocks may show very little outward signs of metamorphism. Burial metamorphism rarely leads to any significant development of metamorphic fabric because pressure is generally lithostatic at the shallow depths in Earth where burial metamorphism occurs.

10.7   Regional metamorphism

Burial metamorphism grades into *regional metamorphism* as temperature and pressure increase. Regional

that are stable under different pressure and temperature conditions. Because different minerals are found in rocks of different compositions, different phase diagrams apply to different kinds of rocks. Figure 10.23a, for example, is applicable to a metamorphosed bauxite deposit. Bauxites, which provide most of the aluminum ore in the world, generally contain a mix of aluminum hydroxide and oxide minerals. This diagram shows that at low temperature, gibbsite ($Al(OH)_3$) may be present, but if a deposit is metamorphosed to higher temperatures, diaspore ($AlO(OH)$) and eventually corundum (Al_2O_3) may replace the gibbsite. Note that both reactions, identified by reaction number, are listed as examples of dehydration reactions in Table 10.2.

Figure 10.23b shows two of the typical retrograde reactions that take place in olivine-bearing rocks. These reactions, also listed in Table 10.2, explain why brucite and serpentine are common minerals in altered ocean-floor rocks, such as the rock shown in Figure 10.22; the two minerals form at low temperature because (high-temperature) forsterite and talc are unstable and react with water. Although the reactions in Figure 10.23b are most significant during retrograde metamorphism, they could be prograde if a rock that

metamorphism, which occurs over millions of years, is not generally due to overlying layers of sediment being deposited and compressing rocks below but is caused by burial and heating associated with mountain-building events (orogenies), such as the one that affected the Adirondack Mountains. The mountain building, which occurs in orogenic belts, may be caused by continental collision or by tectonism at subduction zones and can affect very large areas. As the crust thickens and rocks become buried, the amount of metamorphism increases with depth in Earth. Generally, the greatest burial occurs near the centers of orogenic belts, so we commonly find high pressure-temperature (PT) metamorphic rocks in the centers of mountain ranges, with lower PT rocks near the range margins.

However, we also find vast areas of regionally metamorphosed rocks in places that are not mountain ranges today. During orogenies, when the crust thickens, it naturally moves upward due to buoyancy, creating topographic highs (mountains). Eventually, given enough time, the mountains erode, thus exposing underlying igneous and metamorphic rocks that were once at great depth. Because upward movement and erosion may not be the same everywhere in an orogenic belt, rocks of different metamorphic grade are exposed in different places. The most significant such exposures are in regions called *shields*. All the world's continents contain *Precambrian shields*, areas where ancient mountains, sometimes several billions of years old, have been completely removed by erosion. The mountain roots presently exposed may come from tens of kilometers beneath the surface. The degree of metamorphism and lateral variations in metamorphism that typify shield areas tell us about the size and formation of mountains that are no longer present. Precambrian shields are particularly important because they host many important mineral deposits.

The orange and red colors in Figure 10.24, which circle Hudson Bay, show the *Canadian Shield*. The shield, a large generally flat region, contains most of the metamorphic rock outcrops in North America as well as many different kinds of igneous and sedimentary rocks.

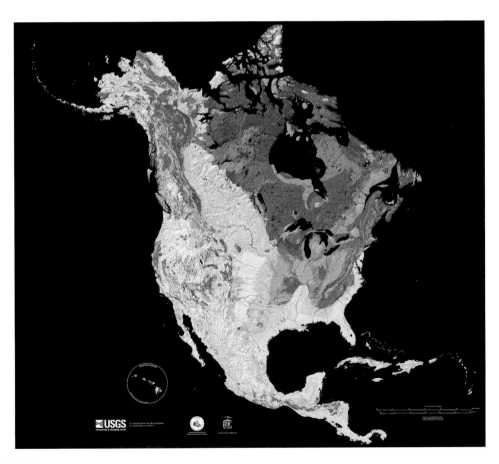

Figure 10.24 North American terranes.
Map from the USGS/Wikicommons.

The darker red colors show regions where rocks that are more than 3 billion years old can be found. The lighter orange colors show regions of 1-billion-year-old rocks. Green, blue, and purple colors, mostly within the United States, show regions where surface rocks are of Phanerozoic age (younger than 600 million years ago). This pattern—ancient Precambrian metamorphic and igneous rocks flanked by much younger sedimentary rocks—is present on all the world's continents.

10.7.1   *The role of the protolith*

Some kinds of rocks change greatly when metamorphosed; others do not. Granite, for example, undergoes few mineralogical changes during metamorphism. Granites are mostly quartz and K-feldspar, and these two minerals are stable at just about all metamorphic conditions. However, if a granite contains mica (muscovite or biotite) or hornblende as accessory minerals, some new metamorphic minerals can form. The new minerals, however, will always be present in small amounts. Consequently, the biggest changes that occur when granite is metamorphosed are textural. Thus, granites may evolve to become orthogneisses if feldspar and quartz concentrate in different layers during metamorphism. Few new minerals will form, but with increasing metamorphism, the overall grain size of all minerals may increase.

Like granite, sandstones made only of quartz, and limestones made only of calcite or dolomite, may recrystallize during metamorphism as they become quartzite or marble, but no new metamorphic minerals can form. Yet, if the sandstone contains some clay, or the limestone contains some quartz or clay, new metamorphic minerals can form. Thus, the mineralogical composition of the protolith is very important.

In contrast with granites, shales (pelitic rocks) and basalts (mafic rocks) change greatly during metamorphism. The primary minerals they contain have limited stability fields and react to produce a series of different metamorphic minerals, depending on metamorphic conditions. And these rock types develop different metamorphic fabrics depending on the degree of metamorphism. Iron formations, too, undergo significant changes during metamorphism because several different iron oxides, amphiboles, pyroxenes, and olivine will form depending on metamorphic conditions. Ultramafic rocks may develop many new minerals (including brucite, talc, and serpentine discussed earlier in this chapter) and textures when metamorphosed.

The discussion of regional metamorphism below focuses mostly on pelites and mafic rocks, and the discussion of contact metamorphism focuses on pelites and carbonates. But pelites, mafic rocks, and carbonates are only examples. The principles that are introduced apply equally to all compositions, and any kind of rock may experience regional or contact metamorphism.

10.7.2   *Metamorphic grade and pelitic rocks*

Metamorphism can occur over a wide range of pressure and temperature. The range depends on the protolith involved. Most rocks will begin to show metamorphic changes at around 200 °C, and some can be heated to temperatures greater than 900 °C or 1000 °C before they begin to melt. Although metamorphic changes can occur in response to either heat or pressure, petrologists usually tend to think of temperature as being more important. Part of this perspective is because heat promotes metamorphic reactions more than pressure does, because adding heat to a system improves reaction kinetics. Additionally, during most metamorphism, increases in temperature and pressure are proportional to each other, and thus it is unnecessary to consider the two variables separately. So, petrologists often talk about *metamorphic grade*, which is a relative scale of metamorphic intensity based on temperature.

Figure 10.25 shows some different minerals that form at different grades of metamorphism in pelitic rocks. Low-grade metamorphic rocks are those that were metamorphosed at relatively low temperature (200–400 °C), medium-grade metamorphic rocks at

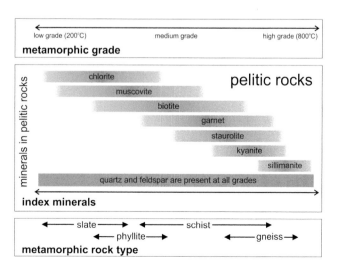

Figure 10.25   Pelitic index minerals and metamorphic rock types.

medium temperatures (350–650 °C), and high-grade metamorphic rocks at high temperatures (> 600 °C). During normal metamorphism, called *progressive metamorphism*, rocks start at low grade and minerals appear in predicable order as grade increases.

The most common way to estimate metamorphic grade is based on the presence or absence of some key *index minerals*. The significance of index minerals was first pointed out by Scottish geologist George Barrow in his classic 1912 study of regionally metamorphosed pelitic rocks of the Scottish Highlands. As shown in Figure 10.25 and listed in Table 10.3, the most important of these index minerals are, from low grade to high grade, chlorite, muscovite, biotite, garnet, staurolite, kyanite, and sillimanite.

The various index minerals have overlapping stability ranges, and the ranges may slide a bit to higher or lower grade because different index minerals appear at different temperatures depending on exact rock composition and metamorphic pressure. Chlorite is found only in low-grade rocks that generally show little metamorphic texture other than, perhaps, slatey cleavage. With increasing metamorphism, muscovite and biotite appear, and rocks may be phyllites or schists. At still higher grades, garnet, staurolite, kyanite, and sillimanite appear. At the highest grades, all micas will be absent and rocks are typically gneisses. Quartz and feldspar are present in rocks of all grades.

Although metamorphic petrologists often focus on temperature, pressure has some effect on the metamorphic minerals that will form. The top diagram in Figure 10.26a shows stability fields for some index minerals that are in pelitic rocks under different PT conditions. The white lines show approximate PT conditions where metamorphic reactions that add or remove a mineral take place.

Besides the minerals listed in Figure 10.26a, quartz is found in all pelites, muscovite is often present, and one of the aluminosilicate minerals (kyanite, andalusite, or sillimanite) may be present as well. The middle diagram, Figure 10.26b, highlights some of the more significant

Table 10.3 Common index minerals in metamorphosed pelites.

	Grade	Mineral	Formula
temperature	low	chlorite	$(Mg,Fe)_5Al_2Si_3O_{10}(OH)_8$
		chloritoid	$(Fe,Mg)_2Al_4Si_2O_{10}(OH)_4$
		muscovite	$KAl_2(AlSi_3)O_{10}(OH)_2$
		biotite	$K(Fe,Mg)_3(AlSi_3)O_{10}(OH)_2$
		garnet	$(Fe,Ca,Mg)_3Al_2Si_3O_{12}$
		staurolite	$Fe_2Al_9Si_4O_{23}(OH)_2$
		kyanite	Al_2SiO_5
	high	sillimanite	Al_2SiO_5

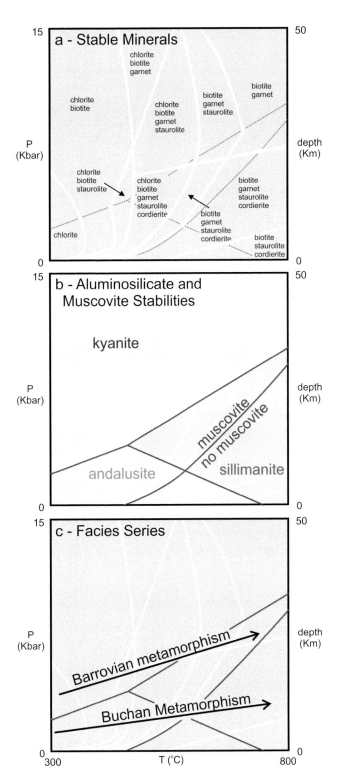

Figure 10.26 Stable minerals in pelitic rocks (a and b), and the differences between Barrovian and Buchan metamorphism (c).

reactions that limit the stability of muscovite and that separate PT regions where different aluminosilicates are stable. Muscovite is only stable in rocks that equilibrated at temperatures to the left of the labeled red line, kyanite only forms in high-pressure rocks (blue region), sillimanite in high-temperature rocks (tan region), and andalusite at low-pressure and moderate temperature (green region). Thus, metamorphic PT conditions for any metamorphosed pelitic rock can be estimated based on the minerals it contains: the index minerals listed in Figure 10.26a, the three aluminosilicate minerals shown in Figure 10.26b, and the presence or absence of muscovite, also shown in Figure 10.26b. Note that the reactions relating kyanite, sillimanite, and andalusite are the same ones depicted in Figure 10.21.

The series of minerals depicted in Figure 10.25 is called the *Barrovian Series*, named in honor of George Barrow. These minerals form when pelitic rocks are metamorphosed along a *Barrovian PT path* such as the one labeled in the bottom diagram, Figure 10.26c. This path is the most common in metamorphic terranes. However, in the Buchan region of Scotland, 100–150 kilometers northeast of where Barrow did his classic study, metamorphism occurred at pressures lower than normal Barrovian metamorphism (shown by the lower black arrow in Figure 10.26c). This kind of metamorphism, called *Buchan metamorphism*, produces many of the same minerals as Barrovian metamorphism produces, but cordierite and andalusite, low-pressure minerals that do not generally form during Barrovian metamorphism, are often present. Additionally, the low pressures of Buchan metamorphism mean that kyanite is absent. The lower-pressure mineral assemblages labeled in the low-pressure part of the top drawing (Fig. 10.26a) are all Buchan assemblages.

10.7.3   Mafic rocks and metamorphic facies

Figure 10.27 shows a diagram equivalent to Figure 10.25, but for mafic igneous rocks instead of for pelites. Just like with pelitic rocks, the presence or absence of specific minerals is a good indicator of metamorphic grade. Table 10.4 gives the compositions of these key minerals. The minerals in mafic rocks have high-temperature (igneous) origins. Consequently, during low-grade metamorphism, new minerals often form by retrograde reactions, which is not generally the case for pelitic rocks. One complication with mafic rocks is that, in contrast with minerals in pelitic rocks, some of the minerals in mafic rocks have quite varying compositions—and thus variable stability ranges. So, the sizes of the red boxes and their order in Figure 10.27 are approximate.

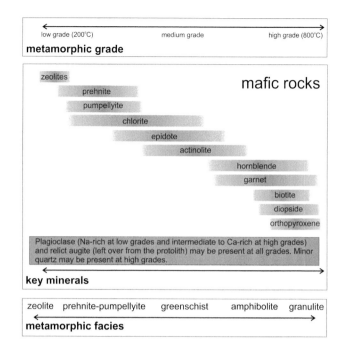

Figure 10.27   Key minerals in metamorphosed mafic rocks.

Table 10.4   Common index minerals in metamorphosed mafic rocks.

Grade	Mineral	Formula
low	zeolites	variable Ca-Na aluminous silicates
	prehnite	$Ca_2Al_2Si_3O_{10}(OH)_2$
	pumpellyite	$Ca_2(Mg,Fe)Al_2Si_5O_{11}(OH)_2 \cdot (H_2O)$
	chlorite	$(Mg,Fe)_5Al(AlSi_3O_{10})(OH)_8$
	epidote	$Ca_2(Fe,Al)Al_2Si_3O_{12}(OH)$
	actinolite	$Ca_2(Fe,Mg)_5Si_8O_{22}(OH)_2$
	hornblende	a complex amphibole
	garnet	$(Fe,Mg,Ca)_3Al_2Si_3O_{12}$
	biotite	$K(Mg,Fe)_3(AlSi_3)O_{10}(OH)_2$
	diopside	$(Ca,Na)(Mg,Fe,Al)(Si,Al)_2O_6$
high	orthopyroxene	$(Mg,Fe)SiO_3$

(temperature — arrow pointing downward at left of table)

At the lowest grades, zeolite minerals, prehnite, and pumpellyite form during diagenesis and the onset of metamorphism of mafic rocks. Often, these first metamorphic minerals fill vugs or fractures, and the overall rock still has the appearance of the protolith. At slightly higher grades, chlorite, epidote, and the green amphibole actinolite form as rocks become greenschists. And, commonly, relict (left over from the protolith) plagioclase and augite persist at low and medium grades. With more metamorphism, garnet and hornblende develop as the lower-grade minerals disappear, and rocks evolve to become amphibolites. The highest grades of metamorphism are characterized by clinopyroxene (diopside of variable composition) and orthopyroxene. Plagioclase is present in rocks of all grades but is generally Na-rich at low grades and more Ca-rich at higher grades. Minor biotite and sometimes quartz may be present, too.

Metamorphosed mafic rocks commonly do not develop the same textures as metamorphosed pelitic rocks, because mafic rocks do not contain large amounts of micas or other platy minerals. Thus, for mafic rocks, the idea of matching rock types including slate, phyllite, and schist with metamorphic grade is inappropriate. But metamorphosed mafic rocks do have distinctive appearances. These distinctions led Penti Eskola, a Finish petrologist, to introduce the idea of *metamorphic facies* in 1920. A metamorphic facies is a general range of PT conditions that produces distinctive-looking rocks, which can be differentiated based on the metamorphic minerals they contain and rock textures. From low grade to high grade, the common metamorphic facies are the *zeolite, prehnite-pumpellyite, greenschist, amphibolite,* and *granulite facies.* Figure 10.27 shows the relationships between facies, metamorphic grade, and the most common minerals that form in metamorphosed mafic rocks.

Eskola, and later Francis Turner in the 1960s and 1970s, recognized that some metamorphic rocks form in conditions that are exceptional. So, they extended the idea of metamorphic facies to include all possible PT conditions at which metamorphic rocks can form. In this diagram, Figure 10.28, conditions in the upper left portion exist nowhere on Earth, because it is impossible to get to the high pressures deep in Earth without substantial temperature increase. In other parts of the diagram, the names of the different facies are the names of different kinds of mafic rocks that form under different conditions or of key minerals that may exist. The boundaries between facies, however, are wide and only approximate. Note that the aluminosilicate reactions (Fig. 10.26b) have been shown on Figures 10.28 and 10.29 to make it easier to compare the different diagrams; aluminosilicates, however, never exist in mafic rocks. The names of the metamorphic facies are often used when talking about rocks that are not mafic, but it can lead to confusion because the names of the facies describe rocks that only form from mafic protoliths. Nonetheless, no matter a rock's composition, geologists find it convenient to use the same facies names to describe the general range of PT at which a rock formed.

10.7.3.1 Barrovian and Buchan facies series

A *facies series* is a sequence of different facies that a rock experiences while being metamorphosed from low grade to higher grade; Figure 10.29 shows several examples. A Barrovian facies series follows a normal geothermal gradient created during crustal thickening related to mountain building. Typical metamorphism progresses from the zeolite facies to the greenschist, amphibolite, and granulite facies. Other rocks may follow a Buchan facies series (low pressure). Buchan metamorphism occurs in places where Earth's crust is rifting, allowing magmas to move upward while carrying a great deal of heat, or in other places where high upward heat flow

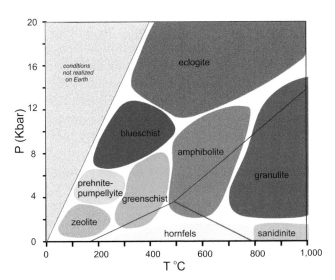

Figure 10.28 The P-T ranges of different metamorphic facies.

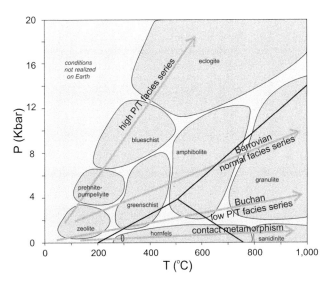

Figure 10.29 Metamorphic facies series.

causes metamorphism at elevated temperature without a great increase in pressure. Barrovian and Buchan are the most common facies series, but as shown in Figure 10.29, some rocks may follow a high pressure facies series or a contact metamorphism facies series. In either case, the minerals that form will be different from those formed by Barrovian or Buchan metamorphism.

Mafic rocks metamorphosed along a Barrovian or Buchan path generally appear similar. Unlike metamorphosed pelites, common mafic metamorphic rocks, excluding very high-pressure blueschists and eclogites, contain no distinctive minerals indicative of higher or lower pressure. Regardless of PT path, metamorphism of mafic rocks begins with the formation of zeolites. Rocks that form within the *zeolite* or *prehnite-pumpellyite facies* generally appear as altered basalt or other mafic rocks that contain distinctive zeolite minerals, or other minerals such as prehnite and pumpellyite. *Greenschist facies* rocks are green and commonly somewhat schistose due to the presence of chlorite, epidote, and actinolite. Amphibolites, which form within the *amphibolite facies*, are distinctive black (hornblende) and white (plagioclase) rocks, generally showing only minor or no foliation, that are dominated by plagioclase and hornblende. And *granulites*, which sometimes look like very high-grade amphibolites, are defined by the presence of orthopyroxene or of garnet with quartz.

10.7.3.2 High-Temperature Facies Series

Rocks of the low-pressure *hornfels* and *sanidinite facies* are generally very hard and fine-grained with no discernible metamorphic fabric. These rocks form when intruding magmas heat crustal rocks that are very near the surface, and thus the rocks undergo *contact metamorphism*. Contact metamorphism that occurs at some depth in Earth grades into regional metamorphism and may produce rocks of the greenschist, amphibolite, or granulite facies. Thus, very low-pressure Buchan metamorphism may be indistinguishable from hornfels or sanidinite facies metamorphism.

Petrologists often divide the hornfels facies into the albite-epidote hornfels facies (low grade), the hornblende hornfels facies (medium grade), and the pyroxene hornfels facies (high grade). Thus, in a field area that contains regional variations in metamorphic grade, zones of the different facies can be mapped based on epidote, hornblende, and pyroxene occurrences in different places. Rocks metamorphosed under sanidinite facies conditions are quite rare and are most commonly found as xenoliths in plutonic rock bodies. Sanidinite facies rocks, characterized by the absence of micas, amphiboles, and other hydrous minerals, often contain unusual minerals. Pelites may contain sanidine (a high-temperature form of K-feldspar), mullite (a rare aluminosilicate), and tridymite (a high-temperature polymorph of quartz), as well as some more common minerals including cordierite and sillimanite.

10.7.3.3 High-Pressure Facies Series

High pressure (labeled high P/T in Fig. 10.29) metamorphism is associated with subduction zones where descending cold and wet lithosphere cools the mantle below (Fig. 10.4). Rock is a good insulator, so temperatures in a subducting slab increase slowly as the slab descends. Consequently, temperatures are not as great as they would be otherwise and metamorphism follows a high P/T path. So, in these environments, metamorphism begins with the zeolite and prehnite-pumpellyite facies and subsequently might continue to the blueschist and eclogite facies (Fig. 10.28). Rocks that form at blueschist and eclogite conditions are relatively rare at Earth's surface.

The kind of rock that forms under blueschist or eclogite facies conditions depends on the protolith. Pelitic rocks metamorphosed at high pressure contain kyanite and garnet, often with chloritoid, zoisite, an Na-rich amphibole, and phengite (a mica similar to muscovite but containing magnesium). Granites may contain the same minerals that pelites contain but in different proportions. Marbles may contain aragonite, a high-pressure polymorph of calcite, although aragonite tends to turn into calcite when high-pressure rocks are brought to the surface. The rock type called *blueschist* forms when basalt and other mafic compositions reach blueschist facies conditions. Blueschists contain *glaucophane*, a distinctive inky blue amphibole (but despite their name do not display schistosity). Figure 10.30a shows a blueschist, partially changed to a green eclogite, from an outcrop on the beach just north of Jenner, California, that contains blue glaucophane, green pyroxene, and red garnet. Glaucophane, an Na-rich amphibole, forms in blueschists because at high pressure, the Na-component of plagioclase reacts to make amphibole. Similarly, at high pressure, the Ca-component of plagioclase is unstable and reacts to form a pyroxene. So, high-pressure rocks commonly contain glaucophane and pyroxenes that are rich in Na and Al and other nonstandard components. The pyroxenes, such as the green ones seen in Figure 10.30a, are called *omphacite*. Many

a. Blueschist from near Cambria, California

b. Eclogite from Almenning, Norway

Figure 10.30 Blueschist from Jenner, California, and eclogite from Almenning, Norway.

Blueschist photo from Jamie Schod; eclogite photo is from Kevin Walsh, Wikimedia Commons.

blueschists also contain lawsonite, a high-pressure mineral with composition equivalent to hydrated anorthite, and epidote.

The photo in Figure 10.30b shows a hand sample of *eclogite* from Norway. Eclogites, which are rocks that contain conspicuous red garnet and a green pyroxene called *omphacite*, have a red-green Christmas-tree appearance. These rocks form when basalt and other mafic rocks reach eclogite facies conditions. The eclogite samples we find in outcrops today originated at great depth within Earth, and the minerals within them are unstable at Earth's surface. Consequently, many eclogites show signs of retrograde metamorphism and have converted partially to become blueschists. The

blueschist seen in Figure 10.30a may have formed this way—from rock that was, at one time, entirely eclogite.

10.7.4 Mapping regional metamorphic terranes

The temperature and pressure at which rocks equilibrated may vary across a geological terrane. So, when petrologists study regional metamorphism, a fundamental undertaking is to map *metamorphic zones*. These zones are regions, small to large, characterized by specific index minerals. Figure 10.31 is a reproduction of one of the first maps of metamorphic zones, created by George Barrow in 1918. The dot on the inset map shows where in Scotland Barrow made this map. In this region, called the Dalradian Region, unmetamorphosed rocks to the south and east of the Highland Boundary Fault (which follows the southern edge of the chlorite zone on the map) give way to zones containing chlorite, biotite, garnet, staurolite, kyanite, and sillimanite to the northwest. These are all pelitic index minerals, and they demonstrate that metamorphic grade increases to the north and west (red arrow).

To make a map such as the one shown in Figure 10.31, geologists do fieldwork to find out where the key index minerals occur. They then can draw lines on their map, called *isograds*, that separate areas with different minerals—and thus separate different metamorphic zones. Metamorphic grade is about the same throughout any zone, and consequently mineral assemblages in rocks of comparable composition are the same. Isograds, which separate zones of different metamorphic grade, may reflect regional variations in conditions at the time that metamorphism occurred or different amounts of uplift and erosion that

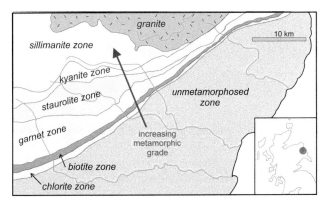

Figure 10.31 Barrovian isograds mapped by George Barrow in the early 1900s.

Modified from a map produced by David Waters.

uncovered metamorphic rocks. Maps such as the one shown in Figure 10.31 are sometimes interpretations based on limited evidence, because finding outcrops containing key minerals may be challenging. The presence or absence of index minerals depends on rock composition, not just on metamorphic grade, and, additionally, many metamorphic terranes have few outcrops to examine.

10.8   Contact metamorphism

Contact metamorphism occurs when intruding magmas or lavas metamorphose shallow crustal rocks at high temperatures but at low pressures. This kind of metamorphism is distinct from regional metamorphism because of the large temperature differences between a magma's temperature and the temperature of the surrounding rock, called the *country rock*. Magmas undoubtedly are heat sources at deep levels, and thus at high pressures, within Earth, where they contribute to regional metamorphism. But rocks undergoing regional metamorphism are generally already at high temperature before intrusions occur, so the effects of magmatic heat are difficult to discern. In contrast, at shallow depths (low pressure), the effects of magmatic heat may be profound.

The entire volume of country rock around an intrusion that is affected by contact metamorphism is called a *contact aureole*. At Willsboro, for example, the aureole consists of three different zones that, together, are about 50 meters thick. Contact aureoles range from being quite small, perhaps centimeters thick, to being many kilometers wide. Part of this variation is due to the size of the intrusion that provides the heat that causes the metamorphism; small intrusions often create small aureoles, and large intrusions may create larger aureoles. But some variation can occur because contact metamorphism often involves hydrothermal metamorphism caused by fluids flowing from the intrusion into the country rock and because the fluids can travel long distances.

Typical contact aureoles contain zones with different index minerals or rock types that wrap around an intrusion or are parallel to the contact between the intrusion and the country rock. Metamorphic pressure is generally not an important factor during contact metamorphism, and the products of the metamorphism depend mostly on the composition of the protolith and on temperature. Basalt, shale, or other fine-grained rocks may become hornfels, sandstones may become quartzites,

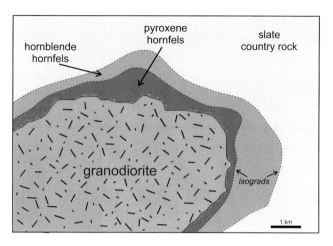

Figure 10.32   Approximate map of metamorphic zones and isograds around the Onawa Pluton, central Maine.

and limestones or dolostones may become marbles, or perhaps skarns.

Figure 10.32, a map of a contact aureole around the Onawa Pluton, a granodiorite body in central Maine, shows two distinct metamorphic zones between the granodiorite intrusion and the surrounding slate country rock. Metamorphic temperatures are always hottest near an intrusion and coolest away from an intrusion, so contrasts in metamorphic grade and mineral assemblages are typical, as shown in this map. Note that the contact aureole around the Onawa Pluton is much larger than the one at Willsboro (shown in Fig. 10.3)—kilometers instead of meters thick. Around the granodiorite pluton, metamorphism produced hornfels. The inner (high-temperature) zone of hornfels contains pyroxene, and the outer (low-temperature) zone of hornfels contains hornblende. The slate country rock, even farther from the pluton, is a product of earlier metamorphism that took place before the Onawa Pluton formed.

10.8.1   Skarns

Some of the most spectacular examples of contact metamorphism develop when magmas intrude carbonate rocks (limestone or dolostone), creating a *skarn*, such as the one at Willsboro. Skarns contain contact aureoles with different minerals in different zones parallel to the contact between the intrusion and the country rock. Calcite (calcium carbonate) and dolomite (calcium magnesium carbonate) dominate limestone and dolostone, but some quartz and clay may be present. So, during metamorphism, reactions can occur that

Figure 10.33 Calcite with monticellite and graphite (top) and marble with garnet (bottom) from Crestmore, California.

Top image modified from Tom Loomis, Dakota Matrix Minerals; bottom photo from Lech Darski, Wikimedia Commons.

produce many different high-temperature silicate minerals, including all those listed in Table 10.1. These minerals have significantly different properties and some are quite colorful, so contact metamorphosed carbonates develop distinctive zones, typically centimeters to tens of meters thick. Skarn minerals often form as large euhedral crystals prized as museum specimens. Figure 10.33 shows two examples. The top photo contains blue calcite, with minor monticellite and graphite near the specimen's bottom. The bottom photo contains striking red garnet crystals up to 2 cm across.

Figure 10.34 shows a vertical cross section of a contact aureole in Crestmore, near Riverside, California. Limestone quarrying for cement manufacturing has uncovered the intrusion and the well-developed skarn that surrounds it. At Crestmore, a granitic magma intruded limestone, creating a contact aureole with four main zones that contain some of the minerals listed in Table 10.1. The thickest zone shown is about 15 meters across at its widest point. The outermost zone of the aureole is a calcite-rich marble that, in places, contains brucite. The calcite has a spectacular blue color that can be seen in the sample of Crestmore calcite in top photo of Figure 10.33. The next zone inward is characterized by monticellite, a Ca-Mg silicate, and a few other relatively rare metamorphic minerals. Still closer to the intrusion, vesuvianite, a Ca-Mg-Fe silicate, appears. Closest to the intrusion, the marble contains diopside,

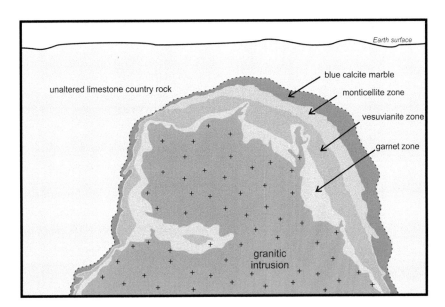

Figure 10.34 Schematic vertical cross section of the contact aureole at Crestmore, California. The drawing depicts an area of about 75 meters across.

Based on a drawing by Burnham (1959).

Ca-Mg silicate, wollastonite, Ca silicate, and garnet, Ca-Al silicate. The bottom photo in Figure 10.33 shows an example of the garnet-bearing rock. The composition of the rock in the four zones at Crestmore is not the same, because significant metasomatism occurred during metamorphism. Thus, the different minerals are found in different zones because of variations in both temperature and rock composition.

10.9   Other types of metamorphism

10.9.1   Cataclastic metamorphism

Cataclastic metamorphism is caused by mechanical rock deformation, for example when rock bodies slide past each other in a fault zone or shear zone. In such cases, grinding of rocks causes them to become crushed and pulverized and may produce fine material called *rock flour* or earthy material called *fault gouge*. Figure 10.35 shows fault gouge, the reddish material behind the backpack, where a fault cuts through the gray rocks to the right and left. In exceptional circumstances, *mylonites* (such as the one in Fig. 10.5), metamorphic rocks created by recrystallization of crushed material, may form, but they generally involve no new metamorphic minerals. The mylonite shown in Figure 10.5, which contains metamorphic garnet, is an exception. Sometimes cataclastic metamorphism produces jumbled breccias called *cataclasites*, which contain angular rock fragments in a rock flour matrix, or *augen gneisses* (Fig. 10.13), in which feldspar or other crystals are stretched in one direction.

Figure 10.35   Example of fault gouge at Tavan Har, the Gobi, Mongolia.
Photo from Qfl247, Wikimedia Commons.

Figure 10.36   Shatter cones at Cap-aux-Oies, 100 km northeast of Quebec City, Canada.
Photo from Jmgas, Wikimedia Commons.

10.9.2   Shock metamorphism

Shock metamorphism, sometimes called *impact metamorphism*, occurs when a meteorite or other extraterrestrial body hits Earth. The impact generates great heat and pressure that can cause mineralogical and textural changes to bedrock. One typical feature is the formation of layers of glassy material, called *planar deformation features*, within mineral grains. If these features are in quartz grains, we call the grains *shocked quartz*. Shock metamorphism can also lead to formation of coesite, stishovite, and other high-pressure minerals that normally would never form at Earth's surface. Additionally, the impact may form *shatter cones* in country rock—curved fracture surfaces similar to what develops when a flying rock hits a car windshield. Figure 10.36 shows shatter cones associated with a meteorite impact site next to the St. Lawrence River in Quebec.

Questions for thought—chapter 10

1. The agents of metamorphism are pressure, temperature, and chemically reactive fluids. What geological processes may cause rocks to be subjected to high pressure? What (two) geological processes may cause rocks to be subjected to high temperatures?

2. Is a mineral that forms at high metamorphic pressures more or less likely to be denser than a polymorph that forms at lower pressures?

3. What kinds of fluids – what are the two major things they might be made of – can cause meta-

morphism? What is metasomatism, and how do fluids cause metasomatism?

4. Besides changing the minerals in a rock, what other changes may result from metamorphism? List and describe some examples.

5. What is the difference between contact and regional metamorphism? Which one affects the largest areas. Why?

6. What causes regional metamorphism and what causes contact metamorphism?

7. Contact metamorphism may create skarns. What is a skarn? And what are some differences between a skarn and a typical marble?

8. Some parts of the world have higher than average geotherms and other places have lower than average geotherms. What is a geotherm? What would cause a geotherm to be higher than normal in a particular place? And what might cause a geotherm to be lower than normal in some other place.

9. In what settings do high pressure, but low temperature metamorphic rocks form? What do we call such rocks?

10. How can high temperature, but very low pressure, metamorphic rocks form? What do we call this kind of metamorphism.

11. Metamorphism occurs over a wide range of temperature but it is difficult to specify the range precisely. Why are there no sharp boundaries for the temperature range of metamorphism?

12. What geological process does metamorphism grade into at low temperature? And, what geological process does metamorphism grade into at very high temperature?

13. What is the difference between lithostatic and differential pressure? And, why is lithostatic pressure generally maintained for much longer times than differential pressure?

14. Some metamorphic rocks contain exceptionally large grains of metamorphic minerals? What are these large mineral grains called? Why are smaller grains generally more reactive than larger grains of the same mineral during metamorphism?

15. Some rocks develop *foliation* or *lineaton* during metamorphism. What do these two terms mean, and how do such metamorphic fabrics develop?

16. Why are phyllites shinier than shales and slates? What happens to metamorphic grains in a phyllite when it metamorphoses into a schist?

17. How does gneissic banding form? What are some of the typical major minerals in both the light and dark layers?

18. Suppose a rock is metamorphosed and becomes a very high-grade gneiss. Then it is heated significantly more. What is likely to happen and what will be produced? What kind of rock is likely to form after it cools?

19. Different protoliths produce different kinds of metamorphic rocks. Why is this? What is the protolith of quartzite? What happens to individual quartz grains during metamorphism that produces a quartzite? What is the protolith for most schists? What is the protolith for most marbles? What is the protolith for most amphibolites?

20. What happens when decarbonation and dehydration occur during metamorphism? What causes them to occur?

21. What is the difference between prograde and retrograde metamorphism? Why are mafic rocks, in contrast with felsic rocks, more likely to be affected by retrograde metamorphism? Name a rock that typically often forms from retrograde metamorphism.

22. Why are clay minerals more affected by heat during metamorphism than feldspars?

23. Why does olivine commonly convert to chlorite, serpentine, or brucite during metamorphism?

24. *Metamorphic grade*, and *metamorphic facies* are two ways to describe the intensity of metamorphism. Contrast and compare the two. What causes an increase in metamorphic grade.

25. What metamorphic minerals present in pelites are usually absent in mafic metamorphic rocks? What about marbles? What minerals are typically present in marbles (name 3) that are absent from other kinds of metamorphic rocks? Why do different minerals occur in different kinds of metamorphic rocks?

What are the major types of foliated metamorphic rocks?

PART III
Surficial Geology and Resources

11 Weathering and Soils

11.1 Mesopotamia

The region between the Tigris and Euphrates Rivers, in modern-day Iraq, once contained the most productive farmland in the world. This area, called *Mesopotamia* ("between two rivers"), is part of the *Fertile Crescent* (in red in Figure 11.1), an arcuate-shaped region that extends from the Persian Gulf north and west to the Mediterranean Sea before curving south and following the Nile River into central Egypt. The first irrigated

Figure 11.1   The Fertile Crescent is the region where modern agriculture began, around 6000 BCE.

Map from Nafsadh, Wikipedia Commons.

societies developed in the Fertile Crescent beginning about 6000 BCE. In Mesopotamia they included, in succession, the Sumerian, Akkadian, Babylonian, and Assyrian empires.

The Tigris and Euphrates Rivers—labeled on Figure 11.1—were key to Mesopotamia's agricultural success. Their headwaters are in the Armenian Highland of Turkey. There, for millennia, copious rainfall caused weathering and erosion of the diverse volcanic and sedimentary bedrock and soil, and the rivers transported eroded materials downstream to deposit them in flat floodplains. The resulting soil was a rich loam containing silt, other fine material, and many nutrients that supported lush grasslands during wet months, but dried and became unproductive during other times of the year.

Early Sumerians overcame the seasonal problems and, by 6000 BCE, had constructed canals and dams to ensure stable water supplies year round. Irrigation soon converted the land to lush green fields, orchards, and pastures. Wheat, barley, and fruit came from the ground, fish from the rivers, and cattle, goats, pigs, and sheep grazed the land. Clay deposits, associated with the soils, provided resources for pottery and ceramics, for buildings and cities, and for tablets used to record the first written records of history. Although the region lacked wood and other key commodities, the agricultural wealth soon led to trade with neighboring people and the growth of a large agrarian society.

Figure 11.2 shows the ruins at Babylon, 80 kilometers (50 miles) south of Baghdad, Iraq. It was the largest city in the world about 1700 BCE, and again between 600 and 300 BCE, but the irrigated society was unsustainable.

A problem with irrigation is that it is a two-edged sword. The water that delivered so much productivity to Babylon and other Mesopotamian societies also delivered carbonates, salts, and other things detrimental to

Figure 11.2   The remains of Babylon, 50 miles south of Baghdad, Iraq, seen from Saddam Hussein's summer palace.
U.S. Navy photo, Wikimedia Commons.

Figure 11.3   Irrigated fields in California's Central Valley.
Photo from the USGS.

agriculture. Before irrigation, natural flooding removed the unwanted chemicals and replenished soil but, once human engineering began, the natural systems were thrown out of balance. Pollutants concentrated in the soil, and infiltrating water carried salt to the water table. The water table rose, eventually leading to waterlogged soil and salt concentrations too great to support plant growth. Good drainage can be a key to help sustain irrigation but, in Mesopotamia, fields were flat and right at river level, so drainage was minimal.

Mesopotamia had times of successful irrigation and times of crises, but by the end of the neo-Babylonian Empire in the 6th century BCE, productivity had declined so much that many of the remaining people abandoned the once fertile crescent and emigrated to Persia. Mesopotamia was the first irrigated society to collapse and, since then, few economies based on irrigating dry lands have lasted for any great length of time.

The same kind of story might be unfolding in California's irrigation districts, such as the area shown in the Central Valley in Figure 11.3. Decades of irrigation have led to prosperity, but the prosperous times may be ending. California's growing population, an agricultural economy that is now in third or fourth place behind technology and manufacturing, and competition for Colorado River water all mean that water for irrigation is becoming increasingly scarce. Additionally, many years of poor irrigation practices have led to significant soil erosion and loss of soil quality. So, currently, an aging agricultural industry is finding it tougher and tougher to maintain profitability. Today (January 2017), a 3-year drought that, at times, seemed catastrophic, may be ending. Yet, California's farmers

are not particularly optimistic, and many worry about when, not if, the next drought will occur, and how bad it will be. And, they worry about continued degradation of soil resources.

11.2 Weathering

11.2.1 Definition of weathering

Over time, geological materials exposed at Earth's surface will break down. Solid material will break into pieces, large pieces will break into smaller pieces, and chemical reactions will occur that change the nature of materials that are present. This process is *weathering*—it was introduced in the previous chapter. Weathering is not limited to natural materials; it also affects human artifacts made from natural materials, including buildings, monuments, and gravestones. Although weathering is different from *erosion*, because erosion involves transportation and relocation of material after it is weathered, the processes are closely linked. Often, weathering produces material transported only a short distance from where it was created, but the limits of weathering and the beginning of erosion are poorly defined. Nonetheless, these two forces—weathering and erosion—are the major forces responsible for Earth's many different landscapes.

Weathering occurs because most rocks and minerals are not formed at normal Earth surface conditions. Instead, most form at elevated pressure or temperature, often buried within Earth. Under such conditions, they develop compositions, textures, and physical properties that are no longer stable once exposed to the environment above ground. Environmental agents of weathering include water and ice, heat from the sun, wind, plant and animal activity, freeze-thaw cycles, abrasion by particles (carried by wind, water, or ice), and chemicals including acids dissolved in water or created by plants.

As discussed in Chapter 10, geologists divide weathering into three types that often work in concert (Table 11.1). *Physical weathering*, also called *mechanical weathering*, produces rock and mineral *clasts* (pieces of rock or individual mineral grains of any size) that are smaller than the original material, but of the same general composition. Figure 11.4 shows a boulder field near the Lang Craigs in Bellsmyre, Scotland. The pieces of rock formed after fracturing weakened nearby cliffs. This is an example of physical weathering.

Chemical weathering changes rock and mineral compositions and produces weathering products with compositions different from the original material. During chemical weathering, many minerals alter to become compounds that are more stable at Earth surface conditions, and the most common such product is clay. Other minerals and rocks simply dissolve.

Figure 11.5 shows a highly weathered Madonna statue in the St. Louis Cemetery, New Orleans. It has undergone intense chemical weathering as the limestone it consists of reacted with acidic rainwater. Instead of turning into new, different, material, the limestone is simply dissolving and washing away.

Biological weathering is not a distinct kind of weathering—it results from combinations of physical and chemical processes—but is a name used for weathering attributed to living organisms. All three kinds of weathering work together, and sorting the effects of one from the others is often difficult.

Because weathering affects materials (termed *parent materials*) of all sorts, and because weathering processes are highly variable, weathering products are diverse and include a wide range of mineral and, sometimes, organic matter. The most obvious product of weathering is usually a heterogeneous layer of broken and decomposed material, termed *regolith*, that overlies solid bedrock.

Table 11.1   The Three Kinds of Weathering

- *Physical weathering*: weathering that produces smaller pieces of the original material.
- *Chemical weathering*: weathering that changes the composition of rock or mineral matter, often involving dissolution (dissolving) of some components.
- *Biological weathering*: weathering caused by plants and animals; essentially, special cases of physical and chemical weathering.

Figure 11.4   A boulder field produced by physical weathering near the Lang Craigs in Bellsmyre, Scotland.

Photo from Lairich Rig, Wikimedia Commons.

Figure 11.5   Madonna statue in the St. Louis Cemetery, New Orleans.

Photo from Infrogmation, Wikimedia Commons.

Figure 11.6 shows several meters of regolith overlain by soil and vegetation near Rio de Janeiro, Brazil. The underlying bedrock—the parent material—is not visible, but is a quartz- and feldspar-rich gneiss. The regolith seen here is relatively fine-grained at its top but coarser below where it is nearer to the gneiss. The coarser material contains conspicuous light-colored feldspar crystals that weathered out of the original parent rock. Rocks in tropical areas, such as Brazil, where rainfall is abundant and temperatures are high, weather much faster than rocks in more temperate climates. Very thick layers of regolith, sometimes up to 100 meters thick, may collect on top of bedrock. In cooler and dryer regions, weathering rates are slow, and the thickness of regolith is less—sometimes there is no regolith at all and bedrock is exposed at the surface.

Weathering and the formation of regolith are not restricted to Earth, but occur on the moon, asteroids, and other planetary bodies. For example, Figure 11.7 shows regolith on Mars. In 2012, the rover *Curiosity*, shown in this photo, sampled weathered material on Mars and found it similar to some regolith on Earth. However, unlike on Earth where there is abundant water and biological activity, weathering on Mars is mostly due to wind and sandblasting and, to a lesser extent, acidic fog. Consequently, the regolith on Mars is mostly very fine grained. Sand dunes are common on Mars, and blowing winds create sand and dust storms that cause most of the weathering that shapes the planet's surface.

Besides creating regolith, weathering produces dissolved material (mostly ions associated with soluble salts or colloidal substances) that (on Earth) is carried away by water. Some dissolved material eventually makes its way to an ocean, but the rest does not. As discussed in Chapter 8, solid regolith or dissolved materials are transported by water, wind, or gravity and deposited elsewhere, the deposits are called *sediment. Clastic*

Figure 11.6   Regolith and soil near Rio de Janeiro, Brazil.
Photo from Eurico Zimbres, Wikimedia Commons.

Figure 11.7   Regolith on Mars.
Photo from the Jet Propulsion Laboratory, NASA.

sediment is material such as pebbles, sand, or silt—solid fragments created by weathering that may evolve to become clastic sedimentary rocks such as sandstone or shale. *Chemical sediment* is formed by precipitation of dissolved components in water and may, after precipitation, become *chemical sedimentary rocks*, such as limestone, rock salt, tufa, or gypsum.

In some flat-lying regions, regolith may remain in place, forming thin to thick accumulations that have essentially the same composition as the parent rocks. In other places, *sorting* may occur when some weathering products are carried away by wind, water, gravity, and other forces. So, some material becomes *transported material*. We call the remainder the *residual material*; it typically does not have the same composition as its parent. In areas with steep topography or cliffs, erosion may remove all weathering products, leaving no residual material and only bedrock or other parent material exposed for more weathering.

Regolith is important to people. Most important, it generally includes upper layers of *soil*, which may be residual or transported, that contain living matter and provide a medium for most plant growth. Residual concentrations in regolith may also concentrate valuable commodities, including clay, gravel, diamonds, iron, manganese, nickel, and aluminum.

11.2.2 Physical weathering

Physical weathering, which involves rocks breaking apart without mineral or compositional change, is most often due to *fracturing* (formation of cracks) or *abrasion* (sandblasting or other eroding by particles transported by wind, water, ice, or gravity). Fracturing produces angular clasts of all sizes from originally hard massive material; abrasion produces mostly smaller and more rounded material with smooth surfaces. Besides water, ice, and gravity, heat, pressure, direct wave or wind action, plants and animals, and less commonly, fires may also cause physical weathering.

11.2.2.1 Ice and salt wedging

Ice, especially when involved in freezing and thawing cycles, is one of the most common causes of physical weathering. When water freezes, its volume increases by 10%. So, when water gets into cracks or holes and then freezes, it causes the crack or hole to enlarge. When ice melting occurs, the water runs out, taking debris with it. Thus, successive freeze-thaw cycles enlarge openings

and the process continues, eventually leading to major fracturing. Even rock that seems very strong may succumb to this relentless process, called *ice wedging* or *frost wedging*, over time. Figure 11.8 shows a *blockfield*, also called a *felsenmeer*, in easternmost Washington. This rubble was created by frost weathering during multiple freeze-thaw cycles.

The effects of ice wedging are especially common in alpine areas where temperatures may be near freezing much of the time. In such places, thick accumulations of *scree* (loose, unsorted debris) or *talus* (piles of moderate to very large rock fragments) accumulate on slopes or beneath cliffs, when gravity aids ice wedging. Figure 11.9 shows a hiker on a talus slope in eastern California. Blockfields, like the one shown

Figure 11.8 Ice-shattered blockfield on Mt. Spokane, Washington. Photo from Eric Brevik.

Figure 11.9 Talus on a slope in the Sierra Nevada Mountains. Photo from D. Perkins.

in Figure 11.8, are different from scree or talus slopes, because the formation of blockfields does not involve significant downslope movement of material under the influence of gravity.

Salt wedging, most common in arid environments, is similar to ice wedging, but results from crystallization of salt as water evaporates. Typical salt minerals include halite, calcite, and gypsum. All of them can precipitate from water on or near rock surfaces or in cracks within rocks. Formation of some clays, called *expanding clays*, may also cause wedging and fracturing like salt minerals do.

11.2.2.2   Thermal expansion and contraction

As rocks warm, they expand; when they cool, they contract. Expansion and contraction cause some rocks to develop *joints*, planes of weakness that may or may not ultimately fracture. Forest fires, for example, may cause jointing and fracturing, because fires heat rocks quickly. Old-time miners used fire to shatter rock as an aid in their excavations. Magma cooling also causes some rocks to joint and fracture. Basalt, for example, commonly develops *columnar joints* when it shrinks during solidification. Figure 11.10 shows columnar joints in a basalt flow at Devils Postpile, near Mammoth Lakes, California. The joints formed when the flow cooled and contracted.

Joints are especially common for rocks that experience cyclical cooling and heating (daily or seasonally), because even the strongest rocks cannot stand up to repeated deformation. This effect can be especially significant for dark-colored rocks in desert regions, because rock surface temperatures may vary by up to 50 °F between day and night. Exposed rocks in mountain terranes also experience major fluctuations in temperature, often making them susceptible to jointing, fracture, and physical weathering. A common result is the formation of large talus fans, or aprons, and talus chutes on steep slopes near cliffs. Figure 11.10 shows talus beneath a basalt outcrop and Figure 11.11 shows a well-developed steep talus slope beneath vertical cliffs in eastern Washington's Channeled Scablands. The steep-sided valleys of the Scablands formed during the last ice age, 10,000–20,000 years ago, when ice dams broke and released catastrophic flood waters that inundated parts of eastern Washington. Topsoil was carried away and the flood waters cut deeply into basalt bedrock. Subsequently, talus slopes developed on valley sides.

Continuous heating and cooling cause some rocks to form internal joints that are parallel to the rock's surface. The outer layers of rock may then peel off in sheets, in a process called *exfoliation*, producing huge slabs like the ones shown in Figure 11.12, on the side of Enchanted Rock near Llano, Texas. On a smaller scale, in desert regions, expansion and contraction lead to *granular decomposition*, especially for coarse-grained plutonic rocks. When this process occurs, constituent mineral grains fall off outcrops but maintain their original grain shape.

Figure 11.10   Columnar joints in basalt at Devils Postpile, near Mammoth Lakes, California.

Photo from the National Park Service.

Figure 11.11   Talus in fans and chutes on a cliff of the Channeled Scablands of Washington.

Photo from Eric Brevik.

48

Figure 11.12 Exfoliation slabs of granite at Enchanted Rock, Texas.

Photo from Wing-Chi Poon, Wikimedia Commons.

Figure 11.13 Exfoliation slabs in Yosemite National Park, California.

Photo from S. Rae, Wikimedia Commons.

11.2.2.3 *Unloading*

Figure 11.13 shows exfoliation slabs on a rock dome near Tanaya Lake in Yosemite National Park, California. Another example of exfoliation in Yosemite is seen in Figure 6.4. When erosion or other processes remove overburden, it causes a decrease in pressure on bedrock and the rock may expand, leading to fractures such as those shown. This process, *unloading*, can occur because of erosion, melting of glacial ice, or other reasons. The fractures that form are typically parallel to the surface and may cause exfoliation, similar to exfoliation caused by thermal expansion and cooling. The two processes, unloading and expansion/contraction, often work together, and water entering fractures in the rock may also contribute to the exfoliation process. Many granite outcrops in Yosemite National Park (like the ones in Figs. 6.4 and 11.3) exhibit exfoliation, producing slabs up to several meters thick.

11.2.2.4 *Abrasion by wind*

Abrasion occurs when material is scraped off and removed by moving particles transported by wind, water, or ice. Abrasion by wind involves mineral particles acting as sandblasting agents. In deserts or semidesert areas, this sandblasting is often very significant. For example, wind abrasion may erode and smooth outcrops or round the corners of blocky rocks. It also may remove material from flat surfaces. Abrasion and removal of material by wind may cause *deflation*, a general lowering of the land surface due to erosion. If localized, deflation may create a depression called a *blowout*, such as the one shown in

the top left photo of Figure 11.14. Blowouts may be small, on the scale of meters or tens of meters, but some are up to kilometers in long dimension.

Sand carried by desert winds can shape and polish desert rocks, removing corners and adding pits, grooves, and other surface features. One common result is the creation of wind polished rocks called *ventifacts*, or *wind-faceted stones*, often displaying polished and reflective surfaces. The top right photo in Figure 11.14 shows one example from the Mojave Desert, California. Small ventifacts can also be seen in the lower right of the bottom photo of Figure 11.14.

In arid and semiarid regions, wind commonly removes medium- and fine-grained sediment while leaving coarser, heavier material behind. This creates *desert pavement* made of relatively coarse material. The pavement acts as a shield and prevents erosion from going any deeper. The bottom photo in Figure 11.14 shows an example of desert pavement in Death Valley, California, that formed because wind carried away all material smaller than pebbles. Desert pavements are more common than sand dunes in most of Earth's arid lands.

The term *ventifact* is not restricted to small individual rocks. It is also used to describe any bedrock outcrop shaped by wind abrasion. Large ventifacts, like the one shown in Figure 11.15, often have bizarre shapes and sometimes appear as toadstools sitting in a sea of sand.

When deflation accompanies wind abrasion, ventifacts may evolve to become linear outcrops or ridges, such as those shown in Figure 11.16. These features, called *yardangs*, can be large or small. The ones in Figure 11.16

Figure 11.15 Ventifact in the Altiplano region of Bolivia.
Photo from El Guanche, Wikimedia Commons.

Figure 11.14 Blowout in the Sand Hills, Nebraska (top left). Ventifact in the Mojave Desert (top right). Desert pavement in Death Valley National Park, California (bottom).

Photos from Scott Lundstrom, USGS (top left), Mark A. Wilson (top right), and Eric Brevik (bottom).

Figure 11.16 Yardangs in the Gobi Desert.
Photo from George Steinmetz/Corbis, Smithsonian Institute.

are exceptionally large; note the vehicle and tire tracks for scale. Yardangs may exist in groups like those shown in this photograph, or they may be individual rock outcrops surrounded by desert sediment.

11.2.2.5 Abrasion at shorelines

Abrasion by water occurs most commonly when powerful breaking waves crash into shorelines or cliffs, removing material and producing cracks. Waves may compress air in the cracks, adding additional stress to the rock and causing it to fracture. Continuous action may lead to undercutting and eventual collapse of overhanging cliffs, forming a *wave-cut notch* like the one shown in Figure 11.17.

Abrasion at shorelines may also create *wave-cut platforms* like the one shown in Figure 11.18. These platforms are nearly flat rock outcrops extending from steep cliffs, often displaying wave-cut notches, to the ocean during low tides. Seawater covers the platforms

Figure 11.17 Wave-cut notch, Wessex Coast, England.
Photo from Jim Champion, Wikimedia Commons.

Figure 11.18   Wave-cut platform.

Photo from Karen8543, Wikimedia Commons.

Figure 11.19   Outcrop showing glacial polish and glacial striations.

Photo from Amezcackle, Wikimedia Commons.

during high tides and, as waves cut notches into the
cliffs, platforms grow in size. If the rock above the notch
falls and is eroded away, the notch may move farther
inland, and the platform will become even larger. The
formation of notches or platforms is limited, depending
on the nature of the bedrock and of the shoreline. Very
solid, hard rock may not abrade or erode sufficiently for
either feature to form. Additionally, platforms cannot
develop if sandy beaches protect the shore from erosion.
Sea levels change over geological time, and if sea level
drops, platforms may become marine terraces perched
above the ocean. If sea level rises, platforms will flood
and eventually marine sediments will cover them.

11.2.2.6   Abrasion by Ice

Figure 11.19 shows an outcrop modified by glacial ice.
Although much of the weathering associated with glaciers is due to freeze-thaw cycles that break off angular
rock fragments, abrasion is also a significant process.
Ice by itself is not generally hard enough to abrade
bedrock, but ice at the bottom and sides of glaciers
incorporates bits and pieces of rock. Consequently,
moving ice acts like scouring powder or sandpaper,
loosening and removing fine rock material called *rock
flour*. During this process, if the bedrock being abraded
is hard, perhaps granite or gneiss, it may become shiny
and obtain a *glacial polish*, as is shown in Figure 11.19.
Sometimes angular rock fragments in advancing ice
scour grooves called *glacial striations* that record the
direction of ice flow—such grooves are also visible in
this photograph.

11.2.3   Chemical weathering

When chemical weathering occurs, the weathering
products include leftover *primary minerals* (minerals
that were present when the rock first formed), *secondary
minerals* created by chemical reaction during the weathering process, and chemicals dissolved in water. The
leftover minerals and the secondary minerals together
make up the *residual minerals*. Residual minerals are
sometimes called the *resistate*, because they resist weathering and remain as solid material after weathering.
Chemical weathering is a gradual process that often
proceeds stepwise. Other things being equal, it occurs
more rapidly in warm environments than in cold environments because warm temperatures speed up reaction
rates. Additionally, the two major kinds of weathering,
physical and chemical, work together. The more a material is fractured or broken up, the more surface area is
available for chemical reaction. So, physical weathering can enhance chemical weathering rates. The opposite is also somewhat true; chemical weathering can
weaken a rock and make it more susceptible to physical
weathering.

During chemical weathering, primary minerals may
react to produce a series of different secondary minerals, changing both the mineralogy and composition
of a rock. The key chemical processes involved include
hydrolysis, *hydration*, *dissolution*, and *oxidation*, and in
different settings, different processes may dominate.
Hydrolysis, hydration, and dissolution involve
water. Sometimes oxidation does, too. So, water is

crucial for chemical weathering. Additionally, chemical weathering is fastest in wet environments and slowest in dry ones.

11.2.3.1 Spheroidal weathering

Chemical weathering occurs at mineral and rock surfaces where solid rock meets the atmosphere or hydrosphere. It progresses from the outside in and is most rapid for small grains, because they have a high surface area to volume ratio. For larger grains and solid materials, chemical weathering is most intense at edges and corners where the most surface area is exposed. So, weathering commonly leads to rounding. For example, chemical weathering is responsible for the *spheroidal weathering* displayed by the 6-inch-wide basalt cobble shown in Figure 11.20. It rounded the cobble, and in the process it produced the *weathering rind*—the thin brown layer seen on the cobble's outside.

On a larger scale, spheroidal weathering can produce large rounded boulders from originally angular blocks. Figure 11.21 shows this process occurring in Joshua Tree National Park, California. In Joshua Tree, as in many desert settings, repeated heating and cooling, ice wedging, and other forces fracture formerly solid rock. The fracturing exposes many corners for chemical attack, so rounding occurs. Simultaneously, granular decomposition removes other mineral material. All the weathering products collect as sand on the desert floor, some of which is blown away, leaving rounded and partially rounded jumbled rocks standing tall in a sea of sand.

Figure 11.21   Granite outcrop in Joshua Tree National Park, California.

Photo from D. Perkins, GeoDIL.

11.2.3.2 Hydrolysis, oxidation, and hydration

Some minerals react with water to produce new minerals, with leftover components dissolved in water. We call the dissolved material the *hydrolysate* and the process *hydrolysis*. Most hydrolysis leads to the formation of clays. Clays are highly variable, and the type that forms depends on the parent rock composition. Granite, for example, always contains much alkali feldspar, and alkali feldspar generally weathers to produce kaolinite (a clay) plus silicon and potassium in the hydrolysate. So, kaolinite is commonly associated with weathered granites. The Joshua Tree granite shown in Figure 11.21 contains abundant K-feldspar that is decomposing by hydrolysis, leaving kaolinite and quartz as residual materials, although the desert dryness means the clay does not persist for long—it decomposes to other minerals. Hydrolysis is also significant for silicate minerals besides feldspar, and because silicates are the most common minerals in Earth's crust, clay minerals are common products of weathering for many kinds of rocks, especially in wet, warm tropical areas.

Oxidation, when iron or other elements (such as magnesium, sulfur, aluminum, or chromium) react with oxygen, often accompanies hydrolysis. Figure 11.22 shows a good example of hydrolysis and oxidation that occurred in a coarse sandstone. Weathering of granite pebbles in the sandstone has produced kaolinite (clay)—the white material in the photograph—by hydrolysis of alkali feldspar. Iron oxides and hydroxides, weathering products produced by oxidation, give this outcrop its overall reddish color.

Figure 11.20   Spheroidal weathering of a basalt cobble.

Photo from Eurico Zimbres, Wikimedia Commons.

Figure 11.22 A weathered sandstone that once contained granitic pebbles.
Photo from James Aber, Emporia State College.

Figure 11.23 Soil from near Fargo, North Dakota.
Photo from Eric Brevik.

Iron and iron oxide minerals (hematite or magnetite) readily combine with water to form *limonite*, the major component of rust. Aluminum combines with water to form *bauxite*, the most important ore of aluminum, and other elements, too, can react to form hydrous minerals. Such reactions, termed *hydration reactions*, differ from hydrolysis because little material is carried away in solution—the only significant change of rock composition is the addition of water. For iron-bearing minerals, hydration is commonly linked with oxidation, and the minerals often weather to produce combinations of hematite, goethite, or limonite (all iron oxide or hydrated iron oxide), giving sediment or soil a yellow brown, reddish, or yellowish color (Fig. 11.22). Less commonly, in low-oxygen environments, different iron and other metal oxides form that give sediment or soil gray, greenish, or bluish hues. Figure 11.23 shows an example of iron-rich soil that formed in a relatively oxygen-poor environment near Fargo, North Dakota.

11.2.3.3 Dissolution

A few minerals, including halite and gypsum, dissolve easily in water. Most, however, have relatively low solubilities, but if water is acidic, mineral solubilities increase significantly for some non-silicate minerals, especially calcite and magnesite. Thus, when carbon dioxide and sometimes organic matter are in water, the water becomes acidic and can dissolve minerals. Calcite, for example, which has a low solubility in neutral water, dissolves in many natural waters. Figure 11.24 shows weathered limestone at Malham Cove, Yorkshire. The rocky outcrop developed fractures that allowed water to

Figure 11.24 Weathered limestone in Yorkshire, England.
Photo from D. J. and F. G. Waters.

penetrate. Local rainwater was acidic enough to cause dissolution of calcite, and it consequently widened the cracks and added pits and holes to the outcrop. Acidic water has also attacked corners and edges, giving all the blocks a rounded appearance. This appearance is typical of many limestone outcrops and is one way that geologists can identify limestone from far away. See also Figure 8.7 and the accompanying discussion.

On a regional scale, dissolution of limestone can lead to a distinct form of topography, called *karst topography*, that includes many sinkholes, caves, and springs. If the calcite later reprecipitates, caves may become adorned with formations called *speleothems*, including *flowstone*, *draperies*, *stalactites*, or *stalagmites*. Precipitation of calcite also accounts for *travertine* and *tufa* deposited at hot springs and in lakes, and for *marl* that sometimes collects on stream bottoms. Speleothems, travertine, tufa, and marl are all different forms of calcium carbonate. Thus, dissolution and reprecipitation can be significant processes, but are restricted to specific kinds of minerals and rocks.

11.2.4 Biological weathering

Rock disintegration, caused by tree roots, started Leopold's atom (Chapter 2) on its lifetime journey. Fracturing of rocks by roots is a form of *biological weathering*. Such weathering involves the decomposition of rock by animals, plants, and microorganisms. For example, burrowing rodents, worms, and insects may create fractures or holes in rocks, allowing tree or other plant roots to grow into openings and pry the rock apart by *root wedging*. In the top photo of Figure 11.25, root wedging by a pine tree is breaking apart a sandstone outcrop. Moss,

Figure 11.25   Root wedging braking apart a rock outcrop (top) and lichen attacking a boulder (bottom).

Top photo from William Vann, EduPic Graphical Resources; bottom photo from Eric Brevik.

fungus, and other organisms can do the same. The bottom photo in Figure 11.25, for example, shows lichen (symbiotic communities containing fungi and algae) that are attacking a boulder at Steptoe Butte, eastern Washington.

While they are physically breaking rocks, some microorganisms and plants produce acids and other compounds that cause chemical decomposition. So, biological weathering is a combination of both physical and chemical processes caused by living organisms.

11.2.5 Weathering products and rates

11.2.5.1 Weathering rates of minerals

Some minerals weather more easily and faster than others do. This *differential weathering* reflects how stable the minerals are. Temperature, pressure, availability of water, availability of oxygen, availability of carbon dioxide (CO_2), acidity, and other factors are key controls of mineral formation. When exposed to the elements at Earth's surface, the rate at which a mineral weathers depends primarily on how different the Earth surface conditions are from the conditions at which the mineral originally formed. High-temperature igneous minerals, including olivine and pyroxene, break down rapidly at Earth surface temperatures compared with igneous minerals that formed at lower temperatures, such as alkali feldspar and quartz. Clays and iron oxides, which form at Earth's surface conditions, weather very slowly or not at all. Gypsum and rock salt, which form in arid environments, weather quickly in wet environments. Limestone, which may be deposited by neutral or basic water, weathers relatively quickly when exposed to acid.

Other factors limit or promote weathering, too. For example, the minerals that break down most easily are often those that are relatively silica poor or those rich in iron. The amount of silica affects the extent to which bonding is ionic or covalent—more silica means more covalent bonds, and covalent bonds are stronger than ionic bonds. Thus, silica-rich minerals resist weathering better than silica-poor minerals. More iron means that oxidation is more likely to occur, which often leads to weathering that produces secondary iron minerals.

Quartz dominates most sand, because it is an extremely common mineral that is very resistant to weathering. However, flowing water and gravity can concentrate minerals of different sorts, commonly separating less-dense quartz from heavier mineral grains. Figure 11.26 shows an 8-millimeter-wide view of sand from Pfeiffer Beach, California. It contains garnet, epidote, zircon, magnetite, spinel, staurolite, and quartz. Some of these minerals are quite stable and will remain in the sediment for a long time. Others will decompose more quickly.

Geologists can compare the relative stabilities of minerals by looking at weathering that occurs in rock outcrops and by studying the minerals present in sediments of different ages. Although there is some variation, typical studies produce lists, like the one in Table 11.2, comparing weathering rates of different minerals. Weathering resistance, however, does not necessarily mean that a particular mineral is abundant in weathered materials. Some of the minerals at the top of the list in Table 11.2 are uncommon compared with others. Zircon, rutile, and tourmaline, for example, are very resistant to weathering, but rarely are major components of sediments because they are only minor minerals in

most parent rocks. Minerals at the bottom of the list are very unstable when exposed to the elements and, consequently, are absent from all but the youngest sediments.

Mafic silicates weather to produce clay minerals and iron oxides; feldspars of all sorts weather to produce clay minerals and dissolved material; quartz is unchanged by weathering; calcite weathers by dissolution producing dissolved ions; and aluminous minerals weather to produce gibbsite or other aluminum hydroxides. Table 11.3 lists the weathering products for the most common minerals. Clays, iron oxides (mostly hematite), quartz, aluminum hydroxide, and a few other minerals dominate the list. While producing these residual minerals, weathering also produces dissolved cations (especially alkali and alkaline earth metals) and anions, which may have a significant impact on water chemistry and quality.

Figure 11.26   Sand from Pfeiffer Beach, south of Monterey, California.
Photo from Siim Sepp, www.sandatlas.org.

Table 11.2   Relative resistance of minerals to weathering.

Primary minerals	Secondary minerals	
zircon	anatase	most weathering resistant
rutile	gibbsite	
tourmaline	hematite	
ilmenite	goethite	
garnet	kaolinite	
quartz	clay minerals	
epidote	calcite	
titanite	gypsum	
muscovite	pyrite	
K-feldspar	halite	
plagioclase	other salts	
hornblende		
chlorite		
augite		
biotite		
serpentine		
volcanic glass		
apatite		
olivine		least weathering resistant

Data from *Soils and Geomorphology*, Peter Birkeland (1999).

Table 11.3   Chemical weathering products of common minerals

Mineral (composition)	Residual minerals	Dissolved ions
halite (Na chloride)		Na^+, Cl^-
gypsum (hydrated Ca sulfate)		Ca^{2+}, $(SO_4)^{2-}$
calcite (Ca carbonate)		Ca^{2+}, $(HCO_3)^-$
quartz (SiO_2)		$(SiO_4)^{4-}$
plagioclase (Ca-Na-Al silicate)	clay	Ca^{2+}, Na^+, $(SiO_4)^{4-}$
alkali feldspar (K-Na-Al silicate)	clay	K^+, Na^+, $(SiO_4)^{4-}$
olivine (Fe-Mg silicate)	limonite (Fe hydroxide), clay	Mg^{2+}, $(SiO_4)^{4-}$
pyroxene (Ca-Fe-Mg silicate)	limonite (Fe hydroxide), clay	Ca^{2+}, Mg^{2+}, $(SiO_4)^{4-}$
amphibole (Ca-Fe-Mg silicate)	limonite (Fe hydroxide), clay	K^+, Mg^{2+}, $(SiO_4)^{4-}$
biotite (K-Fe-Mg-Al mica)	limonite (Fe hydroxide), clay	K^+, Mg^{2+}, $(SiO_4)^{4-}$
muscovite (K-Al mica)	clay	K^+

Modified from table at www.iupui.edu/~geol110/07_Weather_SedRxs/lect02.html.

11.2.5.2   Weathering rates of rocks

The chemical weathering rate of rock depends on the minerals in the rock, the rock texture (grain size, porosity, permeability, etc.), climate, and several other things. Thus, different kinds of rocks weather differently. Figure 11.27 shows a highly weathered outcrop containing what was once a basalt dike cutting through layers of sedimentary rock. Almost all the original mineral content in both kinds of rock is gone, but the basalt, in contrast with the surrounding sedimentary rock, has weathered to a more uniform and finer-grained material. Both kinds of rock have oxidized to a deep rusty red color.

Although mineral content is the most important factor determining rock weathering rates, other factors can be important, too. Coarse-grained rocks generally weather faster than fine-grained rocks, since water has a harder time penetrating the interior of fine-grained rocks. (An exception sometimes occurs if a volcanic rock contains significant amounts of glass, because glass is especially unstable.) Sedimentary rocks contain variable amounts of clays and non-clays, and usually clay-rich rocks weather more quickly than clay-poor rocks. This is because rocks that contain little clay are often hard and massive and thus more impervious to fracture and fluid flow. Additionally, mineral grains in rocks that contain mostly non-clay minerals are commonly held together by silica cement, which is very durable. The final products produced when rock weathers are predictable from the original minerals in the rock; Table 11.4 summarizes them for common rock types.

Figure 11.27   Weathered outcrop containing a mafic dike cutting through sedimentary rocks.
Photo from Eurico Zimbres, Wikimedia Commons.

11.3   Soil

11.3.1   Earth's critical zone and soil

Recent programs developed by the National Science Foundation and other organizations have sought to focus research on what is called Earth's *critical zone*, shown in Figure 11.28. This zone includes Earth's near-surface components, from the tops of the highest vegetation to the lowest levels of groundwater penetration. So, the critical zone consists of the outer surface of our planet, small bits of the adjacent atmosphere and lithosphere, and living organisms (the biosphere). Hence, the zone includes all the habitats, processes, and resources that sustain life on Earth. It is, indeed, critical.

The critical zone is one of the most complicated parts of Earth because it occupies space where rock, water, air, and life interact. It is constantly changing and evolving on time scales from seconds to eons. For example, weather variations cause short-term changes, human actions sometimes cause somewhat longer-term changes, and changing climates and tectonics make changes over millions of years. Perhaps most important, weathering and other zone processes transform geological and biological materials into *soil*. So, interactions between the lithosphere, hydrosphere, atmosphere, and biosphere create the *pedosphere*, the soil part of Earth. All five spheres are crucial to human well-being.

The products of weathering, unless modified, are generally not considered soil—most soil scientists agree that soil formation (*pedogenesis*) must involve the creation of soil layers, commonly called *soil horizons*, of different characteristics. A simple model by Simonson (1959) says that pedogenesis results from, first, accumulation of *parent material* (due to weathering and, in most cases, the transportation and deposition of weathered materials), followed by *horizonation* (layering) caused by material being selectively concentrated, added, or removed or being converted into different material. Soil parent material can come from many different sources. Some forms directly in place from residual materials left over after bedrock weathers, and some forms from thick accumulations of organic debris (e.g., swamp deposits). However, most derives from transported material, including glacial deposits transported and deposited by ice, sediment deposited by flowing water in streams or rivers, debris deposited by gravity in valleys or at the bottom of mountains or cliffs, sediment deposited in lake or ocean bottoms, and dust and other material transported and deposited by wind.

Table 11.4   Characteristics of weathering for some common rocks and typical residual minerals.

Rock	Weathering characteristics	Most common residual minerals
granite	Granite is a hard, massive rock that forms at high temperature and pressure at depth in Earth, so it is unstable when exposed at the surface. Unloading may cause expansion joints or exfoliation. Granite contains subequal amounts of alkali feldspar and quartz. The alkali feldspar weathers at a slow to moderate rate; the quartz does not weather at all.	clay minerals, quartz
basalt	Basalt texture is highly variable. Some basalt is massive and hard, but other basalt may contain vesicles (holes left over from gas bubbles) or have columnar jointing, providing conduits for water and making it more susceptible to chemical attack. Blocky physical weathering is common. Basalt contains primarily calcium-rich feldspar and pyroxene. Both decompose relatively quickly.	clay minerals, iron oxides
sandstone	Some sandstones are loosely cemented and easily break apart during weathering. Others are hard and massive and so resist physical weathering. Sandstone is made primarily of quartz, thus resisting chemical weathering. Sometimes sandstone contains other minerals that weather to produce clay minerals.	quartz, possibly feldspar if it was present in the original sandstone
limestone	Limestone textures are variable, as is limestone weathering. Limestone may fracture and, in some dry environments, may form cliffs or outcrops that undergo physical weathering. Limestone is primarily composed of calcite and does not decompose quickly in neutral or basic water, but water in many environments is acidic, which causes limestone to dissolve. Dissolution is generally focused on fractures and vugs (holes produced by previous dissolution), enlarging them and sometimes producing caverns.	none
shale	Shale does not easily weather chemically since its components (mostly clay) are products of weathering, but it does weather physically, often very rapidly because it is very soft and contains many fractures. Another reason for high weathering rates is that the clays in shale may swell and shrink quickly when water content changes, causing fracturing.	clay minerals
evaporites (e.g., gypsum or rock salt)	Evaporites are sometimes massive in outcrop, but they are quite soft and susceptible to physical weathering. Evaporite minerals have great solubility, dissolving when exposed to water. In dry climates, they may persist for long times; in wet climates, evaporites are absent.	none

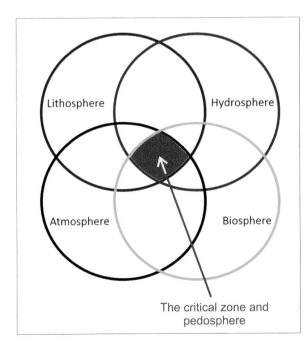

Figure 11.28   The critical zone and pedosphere.
Modified from drawing by Eric Brevik.

11.3.2   *The importance and nature of soil*

There are many different ideas about why soil is important, but the differences are mostly because of perspective. To a *soil scientist* (*pedologist*), soil is the complex system of material at Earth's surface that leads to growth and development of plant life. To many *geologists*, it is part of a thin unconsolidated surface layer, lying above bedrock, that forms by breakdown of geological materials and that makes it difficult to see the bedrock. To an *engineer*, it is the loose material made of mineral and organic material that covers large portions of Earth, has variable properties, and is the site of many engineering projects. Perhaps of most importance to *people around the world*, soil is the crucial layer on the ground's surface that is responsible for crop production and forage. To be sure, soil provides Earth and its inhabitants with many important services. For example, soil:

- provides the minerals and water needed by plants and produces food for animals in the wild.

- is necessary for agriculture, providing the minerals and water needed by plants.
- is a critical component in mining and construction industries.
- absorbs rainwater and releases it later, thus preventing floods and drought.
- cleans and filters water, removing organic material and other contaminants.
- is the habitat for many organisms.
- is a carbon sink, reducing the impacts of global warming.
- is necessary for functioning drain fields and other wastewater disposal.
- provides a medium and base of support for plant growth.

Soil compositions are quite variable, but a typical healthy soil contains about half mineral matter, equal amounts of water and air, and a small amount of organic matter (Fig. 11.29). The United States Department of Agriculture, the Soil Society of America, many textbook authors, and others all have slightly different definitions of soil. Most definitions include some important common characteristics that differentiate soil from other forms of sediment and regolith:

- Soil results from the interaction of weathering and biological activity.
- Soil compositions are highly variable, both spatially and temporally.
- Soil contains mineral and organic matter, gas (generally air), and liquid (generally water).
- Soil has different composition and texture than its original parent material.
- Soil develops horizons, layers with different chemical and physical properties.

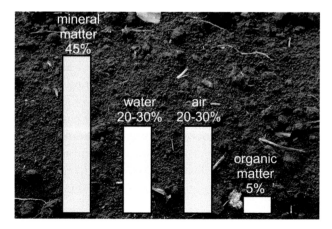

Figure 11.29   Typical composition of soil.

11.3.3 Soil profiles

Soil forms a layer, lying above bedrock, of variable thickness. In some places, thickness is zero (on rock outcrops), and in other places it may be many tens, or even hundreds, of meters thick (especially in tropical areas). *Humus*, a key component of good soils, is concentrated near the surface. It is organic matter formed by decomposition of plant and animal remains that is quite stable in some environments, where it may persist for long times without decomposing. Humus holds moisture and nutrients and provides structure that makes soil cohesive so it resists erosion. Whatever its thickness, the top of the soil layer is generally the boundary between soil and air, or between soil and plant material that has not begun to decompose. The bottom of the soil layer is often gradational, where soil gives way to rock or to earthy material lacking living organisms.

In all but the driest environments, rainwater and snowmelt infiltrate soil and can dissolve soluble soil components and carry them downwards. This process, shown in Figure 11.30, of dissolution and transportation is called *eluviation*. If the dissolved material is deposited in lower soil levels, the process of transportation and deposition is called *illuviation*. Sometimes, flowing groundwater removes dissolved material from a soil column completely, by a process called *leaching*.

So, provided there is enough precipitation, upper soil layers become depleted in soluble material and are often quite porous. Lower layers, in contrast, may be quite

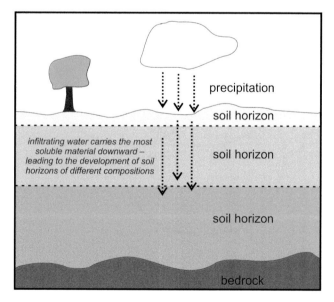

Figure 11.30   Infiltration of water causes dissolution and transportation of material.

hard and compact because illuviated matter fills all pore space. Eluviation, illuviation, and leaching are key processes that help form *soil horizons*, layers of soil parallel to the soil surface, with characteristics that differ from layers immediately above and below.

Soil profiles, vertical cross sections from the surface to the underlying bedrock, show horizons present in a soil and their relative thicknesses. The profiles provide a rapid view of the characteristics that may make a soil desirable for agriculture or other purposes. Soil scientists designate different soil horizons by standard letters (O, A, E, B, C, and R), but sometimes subdivide the horizons in different ways.

Figure 11.31 shows a typical soil profile with the layers labeled with the standard one-letter abbreviations defined. The O horizon is an organic layer, frequently containing greater than 35% organic material, although it can be as low as 25%. It can be very thin, perhaps a thin layer of leaves on a forest floor, or several meters thick, such as the accumulation of organic material in a bog. It is often absent in grassland soils but common in forested soils and wetland environments.

The A horizon is a mineral and organic horizon, usually containing less than 5% organic material, but up to 35% in some settings. It may be somewhat modified by eluviation. The A horizon is the focus of biological activity and the most important source of nutrients for plant growth; it is also easily lost to erosion, which can lead to serious decrease in soil productivity.

The E horizon is a light-colored mineral horizon, characterized by intense eluviation that removes clays and the mineral and organic material that provides topsoil with its dark soil color. It is common beneath forest vegetation.

The B horizon is the zone where most eluviated material accumulates by illuviation. These accumulations can be clays, pigmenting minerals, or several other kinds of materials. The B horizon, like the A horizon, can be important for plant growth, for example, by providing much needed moisture to plants when the topsoil dries out.

The C horizon is altered, or slightly altered, parent material, the geologic material that was present before soil formation.

The R horizon is a layer of consolidated bedrock within the soil profile, generally within a few meters of the surface. If bedrock only exists at greater depth, the R horizon is generally not considered present.

Soils that form in different environments contain different horizons. Most typical soils contain three or four horizons, but in hot, wet, tropical regions where extreme leaching has occurred, some soils may have only one. Grassland soils of the Great Plains usually have well-developed A, B, and C horizons. Soils in a Midwest forest commonly have O, A, E, B, and C horizons. In the southeast United States, characterized by lots of rain and intense weathering, the A horizon may be absent and only O, E, B, and C present. Soils in a river floodplain may only have A and C horizons. Different soil-forming agents and environments explain the different profiles. The difference between grassland and forest, for example, is partly due to different environments, but also due to different soil-forming organisms. Floodplain soils lack a B horizon because new flood deposits cover evolving soils before any B horizon can develop. The presence and thickness of different soil horizons form the basis for some soil classification systems.

11.3.4 Soils and agriculture

Certain soils are better suited for agriculture than others are, and soils can be rated based on their *fertility* and a closely related property, *productivity*. Soil fertility is a measure of the amount of nutrients, organic matter, biota, and other components that are present and promote plant growth. Sixteen nutrients are considered essential, and most of them must come from mineral matter (Table 11.5). Though needs of different plants

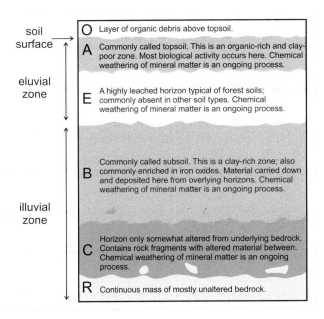

Figure 11.31 The standard horizons in a soil profile.

Table 11.5   The 16 essential nutrients needed for plant growth

Nutrient	Source
carbon	carbon dioxide
hydrogen	water
oxygen	carbon dioxide and water
nitrogen	air, but nitrogen must be processed by microorganisms before it can be used by plants
phosphorus potassium sulfur calcium magnesium iron boron manganese copper zinc molybdenum chlorine	soil mineral content and decomposing organic matter

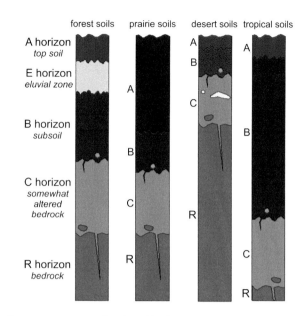

Figure 11.32   Typical soil profiles from four types of ecosystems. Drawing modified from one by Bruce Railsback.

vary, the necessary nutrients, organic matter, and biota are generally present in a soil called *loam*, and loam is considered the closest to an ideal soil for many crops, especially corn and small grains. Soils with a loamy texture have enough pore space to deliver water and provide good root penetration for crop growth.

Productivity, which depends on fertility and other factors, including climate, refers to the size of potential harvests. Figure 11.32 compares soil profiles typical of four common ecosystems, including very productive prairie soil and nonproductive desert soil. The most productive soils, which include prairies and other grassland soils and (to a lesser extent) temperate and coniferous forest soils, have thick, well-developed A horizons. For all soils, the amount of rain and thus the amount of eluviation determine the thickness of the B horizon, because that is where water deposits illuviated materials. Sometimes, in dry climates, dissolved carbonate material may precipitate the C horizon—shown by the light-colored inclusions in the profile for desert soils.

Large amounts of rainfall characterize tropical environments. So, very thin A horizons and thick B horizons characterize most tropical soils. Due to extreme leaching, organic matter is absent except at the surface, even though biological productivity is generally quite high. When plants die, they decompose and release nutrients into the uppermost soil layers, but still-living plants rapidly take up the nutrients. Thus, in tropical rainforests, biological productivity is concentrated in a very thin layer of organic material at the ground surface

and above, much of it in trees. Because no nutrients are present in deeper soil horizons, agriculture in tropical areas is a temporary undertaking; cleared land will only be productive for short times before all nutrients are gone and farmers must move to another plot.

In the United States, as in the rest of the world, soil productivity correlates well with topography and climate. Although productivity is different for different crops, average values, termed the *soil productivity index* (SPI), may be calculated based on soil fertility characteristics. Figure 11.33 is a soil productivity map of the United States. Warm colors (red, orange, and yellow) show areas of low productivity, and dark colors (blue and green) show areas of high productivity. Productivity is greatest in grassland areas of the central United States and northwestern states. Overall, the Central Plains of the United States have the highest SPI. In these regions, thick soils have a readily available supply of organic matter, and precipitation is neither too little nor too much, most of the time. At the other end of the scale, thin soils in mountain areas, and desert soils, are less productive.

Figure 11.34 shows red clay-rich soil from North Carolina. In contrast with U.S. grassland soils, this soil and most soils in the southeastern United States, where rainfall is greater than in the rest of the country, are highly leached. Leaching has removed organic content and nutrients from the A horizon and left iron oxide behind (which makes the soil red), making the soil less productive for corn, wheat, and other small grains that grow so well in grassland regions. Although these red

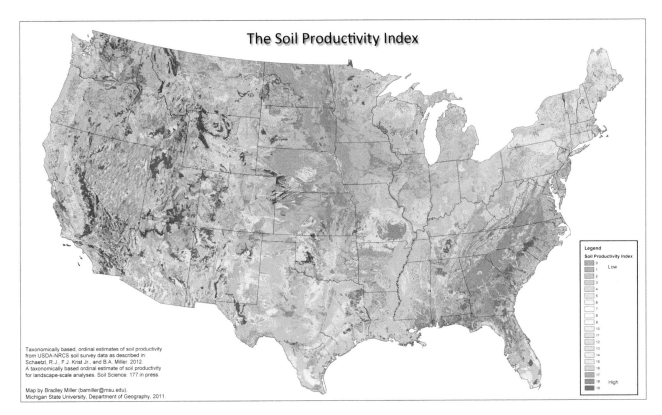

The Soil Productivity Index

Taxonomically based, ordinal estimates of soil productivity
from USDA-NRCS soil survey data as described in:
Schaetzl, R.J., F.J. Krist Jr., and B.A. Miller. 2012.
A taxonomically based ordinal estimate of soil productivity
for landscape-scale analyses. Soil Science. 177 in press.

Map by Bradley Miller (bamiller@msu.edu).
Michigan State University, Department of Geography. 2011.

Figure 11.33   The soil productivity index for soils of the United States.

Map from Schaetzl, R.J., Krist, Jr., F.J. and Miller, B.A., 2012, A Taxonomically Based Ordinal Estimate of Soil Productivity for Landscape-Scale Analyses:
Soil Science, v. 177, no. 4, April, p. 288-299.

Figure 11.34   Red soil in North Carolina.

Photo from David Lindbo (the Department of Soil Science at NC State
University), Wikimedia Commons.

soils do not look like the rich dark soils associated with
very productive farmland of the Great Plains, the red
soils produce large amounts of peanuts, cotton, tobacco,
and other crops. Thus, different crops thrive in different
kinds of soils.

11.3.5 Pedogenesis (Soil formation)

Soils are derived from parent material, which may be
weathered bedrock or unconsolidated sediment. After
the parent material accumulates, horizonation, the
formation of soil horizons, occurs because of physical,
chemical, or biological processes. It may take a long
time. Furthermore, in large part due to the slow rate of
development, some soils develop multiple distinct horizons, while others do not. For example, soils formed
from recently deposited sediment may have no, or only
indistinct, horizons. Eventually, however, most soils
develop distinct and visible horizons.

If soil forms from rock, as shown in Figure 11.35,
the first step in soil formation is weathering. Once rock
begins to weather, fledgling plants may appear, adding
organic matter, and thus humus, to the layers. Soil horizons form, and over time, the horizons usually become
more pronounced. Eventually, in most settings, finer
material begins to dominate and becomes commin-
gled with humus and other organic matter, creating a
rich A horizon, and a healthy soil has been born. Soils
that form from transported (unconsolidated) sediments

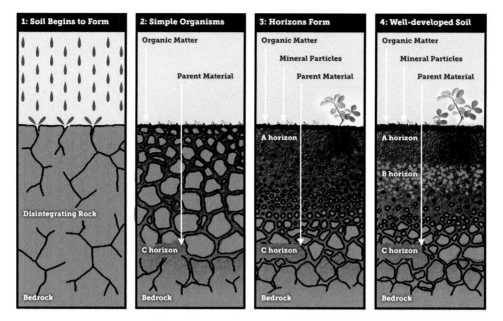

Figure 11.35 Stages of soil formation, starting with barren bedrock. Graphic from Edward Mansouri, Weatherstem.com.

instead of from rock get a jump start on pedogenesis because no onset of physical weathering is needed to get the process going.

No matter what their parent material, soils continue to evolve after their initial formation in response to changes in climate, vegetative cover, human management, and other factors. The entire process of soil creation may take months, hundreds of years, or may not occur at all, depending primarily on five key factors (Fig. 11.36):

- nature of the parent material
- climate
- topography
- time
- organisms (especially native vegetation)

11.3.5.1　Parent material

A soil's *parent material* consists of mineral matter and, less commonly, organic material (e.g., peat) from which a soil forms. Residual or transported quartz, feldspar, mica, calcite, oxides, and other typical weathering products dominate initially, but partially weathered rock, ash from volcanoes, stream deposits, and other components may be present as well. The ratios of the various components depend on many things, including setting (valley, mountain, streambed, etc.), the nature of the bedrock, and climate. Parent material determines overall soil

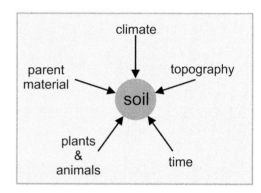

Figure 11.36 Factors that control soil formation.

composition and properties such as color, texture, structure, and permeability. Besides affecting soil composition, the nature of the parent material also influences the rate at which the soil forms. Soils formed from *residual* materials may derive from only one kind of sediment and, therefore, lack some key nutrients. Soils formed from *transported* materials are often more balanced in composition because they commonly include material from more than one source. Overall, the many different sources of sediment and transportation mechanisms lead to a large variety of soil types.

Transported materials arrive at the site of soil formation in several ways. For example, wind can carry dust or silt long distances before depositing them. Such deposits, termed *eolian deposits*, may result in thick, rich soils made of *loess*, a loosely compacted fine-grained sediment. Figure 11.37 shows rolling hills in the Palouse—a major agricultural region in Washington, Idaho, and Oregon. In the United States, other significant loess deposits are found in Iowa, Illinois, Nebraska, and Indiana. The hills shown in Figure 11.37 consist of wind-blown loess deposited during the last ice age. The loess originated as rock powder and other fine material in glacial outwash; wind transported it to where it is today, leaving coarser material behind. In the right climate zones, some of the best soils in the world have developed from loess parent materials, and the Palouse is agriculturally very productive. The fine-grained soil and rolling hills, however, make the Palouse very susceptible to erosion by wind or water.

Water-deposited parent materials may be *marine*, *lacustrine* (lake deposits), or *alluvial* (floodplain deposits). Figure 11.38 shows the alluvial deposits adjacent to the Waimakariri River in New Zealand. The city of Christchurch is in the background. Floodplains such as this one are commonly excellent for agriculture—as can be seen in this photo. Alluvial deposits of this sort are widespread, especially in the United States interior. In contrast, some soils in the Central Valley of California, and along the Gulf Coast and Atlantic Coast of the United States, formed from sediments originally deposited on ocean floors by salty ocean water. And soils near the Great Lakes and near the Great Salt Lake of Utah formed from lacustrine deposits.

Although soils are fine-grained materials, they may derive from coarse parent materials. Soil development from coarse materials, however, may take longer than from fine materials. Glaciers, for example, deposit coarse material in moraines and outwash plains, and

Figure 11.38   A meandering river and flood plain near Christchurch, New Zealand.

Photo from Greg O'Beirne, Wikimedia Commons.

gravity may deposit material in talus cones or on scree slopes that later may become soil. Figure 11.39 shows an outcrop of glacial till above limestone. The till is a jumble of material of many sizes, and the coarser clasts will take a long time to disintegrate. The finer material is, however, available for soil formation, and a thin soil layer can be seen at the top of the photo.

11.3.5.2 Climate

Moisture and temperature are often considered to define *climate*. Most chemical weathering requires water, and chemical reaction rates are faster in warm places than in cold places. So, chemical weathering and soil formation are slow for rocks and minerals of all sorts in dry, cold climates and faster in warm, wet climates. For example, limestone in a typical desert will weather very slowly, but limestone (and other rock) in tropical forest areas weathers quickly. And wet tropical areas often have thick accumulations of weathered material, although much of it decomposes before soil formation.

Climate influences how sediment transportation occurs and what is transported. Climate also limits or promotes dissolution, freezing and thawing, organic activity, and many other things that control soil development. Water and temperature influence the kind of organisms present, and thus the nature of biological weathering, further making the connection among climate, soil, and weathering processes.

Soil horizons develop in different ways in different climates. For example, organic material may persist and

Figure 11.37   Wind-blown soil deposits in the Palouse—a major agricultural region in Washington, Idaho, and Oregon.

Photo from Lynn Suckow, Wikimedia Commons.

Figure 11.39 Glacial till above a limestone outcrop.
Photo from N. Chadwick, Geograph Britain and Ireland.

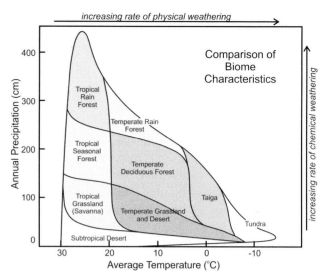

Figure 11.40 Comparing the world's biomes.

be a key to soil horizonation in temperate regions. In tropical regions, where thick soils can form quite rapidly, organic materials decompose relatively quickly and so play a lesser role. Soils that develop in generally moist, seasonal grasslands are rich and productive because of their high organic content and thick A horizons. Soils that form in humid regions are often acidic due to the leaching of alkali and alkali earth elements (Ca, Mg, Na, K) that neutralize acids, and the consequent concentration of Al, Fe, and H that promote acidity. In contrast, soils that form in low-rainfall areas may be alkaline and may contain basic *caliche* (a form of calcite, calcium carbonate) or alkali salts that inhibit plant growth.

Environmental scientists divide the world into *biomes* (ecosystem types) based on the types of vegetation present. As shown in Figure 11.40, differences in vegetation, and thus differences between biomes, are mostly due to differences in amounts of precipitation and average temperatures. Figure 11.40 is a *Whittaker biome diagram*, named after plant ecologist R. H. Whittaker

who first pointed out the logic of such diagrams. Today, biomes are sometimes named or classified in slightly different ways than shown in this figure, but classification always reflects average yearly temperature and precipitation. Chemical weathering is fastest in biomes that receive more rain, such as in tropical rainforests; physical weathering is generally fastest in biomes where freezing occurs, such as taiga or tundra.

The five soil profiles depicted in Figure 11.32 characterize different biomes (tropical rain forest, temperate deciduous forest, temperate grassland, dessert, and coniferous forest) and, consequently, each has a different soil profile. These general profile types are the same in similar ecosystems around the world, although the specific plant and animal species are variable. So, different kinds of soils are consistently associated with the different biomes.

11.3.5.3 Topography

When it rains, water can flow into lakes, streams, or oceans. It may also infiltrate to become part of the groundwater system, or it may evaporate. The proportions that follow each of these three paths vary from place to place and from season to season, but runoff is greatest on steep slopes, and infiltration is greatest on flatter ground.

Topography, therefore, governs how much water is available for weathering, infiltration, and, consequently, soil formation. Thus, topography affects both the rate of soil formation and the development of soil horizons.

Figure 11.41 shows some key relationships. Although not shown in this figure, soils sometime exist on hilltops, where heavy leaching usually makes them light-colored. Soils are generally thin or nonexistent on steep slopes, because the water required for chemical weathering and soil formation runs off too quickly and because erosion rates are high. Soils are often thickest on valley floors where eroded material accumulates, biological activity is high, and erosion is slow. These valley soils are typically dark colored due to high organic content, and they generally contain well-developed soil horizons. However, poorly developed soil profiles are found on valley floors where frequent flooding deposits fresh sediment on top of developing soils, burying them, and restarting the soil evolutionary process. Thick, productive soils also typify many plateau tops if they are characterized by grasslands or prairie.

11.3.5.4 Organisms

Soils contain a large proportion of the world's biodiversity. Yet, soil ecosystems are complicated and poorly understood because most soil biota are microscopic, and much is undescribed. Figure 11.42—a view of the soil food web—gives just an inkling of the many different kinds of organisms that may be present. We know that microorganisms, including bacteria and fungi, promote decomposition of mineral and organic matter, and a single gram of soil may contain thousands of such species. Larger animals, including worms, are more easily identified and studied. They are often important agents that mix soil material, promoting circulation of air and water, when they burrow. Still larger animals, such as moles or gophers (not shown in Figure 11.42) contribute much by keeping soil aerated and providing passages for water and for other plants and animals to navigate. And, at and above the soil surface, birds and other animals consume food generated in soil.

Plant communities, too, are important components of soils. Plants control the nature of humus that is present as they add organic material that helps prevent erosion, and they provide shade that moderates temperature fluctuation. They may help maintain soil moisture or, sometimes, promote moisture loss. They also introduce acids and other compounds that affect soil chemistry and acidity, and some of them harbor bacteria that *fix nitrogen* (change atmospheric nitrogen into ammonia), a key form of natural fertilization. Both plants and burrowing animals are very important agents that cycle nutrients in soil systems.

11.3.5.5 Time

Soil is a work in progress. When initially forming, soil is no more than regolith, but it changes over time. The distinction between regolith and soil is said, by some, to be the point at which the regolith can support life. Yet, even at that point, soil continues to evolve. So, exactly when a soil becomes soil is problematic. Parent materials have a very strong influence on soil properties after soil formation begins, but as time goes by other factors including climate and biological activity tend to become more important and guide further soil development.

Soil lifetimes are finite. Over time, nature may convert soil to other material, such as *bauxite* (a rock consisting of hydrated aluminum oxide), or remove it by erosion. Additionally, new materials, perhaps deposited by a flood or a landslide, which have little similarity to a soil, may cover old soil horizons, forcing pedogenesis to start anew. Most soils, today, are Pleistocene or younger, meaning they formed in the last 2.5 million years—less than 1/10th of 1% of Earth's history. Geoscientists have,

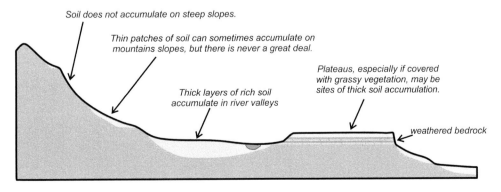

Figure 11.41 Topography and soil development.

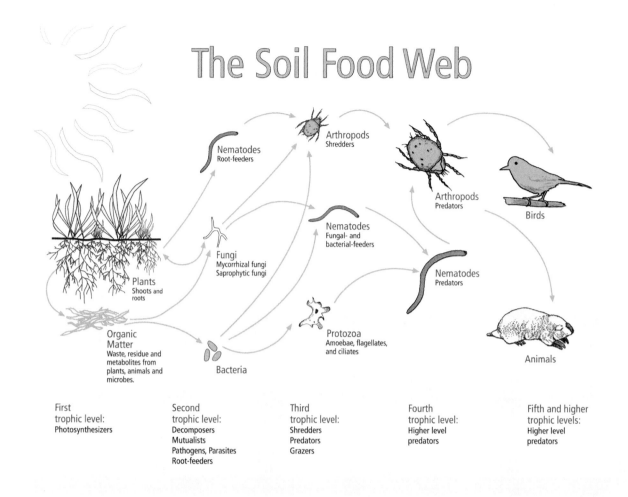

Figure 11.42   Some of the many organisms in the soil food web.
Chart from the Natural Resources Conservation Service (NRCS).

however, identified some fossil soils, called *paleosols*, that date from early in Earth's history. Figure 11.43 shows intriguing 2.2-billion-year-old paleosol in South Africa. This soil is a fossilized *laterite*, a kind of soil that requires intense leaching to develop. So, precipitation and leaching must have occurred quite early in Earth's history.

11.3.6   Characteristics of soils

Different kinds of soils have different properties. Some properties, like density and shear strength, are most important in engineering applications and are discussed in the next chapter. Here, we consider some key properties that help distinguish soil types and that affect agriculture:

- color
- temperature
- texture

Figure 11.43   A 2.2-billion-year-old paleosol from Waterval Onder, South Africa.
Photo from Greg Retallack.

- structure
- consistency
- water content and pore space

11.3.6.1 Color

Soil *color* is one of the easiest ways to tell soils apart and is also the easiest way to see soil horizons (Fig. 11.44). Although color is not a good indicator of all soil attributes, it does provide information about soil composition and moisture content, and about the nature of the parent material. Usually, soil color reflects the presence of different iron oxides, carbonates or other salts, or eluviated humus. Figure 11.44 shows some examples of different colored soils. Highly leached soils, like the North Carolina soil, may have a very red color due to concentration of iron oxides, especially hematite, and absence of other mineral matter. The South Dakota soil is dark colored due to high organic content. The Texas soil contains nodules of caliche (white calcium carbonate). The Rhode Island soil is yellowish due to high goethite (an iron hydroxide) content.

11.3.6.2 Temperature

Soil *temperature* is very important because it limits or promotes seed germination, root growth, microbial activity, development of nutrients, and other things vital to plant growth. Figure 11.45 shows average soil temperatures around the world. Temperatures roughly correlate with latitude because of climatic effects. Soil temperature, however, varies with depth. Climate normally only affects the top few meters of soil; below that, soil temperature increases downwards, in response to heat flowing from Earth's interior. Very cold soils—in permafrost areas—may have surface temperatures at or below freezing; warm soils may exceed 40 °C (104 °F).

During the day, solar energy warms soil surfaces, and conduction moves heat down into the soil. At night, soil radiates heat from its surface into cooler night air, while temperature at depth varies little. The amount of sunlight (solar energy), ambient air temperature, color of the soil (which influences reflection of solar energy), elevation, and thermal conductivity (which controls how fast heat is conducted downward in the soil) are the most important factors controlling temperature.

Wet soils warm slower but hold more heat than dry soils, and evaporating soil moisture causes cooling, so water content affects temperature. Soil texture is also important. Coarse sandy soils are porous and well drained; they may heat up during the day, but space between grains means heat is not conducted downwards efficiently. Sandy soils also cool rapidly at night. Finer-grained soils, and clay-rich soils of all types, generally contain more water and so hold on to more heat, but it takes them longer to heat them up.

11.3.6.3 Texture

The sizes of mineral grains in a soil define its *texture*. Grain size depends primarily on grain mineralogy and on the origin and history of soil parent material. Grain size is important because it affects many soil properties, including compressibility, strength, porosity and permeability, and water content.

The United States Department of Agriculture (USDA) classifies soil texture and categorizes soils using the proportions of sand, silt, and clay that are present (Fig. 11.46). According to the USDA, sand grains range in size form 0.05 mm to 2 mm, silt ranges in size from 0.002 to 0.05 mm, and clays are smaller. Particles larger than sand, such as gravel, are not considered for classification purposes. The USDA classification adds the

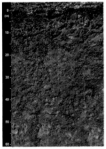   

| North Carolina | South Dakota | Texas | Rhode Island |

Figure 11.44 Soils of different colors.
Photos from the Natural Resources Conservation Service (NRCS).

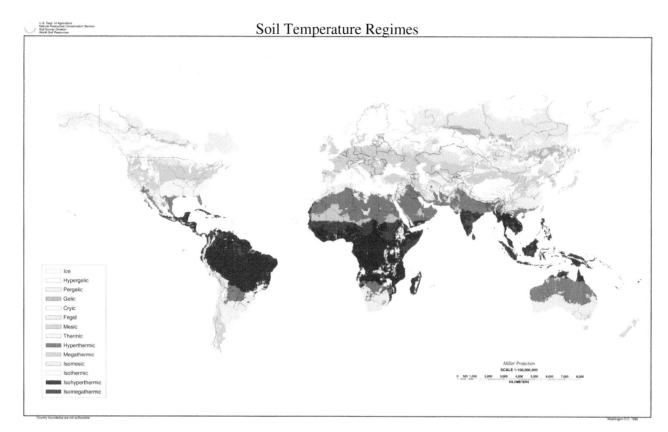

Figure 11.45   Average soil temperatures around the world.
Figure from Plant & Soil Sciences eLibrary.

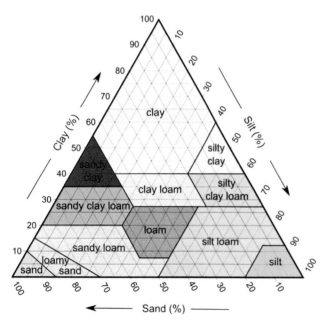

Figure 11.46   The soil texture classification system of the United
States Department of Agriculture.

modifier "organic" to a soil name if the soil contains more than 20 or 30% organic material.

11.3.6.4   Structure

Soil *structure* refers to the shape, size, and amount of *aggregates* (grains combined into larger particles) in a soil. Figure 11.47 shows the aggregation process. In temperate soils, sand, silt, clay, and humus in soil may clump together to form relatively permanent aggregates, called *peds*. Peds are discrete chunks of soil, larger than the grains that make up the soil, separated by space or zones of weakness from surrounding material. Peds are highly variable in shape and are typically described using adjectives such as *platy*, *prismatic*, *columnar*, *blocky*, and *granular*. Peds persist through wetting, drying, and soil deformation and thus provide soil strength and texture, while at the same time allowing water to flow between them. Figure 11.48 shows some good examples of blocky peds. In topsoil where organic material provides

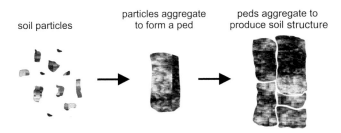

soil particles   particles aggregate to form a ped   peds aggregate to produce soil structure

Figure 11.47 Soil particles and peds.

Figure 11.48 Examples of soil peds in a tropical soil.

Photo from J. Kelley, Soil Science.

the glue that holds peds together, peds may break down and reform regularly, but the presence of peds is overall one of the keys to good soil health. Other kinds of soil aggregates, including *fragments* and *clods*, are similar to peds but form near the surface in response to human or frost actions.

In temperate soils, humus is especially important for aggregate formation because it makes up the glue that holds peds together. In tropical soils, where humus is absent, aluminum and iron may provide the glue that holds smaller grains together. The (tropical) peds in Figure 11.48 are red due to high iron oxide content.

Soil structure is important because it influences pore sizes and thus root penetration, water circulation, water retention, aeration, and drainage. The best soil for agriculture, with good structure and well-developed peds, absorbs enough water to meet plant needs but allows excess water to drain away quickly. Good structure is important for plant growth, especially in fine-grained soils, but is easily damaged. Natural processes that remove soil structure include *slaking* (the breaking down of peds and the redistribution of soil particles by rain impacts and wet conditions) and *cementation*

in subsoil layers caused by iron oxides. Cultivation by people may destroy structure because tilling breaks up aggregates and exposes organic matter to air, promoting its decomposition. Additionally, tillage and compression of soil by tractors and other heavy machinery can lead to loss of pore space and the creation of *plow pans* (hard impermeable layers of subsoil or clay).

11.3.6.5 Consistency

Soil *consistency*, which depends primarily on soil clay content, is a useful property for predicting cultivation or engineering problems. It refers to the way soil sticks to itself (*cohesion*) or to other objects (*adhesion*) and to soil resistance to deformation and rupture.

Consistency, discussed further in the next chapter, is generally measured in three moisture conditions: wet, moist, and dry (Fig. 11.49). Stickiness and plasticity are evaluated for wet soil; resistance to shearing is evaluated for moist soil; resistance to fragmentation and crumbling are evaluated for dry soil; and strength and consistency of cementation, if present, are evaluated in all conditions. Evaluating consistency is subjective and can be done in the field by squeezing and deforming samples by hand, so the terms used to describe consistency are general (Table 11.6).

11.3.6.6 Water content and pore space

Water is key for good soil formation because it is responsible for many of the processes, including eluviation and leaching, that lead to the development of soil horizons. It is important for plant growth because plants are

Figure 11.49 Evaluating soil consistency of a wet sample.

Photo from nrcs.usda.gov.

Table 11.6   Terms used to describe consistency

Soil type	Descriptive terms
dry soil	loose, soft, slightly hard, hard, very hard, extremely hard
moist soil	loose, very friable, friable, firm, very firm, extremely firm
wet soil	nonsticky, slightly sticky, sticky, very sticky; nonplastic, slightly plastic, plastic, very plastic
cemented soil	weakly cemented, strongly cemented, indurated (requires hammer blows to break up)

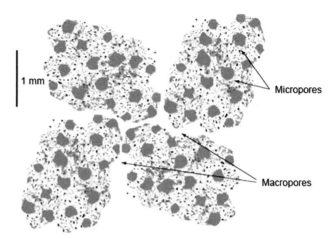

Figure 11.50   Drawing of typical peds with both large pores (mostly between peds) and small pores (mostly within peds).
Photo from Victorian Resources Online.

mostly made of water. Additionally, water is necessary for photosynthesis, delivers nutrients to plants, and is required for plants to remain upright and not wilt. So, the rate at which a soil can absorb water is very important, and the rate depends on soil texture, structure, and also pore space.

Soil *pore space* is space unoccupied by solid (mineral or organic) matter; pores may contain water or air. The water feeds plants, and the air supplies oxygen to microorganisms when they break down organic matter, recycling nutrients for reuse. Pore space also provides opening for flow and storage of water. A good agricultural soil contains about half solids and pores that contain about half air and half water, although not all the water is available to support plant growth. The amount of water available depends on soil characteristics and the type of plant that is growing. Desert plants, for example, can get more water out of soil than can agricultural crops.

Figure 11.50 shows four peds composed of grains of variable sizes and colors separated by small pores. Larger spaces exist between the peds. Overall, larger pores, such as those between the peds in this drawing, are most important for root penetration, aeration, and soil drainage. However, smaller pores within individual peds are often keys to good soil moisture. Very small pores (*micropores*) are less than 2 µm (2 microns) in diameter, and large ones (*macropores*) are up to 0.2 millimeters depending on soil texture and structure (Fig. 11.50). So, small plant roots, about 10 µm in diameter, can grow through most macropores but not micropores. Coarser-grained soils and soils containing more peds tend to have larger pores and, because plants cannot use water in very small pores, light aerated soil with large pores is generally best for plant growth. Good soil structure is the key to having large pores. Tilling is often used to loosen soil and increase porosity. However, the benefits may be short lived because tilling can destroy soil aggregation and structure, and thus lead to a loss of pore space over time.

Plants must apply suction to draw water from soil. The primary mechanism is powered by *transpiration* (evaporation from leaves) that causes more water to move up the stem. When flooded, soils are saturated, water is abundant, and transpiration rates are low. When soil becomes too dry, all available water may be used up. Moisture may remain in the soil but be unavailable because the plant cannot produce sufficient suction to use it. The limiting water content is the *wilting point*.

When soil is oversaturated, drainable water, called *gravitational water*, occupies space between grains. As a soil dries, excess gravitational water drains from the soil until it reaches *field capacity*, the point at which gravitational water is gone and adhesive and cohesive forces stop further loss of water. At field capacity, the smallest pores are water-filled and the larger pores contain water and air. Clay- and organic-rich soils have higher field capacities than sandy or silty soils and so provide less water for plant growth as the soils dry. Although water is necessary for plant growth, excess water can be a detriment because draining gravitational water can remove nutrients and decrease a soil's fertility. Too much water for long times can destroy fertility completely.

11.3.7 Soil classification and naming

Soils have highly variable properties, so a standard way of naming them can be very useful and eliminate the need for lengthy descriptions. The United States Department of Agriculture has developed a hierarchical taxonomy, similar in many ways to the Linnean system used in biology to classify living things (kingdom, phylum, class, order, family, genus, and species). The taxonomy includes orders, suborders, great groups, subgroups, families, and series to describe soils of different types and origins; Table 11.7 describes some of the most important soil orders in the USDA system. The percentage values are the percentage of the world's ice-free land area covered by the soil type.

11.3.8 Soil degradation

Earth contains a finite amount of *arable land*, land that is suitable for growing crops. The numbers in Table 11.7 indicate that mollisols (grassland soils), alfisols (hardwood forest soils), and andisols (volcanic soils), the soil types best suited for agriculture, only add up to about 18% of the available land. But the amount available changes regionally and temporarily, primarily due to human activities. Today the trend is going in a negative direction—we are losing the best arable land at a significant but unknown rate and increasingly relying on chemical inputs and irrigation to produce sufficient food. Simultaneously, the human population continues to grow, and in many places food production cannot keep up with demand. Urban sprawl and related consequences account for some loss of arable land, but loss of soil and loss of soil quality are more significant. The six most significant things contributing to these losses are erosion, compaction, desertification, acidification, contamination, and salination.

11.3.8.1 Erosion

Soil erosion occurs naturally and, over time, has been balanced by soil production. However, a growing human population has caused the rate of erosion to increase, and we are losing 50,000–100,000 square kilometers (19300–38600 square miles) of agriculturally productive land every year. Maintaining soil health is very important. Healthy topsoil contains humus that holds soil together, resisting erosion. Additionally, healthy topsoil supports vegetation that also prevents erosion. Unfortunately, poor agricultural practices, including overcultivation and compaction, can destroy soil structure, leading to accelerated topsoil loss. Other human causes of soil erosion include deforestation, overgrazing, inappropriate construction, and fire.

Figure 11.51 shows farm fields on a farm in Woodbury County in northwest Iowa. Here, terraces, conservation tillage, and conservation buffers are in place to slow soil erosion and improve water quality in the nearby stream. Note the vegetated buffer between the field and the river.

11.3.8.2 Compaction

Earlier in this chapter, we said that a typical soil was about half solid material with the rest being air and water. If soil dries completely, the air and water spaces

Table 11.7 Major soil types (orders) of the USDA soil taxonomy, ordered from most common (%) to least common

Order	%	Description
Inceptisols	17	Form quickly through alteration of parent material. Variable but generally poor for agriculture.
Entisols	16	Do not show any "significant" soil profile development. Minimal soil horizons. Variable but generally poor for agriculture.
Aridisols	12	Form in an arid or semiarid climate. Poor for agriculture except with intense irrigation.
Alfisols	10	Moderately weathered, form under boreal or hardwood forests, rich in iron and aluminum. Good for agriculture.
Gelisols	9	Soils of very cold climates, which are defined as containing permafrost within two meters of the soil surface. Poor for agriculture with limited growing season.
Oxisols	8	Highly weathered soils typically found in tropical environments such as rainforests. Very poor for agriculture.
Ultisols	8	Highly leached soils, often of subtropical climates; dominated by red clays and oxides. Poor for agriculture without significant chemical inputs.
Mollisols	7	Form in semiarid to semihumid areas, typically under a grassland cover. Excellent for agriculture.
Spodosols	4	Typical soils of coniferous or boreal forests. Poor for agriculture without significant chemical inputs.
Vertisols	2	High content of expansive clay. Limited but good for some specific kinds of agriculture.
Andisols	1	Form in volcanic ash and defined as containing high proportions of glass and amorphous colloidal materials. Small extent but potentially good for agriculture.
Histosols	1	Consist primarily of organic materials; equivalent to a peat bog. Poor for agriculture.

Figure 11.51   Terraces, conservation tillage, and conservation buffers on a farm in Woodbury County, northwest Iowa.

Photo by Lynn Betts, USDA Natural Resources Conservation Service.

remain as pores. A soil's *bulk density* is the dry weight of the soil divided by its volume, and the bulk density value reflects the amount of soil pore space. Winter freezing and thawing, and months of rest, tend to cause soil to become less dense. So, bulk density of most soils is lowest in the spring. To grow efficiently, plant roots require space to grow, and if bulk density gets too high, plant root growth becomes restricted and crop yields drop. This is a major issue facing agriculture today, caused largely by modern farming practices that compact soil. For example, bulk density increases when pore space is lost due to tractors and other farm traffic compacting land, or because tilling breaks up aggregates and eliminates space between them. Adding organic matter can sometimes alleviate the problem by decreasing soil density, but usually only near the soil surface.

11.3.8.3   Desertification

Desertification occurs when dry land areas become increasingly arid. Sometimes, lands lose all sources of water and, consequently, vegetation and wildlife. The loss of vegetation leads to significant soil erosion by wind and water, causing the problems to intensify. Desertification is different from drought. Droughts are common in arid lands, but most lands recover after a drought ends. Desertification, in contrast, is a semi-permanent condition, because so much land degradation has occurred that reversing it is, at best, difficult. Although natural climate changes can take their toll, the major causes of most desertification are population pressures and poverty. Contributing factors include overgrazing and cultivation in marginal lands, removal of vegetation for firewood, and irrigation leading to salt buildup in topsoil.

Figure 11.52 shows an example of human-caused desertification. The land to the right of the California fence line has been overgrazed and most of the vegetation is gone. The organic content of the soils is low and erosion by wind is high. Soon the soil may support no life at all. In contrast, the land to the left of the fence, which is also grazing land, has not deteriorated due to better land-use practices. Restoration of desertified land, such as that shown on the right side of this photo, is difficult and costly, and often impossible.

11.3.8.4   Acidification

Soil acidification can lower crop productivity and, by decreasing soil health, increase soil erosion. As discussed previously, natural processes, especially leaching, can lead to soil acidification by removing alkali and alkali earth elements (Na, K, Ca, Mg). Human activities, too, can lead to acidification, for example by growing and harvesting more crops than land can sustain.

Figure 11.52   Healthy and overgrazed land near Fresno, California.

Photo from California Department of Fish and Wildlife.

Nitrogen-based fertilizers contribute to acidification because they may convert to nitrous and nitric acid. Additionally, acidification may occur because of acid precipitation that includes nitric or sulfuric acids.

11.3.8.5 Contamination

Soil contaminants, at low levels, are often within a soil's natural capacity to neutralize or assimilate, but modern municipal and industrial practices sometimes exceed the limits. Drain fields and some other waste treatment processes rely on soil to remove organics and toxics, and if the soil's capacity is exceeded, soil degradation will occur. In metropolitan areas, *brownfields* (abandoned polluted industrial lands) exist where soil contamination by industry makes any development problematic—or impossible. And, in today's modern world, electronic waste, or *e-waste*, is becoming a problem because many discarded electronic devices include lead, cadmium, beryllium, or other toxic components. Proper recycling of old electronics, rather than just tossing them in the garbage, is critical to addressing this problem.

Soil contamination is of special concern because it poses a direct threat to human health. For example, contaminated soil may contain arsenic, lead, cadmium, mercury, or other toxic heavy metals. Although these and other toxic metals sometimes come from natural sources, they are often byproducts of burning fossil fuels, incineration, mining, and other human activities such as landfilling and sewage treatment. Besides metals, toxic organic chemicals can accumulate in soil. They are most often associated with agriculture involving pesticide or herbicide use or industrial activities. Sometimes they are transported long distances by air before being deposited and incorporated in soil.

Cleaning up contaminated soil is not difficult, but can be costly. And, once contaminated soil is removed, disposing of it becomes problematic. So, the only good solution is to avoid contamination in the first place. Figure 11.53 shows an example of soil contamination caused by leakage from underground storage tanks that contained tar. This relatively small site can probably be restored to good condition by replacing all the affected soil. Other cleanup approaches sometimes used include chemical treatment, engineered leaching, bioremediation, sparging, or simply waiting for contaminants to decompose.

11.3.8.6 Salination

Soil *salination* results when salt accumulates in topsoil. Sometimes the salt is visible at the surface; Figure 11.54

Figure 11.53   Soil contamination caused by underground storage tanks containing tar. Encountered during remediation works at a disused gasworks.
Photo from Dumelow, Wikimedia Commons.

Figure 11.54   Salt on rangeland in Colorado.
Photo from Paleorthid, Wikimedia Commons.

shows where this has occurred in Colorado. Salination is a significant problem in arid and semiarid regions where people use irrigation and evaporation rates are high. Even if irrigation waters only contain a small amount of salt, evaporation may eventually lead to salt buildup. Salt may also accumulate if a hydrologic system is thrown out of balance, causing rising water tables to carry salt to the surface. Salination may appear quickly but is difficult or impossible to remove. Typically, it leads to loss of all soil productivity and continuing water quality problems. Many scholars believe that salination was the ultimate cause of the collapse of Mesopotamian society in the 6th century BC.

Questions for thought—chapter 11

1. Weathering and erosion are two related but different processes. Distinguish between the two processes. Explain how weathering can naturally lead to erosion. And, how can erosion naturally lead to more weathering?

2. There are two fundamental kinds of weathering: physical weathering and chemical weathering. Describe these processes. What are the typical products of each?

3. Why is chemical weathering generally more rapid in warm climates than in cold climates?

4. What happens to rocks during physical weathering? What agents cause physical weathering? Can wind be involved? Explain.

5. How does physical weathering promote chemical weathering? And vice-versa.

6. How may plants, animals, and other organisms promote weathering?

7. What is the difference between clastic and chemical sediments? How do each form?

8. Why are the sediments found in deserts mostly sand?

9. How may sandy beaches protect shorelines from erosion?

10. Ice is a very soft substance. Yet, glaciers can scratch and abrade rocks. How can this happen? Explain.

11. What is karst topography? Why does it only form in regions where the bedrock is some kind of limestone? What kind of weathering plays the largest role creating karst topography?

12. What are the differences between speleothems, travertine, tufa, and marl? What do they have in common?

13. What mineral(s) is/are most likely to occur in well-sorted sand? Be specific and explain why.

14. What is the Earth's *critical zone*? What is in the critical zone and why is it important to life?

15. What are he most important characteristics of a fertile soil? List several and explain.

16. What kind of environments are most likely to produce soils with only one or two horizons? Explain why.

17. Why are the soils of the Great Plains excellent for farming many crops?

18. Why may soils that form from transported materials have more nutrients than those that form from residual materials?

19. How do plants protect soils from erosion and degradation?

20. What causes nitrogen fixation and why is it important?

21. Most soils contain *peds*. What are peds and why are they important? What processes can destroy peds?

22. Use names from the USDA Soil Classification System: what types of soils would be found in wet tropical environments, in deserts, in peat bogs, in grasslands, or in coniferous forests?

23. What is desertification and what causes it? Why is it difficult to reverse? How could it best be avoided?

12 Water and the Hydrosphere

12.1 The Salton Sea

The Salton Sea, having a surface area of about 1350 square kilometers, is the largest lake in California. As shown in Figure 12.1, it is in the Imperial and Coachella Valleys, about 70 kilometers north of the Mexican border and 150 kilometers northeast of San Diego. The Salton Sea surface is below sea level, only 1–2 meters higher than Badwater Basin, the lowest part of Death Valley and the lowest place in the United States, but its water depth is generally no more than about 15 meters.

The Imperial and Coachella valleys were the sites of natural desert lakes in the past (collectively referred to as Lake Cahuilla), but the last natural lake disappeared in the 1600s. The present lake was created by accident when a human engineering project went awry in the early 20th century.

Figure 12.2 is a satellite image of the Salton Sea and surrounding region. Irrigation has been practiced there

The Salton Sea and Vicinity

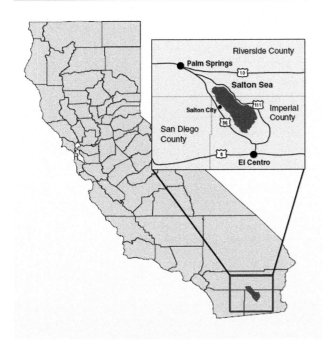

Figure 12.1   The location of the Salton Sea in southern California.
Map from Legislative Analysts Office, California.

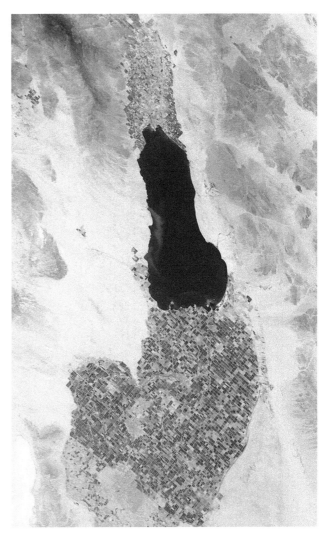

Figure 12.2   Satellite photo showing irrigation districts north and south of the Salton Sea, California.
Image from Bsmuc64ger, Wikimedia Commons.

for more than a century. Today's irrigation projects north and south of the Salton Sea are clearly seen in this image. The Imperial Valley (south of the Salton Sea) and Coachella Valley (north of the Salton Sea) are important food-producing districts today, producing about $2 billion of crops each year. Plans to divert Colorado River water to irrigate the Cahuilla Basin date back to 1859, but it was not until the California Development Company (CDC) was created in 1896 that things moved forward. The first irrigation canal was completed in 1901 and was soon followed by others as farmers dug more than 400 miles of irrigation ditches. Unfortunately, most of the water-delivery canals silted up soon after excavation, and the CDC scrambled to construct new canals and bypass delivery systems. Hasty engineering and heavy rains associated with an early 1905 El Niño season combined to overwhelm temporary dikes, and in February 1905 the entire Colorado River was flowing into the desert. For two years, nearly continuous flooding eroded topsoil and flooded towns and farmsteads, creating the Salton Sea.

Many attempts were made to restore the Colorado to its natural course, and some engineers were skeptical that it could be done. Finally, on February 10, 1907, the breach was closed. The final repair project cost more than $3 million, involved a workforce of 2000 men, and required dike works made from more than 3000 train-car loads of sand, rock, and timber.

Although the Salton Sea has no natural outlet, after the Colorado River was returned to its natural path, Salton Sea water level fell quickly for about 20 years (Fig. 12.3, top). Subsequently, increased irrigation return flows caused the lake surface level to rise again. At the same time, salinity began to increase (Fig. 12.3, bottom) due to evaporation that caused salt concentrations to increase, and the sea is now saltier than typical ocean water. Yet, the Salton Sea is not the saltiest body of water in California, nor is it close to the saltiest sea on Earth. As shown in Figure 12.3, Mono Lake, east of Yosemite National Park, 680 kilometers (425 miles) north of the Salton Sea, has greater salinity, and the Dead Sea, in the Middle East, has salinity nearly 10 times that of the Salton Sea.

At its highest point in 1905, the Salton Sea water surface reached an elevation of 195 feet below sea level, but engineers predicted the sea would dry up after river input was stopped. And, in fact, water level dropped to about 250 feet below sea level over the next decade. However, unexpectedly, it then leveled off and began to rise again (Fig. 12.3). Unaccounted for agricultural

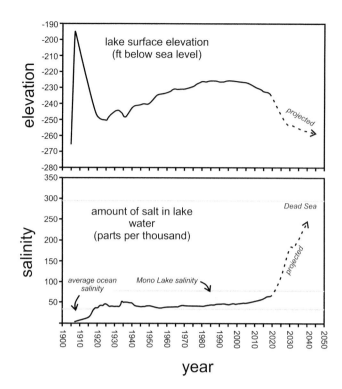

Figure 12.3   Salton Sea lake level and salinity.
Based on a drawing from the Pacific Institute (2014).

runoff, from irrigation projects both north and south of the sea, matched and eventually surpassed evaporation rates, and sea level rose for 90 years. Since about 2000, however, the Sea level has fallen. Population growth in the Southwest has led to increased demand for Colorado River water, and consequently, less water is available for irrigation and maintaining water levels. Lake levels are dropping and lack of freshwater input means that salinities are expected to increase greatly in the next few decades, perhaps surpassing the salinity of the Dead Sea by mid-century (Fig. 12.3).

Some amount of salinity can be good for supporting healthy ecosystems, and several decades after the sea's creation, before the lake became as salty as an ocean, lake water chemistry reached conditions that supported desirable aquatic life and kept the water clear. In the 1950s, resort towns sprang up: Salton City, Salton Sea Beach, Desert Shores, Bombay Beach, and others. A championship golf course and yacht club appeared along its shores. The California Department of Fish and Game stocked the lake with orangemouth corvina and other fish popular with anglers, and tilapia, introduced in 1964, supported a short-lived commercial fishing industry. For 20 years, Salton Sea State Park was one of the most popular destinations in the country, receiving more visitors than Yosemite National Park.

Yet, by the late 1970s, attendance rates were in decline. Increasing salinity, pollution, and flooding took their toll, and within a decade many of the once-prosperous communities were partly or entirely abandoned. Figure 12.4 shows dead and decaying fish on the Salton Sea shore in 2002. The main problem was salinity. Around 1970, the sea's salt content surpassed that of the Pacific Ocean, and by 1980 it was reaching levels that could not be tolerated by many sport fish. Introduction of new species led to some respite but only delayed the inevitable, and massive fish die-offs have occurred since then.

For most of the 20th century, the Salton Sea was a key stopover on the California flyway. Most other California wetlands were drained for agriculture in the mid- to late 20th century, and the sea provided vital habitats for pelicans, quail, orioles, curlews, egrets, cormorants, and many other species. More than 400 bird species were present at one time, equivalent to about half of all bird species in the United States and Canada. Unfortunately, the loss of fish due to increased salinity meant loss of food, and bird populations crashed. Bacteria and disease, along with agricultural pollution, took their toll, too, and bird populations are only a fraction of what they once were.

Salts are not the only things that have become concentrated in the sea. Nutrients (phosphates and nitrates) have increased to high levels, leading to increased plant production. Subsequently, the decay of (too much) organic matter uses up available oxygen, causing parts of the sea to become incapable of supporting life. Algae blooms, accompanied by stinky sulfur smells, are common, and drying sediments lead to serious dust storms, spreading dust, salt, potentially toxic chemicals, and bad odors across vast regions.

Many ambitious plans have been proposed to restore the sea. Some involve removing the salt from the water, others involve diluting the sea with fresh water, and still others involve canal works that would exchange Salton Sea waters for ocean water. All proposals come with a huge price tag, and many are politically and practically impossible.

12.2 Water on Earth

Figures 12.5 and 12.6 show how water is distributed on Earth. Almost all of it—97%—is in the oceans. Small amounts are found in groundwater reservoirs and frozen in glaciers and ice caps. Minuscule amounts are found in other settings. Although 75% of our planet is covered by water, most cannot sustain human life. According to the United States Geological Survey, the total amount of water on Earth is approximately 1.4 billion cubic kilometers, but 97% of it is salt water (Figs. 12.5 and 12.6). Some marine life can survive in salt water, but most of the biosphere cannot.

Just over two-thirds of Earth's fresh water is frozen in ice (Fig. 12.6). Most of the rest is soil moisture or deep groundwater that is difficult or impossible to access. Less than half a percent is available in lakes, rivers, streams, or in shallow groundwater reservoirs—but we rely on those sources to meet most of our needs. These sources are easily overused because their replenishment is generally slow and varies with seasons and longer-term climate changes.

Thus, fresh water—from rivers, lakes, bogs, ponds, and groundwater reservoirs—is but a small part of

Figure 12.4   Dead fish on the shore of the Salton Sea in 2000.
Photo from Jeff T. Alu, Wikimedia Commons.

Figure 12.5   Where is the water on Earth?

Photos from (a) NESCF/NOAA; (b and c) D. J. Bergsma and Neogeolegend, Wikimedia Commons; (d and e) USGS; (f) NASA.

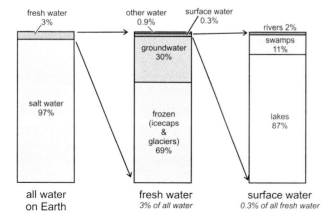

Figure 12.6   Distribution of water on Earth.

Figure modified from the USGS.

Earth's total water supply, but is necessary for the existence of just about all land plants and animals on our planet. Furthermore, many human activities beyond basic survival rely on sources of fresh water, including food production, manufacturing, household maintenance, recreation, and environmental management. Fresh water is also used as a wastewater repository, a mode of transportation, a generator of hydropower, and also has aesthetic value. Yet today, a rapidly growing human population, especially in warm, dry regions, and increased per capita water use mean that fresh water is often in short supply. The uneven distribution of water further exacerbates the problem. India, for example, has less than 10% of the world's fresh water but almost 20% of the world's population.

Some parts of the world do not have abundant supplies of useable water. The shortages affect people directly and often affect food production. Figure 12.7 shows, in red and orange, regions with the worst water-related problems. These problems include insufficient amounts of available water, lack of funds to develop water resources, and in some places high risks of flooding. Note the much of North America is colored in red and orange. Water shortages are already taking their toll today.

Historically, fresh water has been the key to wealth because it is essential for agriculture. Before the ascendency of Babylon 3800 years ago, all people relied on natural systems to deliver crop water when needed and people generally lived near freshwater supplies. Babylonians and now modern humans have become hydraulic engineers—moving water from place to place, adding water to land for irrigation, removing it for drainage, and impounding it for use at a future time. Today, people grow food in many places once thought not arable. Unreliable and insufficient water supplies require engineering projects, including dams, diversions, aqueducts, wells, and saltwater desalination factories. And today, many rivers and lakes, as well as groundwater systems, have inflows and outflows that were engineered, and can be controlled, by people.

Irrigation is by far the largest consumer of water today. As shown in Figure 12.8, irrigation accounts for just over 80% of the total water use in the United States (and in some western states more than 90%). For the world as a whole, irrigation accounts for about 70% of the water consumed, industry uses about 20%, and domestic (household) use is only about 10% on average. Locally, however, there are some exceptions—for example, in some European countries, the bulk of the water is consumed by industry.

As shown in Figure 12.9, water use is greatest in the most developed regions of the world. But demand is increasing in other regions. Demand for fresh water has tripled in the last 50 years, primarily because of population growth, but also because of economic development and increasing affluence. Water consumption in North America and Europe—the two largest bars in Figure 12.9—is mostly due to agriculture. As India, China, and other countries develop more agriculture, demand in Asia will increase and may eventually eclipse North America. Lifestyle changes, too, require more water, biofuels and other manufacturing processes require more water, and some energy development (e.g., hydraulic fracturing to produce oil and natural gas) requires more water. The United Nations predicts that demand for fresh water will increase by 20% in highly developed countries during the next decade; it will double in other countries.

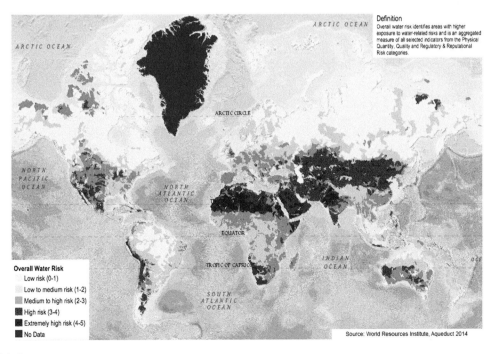

Figure 12.7 Global water stress.
Drawing from Sampa, Wikimedia Commons.

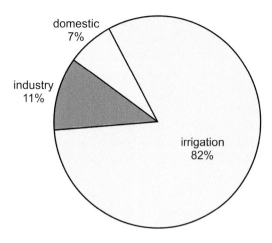

Figure 12.8   Water consumption in the United States.

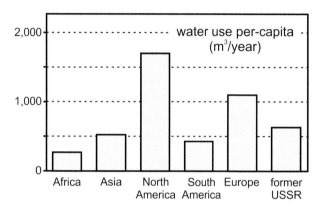

Figure 12.9   The per capita water use in different parts of the world (cubic meters/year).

Figure 12.10   The Aral Sea in Uzbekistan (top) and the Colorado River in Mexico (bottom).

Photos from Audun Kjørstad (top) and Pete McBride (bottom), Wikimedia Commons.

When people use water, much of it ultimately returns to the water system. For example, wastewater is often treated and then returned to rivers, and cooling water from power plants flows back into rivers. In contrast, *consumptive water uses* remove water permanently. Evaporation from irrigated fields and transpiration by agricultural crops, for example, move water from local water supplies into the atmosphere and then to distant places downwind. In many agricultural areas, major rivers used for agriculture do not make it to the sea because of consumptive water use. This problem is especially acute when people try to grow crops in arid land, such as the desert valleys of California.

A fifth of the world's population lives in places where fresh water is being withdrawn from rivers and lakes faster than it is replaced. So, rivers and lakes are drying up; Figure 12.10 shows two examples. In

Uzbekistan (Fig. 12.10, top), river runoff once replenished the Aral Sea yearly, but since 1940, most of the river water has been diverted to support irrigated agriculture. As a consequence, the Aral Sea is today a fraction of what it originally was. In North America, the Hoover Dam, completed in 1935, was the first of a handful of water projects that today divert Colorado River water to agricultural districts and major southwest United States cities. The 1450-mile-long river is, today, generally dry as it nears its mouth at the upper end of the Gulf of California (Fig. 12.10, bottom). The once lush delta is today a desert of cracked earth and salt flats in all but the wettest parts of the wettest years. Groundwater sources, too, are in jeopardy. In Europe, where surface water sources are already used to the maximum extent possible, 60% of the large cities are withdrawing groundwater faster than it is being replaced.

12.3 The water cycle

The total amount of water on Earth and in the atmosphere is constant, but water moves around, sometimes changing from liquid to vapor or to ice as part of the *hydrologic cycle* (the *water cycle*). The ultimate energy source for water circulation is the Sun; without solar energy, there would be no evaporation and no cycling.

Because most of Earth's water is in the oceans, we can, if we wish, think of the water cycle as starting

with evaporation of ocean water. Figure 12.11 shows a standard view of the water cycle, but such depictions are misleading. The implication of Figure 12.11 is that major paths of water circulation are from the oceans to the atmosphere, to precipitation on land, and then back to the ocean via surface or groundwater flow. In reality, most water that evaporates falls back into the oceans where it started, and the much smaller amount of water that evaporates over land mostly falls back onto the land as shown in Figure 12.12. So, almost all water movement is vertical, not horizontal. Most water circulation

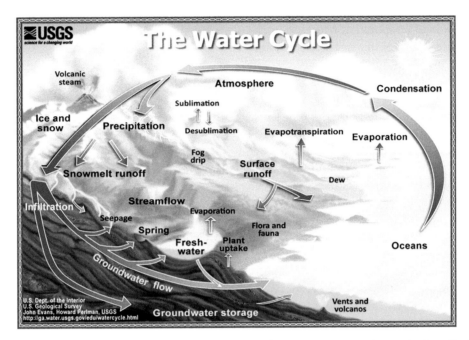

Figure 12.11 A classic view of the water cycle, showing circulation of water from oceans to atmosphere, to land, and back to oceans again. Photo from the USGS.

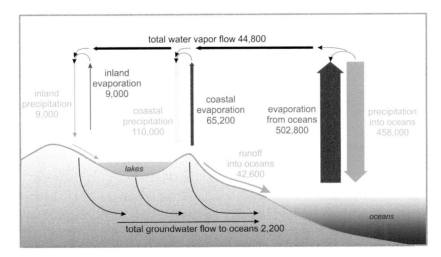

Figure 12.12 A more appropriate view of the water cycle. The numbers are cubic kilometers of water per year.

is up and down over oceans, shown by the arrows in this illustration. Only a relatively small amount is transported from oceans to continental areas and back again. When ocean water evaporates, salts and other dissolved matter are left behind, and fresh water enters the atmosphere. Over time, then, the oceans may become saltier, but the process is a very slow one, because most ocean evaporation (91%) simply falls back into the ocean and (less significantly) because runoff from continents adds water to the oceans.

About 9% of all evaporated ocean water eventually precipitates on land, but this figure, too, is misleading because most of the water in the atmosphere over continents derives from land (evaporation from soil, rock, lakes, rivers, etc.) and a smaller but significant amount from *transpiration* (water emitted by plants directly into the atmosphere). Subsequent precipitation returns water to Earth's surface where it may run off, evaporate, or infiltrate into the ground as it starts heading back toward the sea. In most places, the amount of *evapotranspiration* (evaporation and transpiration together) is greater than runoff and much greater than infiltration, but rates are variable and depend on climate, vegetation type, geology, and, often, human activities.

Some of the water that falls on land becomes *internal runoff*—surface or groundwater flows that never make it to the ocean, but instead collect in closed drainage basins. Evaporation of ocean water does not significantly change ocean water chemistry, but evaporation of inland lakes and seas often causes them to become very salty, much saltier than standard ocean water. Fresh water removed by evaporation may not precipitate in the same watershed, and salt content inexorably increases.

Figure 12.13 shows salt deposits on the shore of the Dead Sea near En Boqeq, Israel. This sea lies at an elevation of 423 meters below sea level; its shores are the lowest dry lands on Earth. Located in the Jordan Rift Valley between Jordan and Israel, the Dead Sea is 55 by 18 kilometers in size, and nearly 10 times saltier than the oceans. Other such water bodies are found on just about all continents. In North America, Utah's Great Salt Lake has variable salinity depending on lake level, ranging from just a bit saltier than the oceans to nearly the salinity of the Dead Sea.

Runoff and runoff rates are of particular importance to people because streams and rivers are conventional sources of fresh water. Of course, the ultimate source of runoff is precipitation, so we can expect places with the most precipitation to have the most runoff available

Figure 12.13 Salt deposits on the shore of the Dead Sea. Photo from XTA11, Wikimedia Commons.

for human use. However, the relationship between precipitation and runoff is complicated because some precipitation may enter the groundwater system, become frozen, or evaporate and reprecipitate before entering lakes or rivers. Additionally, in some places evaporation rates are so high that runoff is limited. In general, this means that runoff rates are lower in warm central latitudes, even if precipitation rates are high.

The amount of infiltration is different in different places, depending most significantly on climate, but also on shorter-term precipitation and, of lesser significance, evaporation rates. Soil texture and structure, vegetation, and soil saturation levels also limit or promote infiltration. Sandy soils have high infiltration rates compared with clay-rich soils. Vegetation and roots can promote infiltration; water may barely infiltrate hard-packed non-vegetated soils. Worldwide, about half the water that falls on Earth becomes surface run off. About a third of the world's total runoff comes from Asia, slightly less from South America, and 19% from North America. The rest of the continents add up to less than 20%. There are many paths, but most water eventually makes its way back to the oceans.

Although we talk about the water cycle, most water does not move quickly and sits in the same general place for long times. Most of it resides in an ocean, with almost all the remainder in ice at the poles or in underground reservoirs. *Residence time*—how long water sits in one reservoir—is a key concept directly related to replacement time (Table 12.1; and see Chapter 2). Groundwater, for instance, can be very old (and is sometimes called *fossil water*), while soil water is always very young.

Table 12.1 Reservoirs and residence times.

	Reservoir	Average residence time
long residence times	ice caps, glaciers, permafrosts	1000 to 20,000 years
	groundwater: deep oceans	up to 10,000 3200 years
moderate residence times	mountain glaciers	20 to 100 years
	lakes	50 to 100 years
	Shallow groundwater	days to years
short residence times	seasonal snow cover	2 to 6 months
	rivers	2 to 6 months
	soil moisture	1 to 2 months
	atmosphere	9 days

From Wikipedia "water cycle" and Physicalgeography.net.

Atmospheric water is typically younger. Young waters turn over (are replenished) quickly and have short residence times. Old waters turn over slowly and have very long residence times. Generally, we calculate residence time by assuming a reservoir has a constant volume and dividing by average recharge rate. It is possible, however, to use isotopes or other geochemical signatures to determine a real residence time (age in years) for any sample of water.

12.4 The ocean systems

Earth's oceans (Fig. 12.14), with an average depth of about 3760 meters and an area of slightly more than 360 million square kilometers, contain 97% of all water. The total volume of ocean water is estimated to be 1.4 billion cubic kilometers, and total yearly water input is about 503,000 cubic kilometers—about 0.03% of the volume. So, dividing these numbers reveals that ocean water residence time is on the order of 3000 years. During this time, however, the water circulates throughout all major ocean basins. Horizontal circulation is slow, typically 0.01–1.0 meters/second, and vertical mixing is much slower. Consequently, the ocean is stratified, with most water movement horizontal.

Figure 12.15 shows some of the major paths of ocean circulation. Surface currents carry a lot of heat and consequently influence climate. In this map, warm currents are shown in red and colder currents in blue. Other important currents that affect climate (not shown here) occur at greater depths.

Although tides produce local currents, global-scale ocean circulation is driven primarily by two forces: wind acting on the oceans' surfaces and variations in seawater

Figure 12.14 The oceans account for more than 97% of Earth's water.
Photo from Jacob Bøtter, Wikimedia Commons.

density due to variations in temperature and salinity. These currents are conveniently divided into *shallow currents* (down to 200–300 meters), also called the *surface layer* or *mixed layer*, and *deep currents*. Shallow currents are driven mostly by wind. The warmer shallow currents, shown by red arrows in this figure, account for only about 10% of all ocean circulation. The colder and more stratified deep currents account for the rest.

Between 60° north and south latitudes, Coriolis forces produced by Earth's rotation cause shallow currents to be deflected right as water flows north or south in the Northern Hemisphere. The opposite occurs in the southern hemisphere: water is deflected to the left as it flows north or south. Additionally, in both hemispheres, continents block current flow. The combination produces *gyres*, circular water movements that are clockwise in the Northern Hemisphere and counterclockwise in the Southern Hemisphere, which are clearly depicted by the arrows in Figure 12.15. Consequently, we get the broad southern-flowing currents in the eastern Atlantic and the stronger, narrower Gulf Stream current in the western Atlantic.

Figure 12.16 shows sea surface temperatures from warm (red) to cool (blue) in the north Atlantic. Clockwise circulation produces the Gulf Stream that originates in the Gulf of Mexico and carries a great deal of warm water north past Florida and up the east coast of the United States. Thus, gyre circulation moves a great deal of heat to high latitudes. Besides moving heat, gyre circulation also leads to upwelling of cold, nutrient-rich deep waters that host great fisheries.

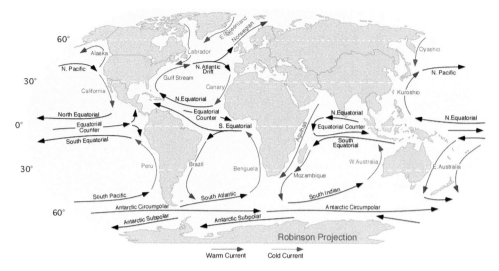

Figure 12.15   The major currents near the oceans' surfaces.
Map from Michael Pidwirny, Wikimedia Commons.

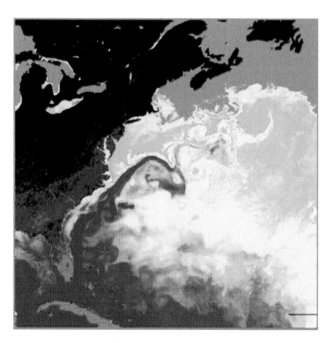

Figure 12.16   Sea surface temperatures from warm (red) to cool (blue) in the Atlantic Ocean off the east coast of North America.
Image from NASA, Wikimedia Commons.

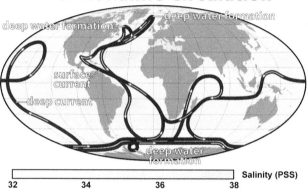

Figure 12.17   Thermohaline circulation in the oceans. PSS is the practical salinity scale.
Map from NASA Earth Observatory, Wikimedia Commons.

Figure 12.17 shows the overall global patterns of ocean circulation, also called *thermohaline circulation*. Although shallow currents (shown in red) are mostly wind driven, deep ocean currents (shown in blue) are largely unaffected by wind energy. They are, instead, driven primarily by water density differences due to variations in temperature and salinity. The density-driven currents reach the deep ocean floors and are a key part of the "global conveyor belt" that transports seawater between ocean basins on a worldwide scale. The deep oceans are stratified, with the densest waters at the bottom, and because it is difficult for dense water to move upwards, most flow is horizontal. Most water in the Atlantic Ocean only returns to the surface every 1000–2000 years, and most water in the Pacific Ocean only returns to the surface every 200–500 years.

The main driving force behind the thermohaline circulation is the formation of dense water masses in the North Atlantic. Evaporation of warm, salty equatorial water that flows into the North Atlantic removes heat from water, decreasing temperature while increasing salinity and density. Eventually the water becomes dense enough to sink. The sinking water mass, called the *North Atlantic Deep Water*, travels to great depth before returning south by deep currents, shown by blue lines in Figure 12.17. Some deep currents travel around Africa and, after warming and some dilution, rise to the surface in the Indian Ocean. Other deep currents flow into the North Pacific before rising. Shallower currents cycle water back toward the Atlantic, warming on the way, so the cycle can repeat.

Figure 12.18 shows ocean-surface temperature imaged by NASA's *Aqua* spacecraft in August 2003. The oceans are warmest near the equator where, at midday on a clear day, seawater absorbs 98% of incoming solar energy. Most heating of Earth's atmosphere occurs near the equator as ocean water evaporates and warm moist air rises and flows north and south in the *Intertropical Convergence Zones.* It cools and descends, and eventually flows back toward the equator at low elevations. As the water-saturated air cools, water vapor precipitates as rain, explaining why there is so much rain in tropical regions. This circulation, extending from the equator to about 30° north and south latitudes, termed *Hadley Circulation*, is the key force driving most of Earth's atmospheric circulation.

Most of the heat trapped on Earth by greenhouse gases ultimately goes into the oceans, which hold nearly 20 times more heat than Earth's atmosphere. Heat carried by ocean currents to high latitudes warms the atmosphere and has significant effect on climate. This is particularly important in some cold regions, where heat emitted from the oceans keeps air and land temperatures warmer than they would be otherwise.

Figure 12.19 shows a garden with palm trees at Poolewe, Scotland. Poolewe is close to the same latitude as Anchorage, Alaska, but the vegetation in Scotland is not at all like that in Anchorage. Heat carried by the Gulf Stream is responsible for the near-tropical climates on the west coast of Scotland during some parts of the year. Some scientists are concerned that the current atmospheric warming may change ocean circulation, altering this very important heat transportation. They point out that large volumes of fresh water entering the North Atlantic can affect ocean salinity and temperature, keys to thermohaline circulation. The bleakest predictions are that the North Atlantic Deep Water will cease to flow, lowering average temperatures in the British Isles and Scandinavia by tens of degrees by the end of the century.

Variations in the ocean's circulation can lead to seasonal or yearly variations in heat transport and to variations in weather patterns. One important variation is the change in the equatorial circulation known as *El Niño* that occurs with an irregular period of two to five years. During El Niño years, a band of anomalously warm current near the equator flows across the southern Pacific Ocean to the western coast of South America. The warm current stops the upwelling of cold nutrient-rich waters near South America, disrupting the food chain and having a devastating impact on aquatic life. Extreme El Niño currents also lead to extreme weather (droughts and floods) in many parts of the world. During *La Niña* years, the opposite occurs—sea surface temperatures are 3–5 °C lower than normal across the southern Pacific.

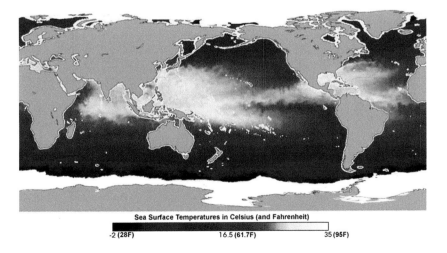

Sea Surface Temperatures in Celsius (and Fahrenheit)

-2 (28F)　　　　16.5 (61.7F)　　　　35 (95F)

Figure 12.18　Ocean-surface temperatures in 2003.
Image from NASA.

Figure 12.19   Garden at Poolewe, Scotland, close to the same latitude as Anchorage, Alaska.

Photo from Vicky Brock, Wikimedia Commons.

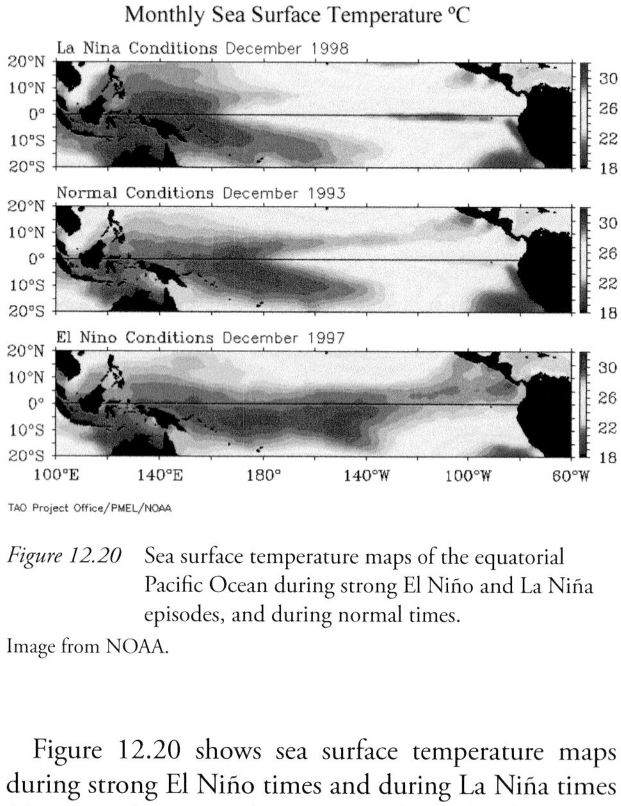

Monthly Sea Surface Temperature °C

Figure 12.20   Sea surface temperature maps of the equatorial Pacific Ocean during strong El Niño and La Niña episodes, and during normal times.

Image from NOAA.

Figure 12.20 shows sea surface temperature maps during strong El Niño times and during La Niña times (the times between El Niño events). Red colors show warm sea water and blue colors show cool. For both El Niño and La Niña, tropical rainfall, wind, and air pressure patterns over the equatorial Pacific Ocean are most strongly linked to sea surface temperatures between December and April. During this time of the year, the effects of El Niño and La Niña conditions are greatest and have large impacts on U.S. weather patterns.

12.5 Freshwater systems

Only 3% of Earth's water is fresh water, and most of that is frozen as ice or snow (Fig. 12.21). The remainder is primarily groundwater. Lakes, rivers, swamps and other wetlands, soil moisture, and atmospheric water only add up to 1.2% of the total. Traditionally, most people have relied on surface sources, rather than groundwater sources, to provide needed freshwater supplies, although this has changed in some parts of the world today. Unfortunately, today, a lack of potable fresh water limits plant and animal populations (including human population) in some parts of the world. Fresh water is arguably the most important resource supporting life on Earth, so its distribution has great social and economic importance.

12.5.1 Rivers and streams

Excess precipitation that does not evaporate or infiltrate to become part of the groundwater system is stored in lakes or runs off in rivers and streams. The amount of runoff depends fundamentally on climate—primarily on the amount of precipitation a region receives and evaporation rates, and secondarily on temperature and humidity. Hot dry regions generally have fewer, and smaller, rivers and streams than cooler wet regions do. However, because infiltration rates depend on geology (bedrock and soil properties, as well as saturation levels), some regions are characterized by high runoff levels and some by low, even if climates are similar. Compared with other parts of the water cycle, rivers and streams move water quickly; residence times average only two to six months.

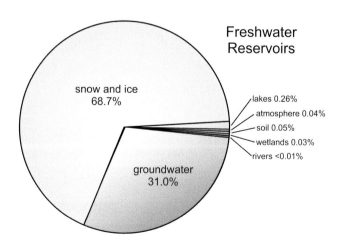

Figure 12.21   Distribution of fresh water on Earth.

Figure 12.22 (top) shows the Mississippi River and some of its tributaries. Rivers vary greatly in size, and size is ambiguous because it can be defined in different ways—most commonly by a river's *length* or *discharge* (the amount of water carried by a river). There are other complications, too. Enumerating river length, for example, depends on unambiguously identifying a river's beginning. Generally, a river's source is defined to be the source of its longest tributary. Using this approach, the beginning of the Mississippi River is at Brower's Spring near West Yellowstone, Montana. The spring flows into the Red Rock River, which flows into the Beaverhead River, which flows into the Jefferson River, and then into the Missouri River at Missouri Headwaters State Park. Eventually the Missouri flows into the Mississippi near St. Louis. This chain of waterways seems clear, yet many cite Lake Itasca in Minnesota as the source of the Mississippi because it is the natural place one ends up if following the course of the Mississippi north from the Gulf of Mexico.

Similar ambiguities arise when trying to determine the longest river in the world. It is either the Amazon or the Nile (shown in the bottom two maps of Figure 12.22), but their lengths are about equal and depend on assumptions about where the rivers begin and end. Although most rivers end at an ocean or large lake, identifying a river's end can be problematic. Where, for example, does the St. Lawrence River end (it becomes increasingly wider over a long distance), and where is the end of a river that dries up in an inland basin? Estuaries and tidal canals are sometimes considered to be parts of rivers, adding length. The Nile River, for example,

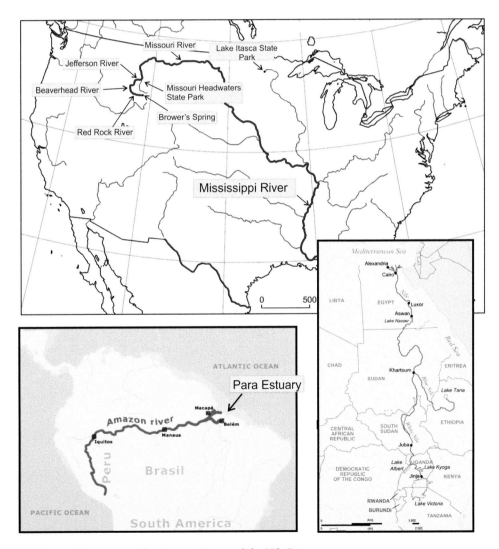

Figure 12.22   The Mississippi River system, the Amazon River, and the Nile River.

is generally said to be longer than the Amazon is, but the Amazon is longer if some of its canals and the Para Estuary are included (Fig. 12.22). Some rivers, such as the Brahmaputra or Ganges, are difficult to delineate because they have many large tributaries and flow into other large rivers.

River size can also be compared by looking at average discharge values or area of drainage basins. Although length and discharge are related (see Table 12.2), they do not correlate exactly because of the influence of climate. Only two of the five longest rivers are also in the top five in terms of discharge (Amazon and Yangtze). The Amazon, Congo, and Nile have the three largest drainage basins. Total yearly discharge from rivers to the world's oceans is about 37,000–39,000 cubic kilometers, which equates to 1.17 million cubic meters/second. So, the discharge from the rivers listed in Table 12.2 accounts for about 40% of the total from all rivers. Of this, a significant proportion comes from the Amazon River alone.

12.5.2 Lakes

Some lakes are closed basins—meaning water collects and does not flow out of the basin—but most are essentially the equivalents of broad, slow-moving rivers.

Table 12.2   Rivers ordered by length and average discharge.

Length rank	Discharge rank	River	Length (km)	Drainage area (km²)	Average discharge (m³/s)	Continent (outflow)	Countries in the drainage basin
1		Nile—Kagera	6650	3,254,555	5100	Africa (Mediterranean Sea)	Ethiopia, Eritrea, Sudan, Uganda, Tanzania, Kenya, Rwanda, Burundi, Egypt, Democratic Republic of the Congo, South Sudan
2	1	Amazon—Ucayali—Apurímac	6400	7,050,000	219,000	S. America (Atlantic Ocean)	Brazil, Peru, Bolivia, Colombia, Ecuador, Venezuela, Guyana
3	4	Yangtze (Cahng Jiang)	6300	1,800,000	31,900	Asia (East China Sea)	China
4	9	Mississippi—Missouri—Jefferson	6275	2,980,000	16,200	N. America (Gulf of Mexico)	United States, Canada
5	7	Yenisei—Angara—Selenge	5539	2,580,000	19,600	Asia (Kara Sea)	Russia (97%), Mongolia (2.9%)
6		Yellow River (Huang je)	5464	745,000	2110	Asia (Bohai Sea)	China
7		Ob—Irtysh	5410	2,990,000	12,800	Asia (Gulf of Ob)	Russia, Kazakhstan, China, Mongolia
8	6	Paraná—Río de la Plata	4880	2,582,672	18,000	S. America (Río de la Plata)	Brazil, Argentina, Paraguay, Bolivia, Uruguay
9	3	Congo—Chambeshi	4700	3,680,000	41,800	Africa (Atlantic Ocean)	Democratic Republic of the Congo, Central African Republic, Angola, Republic of the Congo, Tanzania, Cameroon, Zambia, Burundi, Rwanda
10		Amur—Argun	4444	1,855,000	11,400	Asia (Sea of Okhotsk)	Russia, China, Mongolia
11	8	Lena	4400	2,490,000	16,871	Asia (Arctic Ocean)	Russia
12	10	Mekong	4023	811,000	14,800	Asia (S. China Sea)	China, Myanmar, Laos, Thailand, Cambodia, Vietnam
29	2	Brahmaputra—Ganges—Meghna	2948	1,635,000	42,470	Asia (Bay of Bengal)	India, Bangladesh
58	5	Orinoco	2140	880,000	33,000	S. America (Atlantic Ocean)	Venezuela, Colombia

Water enters primarily by stream flow or precipitation and departs by stream flow or evaporation. Groundwater flow may add water to lakes via springs or remove it by infiltration, but these fluxes are generally small, except for very small lakes or ponds or in karst regions (areas underlain by cavernous limestone).

Table 12.3 lists data for the largest lakes in the world; Figure 12.23 is a plot of volume vs. area for the lakes in the table. Lake depths vary greatly, so surface area and volume of water are not directly correlated (Fig. 12.23 and Table 12.3). The Caspian Sea (371,000 square kilometers) is generally considered the largest lake, but it contains salt water and so is excluded from some lists, and it is not shown in Figure 12.23.

Of the freshwater lakes, Lake Superior has the largest area and Lake Baikal (shown in Figure 12.24) has the largest water volume. Lake Baikal (1642 meters deep) and Lake Tanganyika (1470 meters deep), lakes that occupy rift valleys, are exceptionally deep and thus anomalous. For the other lakes, volume and area are roughly correlated. Lake Vostok, the 16th largest in terms of area, is a subglacial lake in Antarctica and so is often omitted from lists of large lakes. Only a handful of lakes hold most of the fresh lake water on our planet. Lake Baikal, for example, contains a fifth of all non-frozen water on Earth. The total amount of fresh water in the lakes listed in Table 12.3 accounts for about 98% of the fresh surface water on Earth.

Water residence times (and replacement times) are keys for dealing with pollution and water supply

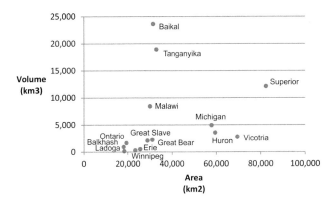

Figure 12.23 Area and volume of the world's largest freshwater lakes.

Figure 12.24 Lake Baikal in Siberia, Russia.
Photo from Sansculotte, Wikimedia Commons.

Table 12.3   Lakes with surface areas > 10,000 square kilometers

Size rank	Name	Countries with shoreline	Area (km²)	Volume (km³)	Average depth (m)
1	Caspian Sea*	Kazakhstan, Russia, Turkmenistan, Azerbaijan, Iran	371,000	78,200	211
2	Superior	Canada, United States	82,414	12,100	147
3	Victoria	Uganda, Kenya, Tanzania	69,485	2750	40
4	Huron	Canada, United States	59,600	3540	59
5	Michigan	United States	58,000	4900	84
6	Tanganyika	Burundi, Tanzania, Zambia, Dem. Republic of the Congo	32,893	18,900	575
7	Baikal	Russia	31,500	23,600	749
8	Great Bear Lake	Canada	31,080	2236	72
9	Malawi	Tanzania, Mozambique, Malawi	30,044	8400	280
10	Great Slave Lake	Canada	28,930	2090	72
11	Erie	Canada, United States	25,719	489	19
12	Winnipeg	Canada	23,553	283	12
13	Ontario	Canada, United States	19,477	1639	84
14	Balkhash*	Kazakhstan	18,428	106	6
15	Ladoga	Russia	18,130	908	50
16	Vostok	Antarctica	15,690	5400 ± 1600	344

*Saltwater lakes.

problems. Average residence times of water in the lakes listed in Table 12.3 is 50–100 years, but some have much shorter or much longer times. Small mountain lakes in wet regions have residence times of hours, days, or weeks. Large and deeper lakes tend to have longer residence times. For example, residence time for Lake Superior is about 200 years, and residence time for Lake Tahoe (California/Nevada) is 650 years. Lake Titicaca (Peru/Bolivia) has water residence time of 1340 years, and Lake Vostok (Antarctica) 13,300 years. The concept of residence time is, however, somewhat misleading for larger lakes. Many of them are stratified, so waters from different levels may not mix, and individual water molecules may remain in lakes for different amounts of time.

12.5.3 Groundwater

Groundwater is, in many places, a convenient potable water source and in some cases the only source of available. Figure 12.25 shows an *artesian well* near Myllylähde Spring in Virttaa, Alastaro, Finland. Artesian wells like the one shown flow naturally without any pumping, because the groundwater reservoir is naturally pressurized.

Groundwater is generally less easily polluted than surface water is, and it is the largest source of fresh water for irrigation, industry, and drinking in the United States and many other places. The total volume of groundwater is more than 30 times that of fresh surface water, but much lies deep beneath the surface and so is difficult or impossible to extract. Groundwater reservoirs are finite and recharge times are long, so groundwater resources

Figure 12.25   Artesian well near Myllylähde spring in Virttaa, Alastaro, Finland.
Photo from MattiPaavola, Wikimedia Commons.

are easily overtapped. Natural replenishment can be very slow, especially in arid or semiarid regions that often have little fresh surface water. Sometimes, in arid regions, water supplies are supplemented with artificial recharge that reroutes rainwater or reclaimed water to underground reservoirs.

Most of the water in the groundwater system originates by infiltration through permeable sediments after precipitation; a very small amount seeps downward to the water table from lakes and streams. A soil's *infiltration rate*, generally measured in millimeters per hour or inches per hour, is the rate at which it absorbs water on its surface. A soil's capacity to hold water is limited by its *porosity*, the amount of open pore or fracture space it contains. If all pores are full of water, the soil is saturated and no further infiltration can occur. If soil is unsaturated, infiltration rate depends on several factors, including the amount of water available, the soil texture and structure, the slope of the soil surface, the temperature, and the vegetation type and cover. A soil's maximum rate of infiltration is its *infiltration capacity*. If precipitation rate is greater than the infiltration capacity, the excess water will run off in rivers or streams, or pool at the surface forming a temporary or permanent lake.

The rate of infiltration also depends on *permeability*, the rate at which water can flow through a soil. Permeability is related to porosity but is not the same thing. A rock with large pores may have low permeability if the pores are not interconnected. Thus, a very porous basalt with lots of open vesicles may be very impermeable. Porosity is generally described as the percentage of a volume of the rock, sediment, or soil that is pore space. Permeability is given in centimeters/second or similar units that describe rate of flow. Relatively impermeable soils become saturated more quickly than permeable ones because water fills pores quickly. So, infiltration may slow or halt completely. Well-sorted sandy soils have high permeability because of interconnected space between grains. Clay-rich soils may have very little permeability because flat aligned clay grains can be barriers to groundwater flow. Vegetation often increases permeability because roots loosen soil and create cracks and other conduits for water to follow.

Infiltrating water will move, by gravity, downward to the water table (shown in Figure 12.26). The water table separates the region above, which contains some empty pore space, from a region below, where water occupies all available pore space. The *unsaturated region* above the water table is also called the *vadose zone*, or the *zone of aeration*. Below the water table, in the *phreatic zone* (also

Figure 12.26 The water table and subsurface hydrology. Drawing from NASA.

called the *zone of saturation*), the soil is saturated, and the water there is called groundwater. The depth of the water table and the size of the phreatic zone may vary depending on season and amount of precipitation. The water table is not a flat surface; in most places, it follows the general topography of the land surface. Water in the saturated zone flows naturally from places where the water table is high to places where it is lower. Flow rates are, however, generally slow—on the order of centimeters per day, except in some coarse sandy soils where flow rates may be meters per day. If river and lake levels are in equilibrium with the water table (which is not always the case), the water table reaches the land surface at river and lake shores (as shown in Figure 12.26).

12.5.3.1 Aquifers

Most groundwater is found in *aquifers*, geological horizons that contain water-saturated permeable material capable of producing significant amounts of water by pumping over extended periods of time. Other horizons may be saturated, but unless they can produce significant water, they are not aquifers. Aquifer water may be extracted from wells in buckets or by pumping, and in some cases may flow naturally at the surface, like the artesian flow shown in Figure 12.25. Good aquifers must be both porous and permeable so they can hold large volumes of water and so water can flow rapidly to replace what is extracted.

Aquifers typically consist of gravel, sand, sandstone, or fractured rock such as limestone. Excellent aquifers produce millions of liters of water a day. Some materials,

including clay or shale, may have a lot of pore space and hold a lot of water, but have low permeability because pores are not connected; they resemble Swiss cheese in texture and do not provide long-term water sources. Such materials are termed *aquitards* (materials with low permeability) or *aquicludes* (materials with no permeability).

Aquifers may be unconfined or confined. In *unconfined aquifers*, groundwater only fills pore space in the bottom part of the aquifer. Unconfined aquifers are also called *phreatic aquifers* or *water table aquifers*, because the upper groundwater surface in an unconfined aquifer is the water table. The *saturated thickness*—a limiter on the total amount of water present—extends from the water table to the (impermeable) base of the aquifer layer. Aquifers are recharged by infiltrating groundwater that flows to the water table and then laterally through the aquifer, as shown by arrows in Figure 12.27. In most settings the water table roughly parallels surface topography. Where the water table reaches the surface, a natural spring will flow (shown near well B in Fig. 12.27). When the water table is below the surface, water may be produced by pumping (wells A and B). Well B need not be as deep as well A, because the water table is closer to the surface at well B.

The right-side of Figure 12.27 shows a *confined aquifer*. In contrast with unconfined aquifers, confined aquifers are overlain and underlain by relatively impermeable material (aquitards) that limit or prohibit groundwater flow (Fig. 12.27). If water in a confined aquifer is pressurized, it may flow upwards in a well or borehole, reaching an elevation higher than the top of the aquifer. The limit of upward flow is the *potentiometric surface*, shown by a blue dashed line in Figure 12.27. If the potentiometric surface is higher than the surface elevation, the pressure is great enough so that water will flow naturally from *artesian springs* or from *artesian wells* dug by people. So, in this illustration, artesian water will flow from wells at locations C and E, but not D. Artesian flow occurs, most commonly, when aquifer recharge is at a higher elevation than a well or spring. Confined aquifers are recharged in regions where the aquifer is exposed at the surface, but recharge areas may be long distances from where water is being produced. In eastern North Dakota, for example, artesian water flows from aquifers that are recharged several hundred miles to the west in the Rocky Mountains.

In some local areas, infiltrating water is unable to make it to the water table because flow is stopped by a low-permeability horizon. In such cases, a small amount of groundwater may become trapped in a *perched*

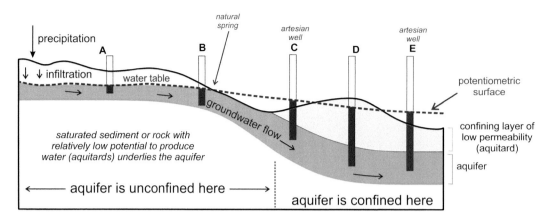

Figure 12.27   Unconfined and confined aquifer, and five well locations (A through E).

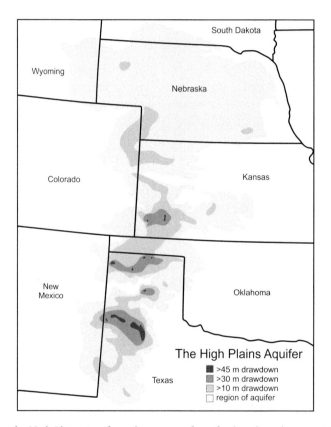

Figure 12.28   Region underlain by the High Plains Aquifer and contours of aquifer drawdown between 1950 and 2003. Map from the USGS.

aquifer lying some distance above other major unconfined aquifers. Perched aquifers may be locally important, but their sizes are typically small and the amount of water they contain varies with season and climate conditions and therefore may be inconsistent.

Figure 12.28 shows the more than 100 million acre part of the Great Plains that is underlain by the *High Plains Aquifer*. This aquifer, which is also called the

Ogallala Aquifer, is the main source of water in one of the most important agricultural regions in the United States that extends from Texas to South Dakota. The aquifer consists of unconsolidated or partly consolidated gravel, sand, silt, and clay. The sediments, commonly coarse, have a wide variation in bed thickness, grain size, and grain sorting. There is much space between clasts and grains, providing the necessary porosity and permeability for an excellent aquifer. Aquifers of this sort most commonly have an alluvial origin but may also derive from valley fill, glacial deposits, or stream fill. They are generally unconfined and range from a few meters to hundreds of meters thick; the saturated thickness of the High Plains Aquifer reaches a maximum of about 300 meters (1000 feet) in Nebraska.

Many important aquifers consist partly or entirely of unconsolidated sediments, permeable sedimentary rocks (including sandstone and limestone). However, fractured igneous (especially volcanic) or metamorphic rocks, may also make excellent aquifers. In many rocks, natural porosity and permeability are low, but joints, fractures, or bedding planes can host water and provide permeability. Some limestones have high porosity and permeability due to the dissolution of calcite after the limestone formed, but many limestones do not.

People have a tendency, once a good aquifer is identified, to withdraw water at ever-increasing rates. For example, Figure 12.29 is a satellite view of irrigated fields in Finney County, Kansas. The circles are 800 to 1600 m in diameter (0.5 to 1 miles). Water for these fields comes from the Ogallala Aquifer. Unfortunately, Ogallala water is being used faster than the aquifer is recharged. So, the depth to the water table has dropped significantly. This process is called *drawdown*, and it has happened, and continues to happen, in the central United States and in many other places around the world.

Since center pivot irrigation began in the Great Plains region in 1950, water withdrawal for irrigation has caused the water table in the Great Plains Aquifer to drop significantly, making water extraction increasingly difficult and expensive and causing some irrigation projects to shut down. As shown by the contour lines in Figure 12.28, drawdown has exceeded 10 meters across much of the region. In areas shown by deepest red color, total drawdown was more than 45 meters (148 feet) between 1950 and 2003. A 2017 report by the United States Geological Survey found that the water table continued to drop from 2003 to the present day and has fallen more than 50 meters (164 feet) in parts of Texas. The rate of drop has slowed somewhat, presumably because less water is being withdrawn. If drawdown continues, more irrigated farms will become

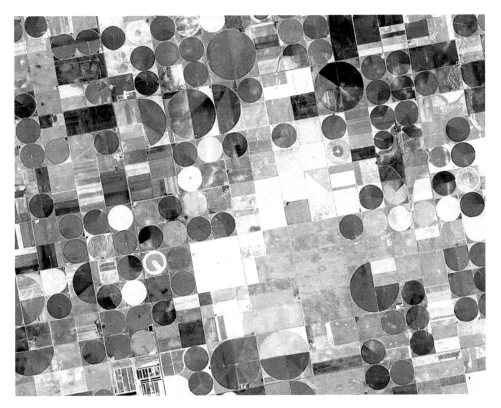

Figure 12.29 Center pivot irrigation in Finney County, Western Kansas. Photo from the USGS.
Map from the USGS.

unprofitable. The easiest solution is to place limits on the amount of water that can be withdrawn from the aquifer in places where the water table is falling, but this solution is difficult to implement in practice for political or economic reasons.

12.5.4 The cryosphere

The *cryosphere* includes all parts of Earth where water is frozen. This means ice or snow and includes glaciers, ice caps, ice sheets, sea ice, ice shelves, icebergs, lake ice, snow cover, and frozen ground. The cryosphere is important because it provides essential animal habitat, it is a key source of fresh water in some places, and its nature and extent record evidence of long-term climate change.

Figure 12.30 shows the major components of the cryosphere, the part of Earth's surface where water is frozen. It thus includes snow cover, permafrost, sea ice, ice sheets (on land), ice shelves (extending into the oceans), sea ice, icecaps on mountain tops, glaciers, and frozen lakes and rivers. In normal continental areas, the cryosphere consists of snow and glaciers at high elevations, seasonal snow at lower elevations, and ice in frozen ground (permafrost). In polar regions, the cryosphere includes ice sheets, ice shelves, and sea ice. In areal extent, when at seasonal, maximum of about 41% of the cryosphere is snow cover, 20% is permafrost, 25% is sea ice, and 13% is ice sheets in Antarctica and Greenland. The remaining 1% is mostly in ice caps, glaciers, and ice shelves. Most of the cryosphere is in the northern hemisphere where, because there is little land near the South Pole compared with near the North Pole, most of the snow cover and permafrost are found. Sea ice is evenly distributed between the two hemispheres. The Antarctic ice sheet is much larger than the Greenland ice sheet and contains eight times more snow and ice. The different reservoirs in the cryosphere have significantly different characteristics. One key one is water residence time. Snow, frozen lakes and streams, and some sea ice may be seasonal; some permafrost, glacial ice, and ice sheets may last for millennia.

Figure 12.31 shows regions where some of the different parts of the cryosphere are found. Snow cover extent shown for the Northern Hemisphere is based on the 1966–2005 February average, and for the Southern Hemisphere is based on the 1987–2003 August average. Sea ice extent for the Northern Hemisphere is the 1979–2003 March average, and for the Southern Hemisphere it is the 1979–2002 September average. Permafrost data for mountain areas and for the Southern Hemisphere are not represented in this map, nor are river and lake ice.

Although during cold months most of the area of the cryosphere is snow cover, most of the cryosphere's volume is in Antarctica and, to a lesser extent, frozen in the Arctic. Antarctica is home to the East Antarctic Ice Sheet, an ice sheet that accounts for most of the ice on Earth, and ice shelves that extend offshore from the ice sheet Near the North Pole, in the Arctic Ocean, sea ice is extensive, and ice sheets that vary in size with season surround Greenland and other land that is covered by glaciers, snow, and permafrost. Away from the two polar regions, some high mountains are covered by snow permanently, or seasonally, although the extent of snow cover has decreased in recent years. In the Northern Hemisphere, winter snow and ice comprise the largest area of the cryosphere during cold months.

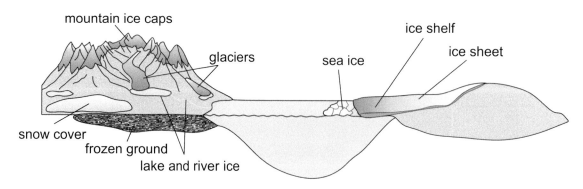

Figure 12.30   The components of the cryosphere.

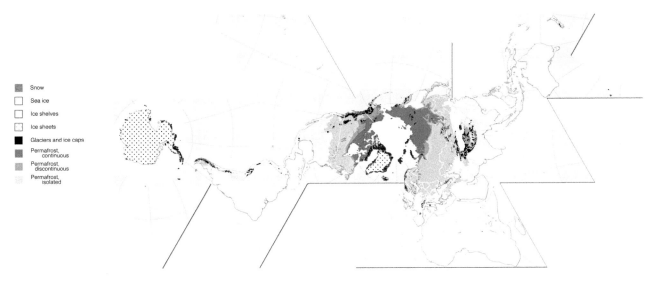

Figure 12.31   The different components of the cryosphere.
From Hugo Ahlenius, Wikimedia Commons.

Figure 12.32   Snow cover in North America on March 5, 2012.
Image from NASA.

12.5.4.1   Snow

Snow is found at all latitudes, but the amount of snow at Earth's surface varies between years and, of course, between seasons. Snow covers ice or permafrost in many places, but melting may expose the underlying materials periodically or seasonally. Figure 12.32 shows snow distribution in March of 2012 in North America. Persistent snow and snow fields are generally at high latitudes (especially near the poles) or at high elevations. Snow is an important part of the water cycle and, in many places, is the source of most runoff and groundwater recharge. Snow is also responsible for much of Earth's *albedo*, the reflectivity that returns some solar energy to space, lessening the rate of planet warming.

12.5.4.2   Glaciers

Glaciers, defined as long-lasting bodies of dense ice that move constantly due to gravity, may be *mountain glaciers* (also called *valley glaciers*), generally of local importance, or *ice sheets*, such as the Greenland Ice Sheet that covers a large area. Table 12.4 compares the distributions of

Table 12.4   Area (km²) of glaciers in different parts of the world.

Region	Square kilometers	Square miles
Antarctica	14,000,000	5,405,426
Arctic Island & Greenland	1,985,500	766,605
North America	124,000	47,877
Central Asia	114,800	44,324
Northern Asia	59,600	23,012
South America	25,500	9846
Central Europe	3785	1461
Scandinavia	2940	1135
New Zealand	1600	618
Africa	6	2.3
New Guinea	3	1.2

Data mostly from the National Snow and Ice Data Center, http://nsidc.org/cryosphere/glaciers/questions/located.html.

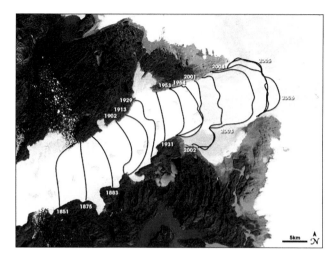

Figure 12.33   Extent of the Jakobshavn Glacier 1851–2006. Image from NASA Earth Observatory.

glaciated regions in different parts of the world. Glaciers are found at all latitudes, but the largest ones are in Antarctica. Antarctica's Lambert Glacier is the largest glacier in the world. It is 515 kilometers (320 miles) long and 64 kilometers (40 miles) wide, although listing glacier sizes is problematic because the distinction between glaciers and ice sheets is not always clear. Not all glaciers are in polar regions; mid-latitude glaciers are especially common in the high Himalaya and Andes Mountains. Surprisingly, some very cold regions, such as Siberia, have few glaciers due to little precipitation.

The extent of glaciation is a reaction to climate and, on a smaller scale, seasonal variations in precipitation and temperature. Nearly all glaciers worldwide are retreating today as a consequence of increasing global temperatures. The Jacobshavn Glacier, pictured in Figure 12.33, in Greenland is a good example. The background image was taken in 2001. The glacier flows from the Greenland Ice Sheet in the upper right to lower left, and has been melting faster than it advances, and thus retreating, since at least 1851. Lines showing the glacier extent and terminus before 2001 were added to the photo based on historical survey data. Lines showing the glacier terminus after 2001 are based on satellite imagery. Since 1851, the glacier has retreated more than 50 kilometers (31 miles).

Figure 12.34 shows two views of Alaska's Muir Glacier in Glacier Bay National Park. The photos were taken in August of 1941 and 2004. During the intervening 63 years, the front of the glacier moved back about 12 kilometers (7 miles) and ice thickness decreased by more than 800 meters (2600 feet). Thus, there has been a huge loss of ice volume. Around the globe, the areal extent of

glaciers has been decreasing, on average, since the end of the so-called *Little Ice Age* (a cool period of time about 150 years ago). Today, about 10% of Earth's land is covered by glaciers; during the last ice age, coverage was about 32%. At present, the rate of glacier melting in polar regions is the fastest it has been any time during the past 10,000 years, and Greenland glaciers are melting faster than any others. Because glaciers contribute substantially to Earth's albedo (reflection of solar energy), as Earth's glaciers disappear, less of the sun's energy will be reflected into space, and the rate of global warming and glacier melting will accelerate. Perhaps the biggest threat posed by melting glaciers is sea level rise. In the past 100 years, melting glaciers caused global sea level to rise by 11 centimeters, and if all the glaciers melt, the world's oceans will rise by an additional 70 meters flooding many large cities.

12.5.4.3   Ice sheets and icebergs

Ice sheets and glaciers can extend into the ocean, where they float on seawater, forming ice shelves in Antarctica and Greenland and, to a lesser extent, elsewhere. Figure 12.35 shows the principal shelves of Antarctica. Antarctica has 15 major ice shelf areas, and the 10 largest appear in this map. They are typically 60–600 meters thick and have persisted for long times. Some of the ice in these shelves derives from snowfall, but most is glacial ice that flows from glaciers on land. Today these shelves are becoming smaller, primarily due to melting caused by warm ocean waters beneath them and by calving that produces icebergs and, in some cases, complete disintegration. For example, two sections of Antarctica's Larsen

Figure 12.34 The Muir Glacier in Alaska in 1941 (left) and 2004 (right).
Image from NASA.

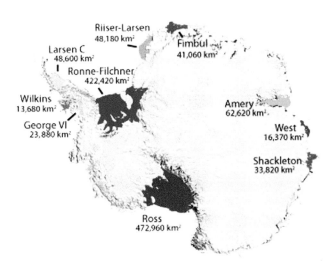

Figure 12.35 Ice shelves of Antarctica.
Image from the T. A. Scambos, National Snow and Ice Data Center.

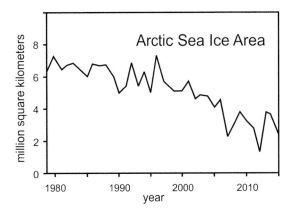

Figure 12.36 Arctic sea ice extent from 1979 to 2011.
Data NSIDC.org.

Ice Shelf, one of the largest ice shelves in the world, shattered into small pieces in 1995 and 2002.

12.5.4.4 Sea ice

Sea ice is frozen ocean water that floats on the ocean surface. During the last century, sea ice has covered as much as 15% of the oceans during cold seasons. The exact amount varies seasonally as it freezes from, or melts to become, ocean water, but today it is disappearing. Sea ice exists in both polar regions, but most significantly in the Arctic. As shown in Figure 12.36, about 4 million square kilometers of Arctic sea ice have disappeared

since 1980. Although there have been a few bumps along the way, the long-term trend is clear: sea ice is disappearing at a rate of about 100,000 square kilometers per year, and between 2003 and 2016 more than 40% of the total volume of Arctic Sea ice disappeared. The data shown in Figure 12.36 have prompted some predictions of an ice-free Arctic Ocean in the near future. The rate of ice loss, however, is highly variable, making precise predictions impossible. Melting sea ice has little direct impact on sea level, but there are other areas of concern. For example, relatively young sea ice contains salt as trapped brine pockets and droplets, but in older ice nearly all the brine is gone. So, as young ice melts and is not replaced, salt is added to the oceans, making the water denser and causing it to move downward—and potentially altering ocean circulation.

Figure 12.37   Stone rings in permafrost, Svalbard Island, Norway. Photo from Hannes Grobe, Wikimedia Commons.

12.5.4.5 *Permafrost*

Ground that is frozen year round is called *permafrost*; Figure 12.37 shows permafrost on Svalbard, a Norwegian island. *Stone rings*, such as those shown in Figure 12.37, are one kind of *patterned ground*, features formed by repeated freezing and thawing, common natural features in some permafrost regions. Earth contains about 60 million square kilometers of permafrost, mostly in polar regions. Small amounts are also found at high elevations in the Rocky Mountains, the Himalaya, and elsewhere. In many places, the uppermost layers of permafrost melt during the summer and can support vegetation and other life. Permafrost is disappearing with global warming, but there is much uncertainty about the rate. Some scientists are concerned that melting of Arctic permafrost could release large amounts of methane that is presently trapped in frozen soils. More methane, a potent greenhouse gas, in the atmosphere could cause the rate of global warming to increase.

12.6   Water chemistry

Water is sometimes called the "universal solvent" because many Earth materials dissolve in H_2O. Rainwater is essentially purified by distillation during a rain-producing process, but even pure water dissolves solids and gases and it travels through the environment. Key characteristics of natural water are the total amount and relative amounts of different dissolved ions. *Total dissolved salts* (TDS) is a numerical value that describes the total amount of ions that water contains. TDS generally increases with time as water travels through soil or rock, or travels across land surface. *Salinity* is commonly (incorrectly) used as a synonym for TDS. It is defined as the total concentration of salts, that are typically found in seawater, in any water sample. So, the term salinity should be restricted to ocean waters.

The relative amounts of major dissolved ions are often good indicators of the material with which water has interacted. Atmospheric gases can also dissolve into water. CO_2, in particular, is very important near Earth's surface because the dissolution of CO_2 into water produces carbonic acid (H_2CO_3), a particularly corrosive acid that can dissolve main minerals in rocks and soils. Table 12.5 lists the most common and important ions that are found in natural waters.

As shown in Figure 12.38, water is classified as fresh, brackish, saline, or brine, depending on its TDS. For examples, average TDS concentrations for some specific water bodies are noted with arrows in this figure. The values are for water at the surface, and some of the water bodies have greater TDS at depth. Note that this chart would have to be five times taller to include the Dead Sea, the Great Salt Lake, and other extremely salty lakes and seas. The units in this figure are parts per thousand (ppt) equivalent to the grams of dissolved material in 1000 grams of water. This unit is sometimes called the *practical salinity unit* (abbreviated PSU or PSS). Average seawater, as seen in this chart, has a TDS value of about 35 ppt.

Very salty water is called *brine*, which, by definition, has TDS greater than 50 ppt. Most brines form when landlocked bodies of water, such as the Salton Sea, evaporate, leading to a buildup of salt content. Brine is also found in deep groundwater associated with petroleum deposits and in brine pools, which are localized bodies of water collected in depressions in the seafloor that have salinities several times greater than normal ocean water. The source of the salt in most brine pools is underlying bedrock salt deposits. Less commonly, in pools near Antarctica, the extra salt derives from brine originally trapped in sea ice. Some rare brines may have TDS values as high as 350 ppt.

Table 12.5   The major dissolved species in water.

Cations	Anions
calcium (Ca^{2+})	bicarbonate (HCO_3^-)
magnesium (Mg^{2+})	sulfate (SO_4^{2-})
sodium (Na^+)	chloride (Cl^-)
potassium (K^+)	

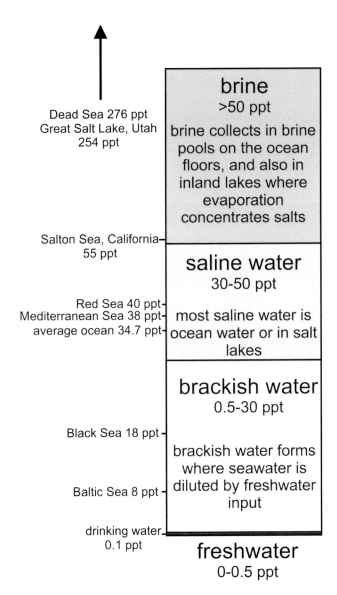

Figure 12.39   Seawater evaporation ponds on the Isle of Rhé, France.

Photo from Ile-de-re-commonswiki, Wikimedia Commons.

Figure 12.38   Water of different salinities (ppt = parts per thousand).

Saline waters (30–55 ppt) are somewhat less salty than brine; they typify the world's oceans and some salt lakes. *Brackish waters*, which usually form when seawater and freshwater mix, have TDS values between 0.5 and 30 ppt. Drinking water generally has less than 1% as much salt as brines, with TDS less than 0.1 ppt. Other forms of *fresh water* may have salinities as great as 0.5 ppt.

Figure 12.39 shows seawater evaporation ponds that produce commercial sea salt. Both saline water and brackish water can be used as a source of salt by processes generally called *desalination*. Historically, salt has been produced by evaporating salty water and collecting the material that is precipitated. Although salt is an important commodity, used for food processing and preservation, a number of industrial processes, and to melt snow on roads, the processing of saline waters and brines to produce drinkable water is also important in regions where fresh water is in short supply. Desalination to provide water is especially important in the Middle East. In Kuwait, almost all drinking water come from desalination; the largest desalination plant in the world is in Saudi Arabia. In the United States, Florida has 150 desalination plants, California and Texas have half to a third as many, and several other states have a smaller number.

12.6.1 Ocean water

Ocean water is, on average, about 96.5% water and 3.5% dissolved salts listed in Table 12.6. The major ions present, shown in Figure 12.40, are chloride (Cl^-), sodium (Na^+), sulfate (SO_4^{2-}), magnesium (Mg^{2+}), and calcium (Ca^{2+}). Other species are listed in Table 12.6 and, in addition to those listed, all other known elements are found in seawater, although in very low concentrations. Although the percentage of dissolved salts in seawater is small, the total quantity is large because of the immense size of the oceans.

Runoff from continents delivers more than 2 billion metric tons of soluble material to the oceans each year. Ocean floor volcanics add additional salt. These inputs are balanced by biological activities and inorganic precipitation of salts. Overall composition of ocean water remains about constant because the rate of salt removal

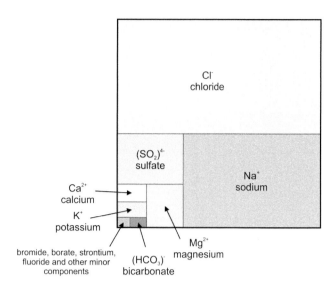

Figure 12.40 Dissolved species in average seawater.
Modified from Wikimedia Commons.

Table 12.6   Composition of average seawater: ion concentrations and total dissolved solids (tds).

Ion	TDS	%Total TDS
Cl⁻, chloride	18.980	55.06
Na⁺, sodium	10.556	30.62
$(SO_4)^{2-}$, sulfate	2.649	7.68
Mg^{2+}, magnesium	1.262	3.66
Ca^{2+}, calcium	0.400	1.16
K⁺, potassium	0.380	1.10
(HCO)⁻, bicarbonate	0.140	0.41
Br⁻, bromide	0.065	0.19
$(BO_3)^{3-}$, borate	0.026	0.08
Sr^{2+}, strontium	0.013	0.04
F⁻, fluoride	0.001	0.00
TDS	34.472	100.00

From www.lenntech.com.

is proportional to concentration, thus buffering the system. For example, the influx of sodium and magnesium from rivers and other waters draining off the continents and into seas and oceans equals the amounts of these elements incorporated into quenching basaltic lavas at spreading zones or infiltrating into the subsurface and becoming incorporated into rocks and magmas. In a few places, high evaporation rates lead to increased salinities; in other places, mixing with fresh water leads to decreased salinities. However, the range of salinities for ocean water is generally quite small, just a few ppt.

The major, and many minor and trace, dissolved ions in oceans are *conservative*, meaning that they have constant ratios to each other and they are present in concentrations proportional to TDS. Conservative elements such as sodium and chlorine are present in relatively large amounts and have long residence times and, consequently, mix throughout the oceans. Salinity may vary in different parts of the open sea, due to freshwater inputs or evaporation, or the melting or formation of sea ice, but the relative proportions of different dissolved salts are nearly the same. Other dissolved substances are *non-conservative*—including gases and nutrients and minor components such as nitrogen, phosphorus, and silicon—because concentrations vary greatly with uptake and regulation by organisms. Nutrients are depleted in warm near-surface water where biological activity is greatest and often concentrate in deeper colder water.

Figure 12.41 shows variation in surface salinity of the world's oceans. The variation is largely due to the extent of local evaporation and precipitation. High salinity (>38 ppt) is shown in red, average salinity in blue (about 35 ppt), and low salinity (about 34 ppt) in purple. Salinity is rather constant over time and is similar in all oceans, but areas in the North Atlantic, South Atlantic, South Pacific, Indian Ocean, Arabian Sea, Red Sea, and Mediterranean Sea tend to be near the high end of the range (red). Areas near the Arctic Ocean, Southeast Asia, and the west coast of North America tend to be a little lower (purple).

When salt water becomes isolated from the main oceans, such as in the Dead Sea, evaporation can lead to very high salt concentrations. Formation of brines due to freezing also can lead to high local salinities, but

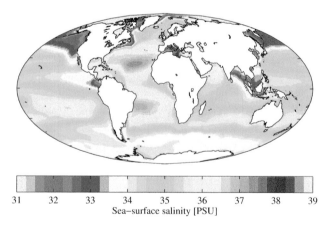

Figure 12.41 The variation in salinity of the world's oceans.
Map from Plumbago, Wikimedia Commons.

marine waters generally have salinity close to 35 ppt. The most saline open ocean is the Red Sea, with a minimum surface water salinity of about 36 ppt at its southern end, where it is slightly diluted by water from the Gulf of Aden, and 41 ppt at its northern end. Because ocean current flow is largely horizontal, some stratification occurs in open oceans. Salinity of the South Atlantic, for example, is slightly greater near the surface than it is at depth (35.5 ppt vs. 34.8 ppt). In closed basins, such as the Dead Sea, stratification may be more pronounced, and salinities are often much greater at depth where denser water accumulates.

Humans have had major impacts on ocean chemistry near ocean margins because of polluted runoff and industrial effluent. On a global scale, however, the oceans contain too much water for human impacts to be seen. An exception to this is a slight increase in acidity; over the past 150 years, the average ocean pH has decreased from 8.25 to 8.14 due primarily to increased CO_2 in the atmosphere, which dissolves in the oceans, producing carbonic acid.

12.6.2 Continental waters

Rivers, lakes, groundwater—any water that is not in the oceans—can be lumped together as *continental water*. Figure 12.42 shows two examples: a mountain stream and the Great Salt Lake in Utah, two water bodies that differ greatly in chemistry and origin. Table 12.7

Figure 12.42   Two examples of continental waters: a mountain stream and the Great Salt Lake, Utah.
Photos from M. L. Smith, Wikimedia Commons, and the USGS.

Table 12.7   Continental surface water chemistry.

	Average seawater	Rain water	River water	River water	River water	River water	Lake water	Lake water	Lake water	Ground water	Ground water	Ground water	Ground water
	1	2	3	4	5	6	7	8	9	10	11	12	13
Ca^{2+}	400	0.65	40.7	1.68	14	22	2.1	10	7.1	144	6.5	3.11	4540
Mg^{2+}	1262	0.14	7.2	0.24	13	17	0.3	4.8	2.8	55	1.1	0.7	160
Na^+	10,556	0.56	1.4	0.16	8	14	2.4	11.9	11	~27	~37	3.03	2740
K^+	380	0.11	1.2	0.31	–	0.5	0.6	3.9	2.2	~2	~3	1.09	32.1
$(HCO_3)^-$	140	-	114	5.4	104	129	3.2	44.5	34	622	77	20	55
$(SO_4)^{2-}$	2649	2.2	36	1.3	4.7	1.3	4.1	2.3	11	60	15	1	1
Cl^-	18,980	0.57	1.1	0.06	8.5	33	3.8	5.9	11	53	17	0.5	12,600
Si^{4+}	–	–	3.7	0.7	24	30	–	11	18	22	103	16.4	8.5
TDS	34,472	4.7	207	10	120	180	–	91	80	670	222	36	20,338
pH	8.1	5.6	–	6.9	7.7	7	5.9	8.4	–	–	6.7	6.2	6.5

Data from www.waterencyclopedia.com/, Van Winkle and Finkbiner (1913), Plisnier *et al.* 1996, Gorham 1955. Samples from (1) www.lenntech.com/; (2) North Carolina and Virginia; (3) Rhine River where it leaves the Alps; (4) Cascade Mountains, Washington; (5) Jump-Off Joe Creek, southwestern Oregon, wet season; (6) Jump-Off Joe Creek, southwestern Oregon, dry season; (7) Gold Mines Lake, Halifax; (8) Lake Victoria; (9) Crater Lake, Oregon; (10) Supai Fm., Grand Canyon; (11) volcanic terrane in New Mexico; (12) spring in Sierra Nevada Mountains, short residence time; (13) metamorphic terrane in Canada, long residence time.

Note: Units are milligrams per liter, which is just about the same as ppm; compositions are for surface water.

compares the dissolved species in seawater, rainwater, and in some examples of continental waters of different kinds; there is a great deal of variation. Although the compositions vary greatly, the main solutes in continental waters are the same as in ocean water. Although there are exceptions in some arid environments, continental waters generally contain much lower (several orders of magnitude) amounts of dissolved ions than seawater, and the ratios of the solutes differ dramatically. For example, bicarbonate (HCO_3^-) makes up less than 0.5% of the salts in seawater but averages more than 50% for many continental waters. Rainwater, which is distilled during evaporation, on average contains even less dissolved material than continental waters do (Table 12.7).

The key factor controlling the composition of fresh continental water is the interaction between the water and soils or bedrock. Soils and bedrock, which may have a wide range of compositions, are the sources of the mineral matter that dissolves in water. For example, water that flows through or over limestone often becomes *hard* (high in Ca^{2+} and Mg^{2+} concentrations), and water that flows through or over sulfide-rich rocks (or mine tailings) may become acidic as it incorporates sulfuric acid. The volume of water and replacement time may lessen or increase the amount of dissolved geological material in water. Large volumes and quick replacement times mean that dissolved components are diluted or removed quickly. Water that has a long time to equilibrate with bedrock tends to have higher concentrations of dissolved material. For example, water #12 (Sierra Nevada spring water) in Table 12.7 is for water with short residence time, and water #13 (slow moving Canadian water), which contains about 1000 times more dissolved material, is for water with long residence time. Climate, vegetation, and topography may also affect freshwater chemistry.

Continental surface water and groundwater are generally fresh water but need not be so. Most lakes contain relatively fresh water, but saline lakes, like the Great Salt Lake in Utah and the Dead Sea, are not uncommon, particularly in arid regions with low rainfall. The Great Salt Lake has TDS ranging from 50 to 270 ppt, depending on lake level and location. Groundwater, too, often has high TDS values, especially if the rock it flows through is soluble limestone, contains appreciable quantities of salt, or is associated with deep petroleum deposits.

12.7 Freshwater quality

Some things that dissolve in water are generally considered beneficial. For example, dissolved oxygen is considered good because it promotes healthy aquatic life. Other solutes, however, are considered deleterious, especially if they are present in excessive amounts, impart bad taste to water, or effect human or animal health. Figure 12.43, for example, is a warning about excessive bacteria and other pathogens in a waterway.

Homeowners are typically concerned about water *hardness*, which is the amount of dissolved Ca^{2+} and Mg^{2+}. Hard water does not pose health risks but does lead to other problems, including the clogging of plumbing and water heaters, stopping soaps from making suds, and depositing scale in teakettles and saucepans. For practical purposes, this means the concentration of calcium and magnesium in the water. Water can be chemically "softened" inexpensively by water softeners that remove calcium and magnesium, replacing them with sodium.

Most water on Earth, including continental waters, is neither seriously toxic nor potable, but the variations are great and determining and ensuring water quality is complex. In its National Water Quality Inventory, the U.S. Environmental Protection Agency reports that the three most important pollutants are dirt, bacteria,

Figure 12.43 Warning about *Escherichia coli* bacteria in surface water.
Image from the USGS.

and nutrients. So, some pollution has a natural origin (although people often make it enter water systems faster than nature would if left on its own). Natural components of water generally come from weathering or decomposition of organic debris, but some may derive from the atmosphere and from interaction with rocks or soil. Natural components include:

- suspended inorganic particulates
- suspended and dissolved organic matter
- dissolved ions (most importantly those listed in Table 12.5)
- dissolved nitrogen, phosphorus, and other nutrients
- dissolved and suspended metals

Suspended matter is often removed when water flows through an aquifer, and this is one reason that groundwater is often a better source of domestic or municipal water than surface water.

Anthropogenic components of water are of many sorts. For example, people often live on shorelines of lakes or streams and, on purpose or by accident, dispose of wastes into water bodies. Figure 12.44 shows two views of Lake Washington, in Seattle. During the 1940s and 1950s, cities adjacent to Lake Washington completed 11 sewer systems that delivered municipal runoff to the lake. Waste disposal of this sort was not a huge problem when the human population was low and waste products not plentiful, because waterways could cleanse themselves by natural aeration, dilution, sedimentation, and filtration, and through natural biological processes. Lake Washington, however, became overwhelmed. Bacteria and nutrient levels increased, the lake became slimy, stinky, and fish died. To solve the problem, sewage treatment was improved, and stormwater runoff was rerouted to the Puget Sound. In time, the lake recovered to be as it is seen here today. Dumping even slightly polluted water into the ocean at Puget Sound does not seem like a good idea, but the volume of the oceans is huge compared with Lake Washington, and so the pollution is diluted and becomes inconsequential.

Today, however, the burgeoning human population and overcrowding, and bad waste management, in many places threaten water supplies. Sewage effluent and stormwater, whether treated or not, pose huge problems. Some pollution comes from *point sources*—a single pipe or drain, for instance, and can often be stopped, regulated, or treated. *Non-point source* pollution, which includes runoff from agricultural fields and municipal areas or dust deposited by wind, pollutes water with sediment, nutrients, toxic chemicals, and pathogens and is more difficult to treat or regulate. Use

1958

2006

Figure 12.44 Lake Washington in Seattle, 1958 and 2006.
Top photo from www.historylink.org/File/1353; bottom photo from saveunionbay.org.

of waterways for transportation and shipping can add still other unwanted components to the environment.

In the more developed part of the world, municipal water supply facilities remove contaminants of many sorts from natural waters to produce water that is pure enough for safe human consumption. Suspended solids, organic compounds of many sorts, viruses, bacteria, algae, dissolved minerals—the list of possible pollutants is very long, but all can be dealt with to produce potable water. In the lesser developed parts of the world, however, water treatment is often inadequate or lacking. And, water treatment is expensive and not affordable for some aquatic systems that have been polluted by human activities.

12.7.1 Water and human health

Figure 12.45 shows a photo of an *Escherichia coli* colony. *E. coli* is a bacterium normally found in mammalian and human intestines. Drinking water is commonly tested for *E. coli* because, although there are many kinds of *E. coli* and not all pose health risks, its presence usually signals contamination by sewage and other kinds of pathogens may be present. Because humans are anthropocentric, we worry more about direct causes of human health problems than about other results of water pollution. So, bacteria, viruses, parasites, protozoa, parasitic

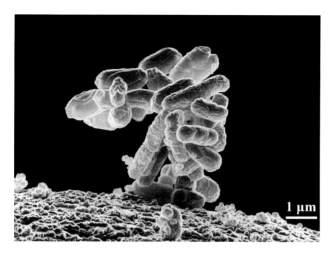

Figure 12.45 Photograph of *E. coli* bacteria.
Photo from Eric Erbe, USDA.

Figure 12.46 Photo of the aftermath of the 2011 meltdown at the Fukushima nuclear power plant in Japan.
Photo from agoracosmopolitan.com.

worms, and other pathogens are considered the most serious of water-borne threats. These pathogens, most commonly originating from improperly treated human and animal wastes, are responsible for cholera, typhoid fever, diphtheria, and other diseases. They are monitored and controlled in most highly developed countries. Many less-developed countries, however, suffer from widespread lack of sewage treatment, inadequate monitoring, and insufficient funds for remediation and cleanup.

Many organic chemicals, including some natural ones like coal, methanol, and cyanide, are found in water and can threaten people and other organisms. Today the natural contaminants are overshadowed by an uncountable number of synthetic contaminants developed for many different purposes. Humans evolved with the natural organics already in the environment and so have some natural defenses, but we often lack defenses to synthetic compounds. Toxic synthetic organics come from many sources, including industrial and household wastes and agriculture. Toxic components of pesticides, plastics, pigments, cosmetics, detergents, oil, gasoline, and others are in waters of all sorts and may concentrate and become more hazardous as they move up the food chain.

Inorganic chemicals and radioactive materials also pose health risks. For example, the 2011 nuclear power plant meltdown at Fukushima, Japan, seen in Figure 12.46, released large amounts of radioactive elements into the air and the Pacific Ocean, most of it due to runoff of water used to quench fires. It is too soon to determine exactly what the effects will be on marine organisms, fisheries, and people. Radioactive elements

that may cause cancer and other health problems in people and animals include uranium, thorium, cesium, iodine, and radon. They all have natural origins, but today, because of processing of ores, emissions from nuclear power plants, and weapons production, testing, and use, they are more widespread and mobile than in the past.

Although the meltdown at Fukushima focused world attention, at least briefly, on the release of radioactive materials into the environment, significant anthropogenic sources of radioactive elements have existed since the 1960s. Other pollutants also pose significant risks to people, more now than in the past. For example, acids, caustics, salts, and metals such as mercury, lead, or arsenic have natural sources (such as rock weathering) but are also manufactured for industrial and municipal use. They are all found in urban and agricultural runoff, mine wastes, and acid rain, sometimes in alarming concentrations.

12.7.2 Ecosystem health

Pathogens and chemical pollutants not only affect people, but they also affect entire ecosystems. Other kinds of pollution do as well. For example, some sedimentation is good for natural systems—e.g., replenishing topsoil in floodplains. Yet, sediment pollution is arguably the most significant type of water pollution that threatens natural aquatic and marine ecosystems. Soil, silt, and salt erosion and deposition can be natural, but humans mobilize these materials much faster than nature does, disturbing land for agriculture, construction, logging, and other purposes. Sediment clogs up rivers, directly affecting aquatic invertebrates and fish by clogging up gills, blocking sunlight, and disrupting reproduction.

12.7.2.1 *Organic matter and eutrophication*

Excessive and reactive organic matter is a major threat to many aquatic ecosystems. It may derive from animal manure, plant debris, sewage, and municipal, agricultural, and industrial runoff. Manure, debris, sewage, and runoff also deliver extra plant nutrients to waterways— especially nitrogen and phosphorus. The nutrients cause rapid and excessive plant growth and algal blooms, adding additional organic particulate matter, which blocks light, clogs waterways, and obstructs animal life. When organic content becomes high enough, the available oxygen may be insufficient to keep up with the decomposition rate of organic debris, and the waterway may become partially or entirely lacking in oxygen, disrupting or killing plant and animal communities. In extreme cases, eutrophication causes waterways to become turbid, choked with green algal slime, and to smell like rotten eggs. This process is termed *eutrophication* and is what happened to Lake Washington, described earlier in this chapter.

Figure 12.47 shows a portion of Lake Erie near Catawba Island at the western end of the lake during a bloom of planktonic algae in 2009. These kinds of blooms are initiated by excessive supplies of dissolved nitrogen and phosphorus and vertical stratification of lake circulation which favor the bloom species. The blooms produce excessive amounts of organic matter

that consume dissolve oxygen and produce toxins that kill off or out-compete other plant species. Invertebrate animals and fish may die, leaving large areas of smelly dead organisms and disrupting ecosystem processes. Small harmful algal blooms can be managed, but large blooms tend to persist until air and water temperatures decrease and water circulation returns at the beginning of the fall. In the past, blooms were rare and largely ignored. However, with increasing human activities, such as an increased use of fertilizers, and climate change, blooms are increasingly common in the United States and worldwide.

Figure 12.48 is a 2003 satellite view of the Caspian Sea. The cloudy bluish water at the sea's northern end is where the Volga River delivers fertilizer and other agricultural runoff that cause algae and other plant life to grow and become present in very large amounts. The vegetation chokes the sea and, after it dies, decaying plant material uses up all the oxygen, leaving none for other aquatic organisms. The problem exists in all parts of the Caspian Sea but is most acute during summer

Figure 12.48 A 2003 satellite view of the Caspian Sea, imaged by the MODIS sensor on the Terra satellite.

Image and caption from Jeff Schmalz, NASA.

Figure 12.47 Eutrophic part of Lake Erie in 2009.

Photo from NOAA.

months. Scant progress has been made to deal with the problem, in part because the shores of the sea are in five different countries: Russia, Iran, Azerbaijan, Kazakhstan, and Turkmenistan, and negotiations to control eutrophication put national needs and goals in conflict.

In the United States, Lake Erie once had problems similar to those of the Caspian Sea. Figure 12.49 shows Lake Erie in 2007. The cloudiness in the western end of the lake is due to silt and other suspended matter delivered by the Maumee and other rivers. The central part of the lake looks cloudy because of an algal bloom. The city in the northwest part of this photo, just west of the roundish Lake St. Claire, is Detroit, Michigan. Although much cleaned up today, Lake Erie, in the 1960s, was highly eutrophic. Excessive algae floated on its surface, blanketed beaches with slime, and eliminated native aquatic species by soaking up all of the oxygen. Decades of international efforts to reduce nutrient runoff have solved some of the problem, and native species have begun to return.

Eutrophication is most easily seen in lakes and rivers, but it is also a common problem in coastal waters. Estuaries, in particular, tend to be naturally eutrophic because natural nutrients become concentrated in confined channels. Human inputs also add to the nutrient and organic load in many estuary and coastal systems.

12.7.2.2 Thermal Pollution

Figure 12.50 shows warm water from a power plant returning to San Francisco Bay. This is an example of *thermal pollution*. Thermal pollution occurs when normal water temperature is changed, either becoming warmer or colder. One of the most common examples of thermal pollution is when water, used as a coolant, is returned to the waterway that it came from, as shown

Figure 12.50 Hot water from the Potrero Generating Station returning to San Francisco Bay.

Photo from Dragonsflight, Wikimedia Commons.

Figure 12.49 Lake Erie in 2007.

Photo from NASA.

here. Some industrial plants and some urban runoff, too, return warm water to lakes and streams. Increasing water temperatures decrease the amount of dissolved oxygen, directly affecting fish. Although many varieties of fish are especially incapable of dealing with temperature variations, plants, plankton, and many shellfish are nearly as intolerant. Warm water also disrupts normal ecosystem equilibria, affecting spawning and promoting or inhibiting chemical reactions and metabolic rates.

Questions for thought—chapter 12

1. Describe several of the most significant ways that people use water. Which human activity uses up more water than any other?

2. Why is the residence time for groundwater in the deep subsurface much longer than water in the atmosphere?

3. Ocean circulation can have a major effect on climate. Explain how this works. And, what are the major things that cause ocean circulation?

4. How might global warming cause cooling in Great Britain and Scandinavia?

5. What causes thermohaline circulation and how is it different than circulation on the surfaces of the oceans? Which type of circulation moves the most water?

6. What are *El Niño* and *La Niña* and how do they affect weather?

7. What is the difference between porosity and permeability? How can a rock be porous, yet impermeable? And – the opposite – how can a rock be permeable but not very porous?

8. What is the water table? Is it always horizontal or can it slope? Explain why. And, what causes the water table to move up and down?

9. Where is the water table, normally, with respect to the water level in a lake or river? What is likely to happen if the water table drops far below a river or lake?

10. What is the difference between confined and unconfined aquifers? Which type of aquifer may be responsible for artesian wells? Why do artesian wells flow the way they do?

11. What kinds of sediments are most likely to produce a good aquifer? Name/describe them.

12. Sometimes basalts or tills can be aquifers. More often, however, they are aquitards. Why?

13. Earth's *albedo* is very important because it means that Earth is cooler than it could be. What is albedo? Explain why and how it affects Earth's climate. Which parts of our planet contribute most to the albedo?

14. What differentiates a glacier from a year-round permanent snowfield?

15. Why do climatologists refer to the Arctic and Antarctica as deserts? Does this explain why there are few glaciers in Siberia? Explain.

16. What is the relationship between times of glaciation and sea levels? Explain how glaciers affect sea level.

17. Earth's cryosphere is shrinking today. What would be the effect on sea level if is disappeared completely?

18. Permafrost is melting at an alarming rate today. Some scientists are concerned that melting permafrost will further contribute to global warming. Explain why they are concerned.

19. What is *salinity* and what compounds are most responsible for making water saline? Where is the most saline water on Earth found? Explain how it came to be so saline.

20. We all know that ocean water is salty. How did the salt get there and become so concentrated. Are there any natural processes that remove salt from seawater?

21. The Mediterranean Sea is generally saltier than most of the world's ocean water. And in the geological past, it has become so salty that it deposited thick layers of halite. Explain why it is so salty, and why it deposited so much halite in the past.

22. What are conservative elements? Give two examples. What causes some elements to be non-conservative in ocean water? Name two non-conservative elements.

23. Different continental water bodies have different salinities. Explain why this is the case and give examples. How do the salinities of continental waters compare with the salinity of seawater?

24. What is the difference between point-source pollution and non-point source pollution? Give examples of both kinds of pollution. Which of the two are most easily controlled or eliminated? Explain?

25. What is *E. coli*? What causes water to be contaminated with *E. coli*? Why do health departments routinely check municipal water supplies to see if *E. coli* are present?

26. What is eutrophication? What causes it, and why is it harmful to aquatic ecosystems? People can cause eutrophication - explain how. Can nature do it without help from people?

13 Mineral Deposits

13.1 Bingham Canyon, Utah

Figure 13.1 is a photograph of the Kennecott Copper Mine 30 kilometers (19 miles) southwest of Salt Lake City (which is in the distant background), Utah. This hole in the ground measures 4 kilometers (2.5 miles) across and is 1.2 kilometers (0.75 miles) deep. At the time of the United States Civil War, before Utah became part of the United States, this hole did not exist. Instead, the area contained sloping hillsides covered with Douglas fir, bigtooth maple, and aspen trees, Western wheatgrass, slender wheatgrass, and fourwing saltbush.

In 1848, two brothers, Thomas and Sanford Bingham, settled in this area and used it as pasture for horse and cattle grazing. The Bingham brothers were among the first Mormon pioneers to arrive in Utah after a long journey west following religious persecution in the eastern United States. Besides using the area for grazing, the Binghams and other early settlers felled timber to build houses and infrastructure. Although the Bingham brothers noticed that some rocks in the area, which people called *Bingham Canyon*, contained copper that was visible to the naked eye, they thought little about it. Eventually, in the early 1850s, they left the area to move closer to Salt Lake City.

Figure 13.1 Bingham Canyon ore deposit, with Salt Lake City, Utah, seen in the distance.
Photo courtesy of Ray Boren.

Logging operations continued after the Bingham brothers left, and loggers noticed that some rocks, besides containing copper, contained lead and silver. These metals were also found in sediments deposited next to streams, and in the early 1860s soldiers from nearby Fort Douglas, a Union Army outpost near Salt Lake City, started to take interest. The soldiers recalled that just about a decade earlier (in 1849), the California gold rush had occurred when gold was found as part of stream sediments near what is today Sacramento. If similar deposits existed in Bingham Canyon, there was much money to be made. This realization kicked off a frenzy of metal prospecting by soldiers excited about the possibility of a new gold rush, this time in Utah.

By the late 1860s, the Utah gold rush was attracting many people. The promise of riches drew miners from Wales, Ireland, and the United Kingdom, some of whom—especially those who arrived first and claimed the largest and most obvious gold deposits—got rich. Many, however, struggled, in part because geographic isolation presented a big challenge to early mining activities. Gold produced during the California gold rush was quickly and inexpensively transported down the Sacramento River to ships in San Francisco Bay. Bingham Canyon, however, was isolated until the golden spike marking the completion of the transcontinental railroad was driven into the ground at Promontory Summit, Utah, on May 10, 1869.

By 1873, a railroad line connected the Bingham Canyon area to the transcontinental railroad, providing a cheap and efficient means to deliver metals to the outside world. As the pace of mining picked up, metals in surface sediments were soon exhausted. Miners then began tunneling underground to extract metal-rich ores. Underground mining was, however, difficult and inefficient, so by the dawn of the 20th century, miners had

started what we now call *open-pit mining* (Figure 13.1). Subsequently, mining operations at Bingham Canyon expanded and attracted many more immigrant workers from around the world.

Figure 13.2 shows the mine and the town of Bingham Canyon, incorporated in 1904, below and to the right of the mine. By the 1920s, 15,000 people lived in the town. About two-thirds of the people were immigrants to the United States from countries that included Austria, England, Finland, Greece, Ireland, Italy, Japan, Norway, Poland, and Sweden, among others.

13.1.1 Why do we need Bingham Canyon?

Bingham Canyon is but one of the many thousands of operating mines in the world today. These mines provide mineral commodities that allow people to live the lives they do. Consider the normal things that you and other people use any day—for examples: the house where you live; the buildings in which you take classes; the coffee shops and restaurants where you eat and drink; the stores where you buy food, clothes, and toiletries; the bicycle, car, truck, train, and plane that you use for travel; the shows that you enjoy watching on television and the music you enjoy listening to while walking to class (hopefully not during class!); the smartphone in your pocket that instantly connects you with people and information all over the globe. All these things, intimate to your daily life, connect you to the giant hole in the ground at Bingham Canyon. Those metals—copper, lead, and silver—which drew settlers and led them to start mining at Bingham Canyon, are in your smartphone, laptop, car, plane, train, and coffee maker. Copper is the conduit for electricity to light your

PANORAMA OF THE PORPHYRY MINES OF THE UTAH COPPER COMPANY AND BOSTON CONSOLIDATED MINING COMPANY AT BINGHAM, UTAH.
Supplement to The Engineering and Mining Journal, September 7, 1907.

Figure 13.2   Bingham Canyon in 1907.
From Walter Renton Ingalls, Wikimedia Commons.

Table 13.1   Some important metals and products that contain them.

Copper	computers, smartphones, photovoltaic solar panels, wind turbines, microwave ovens, refrigerators, air conditions systems, cars, planes, trains, water pipes, electromagnets, electric switches and relays, radiators, brakes, motors, saxophones, boat propellers, power lines
Lead	batteries, solder for electronics, high-voltage power lines, roofing materials, radiation shields, sailboat keels, divers belts, insecticides
Silver	jewelry, mirrors, dentistry fillings, radio frequency connectors, computer keyboards, high-voltage electronics, batteries in hearing aids and watches, saxophones and flutes, catalysts for chemical reactions, medicines to kill bacteria and viruses
Zinc	metal alloys, rust-proof coatings, paint, pharmaceuticals, chemical catalysts, sunscreen, lubrication, food additive, ceramics, rubber manufacturing
Molybdenum	metal alloys, lubricants, chemical catalysts, ink, pigments, electrodes, engines, furnaces

home and recharge your smartphone battery. Lead is in the battery that allows you to start your car. And silver, because of its incredible thermal and electrical conductivity and ability to withstand corrosion and oxidation, is part of electronic switches in both smartphones and cars. Silver nanoparticles are also in your laundry detergent and deodorant owing to silver's anti-microbial properties. The list of things that include copper, lead, and silver is lengthy; Table 13.1 below lists some of the more common products.

13.1.2   Bingham Canyon today

Figure 13.3 shows trucks hauling rock from the Bingham Canyon Mine. The open pit continues to get larger, with about 550,000 tons of rock removed *each day*. The pit today covers an area of 2000 acres (8.1 square kilometers) and is deep enough to stack two Empire State Buildings on top of each other. Most of the land originally occupied by the town of Bingham Canyon has been consumed by the mine, and the last of the town's buildings were razed in 1972.

Since mining began just over 100 years ago, the Bingham Canyon open-pit mine has produced 24 million tons of copper, 790 tons of gold, 6600 tons of silver, and 425,000 tons of molybdenum. To put these numbers in perspective, consider that, in total, the people of the world consume about 20 million tons of newly mined copper, 3000 tons of newly mined gold, 30,000 tons of newly mined silver, and 300,000 tons of newly mined molybdenum *every year*. So, the cumulative amounts of metals mined from Bingham Canyon over an entire century could satisfy peoples' needs for copper and molybdenum for about one year, and silver and gold for only a few months. Thus, to maintain our societies and lifestyles without recycling metals, the world's people need many mines like the one at Bingham Canyon.

Figure 13.3   Trucks transporting ore from the bottom of the Bingham Canyon open-pit mine.
Photo from Hermann Luyken, Wikimedia Commons.

13.2   Why dig such a big hole?

13.2.1   Archaeological periods and mining

The nature of ore deposits and the geologic processes that form them are the focus of this chapter. But before we look at many kinds of ore deposits, it is appropriate to think more about why mines such as Bingham Canyon exist. People mine ore deposits to produce resources used by society, which is not a new phenomenon. In fact, archaeologists and anthropologists define major periods of human civilization based on resources used. The late *Stone Age*, also called the *Neolithic Age*, was followed by the *Chalcolithic Age* from 4500 to 3500 BCE (Before Common Era). During the Chalcolithic Age, humans began using copper (and the name *chalcolithic* is derived from the Greek word *khalkos* for copper) for both decorative and utilitarian purposes. Because copper could be found as malleable pure copper nuggets, people could shape it with available stone tools—a property that no other common minerals possessed. A rise in consumption of copper coincided with the development of a socioeconomic hierarchy, and the wealthy citizens possessed more copper than the proletariat.

Figure 13.4 shows artifacts from the *Bronze Age*, the age that followed the Chalcolithic Age and lasted from 4200 to 1000 BCE. Use of bronze first developed in the Mesopotamian civilization of Sumeria. This was a period characterized by a rapid rise of resource consumption and increasing diversification of products made by metalworking. Perhaps the most significant advancement in metal use was the discovery of how to make bronze, an alloy created by melting and combining the metals copper and tin. Although tin melts at a relatively low temperature (232 °C), copper melts at 1085 °C (a temperature too great to be easily achieved at the time). However, clever metal workers discovered that a mix of one part tin and nine parts copper melts at 950 °C, which was low enough to make bronze manufacturing possible in many places.

Figure 13.5 shows the location of mines of the Bronze Age that provided metals used throughout much of the Middle East. Most of the copper came from the Troodos Mountains of Cyprus, where copper could be found in loose sediments at Earth's surface. Even today, a great deal of copper mining takes place in Cyprus, although operations have moved underground. Some copper and other metals came from mines on the Asian mainland, but tin deposits were generally small or hard to produce. The scarcity of tin in the Middle East and other areas around the Mediterranean Sea meant that tin ores came from as far away as the British Isles, which the Greeks named the *Cassiterides*, which translates to *Tin Islands*.

A key property that allowed humans to work with bronze is that, after pouring the molten bronze into stone molds and allowing the liquid to cool to a solid, the copper-tin alloy could be formed and shaped using hammers at room temperature, a process called *cold working*. And because bronze is much stronger than copper, people could make many improved products, including knives, shields and swords, and tools that led to more productive agriculture.

The *Iron Age* followed the Bronze Age beginning around 1500 BCE, when the Hittite society of ancient Anatolia (modern-day Turkey) discovered how to process iron. Their technological breakthrough was to add a small

Figure 13.4   Bronze Age artifacts.
Photo from Birmingham Museum Trust, Wikimedia Commons.

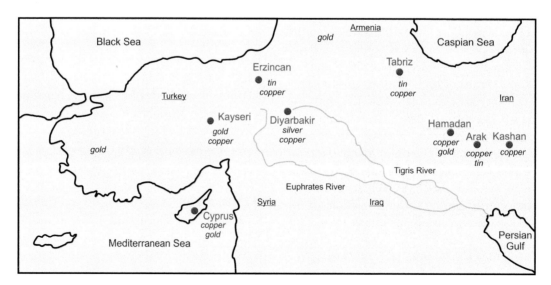

Figure 13.5   Sites of resource extraction in the Bronze Age.
Based on figure at https://vieilleeurope.wordpress.com/.

amount of charcoal (carbon) to rocks that contained iron. Pure iron melts at 1538 °C, but adding carbon results in a carbon-iron mixture that melts at 1170 °C. The Hittites also figured out that iron-carbon alloys could not be cold-worked like bronze, but had to be hammered and shaped while hot. Thus, they invented the art of modern blacksmithing. The iron and alloys produced, once cooled, were much stronger and harder than bronze was.

The source of iron used by the Hittites was metallic meteorites, which also contained a small amount of nickel that improved metal properties. Because iron-rich meteorites were not in abundance, the Hittites carefully guarded their invention of iron metalworking for several centuries. During those centuries, the Hittites exercised military superiority over much of the Middle East and Egypt, where the weaker bronze was used in battle. However, by 1200 BCE, iron metalworking technology had spread across the Middle East, North Africa, and Europe and to Asia; people discovered new sources of iron, and the Hittite empire disappeared.

Mineral resources literally put places on the map of the ancient world. If a region contained abundant amounts of copper, silver, tin, or gold, and later iron, it soon became populated and prosperous. Civilizations established trade routes and developed commercial systems, shipping commodities over increasingly longer distances. If resource supplies became depleted in one location, people sought new sources; thus, early exploration was needed to sustain production and consumption of valuable resources. These same dynamics operate today: when new mineral deposits are discovered, new communities and industries appear. When old deposits become depleted, communities and industries wane. And always, mining companies are exploring to find new sources of economically viable minerals.

13.2.2 Modern society and mining

Copper, tin, iron, and nickel were all important during the early ages of humans, and they are equally important today. Those same metals—and many others—are key parts of a seemingly infinite number of products. For example, Figure 13.6 shows the many minerals that

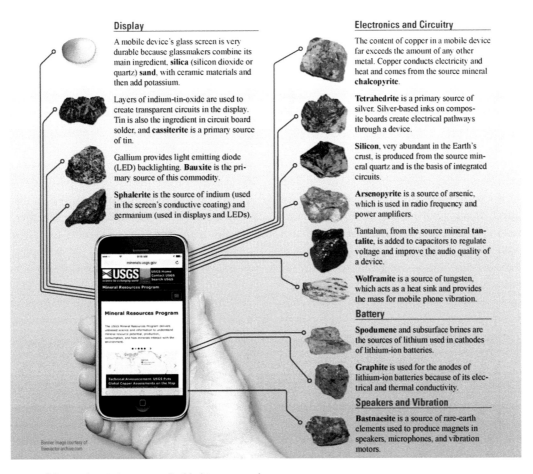

Figure 13.6 Some of the nearly 75 elements embedded in a smartphone. Graphic from the USGS.

provide elements that are in a smartphone. Copper makes up about 10% of the weight of a smartphone, and that copper is the key to moving around the electricity that powers the phone. Tin is used to make the liquid crystal display (LCD) screen and to solder electrical connections that transmit digital information. Iron is combined with the metals neodymium and boron to make magnets that are part of the microphone and speaker. And those are not the only elements embedded in a smartphone; there are about 75 elements in all. Without any one of these elements, smartphones would not exist as they do. Nearly everything that we manufacture contains mineral resources, and the sources for these resources are mineral deposits.

Figure 13.7 Tailings lake near Logan Lake, British Columbia. Photo by Jeffrey Wynne, Wikimedia Commons.

13.3 Mineral deposits and ore deposits

A *mineral deposit*, like the one at Bingham Canyon, is a place in Earth's crust where geologic processes have concentrated one or more minerals at greater abundance than in the average crust. The rocks at Bingham Canyon contain concentrations of copper, gold, silver, and molybdenum that are much higher than the concentrations in average rock. For example, the average concentration of copper in rocks that make up Earth's crust is about 27 parts per million (ppm). The rock mined in Bingham Canyon contains about 0.5 wt% copper, or 5,000 ppm. That means that the rocks at Bingham Canyon contain 200 times more copper than rocks in most places. This concentration of copper allows companies to extract the metal today at a cost that makes mining profitable. We call such a profitable mineral deposit an *ore deposit*. Thus, all ore deposits are mineral deposits, but the reverse is not true. The market value for a particular metal and current mining technology determine whether a mineral deposit can be mined profitably. If markets or technologies change, new deposits may become profitable, or old deposits may no longer make money.

Besides containing ore, most ore deposits contain other minerals, called *gangue minerals* and *waste rock*, that must be separated from ore minerals during processing. The combined gangue and waste make up *tailings* that are sometimes returned to mine pits or tunnels after ore is removed. More commonly, mining companies discard tailings in *tailings piles* or *tailings lakes*. Figure 13.7 shows a tailings lake near Logan Lake, in south-central British Columbia. This 10-kilometer-long lake was created so that slurries containing tailings left over from copper mining could be discarded efficiently.

13.4 The formation of ore deposits

Economic geologists study ore deposits, in particular focusing on the natural processes that concentrate ore minerals sufficiently to make an ore deposit. Understanding these processes is important for developing models that guide exploration for new deposits and ensure a sustainable supply of the resources woven into the fabric of modern society. However, the costs associated with mining are not static because demands for minerals, and the costs of mining, change over time. So, a particular mineral deposit may be economical to mine at one time and not at another. Costs also vary from place to place, and consequently two similar deposits in different countries may not be equally profitable. No matter when and where, however, the greater the *ore grade* (the percentage of ore minerals in a particular deposit), the less that needs to be mined to be economical.

The economics that determine whether a particular deposit is profitable are different for different mineral commodities. Consider iron, for example. The average abundance of iron in Earth's crust is about 5 wt%. Today, to be profitable, an ore deposit must contain about 20 wt% iron. Thus, iron must be concentrated four times over natural abundance to be economical. The ratio of economical ore grade to natural abundance is called the *economical concentration factor*, or sometimes just the *concentration factor*. Table 13.2 compares crustal abundances and economical concentration factors for six important metals. Overall, economical concentration factors are inversely related to crustal abundance. Gold, for example, has an average concentration of 0.0013 parts per million (ppm) in Earth's crust; it must

Table 13.2   Natural abundance, economical ore grade, and concentration factors for some important metals % values are wt%..

Ore resource	Metal concentration in average crustal rock	Minimum metal concentration (ore grade) for profitable extraction	Concentration factor
Aluminum	8%	30%	4
Iron	5%	20%	4
Sodium	2.3%	40%	17
Copper	27 ppm	5000 ppm	185
Silver	0.056 ppm	125 ppm	2200
Gold	0.0013 ppm	10 ppm	7700

be concentrated 7700 times to be profitably mined. In contrast, the more abundant aluminum and iron need only be concentrated four times to be economical.

Although Table 13.2 does not include prices, there is a correlation between the economical ore grades and the price of a given resource. Gold is much more expensive than aluminum, iron, sodium, copper, and silver are, even though the demand for gold is less than for the other five commodities. This price difference exists because the natural processes that concentrate most commonly used metals are much more common than the processes that concentrate gold, so there are many fewer gold deposits than there are of other kinds of deposits.

Many different natural processes can concentrate minerals and create ores. In the rest of this chapter, we will look at some of the most significant of those processes and the kinds of ore deposits they produce. Some processes are physical processes, involving deposition of sediment or of heavy mineral grains. Others are chemical processes that involve elements or minerals dissolving in water before being precipitated somewhere else. Still others are magmatic processes involving minerals crystallizing in a pluton or associated with volcanism. Economic geologists have proposed many classification schemes, but none is perfect because there is great variation between different kinds of deposits, and there is no single, logical way to group them. Thus, the examples below do not include every possible kind of ore deposits, but instead are descriptions of some of the deposits of greatest importance.

13.5 Placer deposits

Placer deposits are a type of ore deposit formed by physical processes and forces: weathering, erosion, water flow,

and gravity. Figure 13.8 shows how such deposits form. The process starts with weathering and erosion that disaggregate rocks to form clastic sediment. In mountainous regions with a significant topographic gradient from high elevation to low elevation, runoff over land and then running water in streams transport clastic sediment from the mountains toward the oceans. When the streams are near their sources, high in the mountains, stream velocities are high, and streams can transport a variety of particles. Clasts of different sizes may be carried and, most important for ore deposits, clasts of different densities may be carried. As streams leave the mountains and slow, they lose some ability to carry coarse and heavy materials, so these materials may be left behind. This winnowing process continues farther downstream as the smallest and lightest clasts continue to be transported while other denser clasts remain behind and concentrate in stream bottoms, beaches, or gravels. *Placer deposits*, also just called *placers*, form when one or more of the minerals concentrated in this way becomes an ore resource. The word *placer* is Spanish for *alluvial sand*. Typically, placer deposits form where a stream's velocity slows on point bars, in braided streams, or in alluvial fans. Other kinds of placer deposits are found in beach sands or gravels on ocean and lake shores, and some placers, although not commonly mined, form in offshore marine environments on continental shelves.

Placer minerals must be both dense and durable to be deposited and remain in place without decomposing.

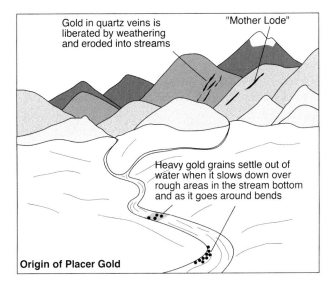

Figure 13.8   Formation of placer gold deposit.
Drawing from *Mineral Resources, Economics and the Environment*, 2nd Edition; Steve Kesler and Adam Simon.

Native metals such as copper or gold, sulfide minerals such as pyrite or pyrrhotite, and oxide minerals such as magnetite or ilmenite are all dense and likely to be found in placers. Metal oxides, especially magnetite (iron oxide), are common and especially dense and durable, and they often dominate such deposits. Gold is dense and extremely resistant to any kind of weathering and so can accumulate in stream and river sediments. Other important minerals found in placers include diamond, garnet, cassiterite (tin ore), ruby, sapphire, monazite, and zircon.

13.5.1 Placer gold

Most of the gold produced (78%) today goes into jewelry. Gold is ideal for this purpose because it is easy to shape and work, it is durable and beautiful, and it does not tarnish. Gold is also used as specie, mostly in the form of gold bars or ingots that remain in banks and national reserves, but also in the form of coins. Additionally, gold is a fundamental metal of the electronics industry. Because it is a good electrical conductor, it is found in many devices, including smartphones, computers, and GPS units. And dentists use gold to make fillings and crowns, because it is inert and does not cause any health problems. Given all these uses, gold is one of the most valuable mineral commodities.

Most of the world's gold comes from placers or rocks that were once placers, and gold placers have been the cause of major gold rushes around the world. Among the most famous gold rushes is the California Gold Rush of 1849–1855, which resulted in California being admitted to the United States in 1850 and later providing a name for a San Francisco football team (the 49ers). When the Gold Rush was in full swing, ships transported thousands of men and women from the eastern United States to California so they could make their fortune mining gold in the Sierra Nevada foothills. Figure 13.9 shows an advertisement intended to get potential miners to purchase ship tickets and join the gold rush. About 50 years later, the Klondike Gold Rush of 1896–1899, in Canada's Yukon Territory, drew an estimated 100,000 prospectors to ports in southeastern Alaska, where they embarked on treacherous journeys over mountain passes to claim territory along the Klondike River. Only about 30,000 to 40,000 miners actually reached the placer gold deposits, with many dying along the Chilkoot Trail and others turning back out of fear and frustration. In Australia, gold rushes

Figure 13.9 Advertisement to convince miners to take part in the California Gold Rush.

Image from G. F. Nesbitt & Co., Wikimedia Commons.

occurred in New South Wales and Victoria during the 1850s and in West Australia (Kalgoorlie) in the 1890s. And the 1886 South Africa Gold Rush resulted in that country becoming one of the world's leading gold producers for more than 100 years.

Gold in placer deposits is found as nuggets (Fig. 13.10), which range from microscopic size to basketball size. The nuggets are never entirely pure, typically containing 2% to 30% impurities that can be separated during melting. The largest nuggets in the top photo of Figure 13.10 are several millimeters in length. The largest placer gold nugget, shown in the bottom photo of Figure 13.10, was discovered in 1869 in Victoria, Australia, by John Deason and Richard Oates. The nugget, nicknamed the *Welcome Stranger*, weighed 216 pounds and was found only 3 centimeters below the ground surface. There was no scale big enough to weigh the nugget, so it was broken into three pieces that were weighed separately. The gold in the Welcome Stranger would be worth about $4 million today (in 2018).

The original sources of placer gold are igneous and metamorphic rocks at high elevation in nearby mountains. Over millions of years, the gold-bearing rocks weathered and eroded, and gold was washed away to become concentrated in placers. Often, the source rocks are long gone and only placers remain to testify to their former existence. Other times, source rocks can be identified, but the rocks may not contain sufficient gold concentrations to make mining them practical. In California, however, the source of gold for the California gold rush was the *Mother Lode*, a long narrow ore zone in the Sierra Nevada Mountains (shown schematically in Fig. 13.8). (The term *lode* refers to any ore-containing

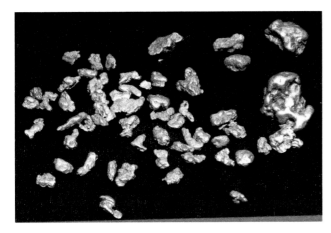

Figure 13.11 Examples of rough diamonds.
Photo from James St. John, Wikimedia Commons.

Figure 13.10 Top: relatively small gold nuggets. Bottom: miners and their wives posing with the finders of the Welcome Stranger nugget (in front of the kneeling person) discovered in Australia.

Top photo from James St. John, Wikimedia Commons; bottom photo from Victoria State Library.

vein or network of veins.) The Mother Lode was, from 1850 to 1967, a productive gold mining formation; today, it is a tourist attraction.

13.5.2 Placer diamonds

Figure 13.11 shows *rough diamonds* (uncut and unpolished) of different sorts. People mine diamonds like these in many places around the world, especially in south-central Africa (South Africa, Angola, Botswana, Namibia, and the Democratic Republic of Congo). The first recorded use of diamonds was in India in the 4th century BCE, when people used the stones for engraving tools and as religious objects. Today, some diamonds are gem quality and appropriate for jewelry, because after being cut they are beautiful and durable. However, society uses most diamonds as abrasives to grind, cut,

and polish softer materials. Many saw blades and drill bits have diamonds embedded in their edges and tips.

Much diamond mining is underground, but productive placer diamond operations are found in Angola, Namibia, and South Africa and on a small scale in several other countries. In southwest Africa, weathering and erosion of diamond-bearing volcanic rock have resulted in diamonds being transported downstream as part of the bed load of the Orange River that forms the border between South Africa and Namibia. Orange River placers produce more diamonds than any other placers in the world. The Orange River eventually flows into the Atlantic Ocean, and there are placer diamond deposits not just along the length of the river, but also in beach sands and shallow marine sediments along almost 322 kilometers (200 miles) of the Atlantic coastline. The modern method of mining the marine deposits uses a vacuum to bring the diamond-bearing sediments up from the bottom of the ocean. Diamonds are then separated from the quartz grains that are the most abundant constituents of the sediment.

13.5.3 Placer tin

As described above, most of the tin used by the ancient Greeks during the Bronze Age came from what is today England. That tin came from tin placers in the modern-day area of Cornwall. During the late Carboniferous and early Permian Periods, from 300 to 270 million years ago, southwest England was an active continental collision zone. One product of that environment was the formation of granitic magmas that contained the tin-bearing mineral cassiterite, SnO_2. Those magmas solidified to form granite plutons in an ancient mountain range that subsequently was subjected to weathering and erosion. Cassiterite, a stable mineral in water at Earth's surface, was physically weathered out of granite and became part of the clastic sediment transported by streams toward the ocean. Cassiterite has three times the density of quartz (beach sand) and more than seven times the density of water. Thus, when the velocity of the streams decreased upon reaching the coast, the cassiterite settled to the bottom of the streams and accumulated to form placer deposits. The Cornwall tin deposits, placer and underground, were mined for more than 4100 years, beginning about 2150 BCE. The last mine, the South Crofty Mine, closed in 1998.

Today, about half the tin produced is alloyed with lead and used as solder. Other major uses include plating (a surface layer) for steel, because tin does not corrode. For a similar reason, tin is in several kinds of alloys. Even though the Bronze Age was 3000–6200 years ago, and the Cornwall tin deposits formed hundreds of millions of years ago, both are relevant today. The screen of a smartphone, such as the one shown in Figure 13.12, for example, consists of several transparent layers, including one that is a mixture of tin, indium, and oxygen, a combination called tin-doped indium oxide. The transparent tin-indium oxide layer is a conductor, and when your finger touches the screen, it changes the electrical state at that point of contact. The resulting voltage change sets in motion all the action that keeps your eyes glued to the screen. So, today, tin ore deposits that powered the Bronze Age literally put the whole world at your fingertip.

Figure 13.12   Tin-indium-oxide screen.
Photo from Adafruit.

13.6 Chemical ore deposits

Chemical ore deposits are deposits that involve ores created by chemical processes. These are secondary processes involving water that concentrates particular elements and minerals sufficiently to make them economically minable. Below we look at two kinds of chemical deposits: laterites and evaporites.

13.6.1 Laterite deposits

Chapter 11 discussed *chemical weathering* of rock. This kind of weathering can lead to the formation of chemical ore deposits. Chemical weathering involves many kinds of reactions that change a rock's composition and mineralogy. The key chemical processes are *dissolution*, *hydrolysis*, *hydration*, and *oxidation*, and thus water is generally involved. So, the effects of chemical weathering are generally fastest in wet environments and slowest in dry ones. Temperature also affects the rate of weathering, because warm temperatures speed up reaction rates. On average, chemical weathering is two to three times faster for every increase in temperature of 10 °C; thus, chemical weathering is more rapid in warm environments than in cold environments. The products of chemical weathering include (1) residual material made of leftover minerals that were in the original rock, (2) secondary minerals created during weathering, and (3) dissolved material that is carried away. The gradual dissolution of the most easily dissolved material is called *leaching*, and even if originally hard rock has been reduced to unconsolidated sediments or to soil, weathering reactions and leaching may continue as the sediment or soil continues to evolve. Thus, eventually, most of the soluble material may be gone and mostly insoluble material may remain. If the residual insoluble material becomes sufficiently enriched in valuable minerals, it may become economical to mine. Important ore deposits of aluminum, nickel, manganese, and iron form in this way.

Chemical weathering and leaching are much more rapid and important in tropical climates, because tropical climates receive lots of rain and have high temperatures year round. Rainwater is slightly acidic, having a pH of 5.7 because of the presence of dissolved carbonic acid. These conditions—hot, slightly acidic water—promote rapid chemical weathering of rocks at Earth's surface, converting rock into soil. In some tropical areas, the *weathering profile* that includes all the vertical layers of weathered material is more than 100 meters thick. As described above, chemical weathering removes soluble material and increases the concentration of insoluble material in the weathering profile.

Consider what happens when orthoclase (K-feldspar) reacts with water. Several stepwise reactions occur, but with enough weathering, the overall reaction can be approximated as:

$$2 \; KAlSi_3O_8 + 2 \; H_2O = Al_2O_3 \cdot H_2O + 6 \; SiO_2 + 2 \; K^+ + 2 \; OH^-$$

2 orthoclase + 2 water = aluminum hydroxide + 6 quartz + 4 dissolved ions

Potassium and hydroxide are very soluble in water and so are quickly removed. The material left behind is generally composed of quartz and a number of different aluminum minerals, including corundum (Al_2O_3), boehmite and diaspore (both different forms of $AlO(OH)$) and gibbsite ($Al(OH)_3$). A reaction, similar to the one above, occurs when albite (Na-feldspar, $NaAlSi_3O_8$) reacts with water, except that Na^+ is produced instead of K^+.

Consider a large volume of granite in a warm, tropical climate that, over a long span of geological time, was subjected to intense chemical weathering and thus developed a thick soil layer above more solid material. The rock was originally composed of quartz (SiO_2), orthoclase ($KAlSi_3O_8$), and albite ($NaAlSi_3O_8$). But the orthoclase and albite reacted with water by reactions such as the ones described above, and sodium and potassium were leached away while the concentrations of silicon and aluminum increased. Simultaneous with the leaching, any iron-bearing minerals present were oxidized, giving the soil a rusty red color.

Typical granites contain about 8 wt% aluminum. Chemical weathering, such as described above, can increase the aluminum concentration to between 25 and 30 wt%—a concentration factor of three to four. Thus, a warm climate and lots of water can transform normal granite into a reddish aluminum-rich soil, like the soil shown in Figure 13.13. This aluminum-rich soil is called *laterite*. Laterites commonly lithify partly, or completely, to become a rock, and some geologists use the term *laterite* as both a soil name and a rock name. Lateritic rocks that are particularly enriched in

aluminum, like the deposit shown in Figure 13.13, are called *bauxite*. Thus, laterites and bauxites are the residue left behind after chemical weathering of granite. And, in bauxite, the aluminum that was originally in orthoclase and albite is in aluminum hydroxide minerals, including primarily boehmite, diaspore, and gibbsite.

Figure 13.14 shows an abandoned bauxite mine in Italy. Bauxite, which also typically contains clay minerals, quartz, and iron oxides, is the main source of aluminum for modern society. When you pick up an aluminum can or drive a vehicle made with an aluminum frame, you are using a resource that nature produced by chemically weathering igneous rocks in a humid, tropical climate.

Figure 13.13   An abandoned bauxite (laterite) mine in Hungary. Photo from VargaA, Wikimedia Commons.

Figure 13.14   Bauxite mine near Otranto in Apulia, southern Italy. Photo from Palickap, Wikimedia Commons.

Historically, people have mined bauxite in many places. Today, because of large deposits that lead to economic efficiency, Australia produces the most bauxite, about a third of total world production. China, Brazil, India, and Guinea also have substantial bauxite industries. The southeastern United States was an important source of bauxite 100 years ago but today accounts for less than 1% of world production.

Granite is not the only kind of rock that breaks down in tropical climates to form laterite. Consider, for example, basalt, a mafic rock that contains high concentrations of magnesium and iron and has undergone intense weathering.

Figure 13.15 shows an example from Vangaindrano, Madagascar. This outcrop is quite a contrast to the photos of fresh basalt (Figs. 7.8, 7.27, 7.34, and 7.46) in Chapter 7, and much of the original mineral and glass material has decomposed.

The essential minerals in basalt are clinopyroxene, $(Ca,Na)(Mg,Fe,Al)(Al,Si)_2O_6$, and plagioclase, $(Na,Ca)(Si,Al)_4O_8$. Many basalts also contain olivine, $(Mg,Fe)_2SiO_4$, or orthopyroxene, $(Mg,Fe)_2Si_2O_6$. Thus, basalt contains six major metallic elements. Of these, the soluble elements sodium, calcium, and magnesium are most easily dissolved and removed. Consequently, the concentrations of iron, aluminum, and silicon increase during weathering. And, the resulting laterite—such as the one seen here—is more iron- and magnesium-rich than one that forms from a granite parent because basalt contains more iron and magnesium to start with. So, weathered basalt contains somewhat different minerals than weathered granite.

Many other kinds of igneous rocks, besides granite and basalt, chemically weather to produce laterite.

Figure 13.15   Weathered basalt in Madagascar.
Photo from Werner Schellmann, Wikimedia Commons.

Different kinds of laterites form, and different metals become concentrated, depending on the mineral composition of the *protolith*, the parent rock from which the laterite was produced. Intense chemical weathering of felsic igneous rocks, such as granite, yields aluminum-rich laterites. Intense weathering of mafic to ultramafic igneous rocks yields laterites enriched in iron, nickel, cobalt, chromium, titanium, manganese, and copper. All these metals are relatively insoluble in water at Earth surface conditions and are not removed during leaching. And all these metals are part of a smartphone. Nickel is a component of the microphone. Chromium is added to iron to make stainless steel; it forms an impermeable surface layer of chromium oxide that prevents oxygen from diffusing into the steel, which could cause the steel to rust or corrode. Cobalt and manganese are part of the lithium-ion battery that stores charge and powers the phone. Titanium is alloyed with other metals to make the back casing of smartphones. Titanium is also used in place of steel in some phones, because it is half as heavy as steel. Thus, larger phones, with more room for batteries, can be made without increasing phone weight.

13.6.2   Evaporite deposits

The previous section of this chapter discussed leaching, removal of material when it dissolves in water, but what happens to the material that is removed? Some of it goes to rivers, lakes, or streams, where it may remain for long periods of time or be precipitated in sediment. But, ultimately, streams and rivers carry most of it to oceans, where it adds to the oceans' dissolved loads. Chapter 8 discusses the formation of *evaporite deposits*, also just called *evaporites*, by evaporation of water. The deposits contain a variety of evaporite minerals, also called *mineral salts*. Halite (NaCl; table salt) is the most common evaporite mineral, in some places forming layers hundreds of meters thick. But ocean water contains many other dissolved salts, which, as discussed in Chapter 8, precipitate in predictable order when seawater evaporates. Calcite, gypsum, and halite are the first to be deposited and are followed by other calcium, magnesium, and potassium chlorides. These are the main minerals in evaporites; however, 80 known minerals can precipitate when seawater evaporates. Evaporite minerals provide the ingredients for many products used by society. For example, people use gypsum to manufacture drywall and use halite to flavor or preserve food and to deice roads in winter.

13.6.2.1 Halite

Although some evaporite deposits form from continental waters, most form in marine basins that have become partially or wholly isolated from larger oceans. For example, as described in Chapter 8, when the Mediterranean Sea and the Gulf of Mexico became isolated in the past, layers of evaporite minerals hundreds of meters thick were deposited. The same thing has occurred in other places.

Figure 13.16 shows an underground salt mine in Romania that is now an underground entertainment center. Underground salt mining has been taking place for centuries. Today, for example, thousands of feet beneath Detroit, Michigan, halite is mined from a layer of salt that extends beneath Lake Huron and Lake Erie into Ontario, Canada. This salt deposit formed when the region that is present-day Michigan was an ocean basin near the equator. About 400 million years ago, the basin was cut off from the main oceans and, during the subsequent 15–20 million years, the basin experienced evaporation and refilling several times. The result was layers of halite, called the *Salina Salt Beds*, that have been mined in the upper Midwest for more than a century. People mine similar evaporite deposits in Kansas, Louisiana, Michigan, New York, Ohio, Texas, and Utah. All the mines together satisfy the demand for halite in the United States.

About half of all halite is used to deice roads to make winter driving safer. Chemical manufacturers also use a lot of halite to make thousands of products, including plastic, paper, glass, polyester, rubber, household bleach, soaps, detergents, and dyes. Additionally, the pharmaceutical industry uses halite to make infusions, medicines, and salines. And a few percent of mined halite is used directly by the food production industry, because halite is an important spice and a preservative. Halite also helps people absorb and transport nutrients, maintain blood pressure, maintain the right balance of fluid, transmit nerve signals, and contract and relax muscles.

13.6.2.2 Sylvite

Sylvite (KCl), the second most abundant chloride mineral, is also mined from evaporites. Sylvite is the world's primary source of potassium used in fertilizer for commercial agriculture, lawns, and gardens. Potassium is an essential nutrient in plants (and humans), ranking third in importance (after nitrogen and phosphorus) for plant growth, because it improves yields and the nutritional contents. So, farmers use potassium fertilizer for growing many crops, including fruits and vegetables, corn, sugar, palm oil, cotton, soybeans, rice, and wheat, among others. Plants deficient in potassium often show yellow discoloration and curling of leaf tips, such as can be seen in Figure 13.17, and overall growth, including root development and fruiting, is diminished.

Historically, potassium fertilizer was known as *potash*, which derives from the Dutch word *potaschen*, translated as *pot ash*. This term refers to the pre-industrial practice of soaking plant and wood ashes in water in a large iron pot. The pot was heated to evaporate the

Figure 13.16 The 17th-century Turda salt mine at Cluj-Napoca, Romania, which is now an underground entertainment center.

Photo from David Stanley, Wikimedia Commons.

Figure 13.17 Soybean leaf showing the effects of potassium deficiency.

Photo from Alandnanson, Wikimedia Commons.

water, which resulted in the formation of a white residue made of potassium carbonate (K_2CO_3), which was used as fertilizer. Some purists today make homemade fertilizer following this recipe. However, for the most part, society mines sylvite instead of making potash.

Sylvite, like halite, is mined from layers of salt that precipitated during evaporation of shallow marine basins. Some mines consist of complex underground infrastructures with elevator shafts and tunnels that allow humans to descend hundreds to thousands of meters below the surface to mining sites. Other operations employ *solution mining* to extract the sylvite, essentially reversing the process by which nature formed the layers of sylvite. For solution mining, miners drill wells into the subsurface to intersect a layer of sylvite. Water, pumped down the well, returns to the surface containing dissolved sylvite, and thus the need for underground mining is eliminated. To recover sylvite, the potassium- and chlorine-containing water is piped into shallow *evaporation ponds* and allowed to evaporate.

Figure 13.18 shows an aerial view of evaporation ponds associated with a potash deposit near Moab, Utah. The Colorado River is on the right side of the photo. Blue dye was added to the water to hasten evaporation. As the water level in the ponds drops, sylvite precipitates. It is then processed to produce potassium fertilizer. Today, people mine sylvite in many places, but the largest production and most significant deposits are in Saskatchewan, Canada.

13.6.2.3 *Gypsum*

Gypsum, a mineral having the composition $CaSO_4 \cdot 2H_2O$, is the most common sulfate mineral. It is often found as individual crystals, which, if clear, are called *selenite*. Photographs of some, especially large and beautiful gypsum crystals in Mexico's Cuevo de los Cristales, are shown in Figure 4.1 of Chapter 4. These large crystals are some of the largest mineral crystals in the world.

Most economic deposits of gypsum, however, are layers of fine-grained rock also called *gypsum*, or *gypsum rock*. Like economic halite and sylvite deposits, these layers of nearly pure gypsum are marine evaporites, and they are commonly found associated with other evaporite deposits that include sulfur or halite. Gypsum is processed and used primarily to make wallboard and other plaster products. Modern homes in developed countries use gypsum sheetrock for interior walls. Gypsum is also used as fertilizer and for manufacturing cement and glass.

Most gypsum is mined from open pits, such as the pit shown in the top photo of Figure 13.19. China is the world's number one gypsum producer, producing almost 10 times as much as Iran and the United States produce, the second and third largest producers. Gypsum mines are

Figure 13.19   Mining gypsum in Fort Dodge, Iowa (top). Gypsum and red beds, Caprock Canyon, Texas (bottom).

Photos from Joe Sutter, The Messenger (top); Freddyfish4, Wikimedia Commons (bottom).

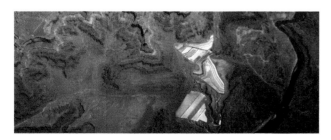

Figure 13.18   Potash evaporation ponds near Moab, Utah.

Photo from Doc Searles, Wikimedia Commons.

found in about 20 U.S. states, but the largest are in Oklahoma, Iowa, Nevada, Texas, and California. A large open-pit mine near Fort Dodge, Iowa—where the bulldozer photograph in Figure 13.19 was taken—may have the largest known gypsum deposits of any mine in the world. Smaller layers of gypsum—generally not ore deposits—are commonly found associated with *redbeds* (oxidized iron-rich sandstones and siltstones) created during arid oxidizing conditions that accompanied evaporite formation. Figure 13.19 (bottom) is an example of such a deposit from Caprock Canyon, near Lubbock, Texas.

13.6.2.4 Continental evaporites

Evaporite deposits have formed in many arid continental regions around the world. Chapter 8 briefly discusses examples of such deposits associated with Utah's Great Salt Lake and California's Searles Lake. Most continental evaporite deposits form when water flows into an inter-montane (between two mountain ranges) valley, creating shallow ephemeral lakes that evaporate quickly. In some places, seasonal variations in rainfall mean that precipitation of evaporite minerals only occurs during dry months. The minerals deposited in continental evaporites are different from those deposited in marine evaporites, because the compositions of continental and marine waters are different. However, calcite, gypsum, and halite are generally dominant in both settings. And many other minerals, especially boron- and lithium-containing minerals, that crystallize during evaporation in closed drainage basins are important resources for society.

Figure 13.20 shows an example of continental salt deposits—halite deposited in Badwater Basin, the lowest place in North America at 86 meters (282 feet) below sea level, in Death Valley National Park. The water that flows into Death Valley and other arid basins or valleys during short wet springs forms shallow playa lakes. These lakes are ephemeral because the annual rate of evaporation is much greater than the rate of water flowing into the basin. As the water evaporates during the dryer months, dissolved salts crystallize as calcite, gypsum, and halite, just as we described previously for marine evaporite deposits that form in ocean basins.

13.6.2.5 Boron deposits

The Death Valley region of California, however, has a storied metal-mining history that includes many booms and busts. Mining operations in mountains around the valley targeted gold and silver, but also antimony, copper, lead, zinc, and tungsten. Unfortunately, most deposits produced for short times, and the last major metal mining ended in 1915.

In the valley bottom, however, mining of mineral salts had a longer and more continuous history. Death Valley is a closed basin, and evaporite minerals were deposited there for thousands of years before people arrived. The first of the salt mines started operations around 1880, and the last, the Billie Mine, closed in 2005. Many different mineral salts are found on the valley floor. Death Valley halite deposits, such as the one shown in Figure 13.20, never supported a robust mining industry, and the most successful mining operations focused primarily on minerals that contained boron.

The most profitable boron ore mineral in Death Valley was *borax*, but many others, such as colemanite and ulexite, were mined as well. Borax and other borate minerals have historically been important as water softeners in laundry detergents. In the late 19th century, mining companies used mules and horses to haul the borax and other ores out of Death Valley—a grueling 165 miles to a train station in Mojave, California, and thence to customers throughout the United States. Figure 13.21 is a photo of a mule team used for this purpose.

Figure 13.20 Death Valley's Badwater Basin.
Photo from Timothy Lavelle.

Figure 13.21 Teams of mules and horses transporting borax in Death Valley, USA.
Photo from National Park Service.

13.6.2.6 *Lithium deposits*

Lithium is a natural component of many springs, and some people believe that bathing in lithium-rich water offers holistic health benefits. The waters of Lithia Springs, Georgia, became quite popular beginning in the late 18th century and remain a popular tourist destination today. In 1929, a soft drink entrepreneur capitalized on lithium's reputation by introducing "Bib-Label Lithiated Lemon-lime Soda" that contained a minute amount of lithium citrate. Figure 13.22 shows one of the original bottles of that product. It was called 7UP because the bottle size was 7 ounces, something that would not do today. Twenty years after 7UP was first marketed, the federal government banned the addition of lithium to soda, but the product is still sold today.

Lithium has many uses, including being a component in greases and specialty glasses, but the main uses today are for lithium-ion batteries that power phones, laptops, and electric vehicles and for medicine to treat individuals with bipolar disorder (because lithium is a mood-stabilizing drug).

Traditionally, lithium has been mined, at the surface and underground, from evaporite deposits, such as those in the Jadar Valley, Serbia. The Jadar Valley is the where the lithium mineral *jadarite*, $(Na_2OLi_2O(SiO_2)_2(B_2O_3) \cdot 3H_2O)$, was first discovered. Similar lithium-bearing minerals are found in evaporite basins in the Andes Mountains of Argentina, Chile, and Bolivia, central Australia, the Qinghai Basin, China, and the Great Basin Desert of Nevada and Utah. Although these deposits were mined conventionally in the past, in the 1990s, mining companies discovered that they could produce lithium from lithium-bearing groundwater more cheaply than mining lithium ores directly.

Figure 13.23 shows lithium salts crystallizing from brine pools in Argentina's Salar de Olaroz Mine. Note the roads and buildings below the left edge of the pools, for scale. Brines are pumped from wells that penetrate the sediments of the desert floor and are allowed to evaporate in large shallow pools on the surface, like the ones shown. The processes produce crystallized lithium carbonate (Li_2CO_3), from which lithium can be purified.

13.7 Sedimentary ore deposits

In contrast with chemical deposits involving precipitation of minerals from water, many sedimentary ore deposits involve the deposition of already-existing mineral grains or other solid material. In these kinds of deposits, ore minerals collect at the time of sedimentation and are later incorporated into rock. Earlier in

Figure 13.22   Lithium was an ingredient in the original 7UP.
Image from Proczach, Wikimedia Commons.

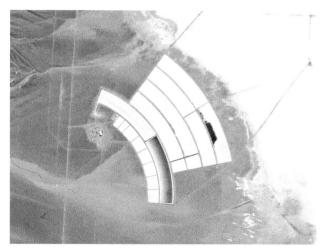

Figure 13.23   Lithium salt crystallizing from brine pools at the Salar de Olaroz Mine in Argentina.
Photo from Planet Labs, Inc., Wikimedia Commons.

this chapter we looked at one kind of sedimentary ore deposit, placers. Most placers, however, remain unconsolidated as sediment. Below we consider three examples of sedimentary deposits that involve sedimentary rock: phosphorites, banded iron formations, and limestone.

13.7.1 *Phosphorous deposits*

Phosphorus, like potassium, is an essential nutrient for all living things and has been used as a fertilizer for centuries. In the 18th century, people used *bone ash* as a source of phosphorus, created by roasting the bones of animals to liberate contained phosphorus. In the 19th century, people turned to *guano*, or bird and bat excrement, as the principal source of phosphorus for fertilizer. Figure 13.24 shows a typical white guano deposit. Similar deposits are found on many ocean islands around the world, where tens of thousands to millions of years of bird and bat excrement has collected. In some places, layers of guano are up to tens of meters thick.

In the 1840s and 50s, guano deposits were developed to produce saltpeter for gunpowder and for fertilizers. Increasing demand, and competition for the best deposits, led the United States Congress to enact the Guano Island Act of 1856. The act allowed U.S. citizens to take possession of any unclaimed island for the purpose of guano mining, if the island was not within the jurisdiction of another country.

Phosphorus collects in oceans after erosion frees it from continental rocks and rivers carry it to the sea. The concentration of phosphorus in ocean water depends on the temperature and acidity. In deep, cold, acidic ocean water, the concentration of phosphate (PO_4) is nearly 0.3 ppm. But ocean circulation causes upwelling of the phosphorus-enriched water along coastlines, where both water temperature and pH increase. These changes cause the water to become oversaturated and, consequently, apatite, a phosphate mineral, precipitates and becomes part of marine sediment that also contains mud, silt, and sand. Additionally, some organisms consume phosphorous, and when the organisms die, their remains become part of the sediment. After lithification, the resulting rock may contain as much as 40% phosphorous.

Figure 13.25 shows a phosphorous mine in Togo, West Africa. Today, instead of coming from bird dung, most phosphorus comes from *phosphorite*—which is what is being mined in this photo—a phosphate-rich chemical sedimentary rock that forms in several different marine environments. The most common depositional environments are in shallow, near-shore marine settings, including beaches, intertidal zones, and estuaries. Over time, once underwater marine phosphorites have been added to continents, today, all significant phosphate mining is on continents. However, offshore mining of phosphate from sediments on the ocean bottom is being attempted off the coast of Namibia.

Figure 13.26 shows the Bou Craa phosphorite mine in the interior of Western Sahara (a disputed territory south of Morocco). A 60-mile-long conveyor belt carries ore to a shipping port on the Atlantic Coast. Morocco and Western Sahara contain 70% of the world's known phosphate deposits today, but China is the number one producer. Some smaller but still significant production occurs in other countries. In the United States,

Figure 13.24 Cormorants on a Guano Island, via Flickr/Gerry Thomasen.
Photo from Danilo Bargen, Wikimedia Commons.

Figure 13.25 Phosphate mining in Togo, West Africa.
Photo from Alexandra Pugachevsky, Wikimedia Commons.

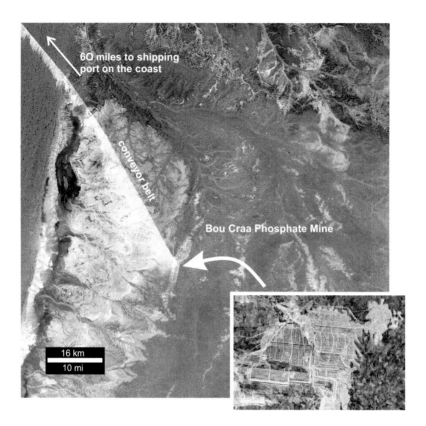

Figure 13.26   The Bou Craa phosphorite mine in Western Sahara and the conveyor belt that carries ore 60 miles to the coast where it is transported for use as fertilizer around the world.

Photos from NASA Earth Observatory.

phosphorite is sometimes mined in Florida, Idaho, North Carolina, and Utah.

13.7.2   Banded iron formations

Figure 13.27 shows *banded iron formation* (left) and the skyline of Dubai in the United Arab Emirates (right). The banded iron formation is a sedimentary rock with repeating layers of black to silver iron oxide (magnetite) and red chert (microcrystalline quartz). The overall red color is because the chert contains inclusions of hematite. The rock shown is from the Mesabi Iron Range of Minnesota where it formed during the Precambrian Eon, 2.1 billion years ago. It is 2 meters high, 3 meters across, and weighs 8.5 tons.

The photo of the Dubai skyline in Figure 13.27 captures the vertical, urban, and economic growth of the United Arab Emirates—from a largely fishing, pearling, and sea trading nation to one of the world's wealthiest nations in only the past few decades. Rapid economic growth in the late 20th century was made possible by production and export of oil and natural gas, although today those commodities account for only a few percent of Dubai's gross domestic product.

What do the two photos in Figure 13.27 have in common? The connection between the banded iron formation and the Dubai skyline lies in the vertical height of the buildings. The buildings can only be that tall because of a steel frame that supports the weight of the structure and the people and things inside it. The tallest building on the skyline is the *Burj Khalifa*, which is the tallest structure on Earth. It is 160 stories high, reaching a height of 828 meters (2716 feet). Typically, wood structures cannot be built higher than five stories, about 15 meters (49 feet), because wood lacks the strength to provide structural integrity required by taller structures. Buildings taller than five stories are built using steel, either steel alone or a combination of steel and concrete, known as *reinforced concrete*. Reinforced concrete is made by building a rebar cage and then filling and surrounding it with poured concrete. This structural

Figure 13.27 Banded iron formation (left) and the Dubai skyline (right).
Photos from André Karwath (left) and Tim Reckmann (right), both at Wikimedia Commons.

strength of the rebar-concrete mixture is much greater than concrete alone. Steel is mostly made of iron, and most of that iron comes from banded iron formations such as the one shown above in Figure 13.27. Thus, these ancient sedimentary rocks play a huge role in our built environment as urbanization drives vertical growth in global cities around the world.

Australia contains many large iron mines; Figure 13.28 shows an iron mine at Tom Price in Western Australia. The bedrock and the soil have the typical reddish color associated with oxidized iron minerals such as hematite and goethite that typify iron formations. Precambrian iron formations are found on all the world's major continents, and where they are found, mining usually occurs. Australia dominates the world iron market, producing 36%, and Brazil and

Figure 13.28 Mount Tom Price iron ore mine in Western Australia.
Photo from Bäras, Wikimedia Commons.

China each produce about 16%. The United States accounts for only about 2%. Most U.S. iron mining is in the Lake Superior region; during most of the last century, 90% of United States iron production came from Minnesota, Michigan, and Wisconsin. While unimportant on the world market, today iron is the third most profitable metal mined in the United States, only bested by gold and copper.

Although steel is dominantly iron, it generally contains a small amount of carbon (0.002 to 2.14 weight %). The carbon content is critical. Too much carbon makes steel brittle, whereas too little carbon makes steel ductile and structurally weak. Other elements are commonly added to steel, depending on its intended use. For example, manganese and nickel are added to increase the ability of steel to withstand tension. Chromium is added to increase the melting temperature and hardness of steel, and vanadium is added to increase the melting temperature and reduce metal fatigue, or structural weakening of steel.

Banded iron formations are massive in scale, in places covering hundreds of square kilometers, and as thick as tens to hundreds of meters. They are all Precambrian in age, and for much of the 20th century, geologists searched for an explanation for their origin. Two key properties of iron are that it can be reduced (Fe^{2+}) or oxidized (Fe^{3+}) and that reduced iron is much more soluble in water than oxidized iron. Geologists believe that these two mechanisms explain the origins of banded iron formations. When Earth's oceans first formed, oxygen levels in the atmosphere and in ocean waters were low, so iron was mostly reduced, and thus the oceans

Figure 13.29 Stromatolite fossils in a glacial erratic found in the Parc des Laurentides, near Laterrière, about 160 kilometers (100 miles) north of Quebec City.

Photo from Silk666, Wikimedia Commons.

contained lots of soluble iron. As long as the dissolved oxygen remained low, the iron remained dissolved in the water. About 2.5 billion years ago, however, the amount of oxygen in the atmosphere and oceans increased, iron slowly oxidized, and iron-rich sediments were deposited on ocean bottoms, later to become banded iron formations.

What caused the increase in oxygen? Fossil evidence shows that at the time that iron formations began to form, *cyanobacteria* had evolved and began producing oxygen through photosynthesis. Figure 13.29 shows an example of such bacteria, fossilized and preserved in Quebec. These fossils were found in a glacial erratic, but came from an outcrop near Albanel Lake 500 kilometers (311 miles) northwest of Quebec City.

The structures seen in Figure 13.29 are *stromatolites*, ancient fossilized algal colonies once composed of cyanobacteria. The oxygen produced by the colonies, such as those in Figure 13.29, is thought to have reacted with dissolved reduced iron in ocean water to form oxidized iron and insoluble iron minerals such as magnetite and hematite, both iron oxides, that settled to the ocean floor. Before the bacteria evolved, there was no chemical or biological mechanism for producing sufficient oxygen to cause iron oxidation, and consequently iron remained dissolved. Some researchers hypothesize that the layering in banded iron formations like the one shown in Figure 13.27 formed seasonally, but other explanations, including daily cycles, have been offered. Banded iron

formations are not presently forming because, about 1.9 billion years ago, atmospheric oxygen levels became so high that today iron easily oxidizes to Fe^{3+} (instead of Fe^{2+}) and is not significantly soluble in water.

13.7.3 Limestone

13.7.3.1 Ancient roads

In 312 BCE, the ancient Romans built the famous Appian Way, known then in Latin as *Appia longarum . . . regina viarum*, translated as "the Appian Way, the queen of the long roads," to connect Rome with Brindisi and other cities in southeast Italy. Figure 13.30 is a map showing where the road—really a network of several roads—went. The road, which stretched a total of 582 kilometers (362 miles), was an engineering wonder when constructed, and it was one of the earliest and most important roads of the Roman Republic. It has proved to be durable and still exists today; the inset photo in Figure 13.30 shows a present-day stretch of the Appian Way in Rome.

Figure 13.30 The dashed line shows the Appian Way, an ancient road in Italy.

Map based on NASA image; inset photo from Livioandronico2013, Wikimedia Commons.

Road construction has evolved significantly over the past few thousand years, and the mass adoption of the automobile in the 20th century was the catalyst for rapid growth of modern road systems. The basic construction methods, however, remain the same today as they did in ancient Rome. *Cement* and *crushed stone*, the building blocks for modern roads, have been used to create 4.2 million miles of paved roads in the United States and nearly as much in India and China. Roads today connect people and allow for the transportation of goods, services, and ideas, just as they have for thousands of years.

13.7.3.2 Quarrying limestone

Cement, and most crushed stone, comes from limestone, a common sedimentary rock composed primarily of calcite ($CaCO_3$) and dolomite ($CaMg(CO_3)_2$). Limestone, because it is relatively resistant to weathering, often forms outcrops at Earth's surface and so is an easy target for mining operations. Most limestone comes from quarries, but a small amount is mined underground. Figure 13.31 shows a limestone quarry on an island in the Mediterranean Sea. Quarrying typically proceeds by sawing blocks or slabs out of limestone bedrock as can be seen here.

Limestone products have many uses. For instance, limestone blocks and slabs are used in construction and for decoration. Egyptians, when they constructed the Great Pyramids of Egypt around 2570 BCE, were

Figure 13.31   Limestone quarry on the island of Gozo, Mediterranean Sea.

Photo from Bogdan Giuşcă, Wikimedia Commons.

among the first people to use limestone for such purposes. Limestone is also processed to produce crushed stone (aggregate), cement, and lime. Lime (CaO) is particularly important because it is a fundamental component of many products, including paper and plastic, and is also used in some chemical, medicinal, and food processing. The greatest use of lime, however, is in steel manufacturing—lime is a flux that removes impurities and improves steel quality.

13.7.3.3 Cement

The ancient Romans made cement by adding water to aluminum-, calcium-, and silica-rich volcanic ash produced by Mt. Vesuvius. Sections of the Appian Way, built using this ancient Roman recipe, remain in use today, testifying to the strength and endurance of the cement. Modern people, in contrast with the Romans, make cement from limestone. To make a type of cement called *non-hydraulic cement*, crushed limestone is heated in a kiln to high temperatures, which drives off CO_2 and produces *quicklime* (CaO). (Unfortunately, this process adds considerable CO_2 to the atmosphere; cement manufacturing accounts for 5%–10% of all anthropogenic greenhouse gases—thus contributing to global warming.) Quicklime, when combined with water, reacts to make *portlandite* ($Ca(OH)_2$), which hardens when it dries and binds well with sand, gravel, and other natural materials. Over time, the portlandite reacts with CO_2 in air and becomes harder as it is recarbonated. Most cement used today, however, contains other ingredients besides quicklime, typically silicon, aluminum, and iron oxides. The hardened product, called *hydraulic cement*, is a mixture of mineral silicates and oxides and thus is harder than non-hydraulic cement.

The most common use of cement is as an ingredient in *concrete*, a mixture of cement with sand, gravel, or other aggregate. Every year, more than three tons of concrete are manufactured for every person on Earth, making concrete the second most consumed material after water. The incredible utility of concrete lies in its plastic and malleable properties when newly mixed and its strong and durable properties when hardened. These qualities explain why reinforced concrete goes into skyscrapers, such as the *Burj Khalifa* in Dubai, and bridges, sidewalks, superhighways, houses, and dams.

13.8   Igneous ore deposits

13.8.1   Porphyry deposits

Typical magmas within Earth are not just molten rock. They are composed of a silica-rich melt (liquid), mineral crystals (solids), dissolved gases such as CO_2 and H_2O, and dissolved elements. Figure 13.32 shows a progression as a typical melt body cools. As it cools, the amount of melt decreases while more crystals form. Simultaneously, the concentration of gases in the remaining melt increases. Eventually, the melt will *exsolve* gas, creating gas bubbles, when the concentration of gas in the melt increases to the point where the melt cannot hold all the gas that is present. The bubbles are mostly CO_2 and H_2O, but also contain other elements and compounds. Some elements, most notably sulfur, copper, gold, silver, and molybdenum, have a strong chemical preference to stay with the H_2O and CO_2, and so they become concentrated in the bubbles. The amount of metal in a bubble is generally quite small, perhaps parts per million to a few weight %, but varies with temperature; higher temperature means more metal.

As shown in the right-hand drawing of Figure 13.32, bubbles in a melt naturally rise, like bubbles in a glass of champagne, because bubbles have a lower density than the melt. So, although they contain metals such as copper, gold, molybdenum, and silver, they are buoyant. Movement may be slow, because melts are very viscous and because mineral crystals get in the way. Still, as shown in Figure 13.32, the bubbles will eventually make it to the top of the magma chamber, hit the roof, and stop rising.

Figure 13.33 shows how rising gas bubbles can move metals from a magma chamber into the crust.

Rising bubbles expand because pressure within Earth decreases moving upwards. As they expand, they put pressure on surrounding material. If the gas pressure becomes great enough, it can cause the magma chamber to explode or the roof to fracture, in much the same way that a cork can explode from a bottle of champagne. The bubbles, accompanied by melt, will move upwards to shallower depths, and lower pressures, in the crust. As they rise, they cool, and eventually the combination of lower pressure and temperature means that the bubbles cannot continue to carry metals. So, the metals precipitate as mineral crystals at some depth between the top of the magma chamber and the surface. Typical minerals that form this way include molybdenite (MoS_2) chalcopyrite ($CuFeS_2$) and bornite (Cu_5FeS_4).

The process described above produces ore deposits called *porphyry-type* ore deposits. As discussed in Chapter 7, the term *porphyry* refers to volcanic rocks that contain large mineral crystals in a sea of groundmass (see Fig. 5.30). This texture is typical of rocks that host porphyry deposits. The Bingham Canyon ore deposit, discussed at the beginning of this chapter, is a porphyry-type ore deposit. It formed 40 million years ago during the final stages of geologic activity by a large stratovolcano. Weathering and erosion removed nearly half the original 13,000-foot-tall volcano, uncovering the porphyry ore deposit that provided the gold, silver, copper, and tin for all the medals at the 2002 Salt Lake City Winter Olympics.

Seven hundred miles south of Bingham Canyon, the Freeport-McMoRan Corporation mines a larger porphyry deposit near Morenci, Arizona. Figure 13.34 shows a satellite image of the mine. Copper minerals were first discovered there by an Army battalion in 1865, and mining began in 1872. Today,

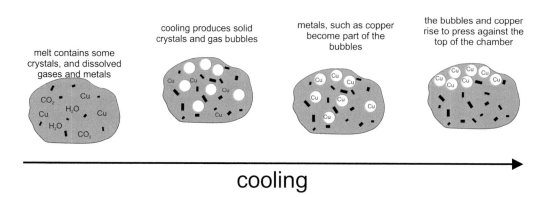

Figure 13.32   The first steps in forming a porphyry deposit.

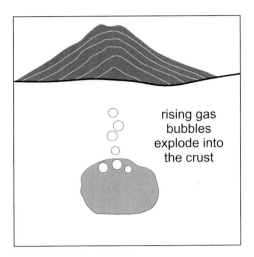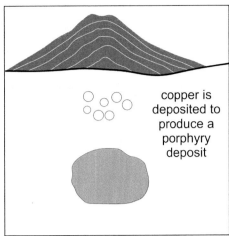

Figure 13.33   Mass transfer of copper from the magma chamber to form a porphyry deposit.

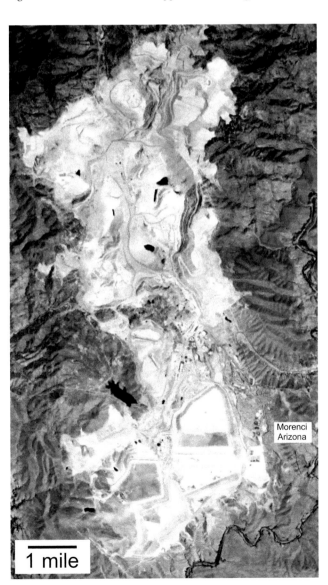

Figure 13.34   View of the Morenci Mine at Morenci, Arizona.
Figure created by D. Perkins.

Morenci, with pits that total almost 130 square kilometers (50 square miles), is the largest copper producer in North America. The mine extends beneath and between several large mountains next to the Morenci town site. Chalcocite and chalcopyrite, both sulfide minerals, are the primary copper ore minerals, but chrysocolla (copper oxide/hydroxide) and malachite (copper carbonate) are found and mined from oxidized ore zones. Although copper minerals are by far the most important ore minerals at Morenci, the mine also produces lesser amounts of sphalerite (zinc ore), galena (lead ore), and molybdenite (molybdenum ore). Table 13.1 lists important uses for all these metals. In particular, zinc is used to coat steel, lead is mostly used in batteries, and molybdenum goes into specialty alloys.

13.8.2   Magmatic sulfide deposits and other cumulates

Mafic and ultramafic magmas, like all common magmas, contain the major elements oxygen, silicon, aluminum, iron, calcium, sodium, potassium, and magnesium. But they typically also contain other elements (minor elements), such as sulfur and nickel and less common metals including platinum, palladium, and chromium. As a mafic or ultramafic magma cools and crystallizes, the first minerals to form are plagioclase, pyroxene, and olivine—all made of major elements. Consequently, the concentrations of sulfur and other minor elements increase in remaining melt. Eventually, sulfur concentration becomes great enough that sulfide minerals begin to crystallize. The sulfide minerals, typically containing iron and nickel, also contain relatively high concentrations important platinum group metals (ruthenium, rhodium, palladium, osmium, iridium, and platinum) and other minor metals.

Small sulfide minerals may form during initial stages of magma crystallization, but with time, the crystals grow larger (Fig. 13.35). These crystals have greater densities than silicate minerals and the mafic or ultramafic melts do. For example, the density of pyrrhotite ($Fe_{1-x}S$) is 4.6 g/cm^3, whereas the density of a mafic melt is about 3.0 g/cm^3. So, the denser sulfide minerals will, over time, sink, as shown in Figure 13.35. Eventually, after more cooling and crystallization, significant deposits of sulfide minerals may accumulate on the bottom of a magma chamber. The deposits, which may form a centimeter- or meters-thick layer called a *cumulate*, are often entirely, or nearly entirely, composed of sulfide minerals. This process produces *magmatic sulfide deposits*, which are the most important sources of platinum, palladium, chromium, and some otherwise rare metals.

Cumulate sulfide ores make up some of the world's most important ore deposits. For example, the Sudbury ore deposit, in western Ontario, yields nickel, copper, platinum, palladium, gold, and other metals. Ore minerals at Sudbury include pentlandite ($(Fe,Ni)_9S_8$), chalcopyrite ($CuFeS_2$), pyrrhotite ($Fe_{1-x}S$), and pyrite (FeS_2). Cumulate sulfide deposits account for almost 60% of the world's nickel production and more than 95% of platinum and paladium production. Deposits of this sort are associated with mafic and ultramafic magmas but not, generally, with felsic magmas, because felsic magmas are so viscous that they cool and crystallize before dense minerals can settle.

Sulfides are not the only kind of mineral that can become concentrated in a cumulate deposit. Oxides—including magnetite (Fe_3O_4), ilmenite ($FeTiO_3$), and chromite ($FeCr_2O_4$), may settle and collect, too. South Africa's Bushveld Complex is the most significant and important example of oxide mineral cumulates; Figure 13.36 shows an example of Bushveld ores. Although the oxide minerals in the ore layer are rich in iron they are seldom mined for iron ore because

Figure 13.36   Chromite layers in the Bushveld layered mafic intrusion, South Africa.
Photo from Kevin Walsh, Wikimedia Commons.

cumulus magnetites generally contain significant amounts of other, diluting, elements, requiring too much processing to extract pure iron. The diluting elements, however, have value, and today, magnetite from the Bushveld Complex is the largest source of vanadium in the world. The Bushveld Complex ores also contain the richest chrome, platinum, and palladium resources in the world.

The Bushveld ores are most prized because platinum, palladium, and chromium have many uses. Platinum, for example, is the substrate needed to synthesize the chemotherapy medication Cisplatin, used to fight several types of cancer, including breast, cervical, bladder, ovarian, lung, esophageal, mesothelioma, brain tumors, neuroblastoma, testicular, and head and neck cancer. The main use of palladium is in catalytic converters in automobiles; palladium is the catalyst driving chemical reactions that reduce smog-forming gases generated during combustion. Palladium is also used to manufacture some food products. Chromium, also used as a catalyst in many applications, is the key component (after iron) in most stainless steel.

13.9   Other kinds of ore deposits

13.9.1   *Volcanogenic massive sulfide deposits*

The Kidd Mine, shown in Figure 13.37, is in eastern Ontario, near Timmins; it is the world's deepest *base metal* (a term referring to industrial metals excluding iron and precious metals) mine, extending to depths of

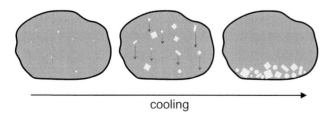

cooling

Figure 13.35   Forming a cumulate mineral deposit.

Figure 13.37 The Kidd Mine near Timmins, Ontario.
Photo from P199, Wikimedia Commons.

Figure 13.38 A black smoker on the ocean floor.
Photo from NOAA, Wikimedia Commons.

more than 3350 meters (11,000 feet). The bottom of the mine is the closest a person can get to the center of Earth. This mine started production in 1966 as an open pit but soon went underground. The moneymaking metals are mostly copper and zinc, but substantial silver, gold, lead, and other metals are produced too.

The Kidd ore deposit formed about 2.7 billion years ago at hydrothermal vents on, or just below, the ocean floor. Ore minerals—mostly iron, copper, zinc, and lead sulfides (pyrite, pyrrhotite, chalcopyrite, sphalerite, and galena)—were deposited when warm, metal-rich hydrothermal waters combined with ocean waters. The resulting deposits are in pods or sheets within sedimentary rock layers that, in places, contain nearly 100% ore.

The Kidd ore deposit is an example of a *volcanogenic massive sulfide deposit.* Most such deposits are small, but the Kidd Mine is one of the larger ones. It is not the largest, however. The Windy Craggy deposit in British Columbia, discovered in 1958, and the Rio Tinto deposit in Spain, discovered in 1972, are both twice the size of Kidd. The quality of the ore in massive sulfide deposits is high, generally greater than 60% ore minerals, so even if small, massive sulfide deposits are alluring mining prospects.

What makes massive sulfide deposits especially intriguing is that they are not all ancient. Figure 13.38 shows a *black smoker* on the ocean floor where hot hydrothermal waters, mixing with ocean waters, create fine particles of sulfide minerals and produce a massive ore deposit. The iron sulfides, which are the most common minerals created at black smokers, are black, hence the name. Black smokers, such as the one shown, are common features at all mid-ocean ridges and are potentially minable. Some prospecting of ocean floors is being done today, and some mining companies have developed plans for mining operations, but to date, the water depth has proven too great for direct mining. However, some areas near black smokers contain sulfide ooze that might, perhaps, be picked up with a vacuum.

13.9.2 Mississippi valley type deposits

Some flat sedimentary rocks in the interior of the United States have strata of limestone that contain ore minerals. These include mineral deposits of the Southeast Missouri Lead District and related deposits in Iowa, Wisconsin, and Illinois. The deposits are especially concentrated in a curved zone called the *Viburnum Trend* in southeast Missouri. Elsewhere, similar deposits are found at Pine Point in Canada's Northwest Territories, in northern England, and in a handful of other places around the world. All of these deposits are called *Mississippi Valley type (MVT) deposits.*

Figure 13.39 shows a museum specimen from an MVT ore deposit in the North Pennines of England. This sample is mostly green fluorite but also contains silver-gray cubes of galena and white and salmon-colored quartz. Primary ore minerals in MVT deposits

Figure 13.39 Galena, calcite, and fluorite from the Rogerley
Mine, County Durham, England.

Photo from Rob Lavinsky, Wikimedia Commons.

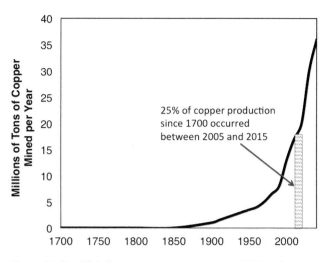

Figure 13.40 Global copper consumption since 1700, and
expected consumption to 2040.

are generally galena (PbS) and sphalerite (ZnS). Fluorite (CaF_2) is commonly present but has little economic value. Weathered or altered MVT ores may contain anglesite ($PbSO_4$), cerussite ($PbCO_3$), smithsonite ($ZnCO_3$), hydrozincite (also a type of zinc carbonate), and secondary galena or sphalerite. Both primary and secondary ore minerals were deposited by flowing groundwater long after limestone formation, but the origins of the groundwater is unknown. The metal-rich ore fluid may have been derived from clastic oxidized sedimentary red beds, according to some geologists.

13.10 Is mining necessary?

Mining is definitely necessary today because it supports just about everything we do and allows us to maintain a high quality of life. So, some people are concerned about ore supplies becoming exhausted. Ore deposits are finite and can run out, and finding new deposits is increasingly more difficult. Thus, it is important to be able to predict how long resource supplies will last so we can plan for the future.

13.10.1 Resources in the future

We usually discuss the question about future resource availability in terms of the rate of consumption of a given resource and the rate at which new ore deposits that contain that resource are discovered. Take copper as an example. Figure 13.40 shows how much copper the

world's people have mined during the past 300 years, projected to 2040. Globally, between 1700 and 2015, about 725 million tons of copper were mined. Nearly half this total, 360 million tons, was mined between 1995 and 2015, and about a quarter was mined between 2005 and 2015. Stated another way, nearly one-fourth of all copper mined since 1700 was mined in only the 10-year period from 2005 to 2015.

This observation causes alarm, especially in the media, because it could mean that the current rate of copper consumption will exhaust the supply. However, this perspective is distorted. The United States Geologic Survey reported in 2016 that the global sum of known copper *resources* is 2300 million metric tons. This estimate is not speculation, but is based on the measured amount of copper that geologists have identified in existing mineral deposits. Of course, some copper resources are too difficult to extract and thus not profitably mined today but that does not mean they will not be mined in the future. Geologists use the term *reserves* to define the quantity of a given resource that can be mined at a profit at the current price of the resource. Thus, the amount of reserves is always less than the amount of known resources. If the market price of a resource increases, though, the proportion of *resources* classified as *reserves* increases.

When calculating the lifetime remaining for a given resource, the known *reserves* are divided by *annual consumption*, excluding recycling. Thus, if one takes the current annual consumption of about 20 million tons of newly mined copper each year, and the global *reserve* of copper of about 715 million tons as of early 2018, we have 35 years of copper left. However, it is important to remember that this estimate of *years of remaining copper* is based only on the amount of copper that we can mine at a profit based on current price. The global *resource* of 2300 million metric tons of copper tells us that there is much more available copper in discovered mineral deposits. If the price of copper increases, some or all of that copper will be mined. Additionally, more copper deposits will certainly be found, mining technology may improve so that unminable deposits become minable, and substitute materials may be discovered.

13.10.2 Recycling of resources

A final consideration when discussing resource availability is the rate of recycling. A metal such as copper can be used many times in a variety of products without losing any of the properties that make copper useful. For example, copper in a smartphone can be recycled and used in a television, recycled again to use in an electric car, and recycled again to be used in a wind turbine. The properties of copper do not change. The same is true for aluminum, gold, platinum, silver, titanium, and other metals in the periodic table.

Figure 13.41 shows the recycling rates for some important metals. Currently, global recycling rates for some metals are very high, meaning that once the metal is mined, it is used over and over again. However, the metals plotted in Figure 13.41 are used in relatively large

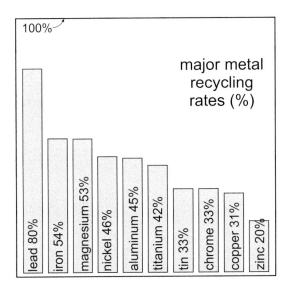

Figure 13.41   Percentage of mined metals recycled and reused in the United States in 2007.
Data from the USGS.

amounts compared to many others. And the recycling rate of some lesser-used metals is very low, even if they are very important to today's society. For example, the rate of recycling of rare earth elements is less than 1%. This low recycling rate is because the costs to recycle exceed the cost to mine virgin rare earth elements. On the other hand, recycling rates for precious metals—gold, silver, and platinum—are very high because of the metals' great value. So, lacking government requirements or other forms of coercion, recycling rates are market-driven. Thus, if resource supplies dwindle, the price will increase and recycling will increase. Additionally, recycling rates may increase if consumers demand increased rates of recycling to reduce the environmental impacts of mining or if manufacturers recognize the benefits of reusing metals rather than mining new metals to replace those disposed of in landfills.

Some recycling possibilities are never considered but perhaps should be. For example, phosphorous and potassium mined in Western Sahara and Canada are shipped around the world to be used as fertilizers. The nutrients enter food products that people consume, and when people urinate, the nutrients go down the drain. Calculations suggest that significant amounts of resources are being discarded in this way. Wouldn't it be better if the phosphorous and potassium were recycled so we did not have to mine as many virgin resources and ship ores long distances? Today, most inorganic fertilizer components end up in landfills or wash off into rivers and streams. Only a small amount is recycled and used as fertilizer.

Finally, we know of many cases of resource use having bad consequences that we have done little to stop. Consider, for example, fertilizer use in North America. In many places, fertilizer runoff from fields and lawns causes *eutrophication* and algal blooms that reduce the level of oxygen in lakes or streams. When concentrations get too low, entire aquatic ecosystems can fail. And where rivers discharge fertilizer into the ocean, *hypoxic zones* result from the low concentrations of oxygen in the water, creating "dead zones."

One of the largest dead zones forms each spring in the Gulf of Mexico (Fig. 13.42) when the Mississippi River delivers 2 million tons of excess fertilizer applied to farms throughout the Midwest. The result is a massive algae bloom, which eventually means that oxygen levels drop, creating hypoxic conditions that suffocate all animals. Figure 13.42 shows oxygen levels on the ocean bottom in the zone; those areas in red are unable to support life, areas in yellow support some life, and

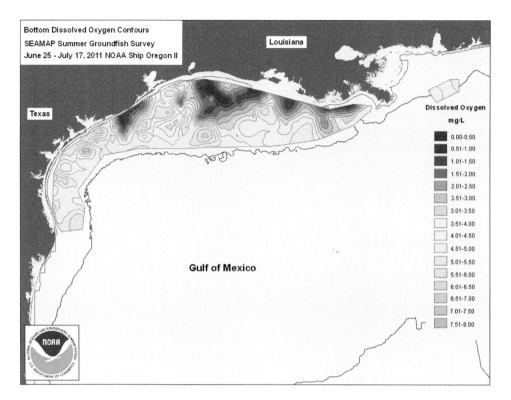

Figure 13.42 The dead zone in the Gulf of Mexico.
Map from NOAA.

areas in green are healthy. Wouldn't it be better if less fertilizer was used in the first place—reducing the cost of agriculture and reducing pollution?

Questions for thought—chapter 13

1. Much of the copper used today is recycled. Yet, recycling is not enough to keep up with human demand. List several reasons why.

2. When were the *Stone Age*, *Bronze Age*, and *Iron Age*? What characterized each? And why did the progression from an earlier age to a later one lead to the expansion of human populations and to prosperity?

3. What is bronze, and what was the breakthrough that allowed people to make it? And, what about iron – what did people figure out that allowed the iron age to begin?

4. What is the difference between a mineral deposit and an ore deposit? Can a mineral deposit evolve to become an ore deposit, or vice-versa? What things might make either of these possible?

5. Tailings piles are commonly associated with mining operations. The chapter refers vaguely to environmental problems associated with tailings but is not specific. So, be specific – what kinds of problem might tailings produce?

6. What is an *economical concentration factor*? Why is this an important concept to mining companies? What may cause economical concentration factors to substantially change over time? There are at least two good reasons.

7. Why is the economical concentration factor of gold much greater than the factor for aluminum?

8. What are *placer deposits*? How do they form? What properties must a mineral have to accumulate in placers? What kinds of valuable minerals are found in placers? There are several.

9. What are *laterites*? What climatic conditions produce them? What is the difference between a laterite and a bauxite? What important metals (name at least two) come from laterites and bauxites? What makes laterites and bauxites often good ore deposits?

10. Why is halite mined near Detroit, Michigan? There are several good reasons.

11. Why are carbon, vanadium, chromium, and other metals often added to iron to make steel?

12. What are some commercial uses for limestone and lime? What properties do limestone and lime have that make them particularly good for these uses?

13. How do magmas evolve to produce porphyries? What kinds of metals are mined from porphyry deposits? List some. Are these deposits generally small or are they large?

14. Cumulate ore deposits are almost always associated with mafic or ultramafic rocks. How do such deposits form? Why are they not associated with granitic or other felsic magmas?

15. Where is the Bushveld Complex? Why is it important? What important metals are produced from the Complex?

16. How do volcanogenic massive sulfide deposits form and what important metals are produced from them? These kinds of deposits are forming today. Where would we have to go to see them forming? Why isn't anyone mining in those places?

17. How do Mississippi Valley type deposits form? What is the host rock of the deposits and what minerals and important metals are typically found in them? In the past, these deposits were mined in open pits, but today most of the mining is underground. Why?

18. What are the major chemicals in fertilizers? Why aren't fertilizers recycled? How might they be recycled? If they are not recycled, where do they end up and what kind of problems does that create? Describe the problems.

14   Energy Resources

14.1   Star wars

If you drive west from Seville toward Cordoba, in the Andalusia region of southwestern Spain, you will come across the Star-Wars-ian scene shown in Figure 14.1. A tall tower, 140 meters (460 feet) high, rises high into the sky, glowing so brightly at its top that you cannot look at it. Thousands of mirrors surrounding the tower are all focusing sunlight on the tower's top. The entire facility, the *Gemasolar Thermosolar Plant*, has been operating and producing electricity for almost a decade. Gemasolar was one of the first of its kind, but others have been built, and are being built, around the world.

Figure 14.2 shows the *Crescent Dunes Solar Energy Project*, completed in 2011, 195 miles northwest of Las Vegas, Nevada. It is in a sandy desert near Tonopah and covers about 2.5 square miles. Crescent Dunes, like Gemasolar, but five times larger, is a *concentrated solar power (CSP)* facility. At Crescent Dunes, 10,347 *heliostats* (reflector units that each contain 1260 square feet of mirrors) follow the sun during the day, focusing light and heat on a *central receiver* at the top of a 195-meter

(640-foot) tall tower. At Crescent Dunes, the total output is about 110 megawatts, which is enough power for a city of 150,000 people. Today, Spain produces half the world's CSP output; Crescent Dunes and other plants in California and Nevada account for most of the rest.

Generating electricity from sunlight is not new, but using conventional solar panel technology, intermittence has been an issue. Standard photovoltaic (PV) panels can generate electricity, but not at night, and not efficiently when the sky is cloudy. And storing electricity for nighttime use has been problematic because, despite major improvements in recent years, batteries are not efficient enough for utility-scale power storage. However, modern CSP facilities have no downtimes and no batteries. Concentrated solar energy melts salt—a mineral resource—in the central receiver at temperatures of 800 °C (1470 °F), and the molten salt provides heat to create steam that drives turbines and creates electricity. Because the salt stores heat, CSP facilities can generate electricity at night or when the sun is obscured—thus,

Figure 14.1   The Gemasolar Thermosolar Plant in Andalusia, Spain.

Photo from Betsy Perkins.

Figure 14.2   Crescent Dunes Solar Energy Project, near Tonopah, Nevada.

Photo from energy.gov.

output remains constant no matter the time of day or weather. Using molten salt to store and move energy is more efficient and less expensive than other technologies. CSP plants are cheaper to construct than new coal plants that meet pollution standards, cheaper than solar installations with photovoltaic panels, and cheaper than nuclear facilities.

As with other kinds of solar power, the electricity generated by CSP is 100% renewable and nonpolluting. For these and economic reasons, CSP technology is spreading. Generating plants such as Crescent Dunes and Gemasolar have, so far, captured only a small part of the solar market compared with photovoltaic installations. But many new and larger CSP projects are planned, especially in China, where government policies promote renewable, nonpolluting energy sources.

14.2   The energy that we use

Most of the energy (99.97%) at Earth's surface comes from the sun. Heat flow from within Earth accounts for almost all the rest, and a very small amount of energy comes from tidal energy caused by the gravitational pull of the moon and sun. Our society obtains its energy from these sources. We can harness solar energy directly with photovoltaic cells or solar furnaces or indirectly by consuming food that ultimately is a product of the sun. And some solar energy has been stored in fossil fuels (coal, oil, and natural gas) for millions of years or for shorter times in wood and other biomass; we can use those fossil fuels, wood, and biomass as energy sources today. Additionally, differential heating of Earth's surface by incoming solar radiation from the sun produces wind, which we can use to generate electricity. The sun also causes evaporation that drives the water cycle and causes precipitation that fills reservoirs so that we can make hydroelectric power. Electricity also comes from uranium, which we mine and use as fuel, along with plutonium, in electricity-generating nuclear reactors. Heat released from volcanic systems produces geysers and hot springs, which we harness to generate geothermal power. And, in some places, we get electricity from tides and ocean waves.

People, especially in the more developed parts of the world, use a lot of energy. The most fundamental use is simply to power our bodies, and the energy for that comes from food. Food consumption is measured in calories, and the more calories consumed, the more energy consumed. Up to a point, more calories

mean good health and more energy available for doing things. Generally, however, when we talk about energy consumption, we are not talking about people; we are talking about energy that powers commerce, industry, transportation, and homes.

Figure 14.3 shows energy consumption in the United States divided into broad categories (economic sectors). About a third (32%) of the energy is used by industries, including manufacturing, agriculture, and construction. The average American uses a myriad of products, and energy was required to make all of them. Of the energy used, 29% is for transportation; no matter how people travel, they use energy. Residential energy (21%) is required to light, heat, and cool our homes, to brew coffee, and use our laptops, tablets, and smart phones. And 19% of all energy use is commercial, for businesses, including stores, offices, schools, and places of worship.

All four of the primary sectors in Figure 14.3 use electricity. Additionally, today's electronic and information age, that requires all the data and communication systems that connect people around the world, is powered by electricity. People want to be connected and they want information quickly, and both desires require energy. Electrical energy allows us to stream movies and to read every book in the United States Library of Congress without leaving our homes. In the United States, today, the electricity used to run the computers belonging to Google, Facebook, and Amazon is almost 5% of the total electricity used for all purposes in the entire country.

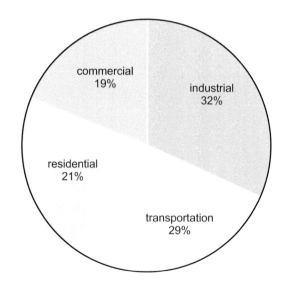

Figure 14.3   U.S. energy consumption by economic sector. Chart from U.S. Energy Information Agency.

Different energy sources became important at different times. Figure 14.4, from the U.S. Energy Information Administration, shows the sources of energy during the past 240 years, with a projection into the future. Before the second half of the 19th century, *biomass*—mostly wood, but also other plant and animal byproducts, including animal fat and manure—was the primary energy source for Earth's people. Beginning in the mid- to late 1800s, with increased manufacturing and the electrification of homes and businesses, fossil fuels—*coal* and *petroleum*—replaced wood as the most important energy source; coal resources were developed beginning around 1850. The first hydroelectric power dams were constructed around 1880. Petroleum first became important at the beginning of the 20th century. Use increased rapidly and, by 1950, the energy produced by burning oil and gas surpassed that produced from coal. And, after World War II, nuclear power plants were constructed using technology originally developed for submarines. Renewable energy sources, notably wind and the sun, have grown in importance during the last few decades. Thus, we have used a mix of energy sources since the beginning of the Industrial Revolution—to sustain our bodies, power lights, take showers, travel, manufacture the products we consume, watch movies, read books, and run our computers, smart phones, and the internet.

Coal, gas, oil, and uranium are *nonrenewable* resources that may someday run out. Geothermal energy, too, may be nonrenewable if produced too quickly. But some of the other sources of energy are *renewable*, or potentially renewable, including solar energy and wind energy; these sources may last forever. Hydroelectric energy is renewable if water reservoirs do not dry up. And hydrogen may provide a renewable source if the hydrogen is produced by other renewable sources of energy. Additionally, people everywhere create municipal and other wastes, forms of biomass, that may be used as sources of energy. Other kinds of renewable biomass, including wood, can also be used as energy sources.

Today, climate change and other problems associated with producing and using fossil fuels cause alarm, most of the appropriate rivers that could produce electricity have already been dammed, nuclear power plants are too expensive to build, and, consequently municipalities and utility companies are moving toward less expensive, nonpolluting alternatives such as solar-powered electrical plants and wind turbines. Yet, the fossil fuel age is not over and will likely persist for many decades; people will continue to rely on coal, oil, and gas as energy sources. Still, the writing is on the wall—and in the Nevada sands near Las Vegas—where a nonpolluting electrical plant uses renewable energy to produce electricity (Fig. 14.2). Fossil fuel use, especially the use of coal to generate electricity, will not sustain the world indefinitely.

14.2.1 Sources of energy

Figure 14.5 shows the sources of energy used by the world's people in 2016. Coal, oil, and natural gas dominated; in fact, for the past two centuries or so, most energy has come from fossil fuels. Energy sources in the United States are similar to sources elsewhere, and

Figure 14.4   Sources of energy used today.

Chart from the U.S. Energy Information Administration.

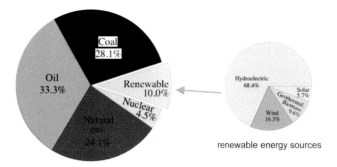

Figure 14.5   Global energy consumption in 2016.

today, a little more than a third of the energy used in the United States comes from oil. A smaller amount (about 29%) comes from natural gas, 14% comes from coal, 9% from nuclear fuels, and 2% from hydroelectric dams. Thus, nuclear reactors and hydroelectric dams have been the most significant alternatives to fossil fuels.

The right-hand side of Figure 14.5 shows renewable sources of energy. These sources are essentially limitless, and humans cannot use them so fast that they run out. Biomass and their byproducts (such as wood and alcohol from grains) and geothermal sources are also potentially renewable, if people are careful and avoid overharvesting and overuse that leads to consumption at higher rates than nature can sustain. Yet, even including biomass and geothermal energy, energy provided by renewable sources is only about 10% of what the world uses today. That percentage is growing, however, as society weans itself from fossil fuels. Figure 14.5 makes it clear that fossil fuels are essential today, and they dominate the left-hand side—the *non-renewable* part of the figure. Petroleum, natural gas, and coal make up 85.5% of world energy supplies, and nuclear fuels account for 4.5%. Unfortunately, people today are consuming fossil fuels faster than nature creates them—much faster. Some scientists think that this overconsumption will become a problem before the end of the current century and that economical alternatives must be found if we wish to maintain peoples' current lifestyles.

14.3   Fossil fuels

Figure 14.6 shows natural seeps of crude oil that reach the surface at the La Brea Tar Pits in downtown Los Angeles, California. Fossil fuels—like the oil shown in Figure 14.6 along with natural gas and coal—are the carbon-rich remains of ancient organisms. Some fossil fuel deposits are relatively young, deriving from organic

Figure 14.6   The La Brea Tar Pits, Los Angeles, California.

material buried roughly 10,000 years ago, but many deposits formed tens or hundreds of millions of years ago. Most fossil fuels are buried deep within Earth, yet people have known about fossil fuels for a long time, because sometimes natural processes bring fossil fuels to the surface. The La Brea Tar Pits are an example, and the history of mining coal from rock outcrops at Earth's surface goes back thousands of years.

14.3.1   Crude oil

In common use, the term *petroleum* refers to its raw ingredient, *crude oil*. The term is also used to refer to products made when crude oil is processed. Unprocessed petroleum consists of both liquid *oil* and *natural gas* (often simply called *gas*). Before the Industrial Revolution, humans did not need petroleum. This changed in the 19th century, when scientists and engineers figured out new ways to use petroleum and coal. Among the pioneers was Benjamin Silliman, a professor of chemistry at Yale University, who was curious about the composition of crude oil. It was well known that crude oil would burn if ignited, but the chemistry of crude oil was poorly understood. Silliman experimented by heating crude oil and causing it to boil. He noticed that, as it boiled, much transformed to vapor, some to watery liquids, and a fraction became a thick substance that had the consistency of room-temperature peanut butter.

Natural gas, which, despite its name, should not be confused with gasoline, is a mixture of gaseous compounds, including those shown in Figure 14.7. Although natural gas consists mostly of methane (CH_4),

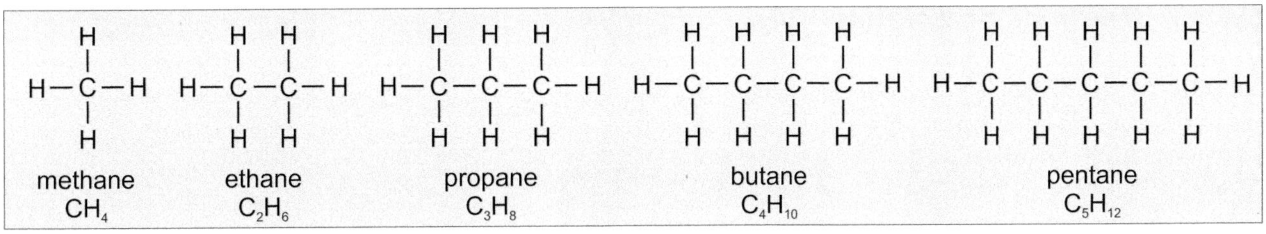

H	H H	H H H	H H H H	H H H H H
H—C—H	H—C—C—H	H—C—C—C—H	H—C—C—C—C—H	H—C—C—C—C—C—H
H	H H	H H H	H H H H	H H H H H
methane CH_4	ethane C_2H_6	propane C_3H_8	butane C_4H_{10}	pentane C_5H_{12}

Figure 14.7 Some of the components in natural gas.

with lesser amounts of ethane (C_2H_6), propane (C_3H_8), butane (C_4H_{10}), and pentane (C_5H_{12}), many other compounds are present. The compounds are different kinds of *hydrocarbons*, compounds made only of hydrogen and carbon. Other hydrocarbons, such as hexane (C_6H_{14}), have molecules larger than the five that are shown in Figure 14.7, and the hydrocarbons that make up liquid oil are even larger and more complex than the ones in gas. Silliman's experiments revealed that it was possible to separate the different hydrocarbons in natural gas and crude oil from each other. This was an important discovery, because crude oil contains a variety of compounds that, when purified, are valuable for different purposes.

Silliman separated individual hydrocarbons from one another using a process called *fractional distillation*, the same process used in petroleum refining today. He isolated products such as paraffin, lubricants and waxes, gasoline, and kerosene, all of which contained different combinations of hydrocarbons. The impacts of Silliman's discoveries, and follow-up work by his son Benjamin Silliman Jr., in the 1850s, were immense, and today, literally millions of products derive from fractional distillation of crude oil.

Petroleum companies distill crude oil (separate it into different components) by heating it and collecting the emissions at different temperatures. Figure 14.8 shows common products from high (bottom) to low (top) temperature distillates. At high temperatures, asphalt is separated from crude oil. At lower temperatures, fuel oil, lubricants and waxes, and other products are produced. Petroleum from different sources contains different mixes, but all the components shown in Figure 14.8 are generally present. Natural gas of various sorts is the lowest temperature distillate. Natural gas—mostly made of the components shown in Figure 14.7—is largely used to heat homes, generate electricity, heat water, and cook food. Gasoline, perhaps the most important component of oil, is also a low-temperature distillate. Gas and gasoline are key fuels fundamental to modern society,

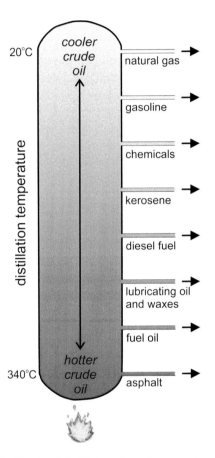

Figure 14.8 Fractional distillation of petroleum produces many different products at different temperatures.

but crude oil contains other fuels including kerosene, diesel fuel, and jet fuel. Besides producing gas and fuels, petroleum distillation also creates lubricants, and very thick oils, called *bitumen,* make asphalt that goes on roads. Additionally, petroleum provides the basic components used to manufacture important chemicals, all plastics, and many other synthetic materials. In fact, several recent studies have concluded that petroleum is too valuable to burn as fuel because of its importance in manufacturing.

14.3.2 Origin of petroleum

Contrary to some popular opinions, petroleum does not come from dead dinosaurs. Most petroleum derives from microscopic plankton and other floating micro-organisms that lived in *photic zones* (near-surface layers where sunlight penetrated) of tropical oceans and large lakes. Life in these shallow-water zones included both *zooplankton* (animal) and *phytoplankton* (plant) organisms. When the organisms died, they sank to the bottom of the water column, settling to the ocean, sea, or lake bottom.

The bottom environments of many oceans and lakes are too oxidizing and have too many scavengers for organic matter to accumulate. However, in some places, the beds of oceans and lakes are stagnant and anaerobic, characterized by low energy, low amounts of oxygen, and generally cool temperatures. Poisonous hydrogen sulfide gas (H_2S) may be present, and the toxic H_2S and generally poor conditions mean that scavengers are absent. In such places, zooplankton and phytoplankton remains cannot, for the most part, decompose; instead, the dead material becomes covered by clastic sediment and buried. Anaerobic bacteria in sediments near the sediment-water interface may convert some dead organic matter into *biogenic gas* (mostly methane), but most of the organic debris persists and is slowly buried to greater depths.

Figure 14.9 shows what happens to organic-rich sediments as they become buried. The deeper they go, the more pressure and heat they experience. Eventually temperatures become too hot for anaerobic bacteria to survive. In most places, the production of biogenic methane ceases once several hundred meters of sediment cover organic debris. With continued burial and heating, diagenesis occurs, causing proteins and carbohydrates in the biological remains to decompose into *kerogen*, a mixture of waxy and solid, chemically complex organic compounds. More burial and heating causes kerogen to change into petroleum, containing both oil and gas, at temperatures of about 40–50 °C. Depending on the geothermal gradient, these temperatures usually occur at burial depths of 1000–2000 meters. The steps that occur as buried biological material becomes petroleum are called *maturation*.

Bitumen and waxes, which are called *heavy petroleum* because of their large and chemically complex molecules, are the first forms of petroleum that develop from kerogen. With increasing burial depths and heating, the kerogen, bitumen, and other organic compounds break

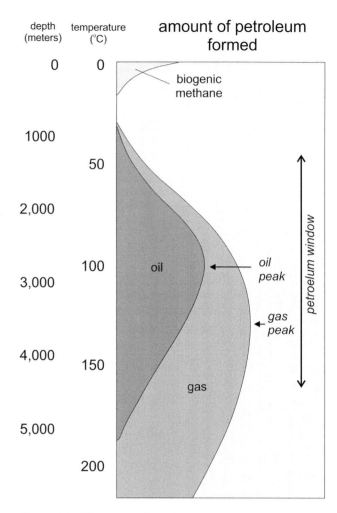

Figure 14.9   The process of petroleum maturation.

down into smaller molecules, including hydrocarbons and other lighter molecular weight organic compounds found in oil and natural gas.

As shown in Figure 14.9, most oil and natural gas forms in a narrow temperature range of 40–180 °C, which we call the *oil window*. Oil production peaks at 100 °C, and natural gas production at slightly higher temperatures. At temperatures above 100 °C, oil substantially decomposes into natural gas. The first natural gas to form is "wet," meaning that considerable ethane, propane, and butane (perhaps 4%–5%) are present with methane. Maximum methane production occurs at about 130 °C. At temperatures above 150 °C, which are usually present at depths greater than 4 kilometers, a process called *metagenesis* occurs, which is the decomposition of bitumen and oil into "dry" natural gas and graphite. "Dry gas" is entirely or almost entirely methane.

14.3.2.1 *Source rocks and migration*

Rocks where petroleum forms, or could form, are called *source rocks*. Most source rocks are organic-rich shales and siltstones, generally of marine origin. Petroleum, however, typically does not remain where first generated because it is less dense than surrounding rock. This density difference causes the buoyant hydrocarbons to *migrate* and flow upward. Additionally, hydrocarbons float on water, so if left unchecked, they will rise to the top of the water table. If the hydrocarbons reach the surface, gas will escape to the atmosphere, and petroleum may create an *oil seep*, such as the La Brea Tar Pits in Los Angeles (Fig. 14.6). Such seeps—which produce oil, asphalt, and tar—are common worldwide and have been energy sources since early times. However, most seeps are small, and unfortunately, as petroleum sits at the surface, it oxidizes or decomposes due to bacterial activity. Consequently, energy value goes down.

Although petroleum migrates naturally upward through many kinds of rocks, migration may stop if the petroleum runs into a *petroleum trap*, a layer of impermeable rock, such as shale. Figure 14.10 shows three examples of. The drawing on the top shows an *anticline trap*—a type of trap involving rocks that folded into an arched shape. Hydrocarbons (shown in red) from depth have migrated upward until encountering the impermeable layer at the top of the anticline—and so the hydrocarbons became trapped and concentrated. The drawing in the middle shows a *fault trap*. Diagonal fault movement, shown by arrows, has juxtaposed permeable and impermeable rocks. So, hydrocarbons flowed upward and became concentrated in the red triangular area with impermeable rock above and to the left. The bottom drawing shows a *stratigraphic trap*. In this kind of trap, impermeable rock (shown in black) surrounds permeable rock that contains petroleum (shown in red). So, crude oil flows into the permeable horizon and cannot escape. Many other kinds of rocks and geometries can trap hydrocarbons, and most are associated with some impermeable *caprock* that stops the upward flow of petroleum.

Natural processes often separate oil and gas. Because of density differences, natural gas often concentrates in pores and fractures just below impermeable cap rocks. Oil concentrates in pores and fractures beneath the natural gas, and water collects beneath the oil. The waters in oil traps, called *formation waters*, are commonly very salty and may contain radium and other potentially hazardous substances. During drilling and extraction,

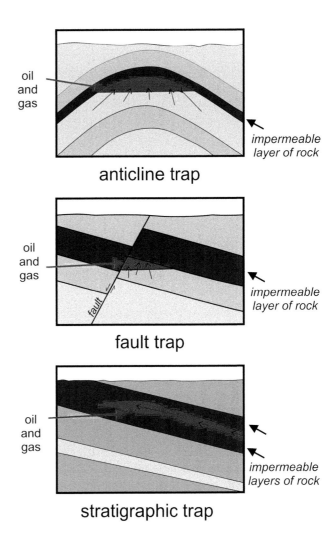

Figure 14.10 Different kinds of petroleum traps.

some formation waters, which unavoidably come to the surface with the oil, can create undesirable pollution. Thus, these waters can create serious environmental, handling, storage, and disposal problems.

14.3.2.2 *Petroleum reservoirs*

The rate at which liquid or gas flows through rock, termed *permeability*, is highly variable depending on rock type. Unfractured granite, for example, may permit flow of only 10^{-9} meters/day, sandstone may allow flow of 10^{-4} meters/day, and highly fractured rocks of all types may allow flow up to 1000 meters/day, sometimes more. Different kinds of rock also have different *porosities*, which refer to the amount of open space between grains. Thus, different kinds of rock have different capacities to hold petroleum. Typical dolomite, for example, contains about 2%–6% pore space, typical shale contains 8%–15% pore space, and typical sandstone contains 10%–30% pore space. Permeability and porosity are related properties but do not always correlate. Some very permeable rocks, such as a granite that contains many fractures, may have low porosity. And some very porous rocks, such as vesicular basalt, may have very low permeability if open spaces in the rock are not connected—like a piece of Swiss cheese that is full of unconnected holes.

Petroleum trapped in the subsurface is an important source of energy, but deposits vary in volume, in productivity, and in quality. *Reservoirs* are volumes of rock that contain significant amounts of hydrocarbons. In *conventional reservoirs*, oil and gas are in porous and permeable rocks, and the petroleum does not migrate away because of traps, such as the anticline, fault, or stratigraphic traps discussed above. In *unconventional reservoirs*, hydrocarbons are in rocks of high porosity and low permeability that may not require an impermeable layer of rock above to keep the petroleum in place. Such rocks, generally termed *tight rocks*, or *tight formations*, include some shales and other fine-grained lithologies. Although oil and gas reservoirs are sometimes mischaracterized as "pools," petroleum does not form a puddle in one place; instead, the petroleum is distributed in pores and fractures between the mineral grains.

The best petroleum production occurs if reservoirs are large and reservoir rocks have both high porosity (so they hold large amounts of hydrocarbons) and high permeability (so hydrocarbons may be extracted easily). Production from such reservoirs is much more easily accomplished than production from tight formations. Very productive oil wells may produce hundreds or even thousands of barrels a day, although such high productivity does not

persist indefinitely. Less productive wells, called *stripper wells*, produce no more than 15 barrels per day.

Crude oil contains many kinds of hydrocarbons and thus is highly variable in composition. Crude oils that have low density are rich in commodities such as gas, gasoline, and diesel fuel—all of great value. Denser crude oils contain higher amounts of fuel oil, waxes, and bitumen—which have less value. Sulfur content, too, is an important characteristic that determines value. "Sweet" oils, oils with low sulfur content, are most desirable because they can be refined more easily and more inexpensively than "sour" oils that contain more sulfur. Therefore, some petroleum deposits have greater value than others of comparable size.

14.3.3 Petroleum exploration and production

Oil and gas exploration and development began in earnest around 1930 or 1940. Since then, most of the easy-to-find and large petroleum deposits have already been discovered and developed. Figure 14.11 shows the amount of oil discovered each year since 1930. The major peaks in the graph are because a few giant oil fields have been discovered at different times. The bumps in the chart correspond to discoveries of the Burgan, Ghawar, Samotlor, Prudhoe Bay, Cantarell, North Sea, and Kashagan fields. These fields have historically accounted for most of the world's oil production. Today, oil companies make many discoveries in remote places (including many offshore locations), but most new oil fields have relatively small size compared with older established oil fields. Shallow wells in easily accessible onshore locations may be drilled for as little

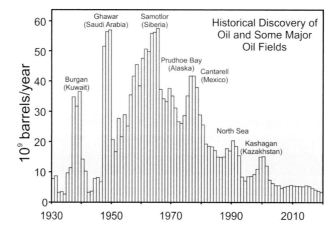

Figure 14.11   Conventional oil discoveries, 1930 to present.

as $100,000. Currently, however, most exploration for, and development of, conventional oil is offshore. The cost of an off-shore well ranges from tens of millions of dollars to 100 million dollars or more, depending on location and water depth. Consequently, most offshore exploration is done by large corporations with deep pockets.

Many conditions must be met for hydrocarbons to form, accumulate, and survive, so finding good oil deposits is a challenge. Exploration is based primarily on the principles of stratigraphy, discussed in Chapter 9, and petroleum geologists long ago identified the kinds of reservoir rocks that are most likely to contain hydrocarbon resources. They also know the kinds of rocks that could be source rocks. So, exploration is focused on some places and never on other places. For example, former sedimentary basins, such as the Michigan Basin, have high potential for oil or gas production, but ancient mountain belts or shield areas that contain igneous and metamorphic rocks do not. Additionally, consideration of tectonic histories and geophysical data permit modeling that may reveal whether petroleum maturation could have occurred.

Although oil seeps or escaping natural gas at the surface provide evidence for hydrocarbons below Earth's surface, most exploration involves geophysical techniques that allow geologists to map and interpret subsurface stratigraphy. Some of the more common geophysical techniques were discussed in Chapter 9. Gravity, magnetic, and seismic surveys are all especially important because they permit mapping of subsurface features that may constitute petroleum traps. Sometimes, natural gas deposits may be detected on seismograms because of the strong density difference between gases and reservoir rock. Oil, however, is usually too dense and too disseminated to be detected with seismic methods. Consequently, exploration often requires direct sampling. Thus, if geophysical data look promising, petroleum companies drill exploration wells to sample hydrocarbons, to create well logs, and to collect core samples.

14.3.3.1   Drilling and recovery of oil and natural gas

Once a promising petroleum deposit is found, focus shifts from exploration to production. Typical oil *production wells* involve a *drill pad* and a *drilling rig* that creates a long borehole. Figure 14.12 shows an example of a well on Bureau of Land Management land near Canyonlands National Park. Below the ground surface,

Figure 14.12   Active oil and gas pad on Bureau of Land Management lands near Canyonlands National Park, Utah.
Photo from the USGS.

a steel pipe called *casing* encloses the hole to keep the hole open and to provide a conduit for petroleum flow. The casing is open near the base of the well, allowing oil to migrate to the well bore. At the surface, a collection of valves, called a *Christmas tree*, regulates petroleum flow during production.

Although the first horizontal oil well, drilled near Texon, Texas, was completed in 1929, before 1990 most oil wells were drilled vertically and one to a location, such as shown by the middle derrick in Figure 14.13. The spacing of wells depended on the richness of the petroleum resource below. Today, however, boreholes may be vertical near the surface but can curve and may become horizontal at depth (left derrick in Fig. 14.13). And, using *directional drilling* (right derrick), multiple wells may be drilled from the same drilling pad—the wells need not be vertical and can be aimed in different directions. Horizontal and directional drilling may extend as far as 10 kilometers (6 miles) from a drill pad, thus reducing the number of wells needed to recover

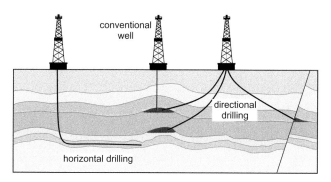

Figure 14.13   Different ways that oil wells may be drilled.

Figure 14.14   Oil pumps in the Kern River Oil Field near Bakersfield, California.
Photo from Antandrus, Wikimedia Commons.

petroleum from different parts of a reservoir. Directional drilling can sometimes tap into more than one reservoir, and such drilling can also be used to recover petroleum from underneath cities and environmentally sensitive areas. Drilling multiple directional wells from a single site can save money and, for this reason, as many as 50 wells have been drilled from a single off-shore drilling platform.

14.3.3.2   Primary oil recovery

Historically, oil companies extracted petroleum using *primary recovery* technology. This technology involves wells that produce oil that flows naturally to the surface and wells where pumping is needed for production. When wells penetrate a petroleum reservoir, the *formation pressure* inside the reservoir is generally greater than the pressure that the overlying rocks exert on the reservoir. Consequently, the pressurized oil and gas will naturally move up through the well casing. In some *naturally flowing wells*, the formation pressure is great enough so that hydrocarbon production may not require pumping, at least initially. Today, oil wells have pressure control equipment, but in the past, *blowouts* sometimes occurred when oil and gas erupted in out-of-control flows at the surface, occasionally resulting in explosions or fires. In the first few decades of the 20th century, blowouts were frequent and were called *gushers* or *wild wells*, but after about 1920, all oil wells had *blowout preventers* and, consequently, gushers were no longer common.

Figure 14.14 shows oil pumps that are producing oil from wells in the Kern River Oil Field near Bakersfield, California. Many petroleum wells, like those in the Kern River Field, were naturally flowing wells when first tapped. Pressure and flow, however, typically fall over time, and some reservoirs never have sufficient

formation pressure to flow naturally in the first place. So, primary production from most wells involves some form of pumping. Primary production, however, has limited extraction potential, either because a reservoir is insufficiently pressurized or because pumps are not powerful enough to extract more than a small amount of the oil from a reservoir. Consequently, primary oil recovery generally produces only 5%–15% of the total petroleum in a reservoir.

14.3.3.3   Secondary recovery

Figure 14.15 shows an example of secondary oil recovery. When primary production wanes and is no longer profitable, *secondary recovery* techniques may be employed. These technologies involve drilling *injection wells* so that water, natural gas, or another gas such as CO_2 may be pumped into the reservoir, artificially increasing formation pressure while also reducing the density of the petroleum that is present. Petroleum will then migrate to the original production well or to additional *recovery wells* that pump it to the surface. Typically, primary and secondary recovery, when combined, will extract 5% to 30% of the hydrocarbons in a reservoir. The drawing in Figure 14.15 shows a mix of CO_2 and water that was injected to force oil to flow to the recovery well on the right.

14.3.3.4   Enhanced oil recovery

Petroleum companies sometimes use other production technologies, collectively called *tertiary recovery*, or *enhanced oil recovery* (EOR), to supplement primary and secondary production. *Thermal recovery*, for example, may involve injecting steam into a reservoir, which

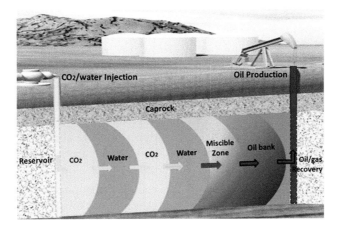

Figure 14.15 The use of carbon dioxide as a secondary recovery technique to remove additional petroleum from a well. Drawing from Los Alamos National Laboratory.

heats petroleum, lowering its viscosity and improving flow rates. Less commonly, thermal recovery involves underground fires that heat oil, with the same goals in mind. *Chemical injection*, another EOR technique, involves injecting chemicals into reservoirs to reduce petroleum viscosity, thus increasing flow to production wells and increasing recovery.

14.3.3.5 Production from unconventional petroleum reservoirs

During the past two decades, as conventional oil deposits have begun to be depleted, oil companies have turned their attention to unconventional oil. The 370-million-year-old Bakken Formation, which underlies 520,000 square kilometers (200,000 square miles) of Montana, North Dakota, Saskatchewan, and Manitoba, provides a good example of an unconventional oil field. Figure 14.16 shows a production site where pumps produce Bakken oil. In the Bakken oil field, companies often drill multiple directional wells from one site to keep costs down. Underground, the wells go in different directions, so oil from a relatively large region is produced with minimal surface impact.

Beginning around 1950, petroleum geologists knew that the Bakken Formation was a productive petroleum source rock. Since it formed 370 million years ago, petroleum from the Bakken has migrated upwards to overlying reservoir rocks that, for many years, have been targets of oil exploration. Although the amount of oil contained within the Bakken is very large, extracting oil directly from the formation has been problematic, due to the low permeability of Bakken rock. Conventional

Figure 14.16 Three wells that produce oil from the Bakken Formation in western North Dakota. Photo from the USGS.

wells drilled into the Bakken have produced for short times and then dried up because petroleum could not flow easily from surrounding rocks to the well. Oil companies solved this problem, beginning around 2000, when they discovered that horizontal drilling, combined with *hydraulic fracturing* (commonly just called *fracking*) could make the Bakken a significant source of petroleum.

Fracking, similar to what is done today, was first tried in 1947. Commercial success soon followed, and, in the past 70 years, petroleum companies have fracked more than 4 million oil wells worldwide. The technology involves injecting pressurized fluids (usually water containing sand and chemicals) into wells to fracture subsurface reservoir rocks, increasing permeability so that petroleum flows more easily to wells. The sand plays an important role because, after fracking creates new fractures, sand can hold fractures open when pressure decreases. Because of the combination of horizontal drilling and fracking, the Bakken is the second most productive petroleum-bearing rock formation in the United States, and North Dakota produces more oil than any other state except Texas.

14.3.4 Tar sands oil

Tar sands, also called *oil sands*, are a type of unconventional petroleum deposit. The sands are generally loose sediments that contain black viscous petroleum (bitumen), which is also called (incorrectly) *tar*. Large undeveloped tar sands are found in many places around the world, although they may not all be producible. The largest deposits are in Canada, and today, only Canada has a significant tar sands industry. Tar sand oil production is the reason that about a quarter of U.S. oil imports come from Canada.

Figure 14.17 shows a tar sands production area in Alberta. Tar sands cannot be pumped because the petroleum they contain is too thick. Consequently, they are, for the most part, mined, and production comes from strip mines and open pits, like the ones seen in this photo. In some places, petroleum companies exploit deeper deposits using *in situ* methods. Steam injection or solvent injection can make the bitumen less viscous or convert it to gas, so the petroleum may be produced through wells. Mining and processing tar sand petroleum requires a great deal of energy, sometimes making it impractical to produce compared with other kinds of petroleum resources.

Figure 14.17   Tar sands oil production in Alberta in 2008.
Photo from Howl Arts Collective, Wikimedia Commons.

Tar sands production and use involve many potential environmental problems. For example, production and burning of petroleum from tar sand deposits create more CO_2 and other pollutants than traditional petroleum resource production. Additionally, the mining process has huge terrain costs, because mining disturbs large areas of land and there is commonly little reclamation. Refining also produces many hazardous byproducts. Water, air, and soil pollution are common, and tar sands production consumes large amounts of water and energy. Additionally, in many places, pipelines to transport tar sand oil are controversial due to concerns about leaks and for other reasons.

Given the rapid increase in demand for oil and gas and in production rates, during the past 60 years, it is not surprising that large petroleum reserves are limited today. But new technologies, including the development of unconventional sources like the Bakken Formation and tar sands in Canada, have made production possible from deposits that could not be utilized in the past. Further innovations are likely as conventional sources continue to decline, but predicting the future is problematic. For now, although the cost of production has increased, unconventional oil means that world production is keeping up with demand.

14.3.5 Coal

People have known about the value of coal, just as they have known about the value of petroleum, for thousands of years. In China, evidence of coal use dates to 3500 BCE, in the Bronze Age. In 15th century America, Aztecs used coal for heating and ornamentation. From the Middle Ages onward, people used coal to heat homes, but today, at least in the more developed parts of the world, that use is rare. Modern society uses coal primarily for two purposes: 85% of the coal mined today, sometimes called *thermal coal*, generates electricity; and about 15%, called *coking coal* or *metallurgical coal*, is used during steel manufacturing.

Coal became especially important during the latter part of the Industrial Revolution, in the 19th century, when it powered steam engines and heated buildings. Burning coal produced much more energy than burning wood, and people could burn coal in places where wood was unavailable. Petroleum began to replace some coal around 1860, but coal continued to be the main source of energy for industry and transportation until after World War II. Today, however, petroleum has largely replaced coal as transportation fuel and as the power source for many industries.

In the late 19th century, several events caused demand for coal to skyrocket. In 1880, Thomas Edison was granted a patent for the first light bulb, and to turn on a light bulb required electricity. The first coal-fired electric station, the Edison Electric Light Station, was in London, England, built by Edison and partners. Shortly afterwards, in 1882, Edison opened the Pearl Street Station in New York City—the first electrical utility company. The station and its electrical distribution system made incandescent lighting less expensive and safer than the gas lighting common at the time. Yet, during through the 1880s, electricity use was relatively limited.

One event may be, more than anything else, responsible for starting the modern age of electricity, which gave a big boost to the coal industry. At the 1893 Chicago World's Fair, U.S. President Grover Cleveland flipped a switch and 100,000 incandescent light bulbs immediately lit up the night sky. Figure 14.18 is a night view of the Agricultural Building at the Fair lit by electric lights. Newspaper reports of that event circulated around the world. People everywhere said, "I want that!" and coal-fired power plants were soon constructed in many places.

Today, there is still much coal that could be mined, and coal energy is relatively inexpensive compared with most other kinds of energy. But mining and burning coal have large ancillary costs that make coal unpopular in many places and countries. Environmental problems include global warming and other forms of air pollution. Coal miner health, the effects of coal burning on people near coal plants, and the destruction of land due to mining activities also are concerns. So, coal is still important, but the era of coal may be moving toward its end. Although coal is still the primary source of electricity worldwide, power generation is slowly shifting to natural gas, nuclear power, and to wind, solar, and other renewable energy sources.

14.3.6 Origin and properties of peat and coal

14.3.6.1 Peat

In most terrestrial environments, organic debris, including dead plant material, disappears when oxidized by reaction with gases in the atmosphere or when consumed by burrowing animals, fungi, and bacteria. However, if organic matter falls and sinks into low-oxygen (anaerobic) water in swamps, bogs, lagoons, or lakes, it will only partially decay because animals, fungi, and anaerobic bacteria are not generally present in low-oxygen environments. The preserved organic matter may accumulate and form *peat*, a soil-like material that forms by diagenesis of ancient grasses, bushes, trees, and other terrestrial plants. Peat deposits, typically accumulating over thousands of years, may, after burial and diagenesis, eventually evolve to become coal. Figure 14.19 shows peat dug from a pit on the Isle of Lewis, the northern part of Lewis and Harris, the largest island in Scotland's Western Isles. People harvest peat to use as fuel in many households of the Western

Figure 14.18   100,000 incandescent lamps light the 1893 Chicago World's Fair.
This photo from the Field Museum Library, Wikimedia Commons.

Figure 14.19   Peat from the Isle of Lewis, Scotland.
Photo from Wojsyl, Wikimedia Commons.

Isles. Because plants did not substantially evolve until the Silurian Period, around 430 million years ago, and were not widespread until the Devonian Period, around 390 million years ago, most of the world's peat and coal are Devonian and younger.

Peat forms in bogs and other wetlands where water flow and continued wet conditions shield dead organic debris from oxygen, thus slowing or eliminating decomposition. So, over long times, huge deposits of organic matter can accumulate and later become peat. Figure 14.20 shows a peat bog on the island of Lewis in Scotland. Many different organisms may become part of peat. *Sphagnum moss*, also called *peat moss*, is the best-known peat component because of its widespread use in horticulture, but other plants—mostly sedges

and shrubs—are commonly present. Areas where peat accumulates are called *peatlands*, and they vary by the vegetation they contain. Swamps (containing trees) and bogs (with few or no trees), common types of peatland, are the most productive sources of peat. Other kinds of peatlands, all quite similar, include fens, mires, moors, and muskegs. Because peat accumulates over thousands of years, scientists sometimes study peat to investigate the recent evolution and development of vegetation and to learn about past climates.

The world, particularly northern Europe and Canada, contains abundant peatlands, shown by the dark green color in Figure 14.21. Lighter green colors show places where peat is found but is not so abundant. Peat deposits are mostly water by weight, but much of the water is squeezed out when peat is compressed. The remaining material can be stacked to dry and compressed further for use as fuel. This is what is taking place in Figure 14.19. In the 1950s, peat was considered an important fuel in many European countries, but it is not as important today. In part, this is because, when burned, peat does not produce a great deal of heat compared to other energy sources, and so has little value on the global market. Additionally, some people and countries are concerned about the effects that peat harvesting has on terrain and about the pollution created when people burn it. Still, peat supplies energy for both home heating and industry in some countries, including most significantly Scotland, Ireland, Finland, and Russia.

Figure 14.20   Peat bog near Callanish, Lewis Island, Scotland. Photo from Bob Embleton, Wikimedia Commons.

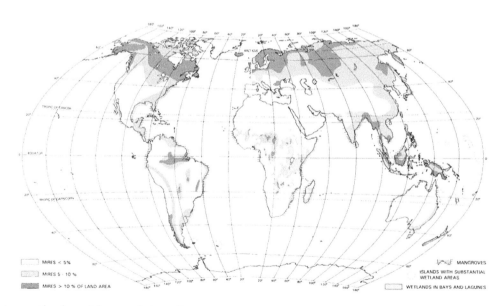

Figure 14.21   Areas with substantial accumulations of peat. Map from the International Peatland Society.

14.3.6.2 Different kinds of coal

Figure 14.22 summarizes the diagenesis of peat that, sequentially, produces different kinds of coal. Peat, the precursor to all kinds of coal, can evolve sequentially to become *lignite, subbituminous coal, bituminous coal,* and then *anthracite.* We classify coal based on rank and grade. *Rank* refers to how much heating occurred as the coal evolved. Heating during diagenesis affects carbon content and other important characteristics. Most importantly, high-rank coal (anthracite) contains less water and has greater carbon content than low-rank coal (lignite). Subbituminous and bituminous coal (medium rank) fall in the middle. *Grade* refers to the quality of a coal based on burning characteristics, carbon content, and impurities. The highest grade, and most valuable coals, are high in combustible carbon, and contain little water, sulfur, or mineral material.

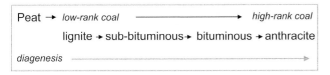

Figure 14.22   Burial pressure, heat, and time cause peat to change into different kinds of coal.

The presence of volatiles, such as sulfur, affects a coal's value. Sulfur content, in particular, is variable and is a significant problem for some coals. When people burn coal, any sulfur that is present produces sulfur oxide gases, which can cause air pollution that threatens human health and contributes to the formation of acid rain. Besides water and sulfur, most coals also contain clays or other minerals that are incombustible and thus remain after coal burning. The *ash content* of a coal refers to the percentage of inorganic materials in the coal that are noncombustible, do not contribute to the heating value of the coal, and create disposal problems after combustion. We much prefer coals with low ash content, because they burn more cleanly and because ash may contain heavy metals or other toxics, making disposal problematic.

14.3.6.3 Low rank coal

When peat is buried to depths of several kilometers, pressure and heat will eventually cause it to become *lignite* by a process called coalification. Lignite, also called *brown coal* (although it is often not brown), is the lowest rank coal. Figure 14.23 shows an example of lignite from Serbia. Although most coal forms from vegetation deposited in ancient tropical and subtropical swamps, any peat is a potential precursor to lignite and other coals. During conversion to lignite, water is lost, organic materials partially decompose, and carbon becomes more concentrated. Commonly, however, lignite contains recognizable organic material, including sticks and leaves, and up to 70 wt% water. Because lignite contains lots of water, and because even when dried lignite does not burn at the same high temperatures as other kinds of coal, transporting lignite long distances is not generally economical. So, most power plants fueled by lignite are close to lignite mines.

14.3.6.4 Higher rank coal

Peat deposits form relatively quickly, but conversion to coal is a slow step-wise process as diagenesis converts lignite into higher rank, and harder, varieties of coal—first subbituminous coal, then bituminous coal, and finally anthracite (Fig. 14.23). These transformations take place over millions of years. Harder coals are most dense, contain more carbon, less water, and fewer impurities, and they burn at higher temperatures, than softer coals. Consequently, harder coals make better energy sources.

Subbituminous coal is a dull black variety used mostly for generating electricity and, less commonly, for heating. *Bituminous coal,* also called *black coal,*

Figure 14.23   Three kinds of coal.
Photos from the USGS.

is darker colored, the most common kind of coal, and the most commonly used to generate electricity. An example is shown in Figure 14.23. Subbituminous and bituminous coals typically form at temperatures of 65 °C to 100 °C over a few tens of millions of years. The rate of formation is fastest at high temperatures.

Anthracite, also called *hard coal*, is denser than other coal varieties and, as seen in Figure 14.23, commonly has a shiny, almost metallic luster (Fig. 14.23). It has greater carbon content and burns hotter with a blue flame, instead of the yellow flame of other coals. It also generally contains fewer impurities. Other ranks of coal are considered organic sedimentary rocks, but anthracite is really a metamorphic rock, because it forms at pressures and temperatures greater than those of diagenesis. When burned, because it contains 90%–98% carbon, anthracite emits little other than CO_2. Anthracite was once used for heating but is not used much today because it is rare

(accounting for only a few percent of the coal in most places) and alternative less expensive energy sources are available. Temperatures of 170 °C to 250 °C are required to convert bituminous coal into anthracite, but very long times are required at temperatures much less than 200 °C. The depth of formation is around 10 kilometers (6 miles). The high temperatures, which are uncommon in most sedimentary basins, and the depth of formation explain why anthracite is rare in mineable near-surface deposits.

14.3.7   Coal production

China produces the most coal of any country in the world, India produces about 20% as much, and the United States slightly less. All types of coal are found in the United States; Figure 14.24 shows where the different coals are found. Anthracite is only in the eastern United States, where it formed when the Appalachian Mountains were created more than 250 million years

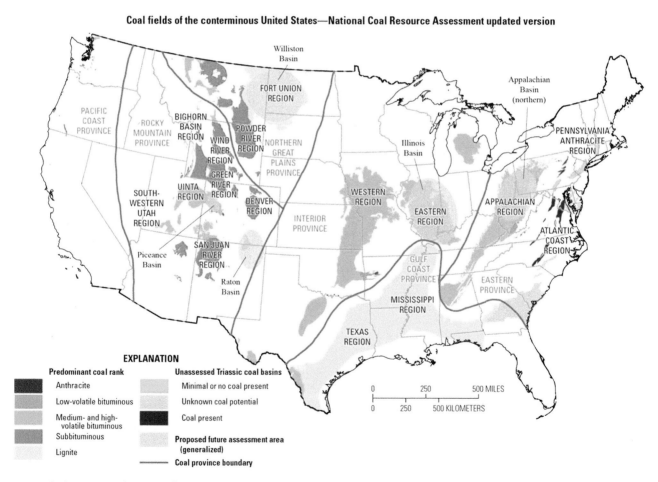

Figure 14.24   Major coal regions of the contiguous United States. Map from the USGS.

ago. Other coal varieties are widespread. Most of the coal in Texas, North Dakota, Mississippi, Louisiana, Alabama, Florida, and Georgia is lignite. Coal in the midwestern and western states is mostly subbituminous to bituminous. Wyoming has some especially important subbituminous coal deposits that contain very low amounts of volatiles, especially sulfur, and thus burn more cleanly than many other coals. So, Wyoming coal is often transported by train to coal-fired power plants hundreds or even thousands of kilometers away from mine sites.

14.3.8 Coal mining

Mining companies produce coal from *coal seams*, dark brown or black banded layers within sedimentary rocks, mined by underground or surface mining techniques. Mining companies use surface mining when coal seams are laterally extensive, flat or nearly flat, and near the surface. Surface mining, which involves removing the *overburden* above a coal seam, is generally more efficient and productive than underground mining because of ease of access and the lack of a need to worry about collapse of mine tunnels. Around the world, surface mining accounts for about 40% of all coal production. In the United States, surface mining, which accounts for about two-thirds of coal production, mostly produces bituminous coal and, to a lesser extent, lignite.

The most common methods of surface mining include strip mining, contour mining, and mountaintop removal. *Strip mining* involves removing layers of sediment and rock (collectively called *overburden*) to expose coal seams below. Shallow strip mines are less expensive than deep ones, and generally, strip mining is economical if the thickness of overburden to the thickness of the coal seam is less than 20:1. Miners use giant shovels, called *drag lines*, to remove and preserve topsoil. The underlying layers of sediment and rock are then removed one by one to expose the coal.

Figure 14.25 shows an example of a coal strip mine, the Garzweiler Mine in western Germany near the Netherlands. This mine is in a productive lignite mining district and, as seen in the photo, coal exists in several horizontal seams. The entire mine covers almost 50 square kilometers (20 square miles) and is expected to produce coal until 2045.

In the past, when mining was no longer profitable, mining companies abandoned surface mine sites, leaving a barren, uneven terrain devoid of soil. Today, in the

Figure 14.25 Lignite strip mine near Cologne, Germany.
Photo from Ekem, Wikimedia Commons.

Figure 14.26 Mountaintop removal to get at coal in Pike County, eastern Kentucky.
Photo from iLoveMountains.org, Wikimedia Commons.

United States, primarily because of the Surface Mining Control and Reclamation Act (SMCRA, passed in 1977), lands must be reclaimed. After mining ends, the different kinds of rock, sediment, and soil are replaced in their original sequence. The mine site is revegetated and restored to its original condition and topography, but at a lower elevation because of the coal that was removed.

Contour mining is similar to strip mining, but occurs in hillier terrain. Mining follows contours around hillsides and ridges instead of accessing a large relatively flat area. Due to steep topography, reclaiming contour mine sites is often challenging, and sometimes never accomplished, after mining is completed.

Mountaintop removal (MTR) involves removing overburden on the tops and sides of mountains and ridges to expose coal seams below. Figure 14.26 shows an example from eastern Kentucky. The overburden is dumped into valleys next to the mountain or ridge being mined. MTR is, today, the main kind of coal mining in the Appalachian Mountains of the eastern United States. Mining companies may remove as much as 120 meters (400 feet) of mountaintops to get at the coal below. After mining produces all the economical coal, no attempt is made to restore land to conditions or topography present before mining. Mountains are gone, and replaced with flattened plateaus covered by jumbles of rock and soil. MTR, which began in Appalachia in the 1970s, is less expensive and requires fewer employees than other forms of surface mining. Today, it is taking place in West Virginia, Kentucky, Virginia, and Tennessee.

Figure 14.27 shows an underground coal mine in Colorado. Underground coal mines, like the one shown and many mines in West Virginia and other parts of the Appalachian Mountains, are more expensive to operate than surface mines but are the only way to mine coal seams that are hundreds of meters below the surface. In the past, underground mining was extremely labor intensive. But today, mining companies use mechanized rotary cutters and conveyor belts, such as the longwall shearer and belt shown in Figure 14.27, making mining more efficient and, consequently, requiring fewer miners. In most underground mines, about half the coal, at least initially, is left in pillars to keep mine tunnels from collapsing. The size and spacing of pillars depend on the depth of the mine and the thickness of the coal seam. Sometimes, the pillars are removed just before a mining area is abandoned.

14.3.8.1 Hazards associated with coal mining

Underground coal mining is one of the most hazardous professions in the world. Historically, chronic diseases

Figure 14.27 Longwall shearer and conveyor built in the Twentymile underground coal mine, in northwestern Colorado.
Photo from the USGS.

have plagued coal miners. For example, miners who inhale rock or coal dust may get silicosis or black lung disease, both of which commonly lead to death. Additionally, toxic, flammable, suffocating, or radioactive gases (such as methane, carbon monoxide, hydrogen sulfide, and radon) may accumulate in mines. This is why, in the past, miners took canaries into mines to detect methane and other odorless, poisonous gases. If the birds died, the miners knew that hazardous gases were present. Today, electronic gas detectors have replaced canaries, and ventilation systems help to remove toxic, radioactive, suffocating, and flammable gases.

Underground mining also poses other risks, including rock bursts, cave-ins, and collapsing ceilings and walls. Such events have killed thousands of American miners over the years. Until the 1960s, not a single decade went by without at least one American coal mine accident killing 100 or more miners. Because of improved technologies and strict safety regulations, the number of mine fatalities in the U.S. has declined from more than 3000 in 1907 to eight in 2017. Of the 16 miners killed in 2014, 10 were in underground mines.

14.3.8.2 Environmental problems associated with coal mining

All kinds of surface mining are controversial because of the impacts that mining has on terrain and ecosystems. Mountaintop removal is especially controversial, because it not only affects the mine site but also the valleys and waterways near the site. Mine runoff, sometimes toxic, is often uncontrolled and may affect people far downstream. Underground mining has its own set of environmental problems. When mining ceases and companies abandon mines, underground mines may collapse and produce sinkholes that damage roads and property.

Figure 14.28 shows sinkholes that formed above the Old Monarch Mine, north of Sheridan, Wyoming. The mine operated from 1904 to 1921. Long parallel depressions lie above areas where coal was removed from below, and the sinkholes developed when water eroded a passage to the surface. Land subsidence and sinkholes are unlikely to be a problem for newer mines because of government regulations that require that mines be shut down responsibly. But many mines were abandoned decades, or more than a century, ago when there were no such restrictions.

Additional environmental problems occur when people burn coal. Burning coal releases CO_2 to the

Figure 14.28 Subsidence depressions and pits above the Old Monarch Mine north of Sheridan, Wyoming. USGS photo, 1978.

Figure 14.29 Acid mine drainage from an abandoned coal mine near North Lima, an hour's drive southeast of Cleveland. Photo from Jack Pearce, Wikimedia Commons.

atmosphere, and coal burning is the number one source of greenhouse gases that are causing global warming. Other pollutants pose problems too. For example, although laws limit the amounts of sulfur and nitrogen oxides that can be created by coal-fired power plants, some of those compounds inevitably escape. The emissions lead to acid rain and contribute to global warming. And, unfortunately, some old coal plants are much more polluting than newer ones that must comply with stricter regulations. Although some coal companies complain about the Clean Air Act and other laws that limit emissions, there is little doubt that the regulations benefit human health and the environment. However, burning coal responsibly increases the cost of electricity for consumers. The additional costs and environmental concerns are why many power plants are switching from coal to natural gas. Natural gas is currently less expensive than coal and produces less carbon dioxide, sulfur, and other undesirable gases.

Abandoned coal mines are also, sometimes, sources of *acid mine drainage*, such as that shown in Figure 14.29. The view shows a mushroom farm where mine drainage is coming from both abandoned surface and subsurface coal mines. Because most coal contains pyrite and other sulfur-bearing minerals, and abandoned mine pits and tunnels are good conduits for water, if the oxygenated water reacts with sulfur minerals, it can create sulfuric acid that flows out of mines and into local waterways, making them as acidic as vinegar or lemon juice. Not only is acidity fatal to aquatic life, but acid will leach toxic elements such as mercury and arsenic from rocks, sediments, and soils—thus contaminating surface water

and groundwater further. Leached metals are the reason that the water in Figure 14.29 has such an orange rusty color.

14.3.9 Coal bed methane

Coal bed methane (CBM), sometimes called *coal mine methane*, is gas extracted from coal beds. CBM is one of the gases that sometimes poses risks to underground coal miners. Figure 14.30 shows, in red, where coal companies produce CBM in the United States. All of it comes from regions where traditional coal mining has occurred or is occurring, shown by the lighter pink colors. CBM is an important energy source in Canada and the United States, as well as in other places around the world, including Australia, India, and the United Kingdom. In North America, production began in the late 1970s.

CBM is mostly an intimate part of the rock (coal) that hosts it instead of occupying open pores like petroleum does. Unlike conventional natural gas, CBM is mostly methane, only containing minor amounts of heavier gases such as propane or butane, and generally very little H_2S. During production, wells drilled into coal seams yield both gas and formation waters. The wells are commonly up to 1500 meters (4920 feet) deep. Once the methane is separated, the formation waters, which sometimes contain undesirable salt or other components, are processed to use as irrigation water or for other purposes. Often, though, the water is simply allowed to evaporate or is reinjected underground—which not

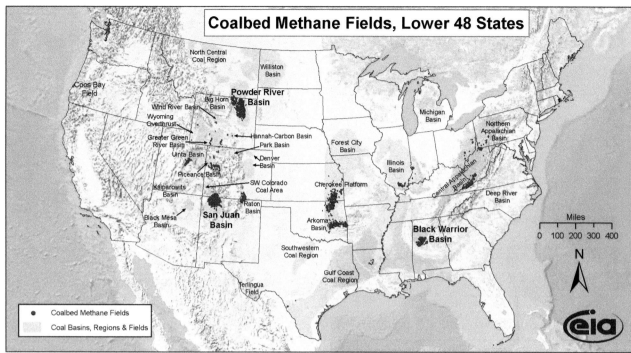

Coalbed Methane Fields, Lower 48 States

Source: Energy Information Administration based on data from USGS and various published studies
Updated: April 8, 2009

Figure 14.30   Map showing where coal bed methane is produced in the United States.
Map from the Energy Information Administration.

only disposes of the water but also helps keep formation pressures, and production, high.

14.4   Alternatives to fossil fuels

Fossil fuels made the Industrial Revolution possible and provided the energy sources that allow people to live the lives they do today. However, we know that continued burning of fossil fuels poses some significant environmental problems. Furthermore, natural processes create oil and natural gas reservoirs and seams of coal much more slowly than people extract and use those resources. Thus, for the good of our planet and all organisms on it, and whether we like it or not, society is at the point at which we must transition from an energy mix based mostly on fossil fuels to one that includes large amounts of nonpolluting and renewable energy resources. The need for this transition is not political. It is climatological and geological. Many other energy sources can replace fossil fuels, such as the concentrated solar power facility described at the beginning of this chapter, but it is unclear which will become dominant in the coming decades. It seems most likely that, in the future, we will continue to use some fossil fuels but will also use significant amounts of nuclear, geothermal, hydro, wind, and solar power.

14.4.1   Nuclear electrical generation

Heat from nuclear reactors powers nuclear power plants by creating steam that turns turbines connected to electrical generators. As of 2019, more than 450 nuclear power plants were operating in 31 countries. Before the Three Mile Island nuclear accident in Pennsylvania in 1979, 236 nuclear reactors were planned or had been built in the United States. After the accident, most building plans were canceled. No new nuclear power plants have been built and gone into operation in the United States since 1978. The number of operating nuclear power plants in the United States peaked at 112 in 1990, but by the end of 2019, there were 98 that together provided about 20% of the electricity consumed in the United States. In comparison, France, which uses more nuclear power than any other country, obtains about 75% of its electricity from nuclear reactors.

Uranium, which fuels most reactors, is a heavy metal that exists mostly as two different isotopes, U-238 and U-235. The isotope U-238 accounts for 99.3% of natural uranium, and U-235 for about 0.7%. Very small amounts of other uranium isotopes, which decay very quickly, may exist for short times. The two principal uranium isotopes have significantly different properties. In particular, U-235 is *fissile*, meaning it easily undergoes fission, making it a good fuel to generate nuclear energy. U-238 is not fissile and thus cannot be used as reactor fuel in conventional reactors. So, nuclear reactors in most electrical plants generate heat from U-235 undergoing controlled nuclear fission. A key reaction during this process involves U-235 that picks up an extra neutron to become U-236, which then splits into isotopes of krypton and barium while emitting heat. Many other reactions take place in reactors too, and the fission products include many different elements and isotopes.

14.4.2 Conventional nuclear power

Commercial power companies use three main kinds of reactors today. Most (75%) are *light-water reactors*, such as the one in Figure 14.31. As shown, fission in a reactor vessel generates heat, which changes liquid water to steam that spins a turbine that in turn spins a generator and creates electricity for the power grid. The uranium used in light-water reactors, the kind of reactors used to generate electricity in the United States, must have a higher concentration of U-235 than in natural uranium ores. So, uranium ores are processed to enrich them and bring U-235 concentration up to 3 to 5 wt% (from less than 1%). Twenty grams of uranium (slightly more than 1 cubic centimeter) contain about one gram of U-235 and produce as much heat as 13.7 barrels of oil.

The first commercial *light-water* power plant, built in Pennsylvania in 1957, operated until 1982. Commercial *graphite reactors*, based on somewhat different technology, make up 5% of the world's electrical generation. They first were built in the United Kingdom and Russia in the late 1950s. And, in 1960, Canadian engineers designed a *Canada Deuterium Uranium (CANDU)* reactor that accounts for about 20% of the world's nuclear electrical generation. Unlike light-water reactors, graphite and CANDU reactors can use unenriched uranium as fuel. The main difference between the three kinds of reactors involves *neutron moderators* that slow neutrons

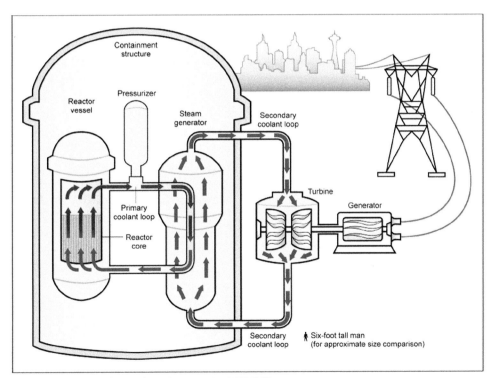

Figure 14.31   Schematic showing the workings of a light-water nuclear reactor.
From U.S. Government Accountability Office, Wikicommons.

so the neutrons are more likely to interact with U-235, ultimately leading to fission decay that generates heat. Light-water reactors—the kind of reactors used in the United States—use normal water as both a coolant and as a neutron moderator. In CANDU and graphite reactors, respectively, heavy water (water in which some or all of the hydrogen, H-1, has been replaced by deuterium, H-2) and graphite are the moderators.

14.4.2.1   Breeder reactors and fuel reprocessing

Figure 14.32 shows the Superphénix breeder reactor in Creys-Malville, France. *Breeder reactors*, in contrast with all three kinds of reactors described above, make use of the much more abundant isotope of uranium, U-238, and can also use thorium as fuel. Nuclear reactions in breeder reactors transform U-238 and thorium into plutonium-239 (Pu-239), which is fissile like U-235 and thus can generate heat and electricity. Breeder reactors are not used for electrical generation today, although the Superphénix reactor once produced commercial electricity. The French government mothballed the reactor in 1996 because it was too expensive to operate and noncompetitive with light-water reactors. Additionally, it had many technical problems that caused it to be shut down periodically.

Breeder reactors make Pu-239, which is good for generating electricity but poses a special problem because Pu-239 is the key component of modern nuclear weapons. Thus, in the United States, the primary use of breeder reactors has been to make Pu-239 for bombs. The Clinch River Breeder Reactor Project was the last

proposed commercial breeder reactor, but the project was canceled in 1977 by President Jimmy Carter because of concerns about nuclear proliferation. Since then, concerns about nuclear proliferation have meant that no new breeder reactor projects have been started.

Pu-239 is, however, still an attractive fuel to those in the nuclear power industry who point out that *spent reactor fuel* (used reactor fuel) can be reprocessed to recover Pu-239 for use in conventional reactors. Originally, this kind of reprocessing was done only to make plutonium for weapons, but later it was used to provide fuel for electrical generation. However, like breeder reactors, reprocessing has been unpopular because it creates weapons-grade plutonium that might fall into the wrong hands. In 2009, the Obama administration stopped plans to construct commercial-scale reprocessing facilities, and no further consideration has occurred since.

14.4.2.2   Uranium ore deposits

Uranium, the primary fuel for nuclear reactors, is much less abundant than copper but far more abundant than gold. The average uranium concentration of Earth's continental crust is 0.91 parts per million (ppm), equivalent to 0.000091%. Uranium ore deposits are of many types. Some are associated with granite and other igneous rocks, some with metamorphic rocks, some with hydrothermally altered rocks, and many with sedimentary rocks, especially sandstones and organic shales. Most uranium production comes from two kinds of deposits—*roll-front deposits*, where uranium occurs in sandstones sandwiched between shale units, and *unconformity-related deposits*, where uranium occurs in sedimentary basins at the interface of sediments and basement rocks. Each of these deposit types accounts for about 1/3 of world production.

The richest uranium ores only contain about 2 wt% of uranium, but that is enough to make mining profitable. The principal ore minerals are *uraninite*, UO_2, *pitchblende* (which is mostly poorly crystallized uraninite), and *carnotite*, $K_2(UO_2)_2(VO_4) \cdot 3H_2O$. Significant uranium deposits exist in Canada, Brazil, Australia, and Niger. Russia and China probably have extensive uranium deposits, but their reserves are largely unknown. The United States also has many uranium reserves in Colorado, Utah, Arizona, Wyoming, and New Mexico. Substantial uranium also occurs in marine organic phosphorite deposits in Florida and Idaho and in black Devonian shales in the eastern United States. However, U.S. uranium consumption has declined over the past several decades because of the end of the Cold War and

Figure 14.32   Superphénix breeder reactor in Creys-Malville, Isère, France.
Photo from Wikimedia Commons. Copyright by Yann Forget.

because no utility companies have built new nuclear power plants for more than four decades.

14.4.2.2.1 Roll-front deposits

Figure 14.33 shows how flowing groundwater produces a roll-front uranium deposit. These deposits exist because natural uranium exists in two ionic forms, U^{4+} and U^{6+}. U^{6+} is very soluble in water, but U^{4+} is not. U^{6+} also is present as soluble phosphate and carbonate complexes in water. Thus, in oxidizing conditions, groundwater may carry significant amounts of dissolved uranium (U^{6+}). But, in reducing conditions, uranium turns into U^{4+}, and uranium compounds containing U^{4+} will precipitate from water. If uranium-containing oxygenated water flows into a reducing (low oxygen) porous and permeable rock (such as sandstone), a boundary, or front, forms between the advancing oxidizing waters and reducing environments in front of the water. As shown in Figure 14.33, the front moves in the direction of groundwater flow toward the reducing zone, but more slowly than the groundwater flow overall. So, dissolved uranium in the oxidizing groundwater eventually reaches the reducing conditions where soluble U^{6+} is reduced to insoluble U^{4+} and potential uranium ore minerals are precipitated. Uranium deposits that form in this way are called *roll-front deposits*, and they commonly occur in sandstones and organic shales. Similar deposits also occur in some Mesozoic stream gravels in New Mexico, Utah, Colorado, and Wyoming.

14.4.2.2.2 Unconformity-related deposits

Unconformity-related deposits are, like roll-front deposits, created by precipitation of uranium minerals in reducing conditions. As fluids flow through *basement rocks* (ancient metamorphic and igneous rocks), the fluids dissolve and incorporate uranium if conditions are oxidizing. When upward-flowing fluids carry uranium-saturated waters from oxidizing conditions in the basement rocks into more reducing conditions associated with sedimentary rocks above, uranium solubility decreases, and ore minerals, typically uraninite (UO_2) or coffinite (a uranium-containing silicate mineral) will be deposited. These kinds of deposits are most commonly of Proterozoic age (Precambrian), but some are younger.

As shown in Figure 14.34, unconformity-related deposits generally occur in lower levels of sedimentary basins where an unconformity (shown in red) separates sedimentary rocks from older deformed metamorphic rocks below. The uranium is most commonly deposited along the unconformity but is sometimes a short distance above or below the interface between sedimentary and basement rocks. The host rocks for uranium are typically quartz-rich sandstones. Unconformity-related deposits, which often contain very high-grade uranium ore compared with other kinds of deposits, are some of the largest and most productive sources of uranium in the world. The most significant occurrences are in Canada's Athabasca Basin, Australia's Kombolgie Basin, and Canada's Thelon Basin. The deposits of the Athabasca Basin are so uranium-rich that they pose health hazards for miners and, consequently, some are mined by robots underground.

14.4.2.3 In situ *uranium mining*

Historically, mining companies have produced uranium ore from large open pits or from underground tunnels. Increasingly today, companies use *in situ* chemical leaching methods that reverse the natural processes

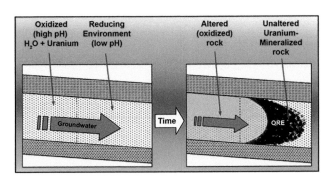

Figure 14.33 Diagram showing how a roll-front uranium deposit may form.

Drawing from the Wyoming State Geological Survey.

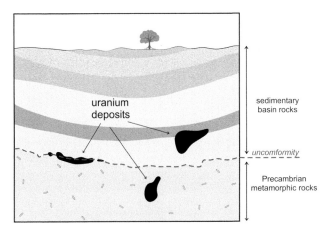

Figure 14.34 Diagram showing where unconformity-related uranium may concentrate.

that create roll-front and unconformity-related deposits. Figure 14.35 (top) shows schematically how mining companies produce uranium this way, and the bottom photo shows wells in a Wyoming uranium recovery field. Companies install wells to pump oxygen-rich water into a uranium-bearing rock below, causing uranium in the rock to oxidize to U^{6+} and, consequently, dissolve in the water. The uranium-containing water then returns to the surface where dissolved uranium is precipitated and concentrated to produce *yellow cake*, a powder that is rich in uranium oxides. Upon completion of mining, wells are filled with concrete and cut off below the surface. Thus, mining occurs without any significant long-term visible changes to mine lands.

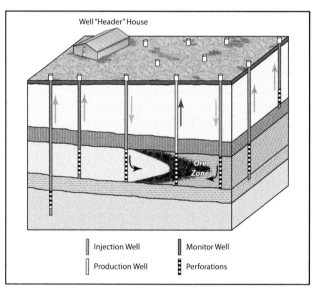

Figure 14.35   *In situ* production of uranium in Wyoming. The barrels in the bottom photo cover wells that have been drilled to intersect the uranium ore at depth. Drawing from the Wyoming State Geological Survey.

14.4.2.4   *What is the future for nuclear power?*

Many believe that nuclear power is an important energy source that we should develop further. Proponents point out that nuclear power plants do not pollute like coal power plants and do not add CO_2 and other greenhouse gases to the atmosphere. They also point out that large uranium deposits remain to be mined, and thus developing more nuclear power would lead to greater energy independence. Opponents argue that radioactive fuel is dangerous, that nuclear plants produce waste that is unsafe for tens or perhaps hundreds of thousands of years, and that many nuclear power plants have suffered catastrophic failures. Some people are also concerned that more nuclear plants could eventually lead to more nuclear weapons. Additionally, after 25–40 years of operation, radiation in reactors weakens the containment structure and plants must be shut down and decommissioned. Decommissioning is expensive, takes years to complete, and produces large amounts of radioactive waste.

The many concerns about nuclear power may be valid. More important at present, however, is that, when compared with other sources of electrical power, nuclear energy is too expensive. Expenses are involved in the mining and processing of the uranium, the construction of the facilities that meet safety requirements, the facility operations, the disposal of high-level radioactive wastes, and the dismantling of facilities after they are no longer profitable to operate. So, in the United States, economics, more than safety or environmental concerns, has stalled the nuclear industry. But economics can change, and at some point, society may have to decide whether we should develop nuclear power further.

14.4.2.4.1   Nuclear accidents

Since 1956 and 1957, when the first commercial nuclear reactors came online at Windscale, England, and Shippingport, Pennsylvania, many potentially serious nuclear accidents have occurred. Most turned out to only have minor consequences, and many went unnoticed by the news media and the public. However, three accidents stand out because they received much public attention: accidents at Three Mile Island, at Chernobyl, and at Fukushima Daiichi. These accidents have shaped public opinion about the safety of nuclear power generation.

On March 28, 1979, at Three Mile Island, Pennsylvania, an instrument malfunction showed that there was too much water in the cooling system. Engineers decreased the water flow, which caused the reactor to overheat and partially melt. Fortunately, no radiation escaped and no people were harmed, because of a safety containment vessel surrounding the reactor. Yet, because this accident occurred in a populated part of the United States and was the first such accident widely reported by major news agencies, it received much attention.

Seven years later, the Chernobyl accident (Fig. 14.36) was much more serious than what had occurred at Three Mile Island. On April 26, 1986, a reactor meltdown and several explosions destroyed a reactor at Chernobyl, which is in northern Ukraine near the border with Belarus. Airborne radiation spread over much of Europe and eventually across the entire Northern Hemisphere. Officially, 31 people died at the time of accident, but as many as 100,000 people may get cancer and die because of exposure to radiation, and many other serious health problems have been reported. Today, much of the Chernobyl area has been cleaned and decontaminated. However, water remains non-potable and valuable farmland will not be useable for more than 100 years.

On March 11, 2011, a magnitude 9.0 earthquake occurred off the east coast of Japan and generated a large tsunami (tidal wave). When the tsunami struck land,

Figure 14.36   Damaged reactor at Chernobyl, Ukraine. Chernobyl, USSR (now Ukraine), 1986.
Photo from the National Nuclear Security Administration, USDOE.

it was 10 meters high and easily went over a seawall at a nuclear power plant in Fukushima Daiichi, Japan, destroying emergency water intake systems and emergency diesel generators. The station lost power because of the earthquake and tsunami and, without emergency power, the reactor cooling system failed and cores overheated and melted, releasing radiation. The Japanese government evacuated 300,000 people from the area around the power plant, but most were exposed to very little radiation and may not suffer long-term consequences. Yet today, cancer, miscarriages, and other health effects for individuals who lived closest to the power plant are major concerns. It is too soon to determine what global impact of the Fukushima accident will be, but the radiation released from Fukushima has spread through all the world's oceans and across the globe. Several days after the Fukushima accident, Germany's Chancellor Angela Merkel, once a proponent of nuclear energy, announced that Germany would shut down all its nuclear power plants because of concerns about safety.

14.4.2.4.2   Radioactive waste disposal and storage

Disposing of spent reactor fuel and other radioactive waste may be the single largest problem facing the nuclear industry today. We classify radioactive waste as low level, intermediate level, or high level depending on how radioactive it is and the isotopes it contains. Nuclear power plants produce mostly intermediate-level wastes that must be stored and isolated for up to 10,000 years before they have decayed enough so that they no longer pose a risk to people. However, the plants also produce some high-level waste that must be isolated from people for at least 200,000 years. These numbers are large and the times are long, so disposing of radioactive waste from nuclear power plants is a big problem that perhaps has no solution. Currently, most of these wastes are stored on-site at nuclear power plants, waiting for a permanent solution.

Beginning in 2001, the U.S. government constructed a potential waste disposal repository for civilian reactor wastes deep under Yucca Mountain, Nevada. Figure 14.37 shows the repository site where, in principle, the government could store high-level radioactive waste in perpetuity. Yucca Mountain was chosen because of its arid climate, impermeable bedrock, and deep water table, which would reduce the possibility of water corroding waste canisters and releasing radionuclides into air and water. Yet, people have raised many concerns about the safety of waste disposal at Yucca

Figure 14.37   Yucca Mountain high-level radioactive waste disposal site in Nevada, USA.

Figure 14.38   Norris Geyser Basin, Yellowstone National Park, USA.

Mountain, especially considering that the waste has to remain secure for hundreds of thousands of years. Furthermore, many people worry about the possibility of truck or train accidents or acts of terrorism, while wastes are transported across the country to the disposal facility. Additionally, even if a moratorium is placed on new nuclear power plant construction, we still have dangerous wastes at many sites around the world. Proponents of Yucca Mountain argue that the wastes must go somewhere, and well-planned and guarded disposal facilities are likely to be better than the current temporary facilities, which are less carefully constructed and quickly reaching their capacities. Federal funding for Yucca Mountain, however, ceased in 2011.

14.4.3   Geothermal energy

14.4.3.1   Sources and uses of geothermal energy

Heat flow from within Earth is about half *radiogenic*, caused by decay of radioactive isotopes, and half *primordial*, leftover heat from the time that Earth formed. Thus, geothermal heat is partially nuclear and partially due to the cooling of Earth since it formed 4.6 billion years ago. Heat from the interior, radiogenic or primordial, moves by conduction and convection to the outside and then into the atmosphere. Convective heat flow involving flowing rock, magma, or groundwater is an efficient way to move heat and has concentrated heat in some places to create significant energy resources.

In active or recently active volcanic areas, magma is within a few kilometers of the surface. The magma can heat groundwater, producing geothermal steam, geysers, and hot springs, such as the Norris Basin geysers in Yellowstone National Park shown in Figure 14.38. People have enjoyed natural hot springs as spas and baths for centuries, and in some places, geothermal waters are suitable for domestic use. Natural hot water heats most homes in Iceland, and Icelandic greenhouses are heated the same way to grow fruits and vegetables. Even in non-volcanic areas, subsurface heat is often great enough to operate heat pumps that provide energy for home heating or cooling.

14.4.3.2   Geothermal electricity

Generating electricity using geothermal energy is technically and economically feasible in many places around the world. Figure 14.39 shows geothermal energy potential for the contiguous United States. Red color shows areas with the most potential, generally in the western U.S., orange colors areas with moderate potential, and yellow or gray colors areas with little potential.

In 1921, the owner of a resort constructed the first American geothermal power plant in The Geysers geothermal field, north of San Francisco, California. Major development, however, did not occur until 1960. Since then, other geothermal power plants have been constructed in the area. Today, The Geysers is the world's largest geothermal field, containing nearly two dozen separate geothermal plants that use steam from 350 wells. Together, the wells supply about 60% of the electricity for California north of San Francisco. Other huge geothermal fields, nearly as large as The Geysers, are in southern California near the Salton Sea and in the Philippines, Indonesia, Italy, Mexico, and Iceland.

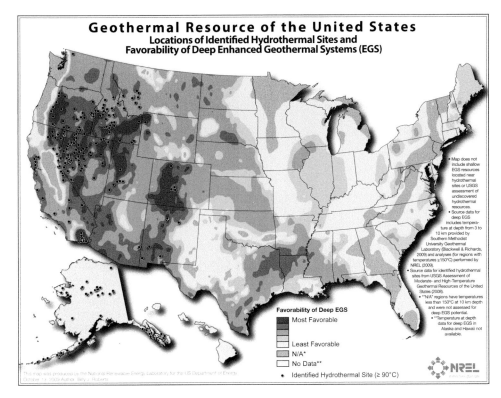

Figure 14.39 Map of geothermal potential for the lower 48 states of the United States.
Map from U.S. Energy Information Agency.

Constructing geothermal power plants involves drilling wells into heated rocks, often to depths of several kilometers or more (Fig. 14.40). Steam or water, produced by production wells, powers turbines that create electricity. The best geothermal sources produce only steam, commonly called *dry steam*, that may be used directly. Many geothermal sites, however, do not produce sufficient steam, and some only produce water. So, heat exchangers concentrate heat from hot water to produce steam at the surface. In most geothermal fields, especially if the produced water or steam is saline, cooled geothermal waters are returned to the heat reservoir, where they may sufficiently reheat to produce more steam. Besides generating electricity and heating buildings, geothermal energy is also used directly for paper manufacturing, drying food, alcohol distillation, and cleaning and drying wood.

Because rocks are generally poor conductors of heat, many geothermal power plants only operate for a few decades before temperatures and resource values decline. In recent years, for example, the geothermal energy available in The Geysers has begun to wane. Several different technological fixes, implemented since 1997, have helped maintain production and even temporarily raised electrical output, but overall heat production has dropped.

Geothermal heat sources last longer in some places than in others. An ideal geothermal site has a heat source at moderate depth, no deeper than several kilometers, and groundwater must be plentiful and in permeable rocks. Thus, in arid and semiarid places, geothermal energy may not be sustainable, because it takes a long time for groundwater to accumulate and infiltrate to replace geothermal waters removed to produce heat or electricity.

14.4.3.3 *Geothermal heat pumps*

The temperature three meters or so below Earth's surface is essentially constant throughout the year, typically 10–15 °C (50–60 °F) depending on latitude. This modest increase of temperature with depth is adequate to provide a valuable energy source. We can harness geothermal energy to heat and cool buildings using *ground source heat pumps* (GSHP), also called *geothermal heat pumps* (Fig. 14.41). GSHP systems are practical anywhere except in areas underlain by permafrost. GSHP systems involve water, or another coolant, pumped

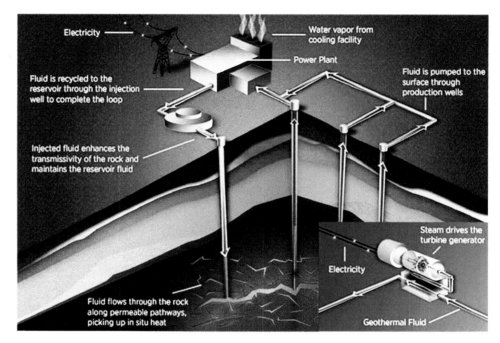

Figure 14.40   Diagram showing generation of geothermal energy.
From U.S. Department of Energy.

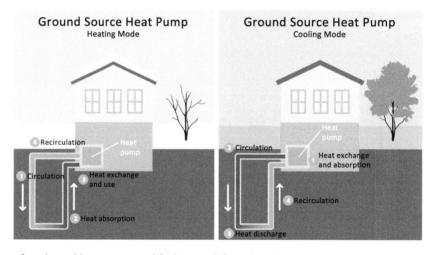

Figure 14.41   Diagrams of geothermal heat pumps used for heating (left) and cooling (right).
Drawings from the Environmental Protection Agency.

through underground pipes in a closed loop system that brings heat from depth to the surface when heating is needed and moves heat from the surface to depth when cooling is needed. The systems include heat exchangers that operate much like air conditioners or refrigerators. Some GSHP installations involve vertical boreholes as deep as 15 to 120 meters (50 to 400 feet), but a less expensive alternative is often to circulate coolant through coils of tubing buried in trenches at depths of one to a few meters (3 to 10 feet).

Installing ground source heat pumps in new homes is more expensive than installing conventional gas or electric systems, but the payback time is typically only 3 to 10 years. Additionally, state or federal tax credits often help cover costs. However, installing ground source heat pumps to heat and cool older homes is not always cost effective. Sometimes, less expensive air source heat pumps, which transfer heat between a building and the outdoors, work as well as ground source systems. They operate much like the box air conditioners that sometimes are in windows, but, unlike with air conditioners, heat flows both into and out of the house depending on need.

14.4.4 Renewable sources of energy

14.4.4.1 Hydroelectric power

Hydropower was, other than animal and human energy, the first source of renewable energy that people used. Before the Industrial Revolution, water wheels supplied energy to grind grain into flour and to saw wood into lumber. Today, large dams provide hydropower that generates electricity, and hydroelectric power is the most productive renewable energy source.

Modern hydroelectric systems include a dam that blocks, or partially blocks, the flow of a stream or river, creating a reservoir behind the dam. Water from the reservoir powers a turbine that generates electricity. For example, the Hoover Dam, seen in Figure 14.42, east of Las Vegas, Nevada, is a hydroelectric facility on the Colorado River on the border of Nevada and Arizona. Water backed up behind the dam, which creates Lake Mead, not only generates electricity, but also provides water for municipal and industrial use and for recreation. Some hydro dams also provide flood control. However, the life span of a hydroelectric dam is only about 50–200 years, because, over time, sediment collects behind dams and they eventually become useless for generating hydroelectric power. Additionally, hydropower potential is limited during times of drought or when upstream demand for water increases. For these two reasons, Lake Mead Reservoir has not been full

since 1983. Water level has dropped more than 40 meters (140 feet) since 1985, and if this continues, Hoover Dam will cease generating electricity in the next few decades. This cessation may be a key reason that the Crescent Dunes Solar Energy Project, introduced at the beginning of this chapter, was built 400 kilometers (250 miles) northwest of Hoover Dam.

Impoundment dams, like the Hoover Dam, stop all, or most, natural flow and funnel water through turbines. *Diversion dams*, also called *run-of-river dams*, only partially block river flow. Diversion dams generate electricity similarly to impoundment dams, but because they do not block the river completely, a constant flow of water continues downstream. *Pumped storage dam* systems, variants of impoundment dams, release water to generate electricity during periods of high demand and pump water back uphill into the reservoir during periods of low demand. So, during the day, when demand is highest, the dam produces electricity by allowing water to flow through turbines and power a generator. At night, when demand is low, electricity from another source is used, so the flow of water is reversed, refilling the reservoir.

Although hydroelectric power is highly efficient and releases no air pollutants, several challenges must be considered when planning a hydroelectric dam. Dams can disrupt ecosystems and human communities both upstream and downstream of dams. For example, reservoirs may flood valuable farmland, wildlife refuges, or villages. Reservoirs may also flood archaeological sites or become contaminated with agricultural wastes, sewage, and other pollutants. Below dams, river flows are severely reduced, which affects groundwater and wildlife habitats. Valuable delta wetlands may be lost because dams prevent the transport of clastic sediments to coastal areas where rivers discharge into oceans. For example, the Mississippi, the Nile, and other large river deltas are presently sinking and eroding. Dams also prevent river navigation for both people and wildlife. For example, the damming of the Columbia River of Oregon and Washington prevents salmon from easily migrating up and down the river and spawning. Consequently, salmon populations greatly declined during the past four decades. To restore the natural habitats of some rivers, the United States government has removed a few old dams and plans to remove more in the future. Still, there are 84,000 dams in the United States that, altogether, affect 600,000 miles (970,000 kilometers) of rivers—equivalent to 17% of rivers in the nation.

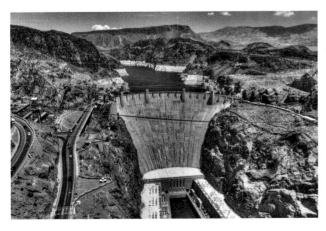

Figure 14.42   The Hoover Dam, a pumped storage dam 29 miles east of Las Vegas, Nevada.

Photo from Shawn Hartley, Wikimedia Commons.

14.4.4.2   Ocean Power: Catch the Wave!

Another way that society harnesses the power of water is by taking advantage of ocean tides. Twice each day, the oceans experience high and low tides because of the gravitational pull of the sun and moon, and the energy from these tides can generate electricity. Figure 14.43 shows the world's largest tidal turbine generating plant, the Sihwa Lake Plant in South Korea. Tidal water flow, when the tide is going either direction, turns turbines that generate electricity beneath the structure that you can see in this photo (see the inset drawing). This plant has a capacity of 254 megawatts, enough power for several hundred U.S. homes and many more homes than that in South Korea.

Tidal power generation relies on several related technologies. One involves constructing a dam across an inlet between the ocean and a bay. Others involve floating turbines or turbines on the ocean floor. When tide rises, ocean water moving toward land and through a dam can generate electricity. During ebbing tides, water flowing away from land generates electricity in the same way. Floating and seabed turbines operate in about the same way, but no dam is involved. Dams that harness tidal energy are effectively impoundment dams and thus limit water flow between oceans and bays, which may affect aquatic ecosystems. In contrast, floating and seabed turbines cause little disruption.

14.4.4.3   Wind energy

For centuries, humans have harnessed the power of wind for many purposes, including sailing ships and turning windmills that pumped water or ground grain. Today, wind energy is also used to power modern wind turbines that generate electricity. Figure 14.44 shows a wind farm near Tehachapi, California, that contains 4731 wind turbines and generates the electricity needed by 350,000 people each year. Wind power in California provides electricity for more than 2,000,000 households, or about 7% of all the electricity used in the state.

Wind velocity is the most important consideration when siting a turbine, because the energy of wind is proportional to the cube of its speed. Thus, if wind velocity doubles, a turbine will generate eight times more electricity. Additionally, the larger the rotor blades, the greater the energy produced. New designs and improved engineering, allowing rotor blade lengths to increase from about 10 meters in 1980 to nearly 40 meters today, have increased electrical production a great deal. And new blade designs with lengths of up to 75 meters will be available in the next few years, further increasing electricity production.

One tremendous advantage that wind energy has over other energy sources is that wind turbines have a very small footprint on Earth's surface. Utility companies can install turbines on farmland with almost no loss of agricultural production. Furthermore, many onshore commercial-scale wind farms are built by siting turbines along property lines and fence lines, areas that are generally undeveloped. Offshore wind farms face their own set of problems, but property lines and fence lines are not issues. Some countries, such as Denmark, get most of their electricity from offshore wind turbines.

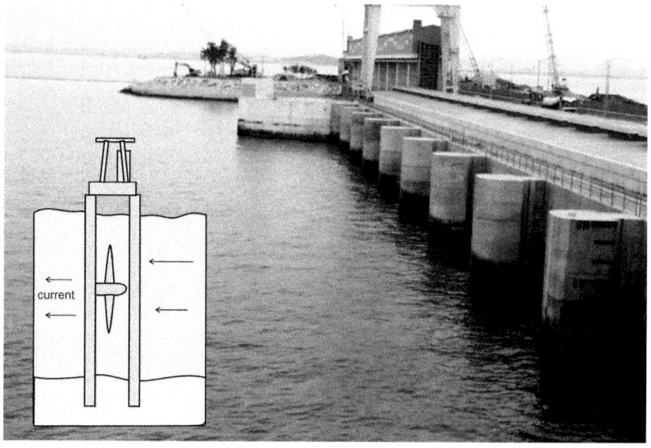

Figure 14.43   Sihwa Lake Tidal Power Station on the east coast of South Korea.
Photo from naver.com, Wikimedia Commons.

Figure 14.44   Wind farm in the Tehachapi Mountains of California.
Photo from Stan Shebs, Wikimedia Commons.

Figure 14.45 shows the average wind speed in different parts of the United States. Because of high wind speeds, the Great Plains have had the most wind energy development, but offshore regions on both the east and west coasts have potential. The Great Plains have an additional advantage: population density is low and there are relatively few competing land-use activities in many places. Today, parts of the Great Plains are centers of coal mining, but in the future, that may shift to wind farming instead.

As wind turbine technology improves, the cost to build commercial-scale wind farms continues to decrease. Today, because wind energy is cost competitive with other energy sources, utility companies and governments are building wind farms that provide nonpolluting, renewable sources of electricity. However, both on land and offshore, siting remains a contentious issue. Many people do not want wind turbines built where they live, because, among other things, they argue that the turbines are ugly and intrusive and sometimes create unacceptable noise. An additional problem facing wind development is that wind does not always blow. Many of the complaints have validity and intermittence is sometimes a problem, but the fundamental value of wind energy is not in dispute.

Overall, wind supplies about 5% of electricity in the United States, but not equally in all parts of the country. Lack of wind or insufficient wind speeds are major limiters to wind resource development. Iowa, South Dakota, Kansas, Oklahoma, and North Dakota all get more than 20 percent of their electricity from wind power, because, in those states, winds blow much of the time and at significant velocities. Other states are not so lucky, but high-voltage transmission lines export wind-generated electricity from states in the Midwest and Great Plains to areas without high wind potential. For example, electricity generated by Wyoming winds travels to California via transmission lines that run through Utah and Nevada.

Today, the most promising sites for new wind farms are probably offshore. Figure 14.46 shows wind turbines near Walney Island in the Irish Sea, but fixed turbines, such as those shown, have in some places been replaced by floating wind turbines. For large coastal cities such as Los Angeles, San Francisco, and Washington, DC, and large inland cities such as Chicago and Detroit on the

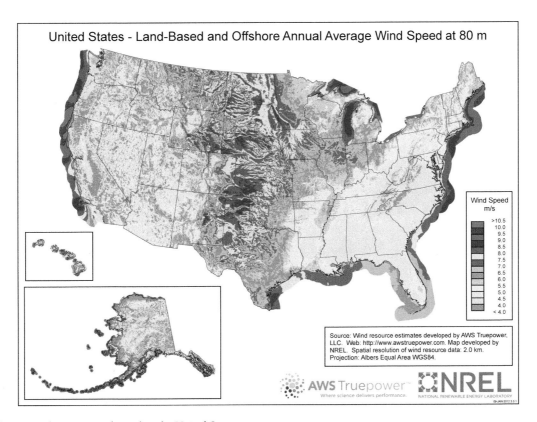

Figure 14.45   Annual average wind speed in the United States.
Map from the National Renewable Energy Laboratory.

Figure 14.46   Offshore wind turbines in the Barrow Offshore Wind Farm, near Walney Island in the Irish Sea.
Photo from Andy Dingley, Wikimedia Commons.

Great Lakes, offshore wind offers incredible potential. Especially important is that studies of offshore wind farms in Denmark and other places have found that wind development has negligible effect on aquatic ecosystems. As technology improves, driving costs down while improving production, it is likely that more cities will embrace wind energy.

14.4.4.4   Solar energy

The solar energy hitting Earth's land surface is greater than all the energy consumed by the people of the world, but it is unevenly distributed. As shown in Figure 14.47, the greatest amount of solar energy hits Earth within 40° of the equator. In the United States, solar energy is greatest in the southwestern states. South America, too, has great potential to use solar energy. But Africa, which to date has not developed many solar resources, may have the greatest potential. Although people can harness solar radiation in several ways, using this energy resource poses some challenges. In particular, the intermittent nature of solar energy, because the sun is not always shining, has meant in the past that other energy sources have always had to supplement solar energy. This problem may disappear over the next few decades as battery technology improves or innovative solar technology, like *concentrated solar power* (*CSP*) facilities discussed at the beginning of this chapter, become commonplace.

14.4.4.4.1   Passive solar heating

Humans have taken advantage of passive solar heating for thousands of years. The technology involves using architecture, windows, landscaping, and other designs to maximize solar heating in the winter and to avoid overheating in warmer months. Figure 14.48 shows some key considerations for a home in the Northern Hemisphere. For example, in North America, windows on the north sides of houses do little to aid heating, and consequently in passive solar homes they are small or absent. In passive solar homes, most windows are on the

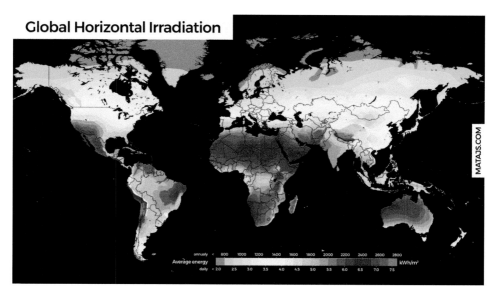

Figure 14.47   The intensity of sunshine around the world.
Map courtesy of V. Matajs, matajs.com.

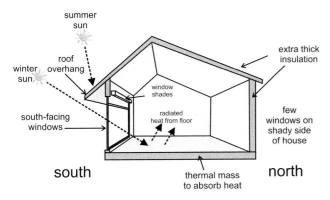

Figure 14.48 Important elements of passive solar design. Modified from drawing at energy.gov.

south side, and because the sun is higher in the summer than in the winter, roof overhangs above south-facing windows shade the windows in the summer but allow solar energy to pass through during the winter. The best passive solar homes are exceptionally well insulated and have a large amount of thermal mass, including a thick cement slab, that will store energy during the day for use at night.

Thinking about passive solar energy is not new. Standard practice for Northern Hemisphere homes built in the early 20th century and earlier was to build them with large south-facing windows and with deciduous trees planted on the south side of the home. During the winter, when all the leaves have fallen from trees, sunlight passes through windows and warms home interiors. During the summer, trees are in full bloom, blocking the sunlight and thus helping to keep homes cool. Unfortunately, those practices seem to be rare today.

Another manifestation of passive solar heating involves constructing walls that absorb heat from the sun and subsequently transfer the heat into a building. For example, because they are poor conductors of heat, adobe and thick stone walls slowly absorb heat during the (warm) day and can re-radiate it into homes in the cooler evenings and nights. And builders make some buildings with walls, rocks, or barrels of water behind windows, and inside the buildings, that absorb and re-radiate heat.

Although passive solar heating usually cannot completely replace furnaces, with proper architectural design and adequate insulation, passive solar systems can substantially decrease energy costs. Unfortunately, the widespread availability of modern heating and air conditioning systems, as well as relatively inexpensive gas and electricity, has led society to forget the architectural

logic of our ancestors. Today, most homeowners and home builders do not think about passive home design.

14.4.4.4.2 Active solar heating

Concentrated solar power (CSP) plants, such as the Gemasolar and Crescent Dunes Solar Energy Projects introduced at the beginning of this chapter, are examples of using *active solar energy*. Active solar energy technologies include CSP, photovoltaic systems, and solar water heating. The difference between passive and active systems is that passive systems use the sun's energy directly and active systems generally convert solar energy into another kind of energy. Active systems also involve transporting the energy to where people need it.

Figure 14.49 shows the basic design of a CSP plant. Thousands of heliostats, composed of mirrors, focus solar radiation, collected over a wide area, on a receiver at the top of a tall solar tower. In the past, receivers contained water that boiled to produce steam. But modern CSP receivers contain salt that melts, providing a high-temperature energy source that makes steam (red line) that ultimately spins a turbine and generates electricity. Condensed steam is recycled (blue line) and reused to generate steam and more electricity. Molten salt has high heat capacity. So, it does not cool quickly and can generate steam at night or when skies are overcast. Thus, CSP plants produce power always and under all conditions. Most CSP facilities are in hot arid or semiarid lands that receive large amounts of direct solar radiation. But, in some places, notably offshore sites near India and Japan, floating CSP solar islands produce electricity, thus not consuming valuable land space.

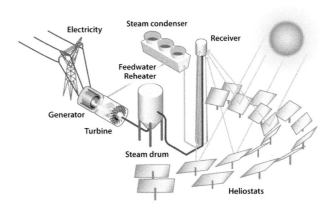

Figure 14.49 The major components in a concentrated solar plant.
Drawing from Energy.gov.

Photovoltaic technology was discovered decades ago and has been used in some specialized applications for a long time. For example, the International Space Station, shown in Figure 14.50, has been powered by photovoltaic (PV) electricity since it was launched into orbit on November 20, 1998. Sixteen large PV panels can be seen in this photo, which was taken by an astronaut in the departing space shuttle *Atlantis*. Many other Earth-orbiting satellites are powered the same way. Photovoltaic cells, also called *solar cells*, convert sunlight into electricity, using special kinds of semiconductors designed for the purpose. The fin-like panels in Figure 14.50 contain thousands of PV cells providing power to the main part of the station in the center.

Most photovoltaic systems involve an array of solar panels that each contain multiple solar cells. In some systems, the panels track (follow) the sun to maximize electrical generation at all times of the day. PV systems generate electricity for immediate use or for storage and future use. Big advantages of PV systems are that they do not pollute and they can be scaled up or down as needed. The main disadvantage is that they do not generate electricity at all times of the day, especially not in the evening and morning when demand for electricity is typically greatest.

Figure 14.51 shows part of the Agua Caliente Solar Project, built in 2014 on 2400 acres in southwestern Arizona near Yuma. Agua Caliente is one of the largest solar panel facilities in the world. The 290-megawatt project delivers electricity to California's Pacific Gas and Electric Co. and provides power for up to 230,000 homes. After hydropower and wind power, PV systems are the third most productive renewable energy sources

Figure 14.51 The Agua Caliente Solar Project near Yuma, Arizona. Photo from Energy.gov.

in the world. In the United States, the greatest amount of PV power is generated in the sunny southwestern states where businesses, many homeowners, and some utility companies have installed solar panels. But even in less sunny states such as Michigan, large arrays of solar panels are providing renewable electricity. Detroit-Edison (DTE), in 2017, completed construction of the largest solar installation in Michigan, a 200,000-panel solar farm on 250 acres of land 60 miles northwest of Detroit that can provide electricity for 11,000 homes. In China and India, PV-solar projects exceeding 1000 MW are operating, with more similar facilities under construction around the world.

14.4.4.4.3 Solar water heating

Figure 14.52 shows two solar water heaters on a rooftop in Israel. Solar water heating (SWH) technologies are relatively inexpensive and generally cost effective. In places with warm, sunny climates, solar energy can be used directly to heat hot water. In cooler places, heat-transfer liquids or gases may collect heat that is subsequently used in an exchanger to heat water. Today, more than 300,000 SWH units provide hot water to homes and businesses in the United States; countless other SWH units heat swimming pools. Typical houses can meet half to two-thirds of their hot water needs using SWH.

Many different configurations are available that, at varying costs, can provide solar water heating in different climates and latitudes. The most common arrangement involves a flat panel or box, like the ones below the tanks in Figure 14.52. The tanks attached to panels store hot water. Many solar water heaters require pumping to move water and thus consume some electricity. But some systems are passive and driven only

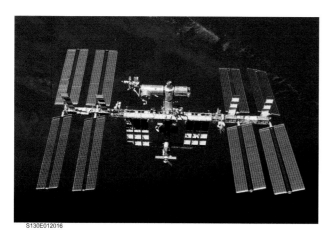

S130E012016

Figure 14.50 The International Space Station in 2009. Photo from NASA.

Figure 14.52 Photovoltaic panels and hot water tanks on a rooftop in Nachalot, Israel.

Photo from Mujadarra, Wikimedia Commons.

by convection. Besides having solar water heaters with water storage tanks, the houses in Figure 14.52 also have photovoltaic panels for generating electricity.

14.4.5 Waste-to-energy

14.4.5.1 Incineration

Municipal solid waste (MSW) composed of trash and garbage can be an energy source using any of several different kinds of *waste-to-energy* technologies. These technologies include incineration, gasification, pyrolysis, and anaerobic digestion.

Dried municipal waste may be burned in incinerators by itself or with other forms of biomass. A big reason to incinerate is that it reduces the waste volume by as much as 95%. So, although it does not eliminate the need for landfills, MSW incineration greatly decreases the space required for landfills. (However, recycling and composting are also effective ways to reduce waste volume and are less expensive than building and using incinerators.)

Sometimes waste is burned just to save landfill space, but burning can also generate electricity when the heat produced by an incinerator is used to make steam that powers an electrical generator. Municipal solid waste has an average heating value that is greater than most peat, sometimes as great as lignite, but less than bituminous coal. So, using waste incineration to make electricity is a common practice in Japan, where space for landfills is limited, and in Denmark, Sweden, and some other European countries. In Denmark, incineration accounts for 5%–14% of total electrical generation.

The environmental impacts of MSW incineration are debated today. Incinerators have high temperatures that convert solid waste into heat, ash, and flue gases, and the ash and flue gases pose potential environmental problems. The gases include CO_2, which contributes to global warming, and sulfur and nitrogen oxides that contribute to acid rain. The ash may contain heavy metals, dioxins, furans, and other toxic elements and compounds. For these reasons, most older MSW incinerators have been shut down. Many modern incinerators avoid the worst of the problems by removing hazardous materials, especially chlorinated plastics and heavy metals, from waste before combustion and by scrubbing gases going up the smokestack. Scrubbing, however, produces toxic waste, posing its own waste disposal problems. Despite the problems, some studies suggest that municipal waste incinerators with the best technology may produce electricity more cleanly than many coal plants.

14.4.5.2 Gasification

Instead of eliminating waste by incineration, a process called *gasification* can convert municipal waste of many kinds, including biodegradable waste and other organic materials, to carbon monoxide, hydrogen, and carbon dioxide. This conversion, which produces *syngas* (synthetic gas), involves reacting organic material with steam in a controlled environment at temperatures greater than 700 °C. Syngas is a better fuel than the original starting waste because it burns at higher temperatures and thus provides more energy. It can power gas engines or may be processed to produce alcohol, hydrogen, or higher-grade synfuels. Gasification, similar to what is done with MSW, is also widely used to create syngas from fossil fuels, most commonly lignite, but that process is not considered a renewable source of energy because lignite is a finite resource.

14.4.5.3 Pyrolysis

Pyrolysis occurs when biomass or other organic materials are heated in anaerobic (oxygen-free) conditions. The material cannot burn and instead converts to charcoal and combustible gases. The gases are then used to make combustible fuels that include mostly *liquid pyrolysis oil*, also called *biofuel* or *bio-oil*, and lesser amounts of syngas. Pyrolysis is sometimes used as a way to create biofuel and syngas from crops such as sugarcane that are specifically grown for that purpose. It is also sometimes used to destroy toxic organic wastes or to remove contaminants from soil. Pyrolysis is commonly the first step in gasification processes.

14.4.5.4   Anaerobic digestion

Anaerobic digestion can convert MSW into fuel just as pyrolysis does. However, anaerobic digestion takes place at low temperature and requires no heating. During this process, bacteria and other microorganisms break down organic material in the absence of oxygen. The products, generally termed *biogas*, comprise mostly methane and CO_2 with a few other minor components including hydrogen sulfide. Sometimes, instead of using waste, crops are grown specifically for biogas production using anaerobic digestion. Anaerobic digestion is also the process responsible for creating marsh gas that causes some swamps to stink and is part of the standard process used to treat municipal sewage.

Many landfills create biogas that can be collected by covering the landfill to trap the gas, typically using wells or horizontal pipe systems. Figure 14.53 (left side) shows how methane produced from rotting garbage in a landfill is captured and used as fuel. The photograph in the right side of Figure 14.53 is of a shut down landfill in Dortmund, Germany where natural gas is being collected by a hood with connected plumbing. Some landfill gases are used directly as fuel; others are processed to remove moisture, CO_2, and hydrogen sulfide to produce high-quality fuel equivalent to natural gas. In either case, the gas may power a natural gas power plant to make electricity or may be used for other commercial or residential purposes.

14.4.6   Energy from other kinds of biomass

For biologists, the term *biomass* refers to the entire mass of Earth's living organisms. For scientists who study energy issues, biomass refers to the biological remains and byproducts of recently deceased organisms, which may be important energy sources. There are many different kinds of biomass that can be used for this purpose; we have already talked about some of them. Figure 14.54 shows the major types of biomass that produce energy today. Unlike coal and petroleum, biomass is not fossilized, and some of it is fresh and green. Biomass includes manure, wood, peat, animal byproducts, forestry byproducts, and crops. These materials generally contain much moisture but, once dried, may be fuels for homes and power plants.

Biomass is classified according to its moisture content. Wet biomass includes sewage, manure, algae, and organic waste waters. Rather than drying and burning wet biomass, it is typically used to make other types of fuels. For example, anaerobic digestion can produce methane from biomass of most sorts. Pyrolysis can convert dry biomass (wood, agricultural residue, and some municipal waste) into other types of fuels or important chemicals, such as charcoal and methanol. Some toilets convert excrement, which is also a kind of biomass, directly to gas for cooking and home heating.

Fermentation of biomass produces ethanol. Ethanol, methanol, and plant oils derived from biomass, collectively called *synfuels*, are potential automobile fuels. Ethanol can also be part of *gasohol*, a mixture of gasoline and alcohol that is a gasoline substitute, and ethanol is an ingredient of *biodiesel*, which can power cars

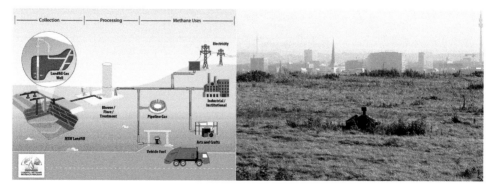

Figure 14.53   A landfill gas system.
Drawing from US EPA. Photo from Frank Vincentz, Wikimedia Commons.

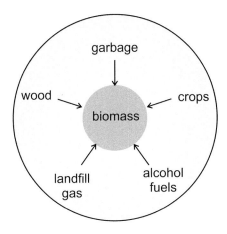

Figure 14.54 Some of the many different kinds of biomass used as energy sources.
Drawing from Energy Information Administration.

14.4.7 *Energy from hydrogen gas*

Figure 14.55 shows a car powered by hydrogen fuel cells. Hydrogen, H_2, is a fuel that we can burn to power cars equipped with special fuel cells. This technology is not new. Engineers invented fuel cells in 1838, and their first commercial use was in the late 20th century when NASA used fuel cells to generate power for satellites and space capsules. Today, hydrogen fuel cells can generate electricity and power motor vehicles and electrical appliances. If the hydrogen comes from clean, renewable energy sources, the environmental impacts of fuel cell use are small, because the only substantial byproduct of hydrogen combustion is water vapor. The heat from burning hydrogen may produce small amounts of nitrogen oxides from nitrogen in the air, but these amounts are trivial when compared with burning fossil fuels.

Today, however, the amount of energy required to make H_2 is about the same as the energy produced when the gas is burned. Hydrogen has, in the past, fueled cars in Iceland, but this use was only economical because abundant geothermal energy was available to make the hydrogen. Today, though, hydrogen production in Iceland is negligible. Nonetheless, as of 2018, people are driving fuel cell vehicles in California, Connecticut, Massachusetts, and South Carolina, where fueling stations are available. The forecast is that fuel cell technology will continue to improve and fuel cell vehicles may penetrate the global auto market. Additionally, engineers are working to develop fuel cells that

and trucks. In some parts of the United States, farmers grow corn specifically for ethanol production. The ethanol, when blended with gasoline, makes gasoline burn more cleanly and completely. Ethanol is also a fuel for some flex-fuel vehicles. Unfortunately, growing corn, fermenting it, and then distilling it to create ethanol requires a great deal of energy, so little, if any, "new" energy is created when we burn corn ethanol. Additionally, recent studies have found that the entire process of growing corn to make fuel for cars is worse for the environment than just using traditional petroleum products derived from crude oil.

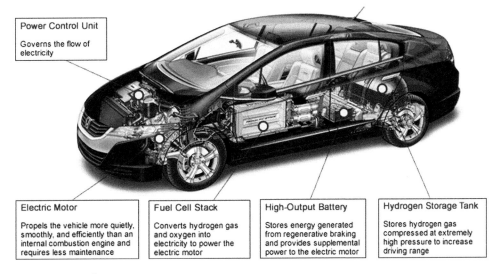

Figure 14.55 A hydrogen-powered car.
Drawing from fueleconomy.gov.

can store solar or wind energy for times when electricity from these energy sources is not directly available.

14.5   Energy sources in the future

A major distinction between different kinds of energy sources is the *energy return on energy investment*, sometimes called *energy return on investment* (EROI), or the *useful energy yield*. The EROI is the ratio of useful energy provided by a resource compared with the energy it takes us to find and make the resource. Finding and producing energy and manufacturing fuel require that some energy be expended. Additionally, once produced, some kinds of energy are more efficiently used than others are. We want energy sources with high EROI values because they provide a great deal of energy and do not require much energy to produce. In contrast, some energy sources, such as most biofuels, have very low EROI.

Figure 14.56 compares EROI values for some energy sources considered in this chapter. In the past, oil from traditional petroleum deposits yielded a great deal of energy when burned, compared with the energy expended to produce and use it. The oil was relatively easy to find and extract. It also burned efficiently, so the first oil resources had very high EROI values and, essentially, set the standard. But, by 2000, oil's EROI had declined significantly, because the world's largest and most easily produced reservoirs were in decline. So, although not shown, the EROI value for oil today is less than in the past, mostly because modern oil production requires much more energy than production did in the past. As shown in Figure 14.56, unconventional sources of petroleum, including tar sands and deposits that require fracking, provide petroleum but have very low EROI values and significantly greater environmental costs.

Relative values of EROI and big differences among EROIs for different resources are important. However, determining precise EROI values for different kinds of energy is problematic because resource qualities vary, technologies change, market values affect production, and for several other reasons. Today, natural gas generally has EROI values comparable to coal. All other major kinds of energy resources lag behind. For example, hydropower and wind power have EROI values that are 40%–60% the value of coal. Nuclear and photovoltaic resources have values less than half as great as hydropower. Newly discovered petroleum fields have lower EROI, and biodiesel and gasohol barely produce any net energy. So, as demand for energy goes up, as we develop energy sources to replace coal, and as petroleum resources continue to decline, the replacement sources of energy will not be as productive. This change will occur over time but ultimately means that the price of energy will increase.

Figure 14.57, from the Energy Information Agency, shows world energy use from 1965 to the present, with extrapolation to 2040. Today, oil and natural gas use continue to climb, but coal use is in decline. Renewable energy use, which is increasing, and nuclear power use, which is not increasing, still make up a small amount of the energy used. In the future, people will surely continue to use energy, and most of that energy will come directly or indirectly from natural sources. Growing world population and economic development mean that the demand for energy will almost certainly increase. However, we might meet the increased demand by developing new energy sources and new energy

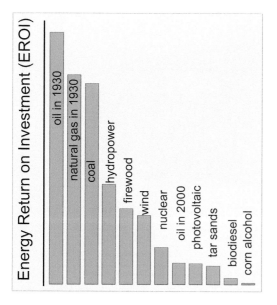

Figure 14.56   Energy return on investment for some different energy sources.

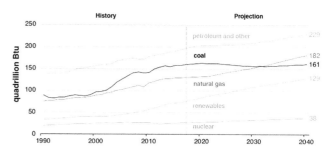

Figure 14.57   World total energy extrapolated to 2040.
Graph is from the Energy Information Agency.

technologies. Alternatively, and costing much less or perhaps nothing, we can meet some of the demand by adopting conservation practices and improving the efficiencies of energy use.

Many of the world's people use far less energy per person than do citizens of developed countries, and at least 1.5 billion people have no access to electricity at all. As the rest of the world seeks to emulate the energy consumers in more highly developed countries, demand for energy will increase. According to the Organization for Economic Cooperation and Development (OECD), an intergovernmental economic organization made up of 36 of the world's most developed countries, during the past several decades, even though energy consumption by highly developed countries has remained relatively flat, significant increased demand has come from the developing parts of the world—from countries not presently in the OECD. Part of this increase is due to improved lifestyles that include access to healthcare and education.

In the future, energy does not have to come from only one or a few types of resources, and sources of energy will evolve; in fact, over the next few decades, the energy mix used to power the developed world will likely change dramatically. The United States and other developed countries may wean themselves from coal and rely increasingly on a fuel mix that includes natural gas, nuclear energy, and renewable energy sources. The dominant renewable source of energy seems likely to be wind, because wind energy is cost competitive with natural gas and less expensive than coal in a free (unsubsidized) market. Replacing all the fossil fuels used today, however, would require many wind turbines.

Because of environmental concerns, many people in the more developed parts of the world argue for an energy mix that does not include any fossil fuels. But, this is unlikely to happen soon because fossil fuels are fundamental to our current lives and economies. And, in lesser-developed countries, coal will likely continue to be a dominant energy source because it is often locally abundant and straightforward to produce and because coal-fired power plants can be built quickly. Additionally, the OECD predicts increased consumption of petroleum products derived from crude oil as developing countries build transportation infrastructure. Although there is tremendous potential for using renewable energy in the developing world, traditional energy sources are almost certainly going to continue to be exploited first.

Questions for thought—chapter 14

1. How may renewable energy resources morph and become unrenewable?
2. When did the Industrial revolution occur and what led to its beginning? How did the Industrial Revolution change how we use energy? What new sources of energy were developed?
3. What is petroleum? What is it made of? List some of the different components and describe how they are separated. Besides fuels, what are other important uses of petroleum?
4. What biological organisms are primarily responsible for petroleum? What conditions are necessary for petroleum to develop? What is petroleum maturation?
5. Why does petroleum often migrate from its source rocks and what might keep it from reaching the surface as an oil seep?
6. What are the differences between conventional and unconventional petroleum reservoirs? Why is petroleum usually easier to extract from conventional reservoirs?
7. Why has much petroleum exploration moved offshore today?
8. Why are petroleum reserves more likely to occur in sedimentary basins rather than in mountains or Precambrian shields?
9. What are *directional drilling* and *horizontal drilling*? What are some advantages of using these approaches?
10. What is hydraulic fracturing? Why is it used today? Has it led to a significant increase in petroleum production?
11. What is the difference between primary and secondary oil recovery? Why do primary methods fail to recover most of the oil?
12. How is petroleum recovered from tar sands? What are the environmental effects of recovering petroleum from tar sands?
13. Coal forms from peat. How does peat form? Why do most dead plants fail to become peat? Why is peat mined?
14. What determines the rank of a coal? What is the difference between coal grade and coal rank? What are the four major ranks of coal? Which is most desirable, and why? There is very little anthracite mined in the United States today. Why?
15. Why are coals lower in sulfur and water more desirable than other coals?

16. Why are most power plants that burn lignite located close to the mines that the coal comes from?

17. Why do coal companies often prefer to do surface mining instead of underground mining? Why isn't all coal mined with surface methods? What types of conditions are optimal for surface mining?

18. Describe the kind of coal ming called mountain top removal. Why is it a good thing, and why is it a bad thing? It is controversial - do you think it should continue?

19. What types of environmental problems result from coal mining? Some problems are associated with mining and some with using coal. Discuss both kinds of problems.

20. Coal is abundant, so why are utility companies switching to natural gas or other alternatives?

21. Why is society in many parts of the world switching from fossil fuel energy sources to alternatives? What are some of the most important alternatives today?

22. Most nuclear power plants in the United States were built over several years during the 1970s. Why did they become less popular after that? What are some good reasons to build nuclear plants to generate electricity? (Some people say there are none!) And, what are reasons not to build any more nuclear plants?

23. How are breeder reactors different from conventional nuclear reactors? Why do some people think they are a good idea? And, why do some people think they are very bad ideas? Why did France shut down their Superphénix breeder reactor in 1998?

24. How long must high-level radioactive waste be safely stored? Why is this a serious problem? In the United States, we have a stalled plan to store nuclear waste in an underground facility at Yucca Mountain, Nevada. Why do you suppose that location was chosen? Why do you think it has never been used? What are the problems associated with not using Yucca Mountain or a similar place for nuclear waste disposal?

25. Describe the three fundamental ways that geothermal energy can be harnessed. Why do most geothermal facilities only operate for a few decades? What are the advantages and disadvantages of hydroelectric power? What are the environmental impacts?

26. What are the advantages and disadvantages of harnessing wind power? Why are the US Great Plains ideal for wind energy? Why do you suppose some North Dakotans think putting up wind turbines is a bad idea?

27. Both concentrated solar power and photovoltaic cells can be used to create electricity from sunlight? Describe the two technologies and how they are different. What are some advantages of concentrated solar power?

28. Describe some of the technologies that are used to harness solar power for heating. What is the difference between passive and active solar heating?

29. What is an advantage of using waste incineration to generate energy? What are the potential environmental problems?

30. There is a lot of corn grown in Iowa that is used to make gasohol. Why is gasohol – sometimes a small amount in gasoline, and sometimes by itself – used as a fuel in motor vehicles rather than pure gasoline? What are some reasons for not growing corn to make alcohol?

31. What are some advantages and disadvantages of using hydrogen fuel cells to generate energy? Why isn't this technology used a lot?

32. What is meant by the "energy return on investment" (EROI)? How do the EROIs of oil in the year 2000, coal, wind, solar photovoltaics, and corn alcohol compare? Why has the EROI of petroleum substantially decreased since 1930?

PART IV
Engineering Properties

15 Soil Mechanics

15.1 The Leaning Tower of Pisa

Figure 15.1 shows the Leaning Tower of Pisa (Torre di Pisa), a bell tower of a cathedral in Italy. The tower was constructed after the Republic of Pisa was victorious in some small, local wars around 1173. Construction took place intermittently, and the project was finally completed 344 years later. There are many tilted structures around the world and, although famous, the Leaning Tower of Pisa does not hold the record for (unintentional) tilt. Its tilt angle is about 4°, and the uppermost floors are offset about 4 meters from the bottom floor. However, a tower in Suurhusen, Germany, holds the Guinness World Record; it tilts at an angle that is 1–2° greater and has greater offset. Additionally, some modern construction has produced buildings that tilt (intentionally) at even greater angles.

The tower began to tip just a few years after the start of construction because the foundations for the tower were too small and the subsoil on one side of the foundations was too soft to support the structure's weight. Construction stalled for a century after that, mostly because the Pisans were in near constant war with one or more of their neighboring republics. During this time, the soil settled somewhat, which probably prevented the tower from tipping over all the way.

Construction began anew in 1272 when builders tried to correct for the tilt by adding levels that were taller on one side than on the other. So, the tower is actually curved. A decade later, construction stopped once more when the Pisans lost a major battle to the Genoans. Finally, in 1319, the top (seventh) floor was added to the tower, but it was not until 1372 that the cupola-like bell tower was completed. Additional modifications were made until around 1517.

The tower's tilt increased during the many years of tower construction and increased at a slower rate afterwards. In 1964, the Italian government, concerned about the tower's ultimate toppling, decided to take steps to prevent further tilting. For tourism reasons, though, the current tilt was to be maintained. An international task force, convening in the Azores, came up with many solutions. One idea that failed to pass scrutiny was to counterweight the tower much like a teeter-totter. Other proposals included removing the tower and installing thicker foundations connected to large

Figure 15.1   The Leaning Tower of Pisa (Torre di Pisa).
Photo from Saffron Blaze, Wikimedia Commons.

pylons. Despite the many proposals, little actual construction took place for more than 20 years.

In 1989, the Civic Tower in Pavia, Italy, collapsed into brick, mortar, and rubble, killing four people and injuring 15 people. To avoid a similar incident, engineering studies were renewed while the Leaning Tower of Pisa was closed to tourists and nearby houses were abandoned. Bells in the upper levels were removed to reduce stress on the foundations and soil, and cables were run from the upper levels to anchors around the tower. Ultimately, the tower was straightened somewhat by the removal of subsoil under one end, and the tower was reopened to visitors. More excavating and soil removal further stabilized the tower, and in 2008, engineers announced that it had stopped tilting for the first time since construction began 835 years earlier.

15.2   Soil mechanics

According to Terzaghi *et al.* (1996):

> *Soil mechanics is the application of laws of mechanics and hydraulics to engineering problems dealing with sediments and other unconsolidated accumulations of solid particles produced by the mechanical and chemical disintegration of rocks regardless of whether or not they contain an admixture of organic constituent.*

Soil mechanics is the part of *engineering mechanics* that deals with soils. It differs from solid mechanics and fluid mechanics because soils are generally much more heterogeneous than fluids or solids.

Soil mechanics, along with rock mechanics, provides the foundation of *geotechnical engineering* and *engineering geology*, subdisciplines of *civil engineering* and *geology*, respectively. Everything people construct has foundations rooted on or within Earth, and usually, this means rooted in soil. The characteristics and properties of soils have fundamental importance but are also important because they are predictors of other engineering properties, such as shear strength, hydraulic conductivity, and compressibility. So, we study the behavior and properties of soil to ensure that our constructions—such as the Leaning Tower of Pisa—are safe for day-to-day use and can withstand extreme events such as floods, earthquakes, or tornadoes. The integrities of buildings, roads, pipelines, dams, and most other human-made structures depend on understanding the properties of soils, including how soils deform and flow and how they interact with water.

15.3   Sediments, soils, and rocks

Chapters 8 and 11 discussed sediments, soils, and their differences. The term *sediment* refers to any granular material that has been formed due to natural erosion, and the general term *dirt* refers to any fine-grained unconsolidated material that comes from the ground. *Soils*, which are more complex and involve interactions between inorganic materials, organic debris, and living material at Earth's surface, are mixtures of four components:

- mineral matter (solid inorganic material including mineral and rock fragments)
- organic material
- water
- air

Soil composition is highly variable, but an average soil contains 45% mineral matter, 25% water, 25% air, and 5% organic matter, as depicted in Figure 15.2. The organic matter may contain living organisms, roots, and decaying debris (humus). The most significant factors controlling soil composition are *parent material* (the bedrock or sediment from which it formed) and *climate*. *Topography*, *biological activity*, and *time* also play important roles. Additionally, soil properties may change over long periods of time due to processes such as leaching. They may also change relatively quickly. For example, air and water, found in pore spaces between mineral matter, change proportions seasonally, or even daily, depending on precipitation and groundwater flow. Clearly, soils and their properties are highly variable and sometimes dynamic.

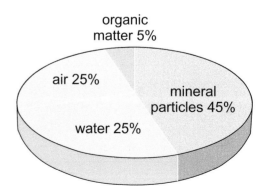

Figure 15.2   The components in soil.

This chapter focuses on the major properties of soils, but most of those properties are shared by unconsolidated sediments, and many are common for other Earth materials as well. Soil, sediment, and rocks of all types share characteristics such as volume, mass, density, and strength, and their properties have important technical and engineering significance. So, although this chapter is about soils, its implications are far ranging.

15.4 Consolidated and unconsolidated materials

Sediments, which contain loose grains, are *unconsolidated materials*. Rocks, in contrast, are *consolidated*, meaning that grains are cemented or otherwise connected to each other to provide rock strength. Rock strength is variable and some rocks, such as poorly cemented *shale*, have little strength. Shale may break and crush easily and at the surface may weather quickly to *punk rock* or *rotten rock* that has properties akin to a sediment. Often, weathered shale can be excavated with a shovel, unlike harder rocks such as *granite*. Yet, at depth in Earth's crust, shale can be a good barrier to fluid flow and can resist deformation, even considering its low strength properties. So, is weathered shale a rock or a sediment? When rocks weather at the surface, they eventually become unconsolidated material and perhaps develop soil horizons, but exactly when the transitions occur is often unclear.

When describing soils and rocks, scientists and engineers may consider an entire *exposure* (e.g., cliff, road cut, trench face, or working face of a mine) as one unit to explain the material's general properties. An aggregated description can be made, and although there may be easily distinguished layers or divisions, considering the entire unit as one mass may be adequate for some purposes. For other purposes, it may be necessary to look at smaller units. Rock outcrops, for example, often include *lithified material* and *discontinuities* between layers that cause the outcrop to not act as one unit. The discontinuities—which may be zones of weakness due to fractures, joints, faults, or bedding planes—may be of fundamental importance for road building or other engineering applications where material strength is important.

When considering unconsolidated material such as soil, we often care about the soil's overall properties rather than the properties of individual components. Each individual particle that makes up a rock or soil mass may have significant strength. If these particles are loosely connected, however, they may behave like a pile of steel marbles. Individual particles are strong and can easily resist deformation due to compression, tension, or shearing. Yet, the entire pile of marbles may have little tension or shear strength and perhaps little compressive strength. So, for many engineering applications, we are interested in bulk properties of the material. Such properties depend on what comprises the soil: mineral content, particle size and shape, density, and other properties termed *index properties*.

15.5 Soil composition and index properties

15.5.1 Soil mineralogy

Soils—mostly made of sand (relatively coarse grained), silt (medium grained), and clay (fine grained)—may have a wide variety of grain sizes, but their mineralogy is surprisingly restricted. Coarse material sometimes includes small rock fragments (from parent material) containing several minerals, but most soil grains are single minerals. Because it is significantly more stable than other minerals and survives weathering, quartz is the dominant constituent in sand and silt. Feldspar, micas, and other minerals may also be present, but generally in subordinate amounts. Clay minerals, including montmorillonite (smectite), illite, and kaolinite, are also common weathering products, but gravity, combined with running water and blowing wind, often does a good job of separating the finer-grained clays from other coarser minerals. So, fine-grained soils may contain significant clay content, but the coarser-grained soils generally do not. This distinction has engineering importance because clays may grow or shrink depending on soil water content, and the presence of clays may have significant effects on soil mechanical properties.

The term *clay* can be confusing because it refers to the specific minerals listed above, but is also used to describe very small grain size. Additionally, silt-sized material is sometimes classified as clay, although not made of clay minerals or having the requisite grain size, if it deforms under stress like many clay-rich materials. Adding more confusion, different classification systems define clay size differently. Fortunately, the distinction between clay (mineral) and clay (grain size) may not matter, because very fine grains in soil or sediment are often made of clay minerals.

15.5.2 Grain shape, roundness, and sphericity

Mineral grains, whether in soil, sediment, or rock, may be characterized by their shape and size, but the individual grains might be difficult to see. In coarse rocks such as conglomerate or coarse sandstone, grains are large enough to see with an unaided eye; for other rocks, sediments, and soils, a hand lens may be required, and a microscope may even be needed. Although individual particles or grains in a soil or rock are sometimes described as resembling a particular shape (Table 15.1), these terms are considered esoteric by many geologists and engineers who instead choose to focus on the properties *roundness* and *sphericity.*

Figure 15.3 shows some quartz grains from desert sand in Egypt. These grains are rounded—meaning that they have no sharp angular corners. So, *roundness*, sometimes called *angularity*, refers to the shapes of corners of a particle or grain. Numerical values for roundness may be calculated based on the radius of curvature of grain edges, but such values are rarely useful. So, roundness is typically described using a six-term scale:
- very angular
- angular
- subangular
- subrounded
- rounded
- well-rounded

Sphericity, a term related to roundness, refers to the degree to which a particle is *equidimensional* (has the same dimension in all directions). As originally defined, it is the ratio of the surface area of a sphere (with the same volume as the given grain) to the surface area of the grain. The theoretical maximum value of sphericity is 1.0 (for a grain that is a perfect sphere), but most grains have a sphericity value between 0.5 and 0.75. Because calculating values for sphericity is tedious, many geologists and engineers do not use numerical values but instead use relative terms—such as low, medium, or high—to describe sphericity. The grains shown in Figure 15.3 have medium to high sphericity.

Figure 15.4 contains drawings showing grains of different roundness and sphericity. Generally, when a geologist or engineer is working in the field or laboratory, shape and roundness are evaluated by comparison to reference charts, such as the one shown. In a laboratory, tools such as an *automated laser diffraction system*, which measures grain shape, size, and roundness, may be used—especially when dealing with large quantities of granular particles.

Table 15.1   Terms used to describe grain shape.

Terms	Meaning
equant	approximately the same dimension in all directions
rod-like, elongate, prolate	long in one direction
flat, oblate	thin in one direction
disk shaped (discoidal)	shaped like a disc
prism shaped (prismoidal)	having multiple sides all parallel to one direction

Figure 15.3   Subangular and rounded grains of sand.
Photo from Wilson 44691, Wikimedia Commons.

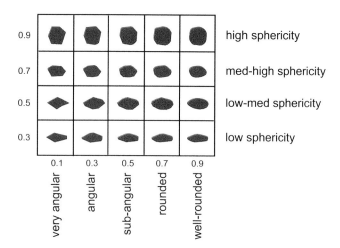

Figure 15.4   A standard reference chart used to estimate roundness and sphericity.

15.5.3 *Grain size*

15.5.3.1 *The Wentworth scale*

Several numerical scales are used today to name different grain sizes, but the most frequently used by geologists is the *Wentworth scale* (also called the *Udden-Wentworth scale*). First developed by Johan A. Udden in 1914, the scale was later modified by Chester K. Wentworth, who published it in 1922. Wentworth compared the many grain size scales used in the early part of the 20th century to determine an optimum size classification system for soils and sediment. The Wentworth scale used today is only slightly modified from his original version (Table 15.2). The scale is useful for describing grains in sandstones, siltstones, and other fine-grained clastic rocks and is applied to other granular material, whether rock, sediment, or soil. A key feature of the Wentworth scale is that it is not linear, but geometric, so the larger grades (e.g., boulder and cobble) incorporate a wider range of grains sizes than the smaller grades (e.g., silt and clay). Additional terms, sometimes added to the Wentworth scale, include *gravel* (for granules and pebbles) and *mud* (to refer to silt and clay). Other grain size scales, such as the International Organization for Standardization *international scale*, use many of the same terms that the Wentworth scale uses, but they are assigned to slightly different grain sizes.

15.5.3.2 *Krumbein phi scale*

The Krumbein phi (ϕ) (pronounced "fee") scale, a modification of the Wentworth scale, is a logarithmic scale mostly used by geologists (Table 15.2). It assigns an integer value to each of the Wentworth grades. In 1937, Krumbein observed that the geometric scale of Wentworth could be made linear by taking the logarithm of the grain sizes and also that the grain sizes in the Wentworth scale were related by powers of two. So, he calculated phi (ϕ) values for each grade:

$$\varphi = -\log_2 D$$

for D, grain diameter, in millimeters. Because the Wentworth scale is mostly used for particles of sand size or smaller, Krumbein included the negative sign in the equation to make most ϕ values positive. Because the phi scale is linear, it is easier to graph and interpolate values when using the phi scale than when using the Wentworth scale.

15.5.4 *Grain size distribution in sediments and soils*

15.5.4.1 *Grain size analysis*

Most sediments and soils contain a variety of grain sizes, and the nature of the variation can be important both to geologists trying to interpret Earth history and to engineers concerned about material properties. Historically, grain size analysis has been done using *sieves* made of metal wire cloth or perforated metal plates. Figure 15.5 shows examples. Indeed, early studies of grain size were limited by the availability of specific mesh sizes due to economic and technologic limitations. Today, sieves are typically used for grains larger than 0.075 mm, and the

Table 15.2 The Wentworth grain size scale and the phi scale.

Size of grain (millimeters)	Size of grain (inches)	Wentworth grade	Phi (ϕ) scale
> 256	> 10.1	boulder	−8
64–256	2.5–10.1	cobble	−6
4–64	0.15–2.5	pebble	−2
2–4	0.08–0.15	granule	−1
1–2	0.04–0.08	very coarse sand	0
1/2–1	0.02–0.04	coarse sand	1
1/4–1/2	0.01–0.02	medium sand	2
1/8–1/4	0.005–0.01	fine sand	3
1/16–1/8	0.002–0.005	very fine sand	4
1/32–1/16	0.001–0.002	coarse silt	5
1/64–1/32	0.0006–0.001	medium silt	6
1/128–1/64	0.0003–0.0006	fine silt	7
1/256–1/128	0.00015–0.0003	very fine silt	8
< 1/256	< 0.00015	clay	> 8

Figure 15.5 Sieves of varying mesh size.
Photo from BMK, Wikimedia Commons.

hydrometer method (based on particle density) is used to measure grain sizes in finer grain materials (0.5 μm to 0.075 mm). Figure 15.6 shows a sketch of a settling column used for the hydrometer method—the method is based on Stoke's law, which says that particle settling rate is a function of grain diameter. The equipment includes settling columns and a blender used to make slurries. Additionally, laser diffraction is used to analyze grain sizes in large samples.

For *sieve analysis*, a known volume of dried soil, with clods broken down to individual particles, is put into the top (coarsest opening size) of a stack of sieves. Shaking and tapping the sieves causes the grains to pass through some sieves until they are stopped by a sieve with openings too small for the grains. Different grain sizes collect on different sieves, and the separate size fractions can be weighed to determine percentage of each in the sample. This method works reasonably well for particles in the sand and gravel size range, but finer particles tend to stick to each other and sieving can be problematic. Water, running through the sieves along with the sample, sometimes overcomes the problems. But sediments with considerable salts and other water-soluble materials must be dry sieved. Dry sieving separates materials based on their intermediate dimension, not their largest dimension. This is why it is possible for long skinny wires to pass through a fine sieve.

A variety of sieve sizes is available; the standard ones are listed in Table 15.3 and Figure 15.7. Sediment and soil generally contain a range of particles sizes, and naming the sediment or soil is often based on determining the sieve size that lets half the material pass through. Table 15.3 compares some of the more commonly used systems. Petroleum engineers have their own naming system, as do practitioners in other fields. Some countries also have their own definitions for mesh openings.

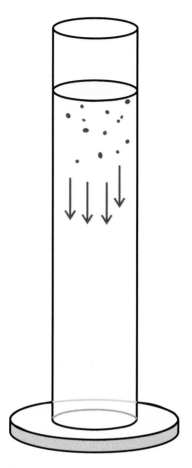

Figure 15.6 Sketch showing a settling column used to determine grain size.

Table 15.3 Sieve sizes and comparison to naming and classification systems.

Sieve number (openings per inch)	Opening size (mm)	Phi (φ) units	Name if it passes through sieve (Wentworth)	Unified Soil Classification System	British Standard System
2	9.500	–3.25		gravel	
4	4.750	–2.25	pebble		
6	3.350	–1.75			
8	2.360	–1.25			
12	1.680	–0.75			
16	1.180	–0.25			
20	0.850	0.25			
30	0.600	0.75		sand	sand
40	0.425	1.25			
50	0.300	1.75	sand		
60	0.250	2.0			
80	0.180	2.5			
100	0.150	2.75			
140	0.106	3.25			
200	0.075	3.75			
270	0.053	4.25	silt	silt and clay	silt

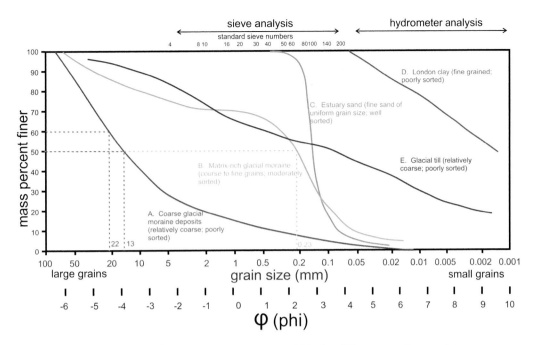

Figure 15.7 Particle size distribution curves show variation in grain sizes within five different granular materials.

15.5.4.2 Grain size distribution

Grain size distribution is a key characteristic that affects soil and sediment properties. For example, if all of a material's grains are the same size, like a box of marbles for example, the material will have great *porosity* and *permeability* and may *deform* quite easily. If grain size ranges from very fine to marble size, there may be almost no porosity and permeability and much more strength. Grain size distribution also affects other properties such as *load-bearing ability* and *expansivity*.

Once grain size data have been collected using sieves and hydrometer studies, the data are generally plotted on a semilog plot such as the one in Figure 15.7. The horizontal axis is grain size (on a logarithmic scale) and phi value (on a linear scale), and the vertical axis is the percentage of grains smaller than a given grain size. For example, in sediment from a glacial moraine (Fig. 15.7; sample A), 60% of the material is smaller than 22 mm in diameter, and 50% of the material is smaller than 13 mm. The horizontal axis for grain size uses a log scale because most sediment and soils are dominated by fine material, and the log scale distinguishes fine sizes better than a linear scale would. Note, however, that the phi values are linear across this diagram. In diagrams such as this, if curves have vertical or near-vertical sections, like the estuary sand (C) it means that the material is well sorted (poorly graded)—grains are of near-uniform size.

The five examples in Figure 15.7 all show distinctly different grain size distributions. The estuary sand (C) contains grains of near uniform size, about 0.1 mm in diameter. The London clay (D) contains only very fine material but is less uniform. The glacial till (E) is a mix of grains of all sizes. The two moraine samples (A and B) contain the coarsest material, but one is coarser than the other. A standard way to compare the sizes of two sediments, such as the two moraine sediments, is to compare the median particle diameter of each: in this case 13 mm and 0.23 mm. The finer-grained moraine sample (B) contains a great deal of fine matrix material between coarser clasts.

15.5.4.3 Sorting and grading

The degree to which a particular soil, sediment, or rock has a uniform grain size is termed *sorting*. Figure 15.8, for example, shows two sand samples. To a geologist, the sand on the left is well sorted because the grain sizes are very uniform. The sample on the right of Figure 15.8,

Figure 15.8 Examples of well-sorted and poorly-sorted materials. Photo by J. Schod.

however, has a variety of grain sizes and would be considered poorly sorted. However, geologists and engineers use different terms to describe sorting. A geologist would say that the estuary sand (C in Figure 15.7), which has nearly uniform grain size, is *well sorted*; an engineer would describe it as *poorly graded*. The coarser glacial moraine sediment (A in Figure 15.7) is *very poorly sorted* (geologist) or *well graded* (engineer). The finer glacial moraine sediment (B in Figure 15.7) is *moderately sorted* or *moderately graded*, because it contains a large fraction of material with a diameter of 0.1–0.2 mm. Sorting is important to geologists because it provides information about the energy and duration of deposition as well as about the environment of deposition (river, debris flow, wind, glacier, etc.). Gradation is also important in an engineered design because specific kinds of gradation are an indicator of important engineering properties such as *shear strength, hydraulic conductivity, packing,* and *compressibility*. Poorly graded soils and fill are desirable when good drainage is desired because they have good permeability. Well-graded materials, in contrast, with lower permeability and porosity and greater strength, provide better material to support foundations of buildings or roads.

15.5.4.4 Representative grain sizes

For some purposes, it is convenient to know *representative grain sizes*, symbolized by D_x where x is some percent value between 0 and 100. Figure 15.9 shows the representative grain size distribution for a typical soil. The horizontal axis is labeled with an exponential grain size scale and the linear phi scale. The curved line shows the mass % of material that is smaller than a particular grain size. For any soil, representative grain sizes, such as those listed on this figure, can be read from graphs of this sort. D_{90}, for example, is the diameter of a grain that is larger than 90% of the grains in a sample. D_{50} is the median grain diameter (a diameter greater than half the grains), and D_{10} is the diameter of a grain that is larger than only 10% of the sample.

15.5.4.5 Hazen approximation

One use of representative grain size values is for the *Hazen Approximation* for *hydraulic conductivity*. Hydraulic conductivity (K) approximates the velocity at which a fluid can travel through a material. The Hazen Approximation says that hydraulic conductivity of unconsolidated material can be estimated from D_{10}:

$$K \approx 100 \left(D_{10}\right)^2$$

With D_{10} in units of centimeters, hydraulic conductivity will be in units of cm/sec. The Hazen Approximation provides reasonably accurate values for many non-cohesive granular materials because hydraulic conductivity depends mostly on the size of the fine material present, which is quantified by D_{10}. Table 15.4 compares measured values of hydraulic conductivity with

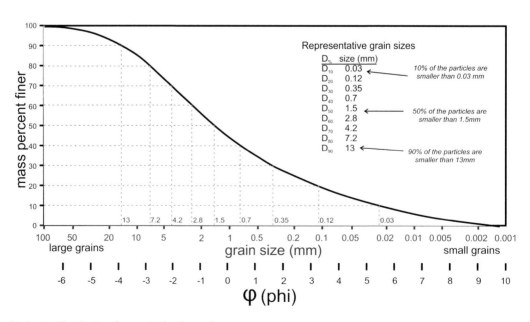

Figure 15.9 Grain size distribution for a typical soil sample.

Table 15.4 Comparing some measured values of hydraulic conductivity with those calculated with the hazen approximation.

Material	D_{10} (mm)	K (cm/s) (measured)	K (cm/s) (calculated using the Hazen Approximation)
uniform coarse sand	0.6	0.4	0.36
uniform medium sand	0.3	0.1	0.09
clean, well-graded sand/gravel	0.1	0.01	0.01
uniform fine sand	0.06	0.004	0.0036
well-graded silty sand and gravel	0.02	0.0004	0.0004
silty sand	0.01	0.0001	0.0001
uniform silt	0.006	0.00005	0.000036
sandy silt	0.002	0.000005	0.000004
silty clay	0.0015	0.000001	0.000002
clay-rich sediment (30%–50% clay)	0.0008	0.0000001	0.0000006
colloidal clay (50% < 2 μ)	0.000004	0.000000001	0.00000000001

Table 15.5 Coefficient of uniformity (C_u) and coefficient of curvature (C_c) for samples depicted in figure 15.7.

Sample	C_u	C_c
A. coarse glacial moraine	55.0	3.4
B. matrix-rich glacial moraine	4.2	1.5
C. estuary sand	2.5	2.2
D. London clay	cannot be calculated because too fine grained; D_{30} and D_{10} unknown	
E. glacial till	cannot be calculated because too fine grained; D_{10} unknown	

those calculated using the Hazen Approximation for some natural materials. Agreement is excellent for all but the finest-grained materials, which is not surprising. In fine-grained materials, *cohesion* (grains sticking together) often affects material properties.

15.5.4.6 *Coefficient of uniformity and coefficient of curvature*

Although adjectives such as *well*, *moderate*, or *poor* can be used to describe sorting (gradation), it is sometimes desirable to have numerical values. The two values most commonly used are the *coefficient of uniformity* (C_u) and the *coefficient of curvature* (C_c) (Table 15.5), calculated as follows:

$$C_u = D_{60}/D_{10}$$
$$C_c = \left(D_{30}\right)^2 / \left(D_{10}\right)\left(D_{60}\right)$$

C_u provides a numerical value for grading (sorting)—the smaller the value, the more uniform (better sorted) a sample is. On a grain size distribution plot, the curves for nearly uniform samples, such as the estuary sand (C in Figure 15.7), have steep (near vertical) sections; C_u for the sand is 2.5. The curve for the matrix-rich moraine material (B in Figure 15.7) reflects moderate sorting and has a less pronounced steep section; C_u for

that sample is 4.2. In contrast, the coarse glacial moraine (A in Figure 15.7), like most moraine deposits, is very poorly sorted and has a C_u value of 55.

If only one grain size is present, $C_u = 1.0$, and some sands, such as the Ottawa sand from Illinois, have values very close to 1. Beach sands, sorted by waves and wind, typically have values between 2 and 6, glacial till typically has values 10–40, and moraine deposits may have values approaching 1000. C_u cannot be calculated for most very fine-grained samples, such as samples D and E in Figure 15.7, because accurate grain size distribution data are generally unavailable for significant portions of the material.

The coefficient of curvature, C_c, sometimes called the *coefficient of gradation*, describes how grain sizes are distributed around the average grain size. More plainly, the value describes the general symmetry of a grain size distribution curve. Values are typically 1–3, with a uniformly (well) graded gravel > 6.0.

15.5.5 *Soil phase relationships*

Figure 15.10 shows the materials that make up a typical soil. Soils contain three fundamental types of materials: gas, liquid, and solid. These three *phases* vary in proportion depending on the type of soil. In all normal, unpolluted soils, the gas phase is air and the liquid phase is water. The solid phase may contain any of several different inorganic (mineralogical) or organic components. So, the major characteristics differentiating one soil from another are the proportions of the three phases and the composition of the solid phase. Gas and liquid occupy pore space between mineral grains. In tight, compact soils, the amount of gas and liquid may be very small, but in some clay-rich or organic soils, gas and liquid may account for more than half of the volume. The soil depicted in Figure 15.10 contains 72% solid

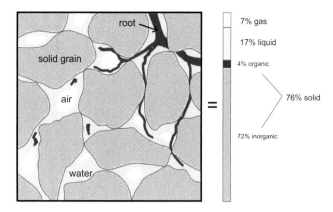

Figure 15.10   Composition of a hypothetical soil (volume %s).

Table 15.6   Void ratios of some soil types and rock.

Material	Void ratio (e)	Porosity (n) %
soft organic clay	2.5–3.2	71–76
soft clay	0.9–1.4	47–58
loess	0.9	47
poorly graded loose sand	0.8	44
loose sand with angular particles	0.65	39
stiff clay	0.6	38
well-graded compacted sand	0.45	31
compacted sand with angular particles	0.4	29
glacial till	0.3	23
intact rock	0.005–0.65	0–65

inorganic material, 4% organic matter, 17% water, and 7% air.

15.5.5.1   *Void ratio and porosity*

The *void ratio* (*e*) of a granular material is the ratio between the volume of voids (also called the *void space* or *pore space*) and the volume of the solid material. So,

$$\text{void ratio} = e = \text{volume}_{voids}/\text{volume}_{solids}$$
$$= (\text{volume}_{air} + \text{volume}_{water})/\text{volume}_{solids}$$

Soils that are well sorted (poorly graded) have high void ratios due to unfilled space between grains. Soils rich in organic matter or clay also have high void ratios because organics and clays have structures that give soils a fabric with a lot of porosity. Void ratios for some low-density organic-rich soils may be 5 or greater. Void ratios for most granular soils, especially those that are poorly sorted, are lower (compare the values in Table 15.6). Void ratios for solid rock vary greatly depending on rock type.

Void ratio is an important property for a number of reasons. For example,

- If void ratio is high, a soil will often have both high permeability and high porosity. If it is low, the opposite is true.
- If the void ratio is high, particles can move around. If low, they cannot, especially in poorly sorted materials.
- If the void ratio is high, the ratio may decrease significantly under loading as grains move closer together. If the void ratio is low, loading may lead to deformation of solid particles.

Porosity is closely related to void ratio but calculated slightly differently:

$$\text{porosity} = n = \text{volume}_{voids}/\text{volume}_{total} \times 100\%$$

and porosity can be related to void ratio:

$$\text{porosity} = n = e/(1 + e)$$
$$\text{void ratio} = e = n/(1 - n)$$

Although percent is a standard way to report porosity (Table 15.6), it is sometimes reported as a ratio or decimal instead. Some petroleum engineers and geologists use the Greek symbol phi, φ, for porosity, the same symbol used for a strength parameter of materials known as the *internal angle of friction*, and also for the Krumbein phi scale for grain size. To avoid confusion, most other engineers and geologists use *n* for porosity instead.

The original porosity of a rock, sediment, or soil depends on several factors—including grain size and shape, sorting (grading), and how grains are packed together. Additionally, porosity may increase or decrease due to diagenesis or fracturing. Values for void ratio or porosity do not give any information concerning the origin of the porosity, variability of pore sizes, or pore connectivity.

Porosity and permeability both depend on opening between grains but do not always correlate. A bucket of marbles, for example, may have the same porosity as a piece of Swiss cheese, but the marble bucket has high permeability and the cheese does not. Many carbonate rocks have high porosity due to vugs formed by dissolution. Often, however, they have quite low permeabilities. Porous materials only have high permeabilities if their pores are well connected. So, some engineers, especially petroleum engineers, define different kinds of porosity, such as:

- *total porosity*—as defined above
- *primary porosity*—porosity of the rock resulting from original depositional processes
- *secondary porosity*—porosity resulting from diagenesis
- *effective porosity* (also called *connected porosity*)—ratio of the connected pore volume to the total volume
- *fracture porosity*—porosity resulting from fractures in the rock at all scales
- *vug porosity*—porosity associated with vugs, commonly in carbonate rocks

15.5.5.2 Effective porosity

Soil or other material may have high porosity, but unless the pores are connected, the porosity may not lead to high rates of fluid flow (high hydraulic conductivity). Additionally, fractures may increase the void space in a given sample, providing non-pore space for water and air to occupy. So, engineers define *effective porosity* (generally only for rocks, not for unconsolidated material), which is the porosity that contributes to fluid flow. Effective porosity is a very important property when considering production of oil or water from wells, because it does not matter what the total porosity is if there is no flow. For example, the material depicted in Figure 15.11 has much interconnected pore space, but the fracture ends create stagnant zones with little to no fluid flow. These zones may significantly affect important properties such as reservoir production, and the effective porosity may be considerably less than the total porosity.

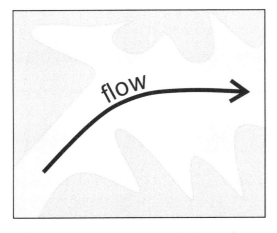

Figure 15.11 Example of a material with high overall porosity relatively slow rates of fluid flow.

15.5.5.3 Saturation and water content

A material (soil, sediment, or rock) is *saturated* if all available voids are filled with water. If a soil contains more water than needed for saturation, it is *oversaturated*; there is so much water present that some grains are held apart by water between them. If soil contains less water than needed for saturation, the material is *unsaturated*. The percentage of water volume compared to the maximum possible is the *degree of saturation* of the soil or rock mass:

$$S_w = \text{volume}_{water} / \text{volume}_{voids} \times 100\%$$

The degree of saturation is called the *total saturation* if the total porosity of the material is due to pore space only. It is called the *effective water saturation* if the effective porosity differs from total porosity because fractures or some other void space may add potential water storage volume to the material.

A completely dry soil has a saturation of 0%. A fully saturated soil (water in all available space) is 100% saturated. Natural systems rarely reach these extremes; (very) small amounts of water are found at shallow depth even in some of the most arid environments, and a small amount of trapped air in sediments and soils is usually associated with wet environments.

The *water content* (sometimes called the *moisture content* or, for soils, the *soil moisture*) is the ratio of the mass of water to the mass of solids of a given material:

$$\text{water content} = w = \text{mass}_{water} / \text{mass}_{solids} \times 100\%$$

Because water content is calculated as a ratio of the mass of water to the mass of the *solid portion* of the material, instead of to the *total* mass of the material, water content can have values greater than 100%. Organic-rich clays may have water contents greater than 300%—which means that the mass of the water is three times greater than the mass of the mineral material present. For most purposes, water content is calculated using mass (not volume) ratios, even though saturation is calculated by using volumes. Determining water content is not difficult. A sample is weighed, dried, and weighed again to determine the total weight and the weight of water present.

15.5.6 Atterberg limits

Atterberg limits, shown in Figure 15.12, describe behavior of fine-grained soils in response to changing water

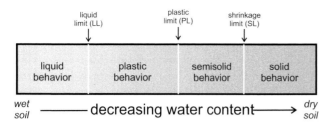

Figure 15.12 Atterberg limits for soil.

Table 15.7   The four possible states of soil.

State	Characteristics
liquid	Soils behave as thick muddy water or a slurry; they may be very viscous but flow like a liquid. Soil volume changes in response to changes in water content.
plastic	Soils change shape but not volume continuously under the influence of stress (pressure) and hold their shape when the stress is released. Soil volume changes in response to changes in water content.
semisolid	Soil does not deform plastically; instead, it crumbles or breaks apart in response to stress. Soil volume changes in response to changes in water content.
solid	Soil is unsaturated and may be crumbly. Changes in water content cause no change in volume.

content, in particular specifying the degree of saturation that causes a soil to change its properties significantly. These limits are important because

- saturated soils swell and shrink in response to changes in water content.
- soil strength is less for very wet soils than it is for dry soils.
- very wet soils may flow like liquids; drier soils do not.

The properties of soils change significantly when water is added or taken away. Very wet soils behave similarly to a liquid. As shown in Figure 15.12, when they dry, they eventually reach the *liquid limit (LL)*, and soil behavior changes from liquid to plastic. Further drying and shrinking takes the soil to the *plastic limit (PL)*, where behavior changes from plastic to semisolid. With more drying, the volume and water content decrease until the *shrinkage limit (SL)* is reached. After that, the soil continues to dry and the water content continues to decrease, but the soil volume remains constant. So, depending on soil moisture, a soil will be in one of four *states*: liquid, plastic, semisolid, or solid (These terms are defined in Table 15.7). Liquid, plastic, or semisolid soils are supersaturated with water; solid soils are unsaturated.

The three moisture contents that correlate with change in soil properties—the *shrinkage limit*, the *plastic limit*, and the *liquid limit*—are called the *Atterberg limits*. They can be measured relatively rapidly by inexpensive testing. These limits have numerical values, termed *gravimetric water contents* (GWC), which are the percentages of water compared with the original mass of dry soil. Thus, a GWC of 100% means the weight of water in the soil is the same as the weight of the solid material. A GWC of 150% means that the weight of water is 50% greater than the weight of the dry soil. Water content (GWC) values may range from 0% to very large values over 100%, but most soils have a *liquid limit* with GWC below 100% and a *plastic limit* with GWC below 40%.

The difference between the liquid limit and the plastic limit is the *plasticity index*:

plasticity index = PI = LL – PL

The primary use of the plasticity index is to quantify plastic behavior and allow soil classification with the *Unified Soil Classification System*. Physically, the plasticity index is the range of moisture content over which the soil remains plastic. This is practical information for an engineer on a construction site when compacting soil or remolding it in any way, and it assists in understanding the expansive properties of the soil. Soils with low PI values tend to have less cohesion and less clay binding the soil matrix together. However, soils with large PI values may indicate that they are expandable when wetted. A volume increase in response to wetting can cause stress in the soil and on surrounding infrastructure.

15.5.6.1   *Liquid limit*

Figure 15.13 shows the original method (the *Casagrande test*) used to determine the liquid limit (LL), the minimum water content at which a soil's behavior changes from plastic to liquid and the soil begins to flow. A Casagrande test involves placing wet soil on a dish of a specialized device, making a groove through its center, and then banging the dish to see how fast the groove would close. Figure 15.13 shows a sample prepared for analysis in this way. The liquid limit is defined as the moisture content when it takes 25 standardized bangs to close a groove that is 13.5 mm across. LL is below 100 for most soils. The Casagrande test is still widely used today, but other methods are available.

Figure 15.13 A soil sample prepared for a Casagrande test.
From E smith2000, Wikimedia Commons.

Figure 15.14 Determining the plastic limit of a soil sample.
Photo from L. Yarbrough.

Figure 15.15 A soil whose moisture content on the surface is less
than the shrinkage limit.
From Plogeo, Wikimedia Commons.

15.5.6.2 Plastic limit

Moist soils have high cohesion and behave more like gooey masses than do dry soils; this behavior is the basis for the *plastic limit*. The plastic limit (PL) is defined as the minimum moisture content for a soil to behave as a plastic material and deform without crumbling. PL value is below 40 for most soils. The plastic limit is determined by rolling out a thread of the fine portion of a soil. Figure 15.14 shows this being done. If the soil is plastic, the thread will hold together down to very small diameter. If the moisture is less than the plastic limit, the thread will crumble. To determine the plastic limit, an initially plastic sample is rolled and re-rolled multiple times, which causes it to dry. Eventually the plastic limit will be reached—defined as the moisture content at which a long thread of 3.2 mm (1/8 in.) diameter just begins to crumble. So, the rolling is repeated multiple times until crumbling occurs, and then soil moisture content is measured. A soil is considered *non-plastic* if a thread cannot be rolled out down to 3.2 mm at any moisture content.

15.5.6.3 Shrinkage limit

Figure 15.15 shows a soil that cracked, forming mudcracks, as it dried. Desiccation cracks, or mudcracks, form when the surface of a fine grain granular soil dries more quickly than underlying layers. The shrinkage limit is the moisture content when the soil can dry no further without cracking. The cracks form due to low tensional strength of the soil. The water content of a cracking soil, like the soil shown in Figure 15.15, is less at the surface than the shrinkage limit. Water content at depth, however, is generally greater than at the surface.

In oversaturated soil, water occupies space between mineral grains, and if the soil dries, it shrinks (loses volume) without cracking as water is given off. The water content of the soil at which no further volume change occurs with further loss of moisture is the *shrinkage limit* (*SL*). The shrinkage limit is the moisture content of fully saturated soil (S_w = 100%) at just the point at which individual grains are touching, so removal of more water causes no more shrinkage and may lead to cracking instead (Fig. 15.15). One method to determine the shrinkage limit is to add water to a soil to make a slurry and then to place the slurry into a small mold and allow it to dry slowly. After a time, depending on the temperature and humidity, the sample will reach the point at which it begins to form surface cracks. This point is the shrinkage limit.

15.5.6.4 Density

The composition of a soil may be described in terms of volume or mass of its components (Fig. 15.16). Typically, those components are air, water, and solid material. Sometimes the solid material may be further broken down into different kinds of materials. For some purposes, soil composition is expressed in terms of mass fractions (or percents) of each component, and for other purposes, volume fractions are used. The two are not the same.

Figure 15.17, on the left side, shows the composition of a hypothetical soil—equivalent to a good agricultural soil—containing 46% minerals, 26% water, 24% air, and 4% organic material by volume. The right side of the drawing shows composition by mass %. The density of common soil minerals is 2.6 times that of water, 3.5 times the specific gravity of organic matter, and over 2000 times the specific gravity of air (Table 15.8). So, when the composition of the hypothetical soil is converted to mass %, the value for minerals increases significantly, the values for water and organic content decrease, and the value for air becomes minuscule and doesn't show on the right-hand pie diagram. The *density* of the hypothetical soil,

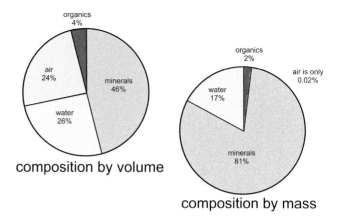

Figure 15.17   Comparisons of volume % and weight % for a hypothetical.

Table 15.8   Densities of soil components and a model soil.

Material	Density (g/cm³)
soil minerals	2.65
water	1
organic matter	0.75
air	0.0012
soil (containing 46% minerals, 26% water, 24% air, and 4% organic material by volume)	1.51

more properly termed the *mass density* by engineers, is about 1.51 gm/cm³:

$$\text{mass density} = \rho = \text{total mass/total volume}$$
$$= M_{total}/V_{total}$$

In science and engineering, common units for mass density include grams per cubic centimeter (g/cc or g/cm³), grams per milliliter (g/ml), and kilograms per cubic meter (kg/m³). Engineers may also use English units, such as pounds/square foot (lb/ft²).

When considering the properties of soils, especially of fine-grained soils, absolute density is often not as significant as relative density. *Relative density*, the ratio of in-place (natural) density to the maximum possible density of a material, is a measure of what could happen to a soil under certain conditions. It is defined as:

$$\rho_r = \frac{\left(e_{max} - e_0\right)}{\left(e_{max} - e_{min}\right)} \times 100$$

Where e_{max} is the void ratio of a soil in its loosest (least dense) condition; e_{min} is the void ratio in its densest condition; and e_o is the void ratio in its natural state. The

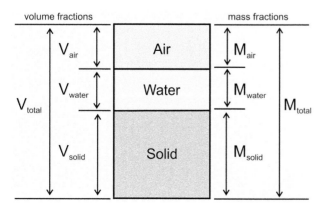

Figure 15.16   Volume fractions and mass fractions of soil.

relative density is an indication of possible increases in density or compaction when a load is applied to soil. Compaction of soil under load is referred to as *settlement*. Settlement is important because it causes an increase in soil density, a loss of void space, a loss of water, a disruption of foundations and other structures, and an increased chance of flooding.

Some scientists and engineers prefer to talk about *specific gravity* (S.G., SG, s.g., or G) instead of density. Specific gravity is the ratio of density of a material to the density of a reference material, typically water:

$$\text{specific gravity} = SG = \rho_{sample} / \rho_{water}$$

Because water has a density of about 1 gm/cm^3 at Earth-surface conditions, the density of any substance in gm/cm^3 is about the same numerically as its specific gravity relative to water.

Some engineers, especially petroleum and mining engineers, report values for and do calculations using *specific weight* (γ), also called *unit weight*, instead of density or specific gravity. It is calculated by multiplying density by the acceleration of gravity and, because gravity is not the same everywhere, it is not an absolute value. It depends on location:

$$\text{specific weight} = \gamma = \rho g = \text{mass/volume} \times g$$

The specific weight of water, a commonly used engineering value, may be calculated in kilonewtons per meter squared (kN/m^2) or pounds per foot squared (lb/ft^2) as follows:

$$\gamma_{water} = \rho_{water} \times g = 1 \text{ g/cm}^3 \times 9.8 \text{ m/s}^2 \text{ and correcting}$$

units:

$$\gamma_{water} = 9800 \text{ kg m/s}^2 = 9.8 \text{ kN/m}^2 = 62.4 \text{ lb/ft}^3$$

Other materials have densities and specific weights that may be several times that of water (Table 15.9).

15.6 Soil classification

Several soil classification systems see widespread use today. They provide a way for engineers, soil scientists, and others to describe various kinds of soils that may be appropriate for different uses. These systems reflect different purposes and provide different information. The most widely used systems in North America are the *United States Department of Agriculture (USDA) Soil Classification System*, the *USDA Soil Taxonomy*, and the

Table 15.9 Densities, specific weights, and specific gravities for some common soil materials.

Material	Mass density (ρ) kg/m^3	Specific weight (γ) lb/ft^3	kN/m^2	Specific gravity (SG) unitless
water	1000	64	9.8	1
basalt (rock)	2900	185	28.44	2.9
bentonite	593	36.6	5.82	0.59
fine dry gravel	1682	103	16.5	1.68
dry sand	1602	98.9	15.71	1.6
wet sand	1922	119	18.85	1.92
dry clay	1089	67.2	10.68	1.09
wet clay	1826	113	17.91	1.83
fresh water	1000	62.4	9.81	1.00

Unified Soil Classification System (USCS). Some government agencies in the United States have established their own classification systems, as have many foreign governments and agencies. The USDA Soil Classification System and Taxonomy were discussed in the Chapter 11. Here, we look at the USCS system because it has more implications for soil properties and engineering.

15.6.1 The USCS

Engineers desire to classify soil according to engineering properties that affect the soil's ability to act as building material, road bedding, foundation support, or for other purposes. They also desire classification systems that are easily used in the field. The Unified Soil Classification System, first developed in 1948 for military construction projects and revised subsequently for broader uses, meets both needs. The system, which can be applied to just about any unconsolidated material, is focused on engineering properties and applications. It may not be useful to fully describe a soil for soil science, agricultural, or some geological projects.

Figure 15.18 shows the USCS system. It is divided into two parts, coarse-grained soils on the left and fine-grained soils on the right. The USCS system considers both grain size and texture, recognizing that, for engineering purposes, coarse-grained soils are best classified based on grain size distributions, while the properties of fine-grained soils are primarily related to their plasticity. *Coarse-grained materials*, those with more than half the material retained by a No. 200 (< 0.075 mm) sieve, are further divided into *sands* and *gravels*. Each soil type is represented by a two-letter symbol with specific meaning—they are defined in Figure 15.18.

Fine-grained materials (the right-hand side of Figure 15.18), those with less than half the material

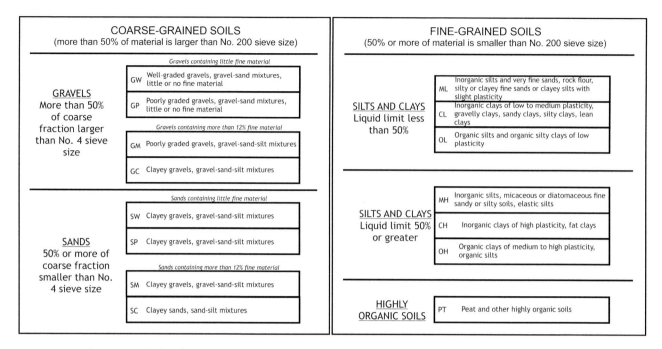

Figure 15.18   The USCS Soil Classification System

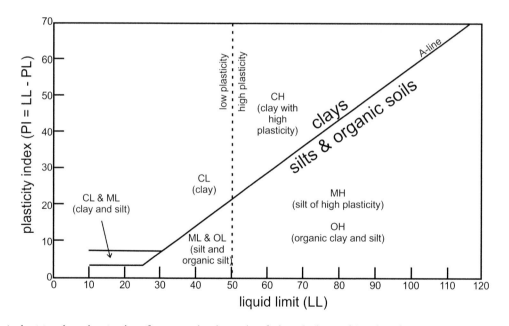

Figure 15.19   A plasticity chart showing how fine-grained soils are classified on the basis of Atterberg limits.

retained by a No. 200 sieve, are further divided into high or low plasticity categories based on Atterberg limits. Figure 15.19, called a plasticity diagram, better shows how this is done. A vertical line (dashed) divides soils into those that have low plasticity and those that have high plasticity. And, a diagonal line, called the A line, separates clays from silts and organic soils.

15.7   Soil strength

15.7.1   Stress

The safety of many engineering endeavors depends on soil strength, because if soil fails, any structures built on or in it will fail too. To evaluate soil strength, engineers must be able to predict the effects of forces acting on a soil and determine whether a soil will support any proposed activities. When a force is applied to a volume of soil, rock, or any other material, the effects depend on how intense the force is. A moderate amount of force applied to a very large surface may cause insignificant effects, but the same amount of force applied to a very small area may cause major deformation. The key consideration is ratio of force to area, called *stress*:

stress = force/area

When enough stress is applied to any material, it will deform. The amount of deformation caused by stress is termed *strain*. Stress, strain, and the relationships between them are important considerations in many geology and engineering subdisciplines, especially *geomechanics*, a specialized area of engineering involving application of classical mechanics to geological materials.

Stress-strain relationships, which are different for different kinds of soils or other materials, are commonly depicted as a curve on a stress-strain diagram, such as the one in Figure 15.20. The shape of the curve shown is typical for a soil subjected to shearing forces but is also similar for stress of all kinds. As seen in this figure, when stress is low, the soil deforms gradually—the more stress, the more deformation. Under these conditions, soil strength plays a small role resisting deformation but cannot stop all deformation. Continued stress causes deformation to increase until it reaches a peak value and soil strength is exhausted. At this point, the soil cannot handle any more stress. So, it deforms rapidly to reduce stress levels, eventually reaching a critical state at which it flows and behaves like a liquid and stress remains constant.

Stress is closely related to pressure, another kind of force that may act on a material. Both are measured in the same units, typically newtons per square meter (N/m^2 = Pascal) or pounds per square inch (lb/in^2 = psi). Pressure increases with depth in Earth, and we use the terms *lithostatic stress* or *lithostatic pressure* to refer to pressure applied to a mass due to the weight of overlying material. Lithostatic stress, like other kinds of pressure, is a scalar (non-directional) quantity; the forces of lithostatic stress are the same in all directions.

Figure 15.21 depicts the different kinds of stress that may affect soil. *Directed stress*, in contrast with lithostatic stress, is a vector (directional) quantity and can be one of three types: *compression* (*compressional stress*), *tension* (*tensional stress*), and *shear* (*shear stress*) (Fig. 15.21). Compression and tension (both symbolized by the Greek letter sigma, σ) act perpendicular to surfaces of a mass and so are called normal stresses; they cause compression or expansion in one direction. Shear stress (symbolized by the Greek letter tau, τ) occurs when differential stress is applied to a mass and causes sliding or slippage along generally planar surfaces.

When subjected to lithostatic stress, soils are compressed into smaller volumes. When subjected to tension, soils, unlike rocks, simply pull apart because they have little tensile strength. When subjected to compression or shear stress, soils initially strain and deform *plastically* (meaning they will hold their shape if the stress is removed). If stress and strain continue, the soil will eventually reach its peak shear strength—its critical state—and subsequently behave as a fluid (Fig. 15.20). In effect, the soil strength is exhausted by all the stress

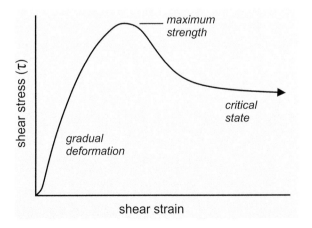

Figure 15.20   Stress-strain relationships for a typical soil.

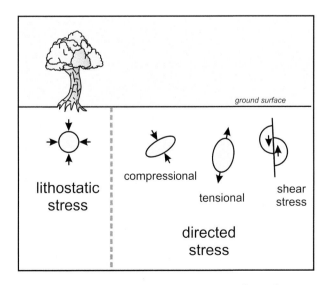

Figure 15.21   Different kinds of stress that may affect soil.

and the soil gives up. This process is irreversible because the soil structure that once gave soil its strength no longer exists.

15.7.2 Pressure, stress, and depth

Directed stress is most common at or near Earth's surface. For instance, gravity causes directed stress on slopes, pulling more strongly downhill than in other directions. Humans are also good sources of directed stress through excavations and other engineering projects. Deeper in Earth, however, most stresses are lithostatic, or nearly lithostatic. Scuba divers know that pressure increases as they go deeper, increasing by 1 atmosphere every 33 feet. So, at a 33 foot depth, the total water pressure is 2 atm (1 atm air pressure at the surface plus 1 atm due to weight of water); at 66 foot depth it is 3 atm; etc. Most soil and rock material is denser than water, so lithostatic pressure increases more quickly with depth in solid Earth than water pressure does under oceans—about three times more quickly, because rock material averages about three times the density of water.

Generally, we think of soil or other material deforming *isotropically* (the same in all directions), so the weight of overlying material causes a general increase in lithostatic pressure and compaction with depth. If, however, the material is *anisotropic* (having different properties in different directions) and has enough strength, deformation may not be uniform in all directions and directed stresses may be present. In rock or soil containing water, the water is under *hydrostatic pressure*, generally very close in value to lithostatic pressure. Large amounts of water often cause pressure to equalize in all directions and so may minimize directed stresses.

15.7.3 Soil compaction and consolidation

Soil density naturally increases with depth because of compression due to increased lithostatic pressure, but humans can cause density increases too, for example, by driving heavy machinery over loose soil or even just by walking on it. In farm fields, continued tilling can produce *tillage pans*, hard compacted layers just below the depth of tillage. Limited crop rotation also commonly leads to compaction, because crop roots access the same horizons in the same ways every growing season and because limited cropping often correlates with more early-season use of heavy equipment.

Figure 15.22 shows what happens when lithostatic or directed stress causes a soil to compact. Pores, formerly occupied by air or water, disappear as the soil loses pore space. Soil particles may become flattened and may align with one another, thus reducing permeability. At the same time, the soil may become stronger and more stable. In the narrowest terms, *compaction* refers to the increase in soil density when air is forced from pore spaces around soil particles. *Consolidation*, a closely related process, refers to density increases due to water being forced out. Often, however, the term compaction is used to refer to both processes, and often the two processes are inseparable. Both compaction and consolidation contribute to soil settlement.

Soil compaction is essential for many engineering purposes. Among other things, compaction
- increases soil strength
- decreases permeability
- reduces settlement of foundations
- increases slope stability of embankments.

So, intentional soil compaction is often a necessary first step for many construction projects involving foundations, roads, walkways, earthen dams, or other containment structures. It is especially desirable in areas that have been filled or backfilled to provide land for development. Proper compaction, for example, might have stopped the Leaning Tower of Pisa from leaning.

Although compacted soils offer some engineering benefits, compaction of agricultural soils is generally considered bad because it reduces crop yields. Large pores are needed for good circulation of soil water and air—essential for healthy plant growth—and compaction removes large pores quickly. Compaction also means less infiltration of rainwater, reducing the amount of water available to support plant growth, decreasing the rate of groundwater recharge, and sometimes leading to more soil erosion at the surface. In compacted soils, roots have a hard time propagating, and burrowing

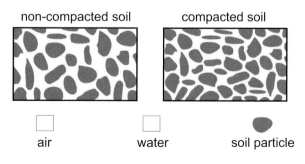

Figure 15.22   Non-compacted and compacted soils

animals, including earthworms and desirable insects, have a difficult time burrowing.

Once compacted, some soils recover naturally if left alone. For example, soils that contain lots of expandable clays may regain their original structure as the clays shrink and swell with varying moisture supplies. In some climates, freeze-thaw cycles help loosen soil, but the effects are near the surface only. Many soils, however, including some containing non-expanding clays, will never recover without human, or possibly burrowing animal, interventions to loosen them up. But a practice like tilling, which can be effective, is a two-edge sword because it can also destroy soil structure.

15.7.4 *Shear strength*

Figure 15.23 shows a slope where a landslide might occur and cause slope failure. Gravity is directed downward, not parallel to the slope. A portion of the gravitational force—called the *shear stress*—will cause sliding if the material that makes up the slope is not strong enough. *Shear strength* is the greatest amount of shear stress that a soil (or rock) can maintain before it gives way and shears (slides or fractures). It is an important property that determines, among other things, whether a landslide will occur and the safety of structures built on soil.

As seen in Figure 15.24, gravitational force can be divided into two components, the shear stress (τ) and the normal stress (σ). The shear stress is parallel to the slope and promotes sliding; the normal stress is perpendicular to the slope and is responsible for friction between layers that resists sliding. If the total gravitational force is G, then

$$\text{shear stress} = \tau = G \sin(\theta)$$
$$\text{normal stress} = \sigma = G \cos(\theta)$$

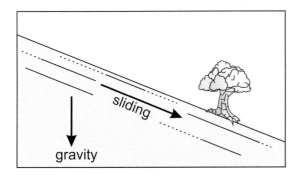

Figure 15.23 Gravity creates shear stress that can lead to landslides.

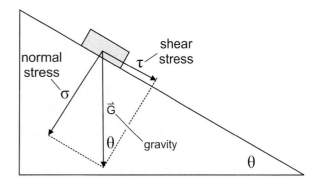

Figure 15.24 Gravity produces both shear stress that causes sliding and normal stress that resists sliding.

Figure 15.25 Sand piling up at its critical angle of repose.
Photo from Peter Craven, Wikimedia Commons.

The relative magnitudes of the two stresses depend on the steepness of the slope. These relationships tell us that for shallow slopes (small θ), shear stress is low and normal stress is high. So, friction is high and slippage is unlikely. For steep slopes (large θ), shear stress is high and normal stress is low, so slippage will occur.

Figure 15.25 shows sand piles being created. An angle, called the *critical angle*, is the lowest slope angle at which sliding of sand grains will occur. If we add sand to the top, it will naturally slide down the sides creating a conical shape. We could do this 100 times and the result would always be the same. The angle of the slope is the *critical angle*, also called the *angle of repose* (Fig. 15.25). For dry material, the angle of repose increases with grain size but usually is between 30° and 40°. The angle of repose for a dry, well-sorted sand is approximately 34°, whether it is poured on a table or piled up outdoors (Fig. 15.25).

Different soils have different shear strengths, depending on factors such as soil type, texture, structure, and consistency. Consider, for example, the two layers of soil shown by brown rectangles in Figure 15.26. For any value of normal stress, some minimum value of shear stress will cause sliding of the top layer over the bottom layer; this minimum value is the shear strength. If normal stress is greater, a larger shear stress is needed to cause sliding. The plot on the right side of Figure 15.26 shows the combinations of shear and normal stress that may or may not lead to slippage. The solid line divides the space into conditions at which slippage will and will not occur, and the slope of the line is the internal angle of friction (φ). Table 15.10 lists some values of the internal angle of friction for some different materials. For well-sorted sand, this angle is about 34°, and for any granular, non-cohesive soil the internal angle of friction is the same as the angle of repose.

The relationship depicted in Figure 15.26 is not quite the same for all geological materials. Some materials are *cohesive*, meaning they have an intrinsic shear strength independent of normal stress. Fine-grained soils, moist soils and sediments, and those that contain clays all exhibit cohesive properties, and the line separating slippage from non-slippage moves up in the diagram, as shown in Figure 15.27. This cohesion explains why you can hold clayey soil in your hand without it running out between your fingers the way sand does.

The graph in Figure 15.27 is one example of a *Mohr-Coulomb diagram*. Diagrams of this sort apply to most materials, like soils, for which compressive strength is much greater than their tensile strengths. These diagrams provide pictures of how a soil or rock perform in a specific stress state. The diagonal lines in these figures are referred to as *failure envelopes*. If a soil is exposed to normal stress and shear stress that plots below the failure envelope, the soil will not fail (slip). However, if the soil plots above the line (outside the failure envelope), then the material will likely fail.

Why do different materials have different shear strengths and failure envelopes? Consider, for example, the conical pile of dry sand shown in Figure 15.25. If the sand particles were very angular, the grain-to-grain friction would be greater than for well-rounded grains. Thus, the angle of repose would be larger, the slope of the line in a Mohr-Coulomb diagram would be steeper, and the failure envelope would be larger. Thus, as shown in Figure 15.28, a Mohr-Coulomb diagram

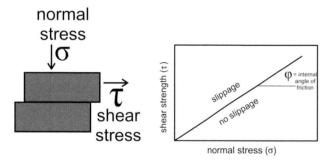

Figure 15.26   The drawing on the left shows normal and shear stresses associated with two layers of sand. The drawing on the right is a stress-strain diagram that reveals the internal angle of friction.

Table 15.10   Internal angle of friction for some geological materials.

Material	Angle of internal friction (φ)
rock	30°
sand	34°
gravel	35°
silt	34°
clay	20°

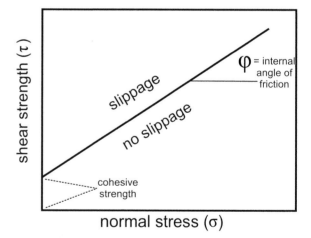

Figure 15.27   Mohr–Coulomb diagram for a cohesive material. A fundamental strength, the cohesive strength, must be overcome before sliding can occur, regardless of normal stress.

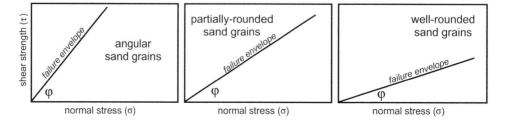

Figure 15.28   Comparing the failure envelope for sands with grains of different roundnesses. Angular grains do not slide past each other as easily as rounded grains and, consequently, have a smaller failure envelope.

for angular sand grains has a steeper slope than for partially rounded grains or well-rounded grains. The sand composed of more angular fragments has greater shear strengths.

15.7.5 Effect of water on strength

Figure 15.29 shows a sandcastle at Castle Beach in East Coast Park, Singapore. Sand castles have just the right amount of soil moisture to hold together the sand grains and create vertical surfaces far exceeding the material's dry angle of repose. So, water may help hold soils together, but it may also lead to failure. Slightly wet, unconsolidated materials exhibit a very high angle of repose because grain-to-grain contact, surface tension, and adhesion between the water and the grains tend to hold the grains in place. This is why kids learn, at an early age, to use damp sand when they make sand castles.

However, if material is saturated with water, the water gets between the grains and eliminates grain-to-grain contact. Although total normal stress may be higher, the *effective stress* holding individual grains together may be small. Effective stress is equivalent to total stress minus the pore water pressure holding grains apart. In some saturated soils, pore pressure may become as high as or higher than total stress, reducing shear strength to zero.

Figure 15.30 shows a warning sign advising swimmers to keep away from some sandy piles surrounded by quicksand. Quicksand forms when loose sand that is saturated with water becomes agitated—perhaps when someone walks on it. The water in the sand cannot escape, so effective stress is reduced and the sand flows like a liquid and cannot support weight. Quicksand is often associated with springs because upward flowing water can keep sand particles in suspensions, but quicksand also forms in standing water.

Figure 15.31 shows a car sinking through a road and into the ground when the ground lost all strength as it became saturated with water. This loss of strength is termed *liquefaction*. Liquefaction is a special case of quicksand commonly associated with earthquakes.

Figure 15.30 Quicksand warning sign.
Photo from Hughesdarren, Wikimedia Commons.

Figure 15.29 A sandcastle at Castle Beach in East Coast Park, Singapore.
Photo from William Cho, Wikimedia Commons.

Figure 15.31 Car sinking into the ground because of liquefaction during an earthquake in Christchurch, New Zealand in 2011.
Photo from Martin Luff, Wikimedia Commons.

Shaking by earthquake tremors mobilizes water in loose soil as it is compressed, so pore pressure increases and, consequently, soil strength disappears. Buildings or other objects on the surface may tip over or sink into the ground. Today, building codes in many countries require engineers to consider the effects of soil liquefaction when designing new structures, but this has not always been the case. Even with some of the strictest building codes in the world, liquefaction caused significant damage during San Francisco's Loma Prieta earthquake in 1989.

Questions for thought—chapter 15

1. What is the difference between sediment, soil, and "dirt"?

2. Why did the Leaning Tower of Pisa lean? What was done to remedy it and make it safe for tourists again?

3. What are the four most important components of soil? Are all components present in all soils? Explain your answer.

4. What are *consolidated materials* and what are *unconsolidated materials*? Name examples of each. If you had your choice, which would you prefer as the foundation for an apartment building you are erecting? Explain your reasoning.

5. Why are the bulk properties of soils often more important to engineers than the properties of individual grains within the soils? Why is (bulk) soil strength important in engineering?

6. What minerals are most common in soils? Explain why.

7. What are the two meanings of the term *clay* when used to talk about soils?

8. Some soils contain predominately fine-grained material. Other soils contain lots of coarse grained material such as pebbles ore even larger rocks. Choose one example of each and describe how it may form.

9. What types of soils and sediments provide good drainage? Describe the properties that make them good for drainage.

10. What types of soils and sediments provide good support for buildings and roads? Describe the properties that make them good for this purpose.

11. Why is effective porosity usually more important than total porosity?

12. Sometimes soils settle. What causes this and why can it be an important consideration? What are bad consequences of settlement.

13. Why would engineers have their own soil classification systems (the USCS system) rather than using the one from the USDA that was described in Chapter 11?

14. What typically happens to a soil under compaction? What about when it experiences increasing shear stress? What happens to most soils under tension? How does a soil's response to shear compaction, shear stress, and tension, differ from a rock's response?

15. What may cause a soil to become compacted or consolidated. They are very similar processes, but what is the difference? What problems may arise when a soil is compacted or consolidated? Why do engineers often favor soil compaction and farmers avoid it?

16 Rock Mechanics

16.1 Problems in Coeur d'Alene

16.1.1 Rock bursts

Figure 16.1 shows a map of Idaho's Coeur d'Alene Mining District, also called Silver Valley. The district has been producing lead, zinc, silver, and copper since about 1900, and more than a dozen mines, many shown on the map, have operated in the district. Most ores come from steep, narrow veins, so for maximum production, mines go to great depth. The mining district is in a minor natural earthquake zone, and since mining began in the area, mining activities have caused additional earthquakes to occur. Most of the mining-induced earthquakes are small, rarely exceeding Richter magnitude 3, and seismometers record many tremors not felt by people on the surface. The mining-related earthquakes are the result of unstable, irreversible rock deformation that includes fracturing of brittle rock and sliding along weak rock planes.

The smallest of the anthropogenic tremors— sometimes called *minor seismic events*, *microseismicity*, *bumps*, or *knocks*—are due to minor fracturing and readjustment within rock masses. Larger tremors are the products of *rock bursts* associated with mine tunnels and rooms. Figure 16.2 shows the results of a rock burst that produced a Richter magnitude 3.4 earthquake. Rock bursts occur because mine wall rocks, under a great deal of pressure, may release that pressure by exploding into open spaces where no rock provides confinement. Rock bursts, such as the one shown here, involve a sudden and short-lived violent fracturing of rock that releases accumulated energy. The Coeur d'Alene district is one the few metal-mining districts in the United States where sizable rock bursts occur.

Courtesy of Petur Hemmingsen, Slideshare.net. During the last 70 years, rock bursts have accounted for more than two dozen fatalities around Coeur d'Alene, including deaths at the Sunshine, Star/Morning, Galena, Lucky Friday, and Coeur Mines (Fig. 16.1). Some deaths were caused by flying or falling rock, but many resulted from burial and suffocation. The number of rock bursts has increased over this 70-year period because mining activity has increasingly gone to deeper levels, and thus mine walls are subjected to greater pressures. Additionally, the more recent mining activity has focused on ores in brittle quartz-rich rocks not exploited before. Rock burst deaths in the Coeur d'Alene mines are not unique.

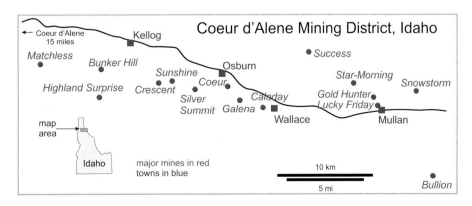

Figure 16.1   Major mines in the Coeur d'Alene Mining District, Idaho.

Figure 16.2   Collapse of a mine tunnel due to rock burst.
Photo from Peter Hemmingsen.

In South Africa, for example, rock bursts kill about 20 miners every year.

Rock bursts are more common in deep mines (where pressure is greater) than in shallow mines and in large openings than in small openings. So, the possibility of rock bursts puts limits on mine depth and mine design. The three main types of rock bursts are *strain bursts, pillar bursts*, and *slip bursts*. All three types, which may vary in volume and explosiveness, expel fist-sized to smaller debris into mine openings. The primary controls on the size and intensity of a burst are the nature of the bedrock and the pressure on the rock.

16.1.1.1   Strain bursts

Mine tunnel or room walls may bulge or buckle when under a great deal of pressure. If pressure gets great enough, *strain bursts* can occur when the walls fracture suddenly by spalling (breaking into small flakes along the wall surface) or shattering completely. Tunnel and room ceilings, too, may fracture the same way. During strain bursts, most of the energy is released perpendicular to the rock surface. Sometimes only a relatively thin layer of rock shatters, but if pressure continues to build, additional bursts may occur. Other times, larger volumes of rock are involved, as seen in Figure 16.2.

16.1.1.2   Pillar bursts

Many mines have pillars that hold up the ceilings of rooms. If the downward force on the pillars becomes too great, *pillar bursts* may occur. Energy and shattered debris go in all directions, and sometimes exploding pillars cause other nearby pillars to collapse violently as

well. Rock fracturing, sliding, and buckling are involved in pillar bursts, but because of great vertical forces, the effect is sometimes as if the rock has been crushed. Like strain bursts, rock distortion may precede pillar bursts.

16.1.1.3   Slip bursts

Slip bursts mostly occur because of forces along fault or joint planes, so energy is concentrated on those planes as rocks on either side move past each other. The planes could be longstanding fault traces or new planes of weakness created by mining activities. Less commonly, the planes involved are bedding planes. This kind of burst generally occurs because mining activity releases pressure that is holding both sides of the plane of weakness together. Slip bursts are sometimes called *fault-slip* or *stick-slip* types of rock burst. A small amount of distortion or creep may take place before a slip burst occurs, releasing energy incrementally, but when a slip burst finally occurs, it releases the remaining energy all at once. Slip bursts can affect large areas, and the tremors the bursts create are more often felt by people than the other two kinds of bursts are.

16.1.1.4   Hazards and engineering

Of the three types of bursts, pillar and strain bursts are the more hazardous to miners, because these bursts more often occur where miners are working. Slip bursts, however, produce the largest earthquakes. All three are important to mining activities, because the risk of rock bursts influences mine planning. Thus, mining engineers and others want to know rock properties and be able to predict how rock will behave under different physical conditions. Evaluating risks is not a simple task, because there are many kinds of rocks, many different geological settings, and physical conditions. Investigations of rock properties and behaviors comprise the field of study that we call *rock mechanics*.

16.2   Forces and stress

Forces that act on a rock may be *body forces* or *surface forces*. Body forces are the same for all parts of a rock body. Gravity, magnetic fields, and electrical fields are examples of body forces; they are imposed from the outside and affect all parts of a rock body equally and simultaneously. Surface forces, in contrast, are applied only to the outsides of a rock body. Examples of surface forces include force that is applied to a rock by squeezing it or by trying to pull it apart. Surface forces can also be applied to planar surfaces within a rock. Surface forces, however, will affect an entire rock body (unless the rock fractures). For example, if a rock is squeezed from the outside, the strength and rigidity of the rock mean that the force is transmitted to the inside of the rock as well.

A rock becomes *stressed*, or is in a *state of stress*, when force of any kind is applied. Although some stress can be generated within a rock or shape, in this chapter we focus mostly on stress that is applied from the outside. Such stresses may cause a rock's volume to change. These changes may be reversible, if removing the stress would allow the rock to return to its original volume and shape. We call this reversible process *elastic deformation*. Alternatively, changes in volume and shape may become permanent if *plastic deformation* (permanent deformation, without fracturing) or *brittle failure* (near-instantaneous fracturing or collapsing) occur. When rock bursts occur in the mines at Coeur D'Alene, they are commonly preceded by elastic and plastic deformation before brittle failure (the rock burst) occurs.

16.2.1 The importance of stress

If a force is applied over a very large area, it may have small effect. But if is applied over a small area, it can be quite significant. Consequently, the force per unit area is often more significant than the total force. Force per unit area is called *stress*, and a numerical value for stress is calculated by dividing the total force by the area over which it is applied.

The standard symbol for stress is σ (sigma) and thus

$$\sigma = force/area = F/A$$

In standard SI units, force is measured in newtons (1 newton = 1 N = 1 kg•m/sec^2) and area is measured in square meters. Thus, stress is given in newtons per square meter. To simplify things, 1 N/m^2 is called a pascal (Pa). So in SI units, stress is generally expressed in pascals. In imperial units, stress is typically measured in pounds per square inch (psi).

Consider a 100-pound weight sitting on top of two different columns of rock, as shown in Figure 16.3. The diameter of the pillar on the left is 1 in. and the diameter of the pillar on the right is 4 in. The vertical stress is clearly much greater on the left pillar than on the right

pillar, which means that the left pillar is much more likely to collapse than the right pillar is. Engineers use considerations of this sort when they are designing pillars to hold up the ceiling of rooms in a mine.

The area of the pillar top on the left (0.79 in^2) is much smaller than the area of the pillar top on the right (12.57 in^2). So, the calculated stresses are 100/0.79 and 100/12.57, or 127 pounds per square inch for the pillar on the left and only 8 pounds per square inch for the pillar on the right. The stress on the skinny pillar in Figure 16.3 is 16 times the stress on the fat pillar, although the pillar sizes only differ by a factor of 4.

Figure 16.4 shows the three fundamental kinds of stress. *Normal stress* is stress that is perpendicular to a surface, symbolized by σ or sometimes σ_n. *Shear stress* is stress that is parallel to a surface, symbolized by the Greek τ or sometimes σ_s. If normal stress is pulling an object apart, we call it *tension*, or *tensional stress*. If normal stress is squeezing an object—pushing it together—we call it *compression*, or *compressional stress*.

Pressure and stress have the same units (force per area), but they are not quite the same thing. Pressure is commonly considered when talking about fluids (gases or liquids), and in fluids, pressure is always the same in

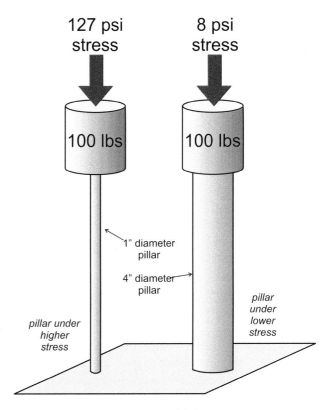

Figure 16.3 Stress on two pillars of different sizes.

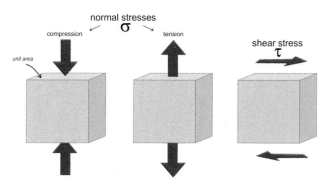

Figure 16.4   Different kinds of stress.

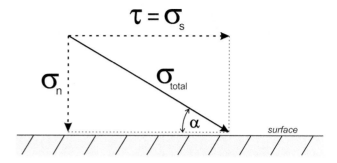

Figure 16.5   Dividing a stress vector into normal and shear components.

all directions. In contrast, if a solid (non-fluid) object is under stress, it means that force is being applied to the object's surface, and the resulting stress may not be the same in all directions within the object. Fluids and solids behave differently in response to forces, primarily because fluids cannot completely resist all shear force applied to them; any shear stress is fleeting because fluids flow until the force is reduced to zero. Any resistance a fluid has to stress is due to viscosity, a variable property that changes with temperature and pressure. Additionally, fluids are *isotropic*, meaning that they have the same properties (strength, viscosity, etc.) in all directions, while many solids are anisotropic. Therefore, the effects of stress do not depend on stress orientation in fluids.

Stress, in contrast with pressure, is a vector property that can be parallel or perpendicular to an object's surface or at any angle between. Stress affects an entire object, not just the surface of the object. Because it is a vector, we can describe stress using vector components, perhaps a north-south, an east-west, and a vertical component or a normal stress component and a shear stress component. Additionally, stress vectors may be added. If stress comes from two directions, the overall stress on an object will be the sum of the two vectors, oriented somewhere between the two separate components.

As shown in Figure 16.5, if a stress is inclined to a surface or a plane within an object, it can always be resolved into normal and shear components. This resolution is commonly done when thinking about rock strength and properties, because normal stresses compress rocks, sometimes making them stronger, while shear stresses may cause them to deform or break apart. Many rocks can resist normal stresses much better than they can resist shear stresses, and thus shear stresses may be more important considerations when thinking about rock strength. This weaker resistance to shear stress

is why a spinning drill bit is more effective than one pounded into a rock. The spinning bit applies shear stress, and the pounded bit applies normal stress.

One standard simplification is commonly made when modeling or analyzing stress. In principle, if we wish to describe the total stress acting on a volume of rock, we should specify the *stress field*, which is the stress at every point within that volume, because stress may not be the same everywhere. But although there may be variations within the rock body, to simplify, we often assume that stress is homogeneous—the same at all points. This assumption is certainly more valid for smaller rock bodies than for larger ones. However, the assumption is invalid if a volume of rock contains fractures or other planes of weakness. A highly fractured rock body, with properties more dependent on the fractures present than the rock itself, is called a *rock mass*.

16.2.2   Stress on rocks within Earth

Rocks within Earth are under pressure due to the weight of overlying rock. We call this pressure *lithostatic pressure*, which is analogous to water pressure that increases with depth in an ocean. Water pressure and lithostatic pressure are isotropic—the same in all directions. *Lithostatic stress*, or sometimes called *confining stress* or *confining pressure*, strengthens rocks. Rocks under greater confining pressure have more strength than less confined rocks do. Pressure pushes mineral grains together in confined rocks, closing small cracks and making grains interlock, thus increasing the viscosity of the rock which hinders deformation. In principle, with enough confining pressure, a rock becomes virtually indestructible.

However, many rocks within Earth experience *differential stress*, also called *deviatoric stress*; their state of stress includes stresses that are greater in some directions than in others. For example, they may be under great stress vertically, which is typical, but much lesser stress in other directions. Thus, there are commonly directions of maximum stress and minimum stress. For a small volume of rock, the orientation of *maximum normal stress* is always perpendicular to the orientation of the *minimum normal stress*. Additionally, a third stress, called the *intermediate stress*, is perpendicular to the first two. Thus, the directions of maximum, minimum, and intermediate stresses form an orthogonal axial system. We designate the maximum stress σ_1, the intermediate stress σ_2, and the minimum stress is σ_3. Note that if all three stresses are equal, the overall state is one of lithostatic stress.

Figure 16.6 shows the three normal stress vectors for a compressional stress state and for a tensional stress state. Most rocks within Earth are in a compressional state, and σ_1 is generally vertical—because the greatest stress is vertical. Within Earth, the minimum normal stress, σ_3, is never zero because there is always some confining pressure that contributes stress in all directions. And although not shown in the drawings, in addition to normal stresses, there are commonly shear stresses that operate in the planes of the rock volume surfaces.

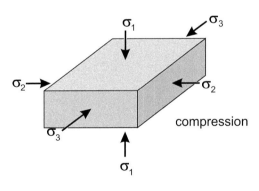

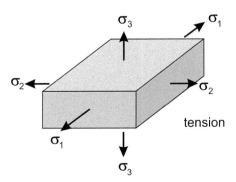

Figure 16.6 Three perpendicular stress components for compressional and tensional stress.

16.3 Strain—rock deformation in response to stress

If a rock body becomes stressed, it may deform. The deformation is called *strain* (symbolized by epsilon, ε), and different kinds of rock have different properties and thus strain in different ways. Like stress, strains may be of three types: compressional strain, tensile strain, or shear strain (Fig. 16.7). Compressional strain makes a rock shorter, tensile strain stretches a rock, and shear strain deforms a rock by causing different parts of the rock to move in different directions. Numerically, strain is quantified by dividing the amount of change in a dimension by the original length of the dimension. Thus, values for strain have no units and are reported as a number or sometimes a percentage.

Some fundamental laboratory measurements, called *uniaxial stress tests*, involve applying force in one direction, usually vertical, to a sample, with no force applied in other directions. Force is gradually increased, and the amount of strain for different stresses is recorded. Eventually the sample fractures; the stress that causes fracture is called the *ultimate strength*, sometimes simply called strength. Uniaxial tests may involve tensional or compressional forces, but uniaxial compressional measurements, often called *unconfined compression tests*, are more easily carried out. More complicated tests, called *triaxial tests*, involve applying stress in more than one direction to a sample simultaneously. Triaxial tests include a confinement pressure around the outside of samples, thus mimicking the conditions that a rock experiences at depth in Earth.

Figure 16.8 shows the geometry of a uniaxial stress test on a cylinder of rock compressed vertically and, consequently, bulging outward near its center. Values for strain in the vertical direction, called *axial strain*, and strain in the horizontal direction, called the *radial*

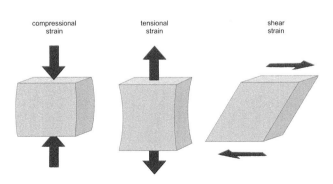

Figure 16.7 Different kinds of strain.

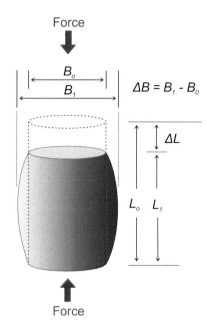

Figure 16.8   The geometry of a uniaxial stress test.

strain, or *lateral strain*, can then be calculated based on measurements of cylinder width and length before and after compression:

$$\varepsilon_{axial} = \frac{(L_1 - L_0)}{L_0} = \frac{\Delta L}{L_0}$$

$$\varepsilon_{lateral} = \frac{(B_1 - B_0)}{B_0} = \frac{\Delta B}{B_0}$$

We call the ratio of lateral strain to axial strain *Poisson's ratio*, named after the French engineer and physicist Siméon Dennis Poisson. Poisson's ratio, commonly used in material science to describe the elastic behavior of different materials when under stress, is symbolized by the Greek letter nu (ν).

Thus, if a perfectly elastic material is compressed along any axis, it will bulge perpendicular to that axis and its total volume will not change. When stress is released, the material will return to its original volume and shape. Poisson's ratio, ν, is a constant for such elastic materials.

Rubber is an exceptionally elastic material and has a Poisson's ratio of about 0.50. Very rigid materials have small values, sometimes approaching 0, of Poisson's ratio. Rocks and minerals are not very elastic and generally have Poisson's ratios between 0.08 and 0.40, most commonly values between 0.25 and 0.30. Such materials do not deform (change shape) greatly when under

stress, and radial strain will be only 25%–30% of axial strain. However, even the most elastic rocks and minerals are neither isotropic nor homogeneous. Thus, Poisson's ratio typically varies from one sample to another, even if those samples come from the same rock mass. Furthermore, the degree to which a material is elastic is commonly different for different states of stress.

16.3.1   *Different kinds of deformation*

Figure 16.9 shows a typical stress-strain diagram for a rock. If stress is applied to a rock, the rock will strain. The strain may be elastic and reversible, but rocks under relatively high levels of stress often respond by breaking or by bending or deforming in other ways. If a rock bends, deforms, or flows like clay or putty, then *ductile deformation* has occurred. If a rock breaks, *brittle deformation* has occurred.

Consider a rock that is experiencing stress. Initially, if stress increases, rock strain will increase. But the rate of increase is variable for different rocks and may not be proportional to stress under some conditions. Most rocks exhibit *elastic deformation* when stress is low, deforming like a spring or a piece of rubber that is being compressed or stretched. During elastic deformation, stress (σ) is proportional to strain (ε). This proportional relationship is called *Hooke's Law*, and the slope of the stress-strain curve is *Young's modulus*, also called *the modulus of elasticity*, denoted by E. When the load is reduced or removed, the deformation disappears as the rock returns to its original shape.

However, there is a limit for elastic deformation, and if stress increases to a level called the *elastic limit*

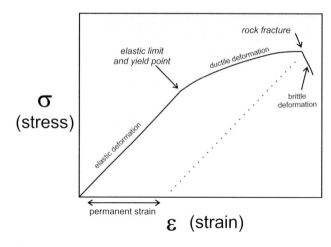

Figure 16.9   Idealized stress-strain curve.

(Fig. 16.9), materials become non-elastic. With a further increase in stress, materials reach the *yield point* that marks the beginning of material flow and *ductile* behavior. The difference between elastic limit and yield point is, for most materials, generally so small that it can be ignored. *Ductile deformation*, also called plastic deformation, in contrast with elastic deformation, is partly or entirely irreversible; if the stress causing the deformation is removed, the material will not return to its original shape. As shown in Figure 16.9, some permanent strain will remain.

At stress levels beyond the yield point, rocks will, up to a point, deform ductilely with increasing stress. Eventually, however, all materials, if placed under high levels of stress, will reach a point of *failure*. Failure occurs when a rock loses coherent strength, and if more force is applied, the rock deforms, fractures, or flows so that no increase in stress is possible. The limiting value of stress is a rock's strength. The deformation or fracturing that occurs when a rock's strength is exceeded constitutes *rock failure* because the rock *fails* to provide any additional strength.

Failure may mean fracturing that, on a large scale, results in faults. It also sometimes includes bending and crushing and wholesale displacement of individual mineral or rock fragments. Most rocks at Earth's surface fail by *brittle failure*, especially when stress is applied relatively quickly, and thus fracturing is the most common type of rock failure we see. Other rocks flow or deform instead of fracturing, thus demonstrating *ductile failure*. The major difference between the two kinds of failure is that brittle failure involves almost complete loss of rock strength at one time, because cracks spread rapidly with little accompanying plastic deformation. Ductile failure is more gradual—cracks develop slowly and are generally accompanied by significant plastic deformation.

A rock responds to stress differently depending on pressure and temperature and the mineral composition of the rock. But, no matter the rock composition, when rocks are under high confining pressure, they do not tend to fracture because confining pressure holds rocks together and hinders the formation of cracks. Additionally, rocks at higher temperature are less rigid and more malleable than rocks at low temperatures, and we know that temperature increases with depth in Earth. For these reasons, brittle deformation (rock breaking or fracturing) occurs in many rocks near Earth's surface but gives way to elastic (reversible) and then to ductile (nonreversible) deformation at depths where confining pressure and temperature are greater. At such depths,

rocks flow and deform without fracturing. Thus, earthquakes do not occur at depths greater than 750 kilometers (400 miles) in Earth—rocks flow instead of fracturing and creating tremors at greater depths.

Figure 16.10 shows stress-strain functions for two kinds of rock, granite and marble. Materials, like granite, that exhibit steeper slopes (and thus have a greater value for Poisson's ratio) in response to elastic deformation are considered *stiff*, while materials such as marble with lower slopes have lower stiffness.

The example granite displays high stiffness and a brittle failure; the marble has relatively low stiffness with a ductile failure. Materials considered *brittle* exhibit elastic behavior, but once the elastic limit is reached, there is only a relatively small region of ductile behavior before they fracture. In contrast, materials considered *ductile* have a smaller region of elastic behavior and a larger region of ductile behavior before they fail. Note that both granite and marble have *ultimate strengths* at which they fail. Figure 16.10 shows the response of rock to compressive stress. A different chart could show rock response to tensile or shear stresses; rocks have different stress-strain curves and strengths for compression, tension, and shear.

Table 16.1 compares physical and mechanical properties of some common materials and rock types. The entries are ordered from lowest compressive strength (concrete), at the top of the table, to greatest compressive strength (quartzite), at the bottom of the table. The correlation between compressive strength and tensile strength is high—compressive strength is 10–20 times

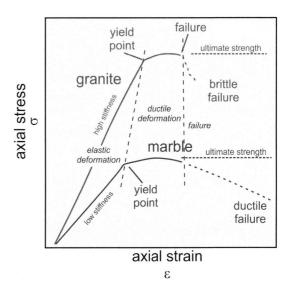

Figure 16.10 Stress-strain curves for two rocks: granite and marble.

Table 16.1   Properties of common Earth materials, concrete, and steel. Data from several different publications.

Lithology	ρ Density (g/cm³)	ν Poisson's ratio	E Modulus of elasticity (x 10⁹ N/m²)	S_c Compressive strength (x 10⁶ N/m²)	S_t Tensile strength (x 10⁶ N/m²)
Concrete	2.7–3.2	0.15	2.1–1.0	0.41–0.21	0.04–0.02
Ice	0.93	0.31	87.0	5–25	0.7–2.1
Rock Salt	2.20	0.33	4.64	35.5	2.5
Limestone	2.45	0.25	50.4	58.7	4.0
Sandstone	2.20	0.24	15.3	107.0	8.2
Marble	2.71	0.23	46.3	110.0	8.7
Schist	2.89	0.12	42.4	129.6	5.5
Basalt	2.73	0.25	42.6	130.3	10.3
Shale	2.75	0.08	13.7	165.8	8.4
Gneiss	2.75	0.21	58.6	192.0	10.5
Granite	2.64	0.23	59.3	226.0	11.9
Amphibolite	2.94	0.29	92.4	278.0	22.8
Steel	7.85	0.29	200.0	365.0	365.0
Quartzite	2.55	0.15	70.9	629.0	23.4

greater than tensile strength is, except for steel. Compressive strength also correlates, but not as well, with Young's modulus of elasticity. Other correlations are absent.

Considering only the rocks listed above:

• This table reveals that rocks vary greatly in strength, and the difference between the strongest and the weakest is a factor of 15. Rock salt and limestone are the weakest, and quartzite and amphibolite are the strongest. Other rocks are between.

• Values of Poisson's ratio $\nu = \left| \dfrac{\varepsilon_{lateral}}{\varepsilon_{axial}} \right|$ are indicators of how isotropic a rock is. Of those listed, ice and rock salt are the most isotropic, while shale and schist are quite anisotropic.

• Values of Young's modulus, the slope of a stress-strain curve, are indicators of stiffness. Of the materials listed, rock salt, sandstone, and shale (relatively soft sedimentary rocks) have the least stiffness, and the other, harder, metamorphic and igneous rocks have greater stiffness. Greater stiffness generally correlates to greater brittleness. Thus, although they are stronger, when they fail, stiffer rocks often fail by fracture. Recall that stiff quartz-rich rocks were, in part, responsible for the rock bursts in mines near Coeur d'Alene.

16.3.2   What determines brittle or ductile behavior?

Bonds in some minerals—including quartz, the feldspars, and olivine—are strong and about the same strength in all directions. In general, this means that these minerals are brittle and have no preferred directions of weakness. Other minerals, including the carbonates that make up limestones and clays that make up many other sedimentary rocks, are more ductile. These minerals are bonded tightly in some directions but not so tightly in others, allowing fracture and gliding along planes of weakness and thus plasticity. Consequently, the mineral makeup of rocks has a large influence on rock mechanical properties.

The amount of water, and occasionally other fluids, that is present in a rock is also important. As pore pressure (the amount of water) increases, the forces holding constituent grains together are weakened. Consequently, saturated rocks can strain relatively easily because mineral grains move more easily within them. Thus, saturated rocks behave more ductilely than dry rocks do.

The rate at which stress is applied (called the *strain rate*) is also significant. Rocks experiencing low stress over long times may bend or deform in other ways that are not catastrophic. For example, rock layers that seem hard and strong can be bent into folds if stress is applied slowly. In contrast, rocks that experience great stress over short times may fail. Thus, high strain rates favor brittle behavior and low strain rates favor ductile behavior. These considerations are important when testing samples in a laboratory; slow strain rates allow a sample to adjust to stress and fast strain rates may cause the sample to fracture, even if it would behave elastically or ductilely under slower rates. Natural processes are generally quite slow, but sometimes stress is applied relatively quickly, which affects the nature of deformation or failure.

Finally, as alluded to above, the temperature and pressure regimes influence the nature of rock deformation. At high temperature, bonds within minerals are more elastic, and thus minerals will behave in a more ductile manner than when at low temperatures. The forces holding mineral grains together are more elastic too. And, at high pressure, rocks are less likely to fracture than if they were at low pressure, because the pressure hinders the formation of fractures. Thus, at high pressures, rocks will be less brittle and more ductile than at low pressures.

16.3.3 Different kinds of rock failure

Figure 16.11 shows some different ways that rocks may fail when subjected to tension or to compression. Brittle behavior is depicted on the left and ductile behavior on the right for both tensional and compressional states of stress. An undeformed column of rock is shown in drawing a. If that column of rock is subjected to tension, it may fail brittlely by fracturing (drawing b), it may fail ductilely by becoming irreversibly deformed (drawing d), or failure may be a combination of both brittle and ductile behavior (drawing c). Similarly, if a rock is compressed, it may fail by fracturing (drawings e and f), by deforming (drawing h), or by some combination (drawing g). The nature of failure depends on the nature of the stresses involved and on rock properties. Laboratory experiments reveal that, when under compression, the fractures that first occur in most materials are at an inclined angle to the direction of stress, as in drawings f and g.

Predicting or modeling stress-strain relationships and rock failure may be complicated. The simplest models are based on assumptions (1) that rock is isotropic and homogeneous, (2) that stresses are applied uniformly and at a constant rate, and (3) that intermediate stresses are unimportant. All these assumptions are invalid to some extent for natural systems, but the models and predictions are acceptable approximations for most systems most of the time. However, in some rare cases, simple models do a poor job of predicting rock behavior.

Consider a cylinder of typical rock, having area A on top, that is being compressed vertically, as shown in Figure 16.12. Vertical force is applied—so σ_1, the maximum principal stress, is F/A in the vertical direction. A potential plane of failure, labeled P, is shown as a black diagonal line. The force acting on plane P can be separated into normal (F_n) and shear (F_s) vector components (drawing B in Fig. 16.12), and, as shown in drawing C, the magnitude of the normal and shear components depends on the angle the plane makes with the vertical (the direction of stress, σ_1). Note that if the plane is nearly vertical, $F_s \gg F_n$, and if the plane is nearly horizontal, $F_n \gg F_s$.

In Figure 16.12, the normal and shear forces on plane P act, somewhat, in opposition. The shear force, if great enough, will induce strain and failure along the plane. However, the normal force on the plane pushes the grains in the rock together, making it more difficult for slippage to occur. Still, if the force being applied is great enough, the rock will fail and fracture despite the normal force. Whether failure occurs, however, does not depend on the strength of shear and normal *forces*. It depends, instead, on the strength of shear and normal *stresses*, which are force/area.

Stress changes if the angle, α, changes. For a horizontal plane, the area being affected by force is the area (A) of the cylinder top. As the angle α increases, the plane slopes, and the area increases. For a (near) vertical

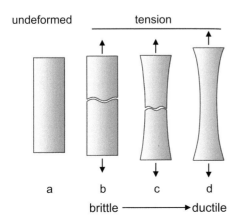
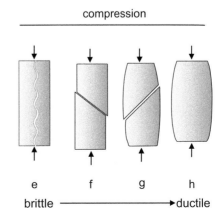

Figure 16.11   Different kinds of rock failure.

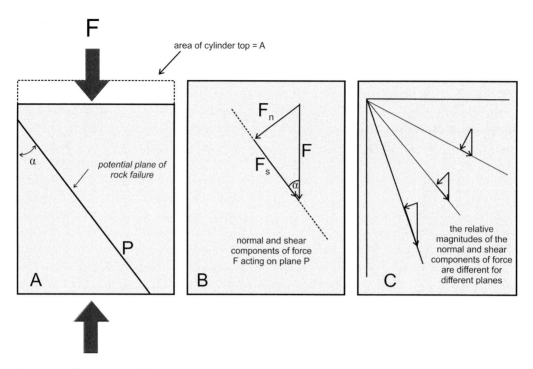

Figure 16.12 The normal (F_n) and shear (F_s) components of a force acting on a plane.

plane (α approaching 90°), the area is (near) infinite. Thus, the area over which the forces act is different for different angles of the plane, and calculating the stresses is not as simple as the geometry shown for forces in the middle drawing of Figure 16.12.

Consider a cylinder of rock, similar to the one considered above but with stress being applied in two directions (Fig. 16.13). σ_1, the maximum stress, is in the vertical direction. σ_3, the minimum stress, is perpendicular to σ_1. For a plane, P, oriented at an angle α to the direction of maximum stress, the values of normal and shear stress, σ and τ, may be calculated as:

$$\sigma = \frac{\sigma_1 - \sigma_3}{2} - \frac{\sigma_1 + \sigma_3}{2}\cos 2\alpha$$

$$\tau = \frac{\sigma_1 - \sigma_3}{2}\sin 2\alpha$$

The origins of the two equations above are not intuitive, and their implications are not immediately clear when looking at the equations. Their derivations, based on fundamental trigonometry and vector concepts, can be found in many books and on the internet, and their implications are best seen by examining diagrams that show variations in normal and shear stresses on a plane depending on plane orientation.

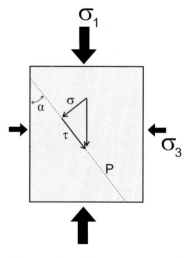

Figure 16.13 The normal and shear components of stress acting on a plane.

16.3.3.1 Constructing a Mohr's circle diagram

To depict the normal and shear stresses on a plane and to better visualize their implications, materials scientists and others use *Mohr's circle diagrams*. Figure 16.14 shows how such diagrams are constructed. The diagram has a vertical axis and a horizontal axis that intersect at the origin. The vertical direction represents shear stress (τ), and the horizontal direction represents normal

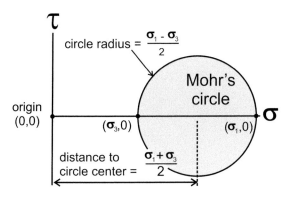

Figure 16.14   Constructing a Mohr's circle diagram.

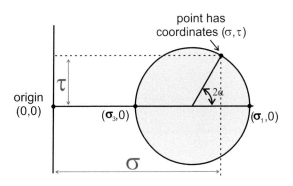

Figure 16.15   Interpreting a Mohr's circle diagram.

stress (σ). A circle, centered on the horizontal axis, intersects the axis at points (σ_3,0) and (σ_1,0) and thus has radius of ($\sigma_1-\sigma_3$)/2. The distance from the origin to the circle's center is ($\sigma_1+\sigma_3$)/2. A standard Mohr's circle diagram, like the one shown, only considers two dimensions—the directions of maximum and minimum stress. Modifications can be made, however, to consider the direction of intermediate stress (σ_2).

16.3.3.2   Interpreting a Mohr's circle diagram

Figure 16.15 shows why Mohr's circle diagrams are useful. For an inclined plane, such as the ones shown in Figures 16.12 and 16.13, that makes an angle α with the direction of maximum stress, we can use a Mohr's circle diagram to determine the normal (σ) and shear (τ) stresses on that plane. A radius from the circle's center, at an angle of 2α to the horizontal, intersects the circle at a point with coordinates (σ,τ). We could also calculate these coordinates using the equations given above. Note that if $\alpha = 0°$ (a horizontal plane), the normal stress on the plane is equivalent to the maximum principal stress (σ_1), and there is no shear stress ($\tau = 0$). Similarly, if the plane is vertical ($\alpha = 90°$), the normal stress on the plane is equivalent to the minimum principal stress (σ_1) and there is no shear stress ($\tau = 0$). Between these two limits, the maximum shear stress, which has a value equal to the circle radius, ($\sigma_1 - \sigma_3$)/2), is when $\alpha = 45°$. At this angle, the normal stress has the value ($\sigma_1 + \sigma_3$)/2. Although the maximum shear stress is reached for a plane at 45° to the maximum principal stress, experimental studies of rock failure reveal that rocks break at angles that are considerably less. This lessening of angles is because the normal stress, as mentioned above, works against shearing forces. Thus, the

effective stress promoting rock failure is equivalent to the shear stress minus the effects of normal stress. The result is that most rocks break along a plane oriented with angle $\alpha \approx 30°$—at an angle that is 30° from the direction of maximum stress.

16.3.3.3   Failure criteria

Rocks fail if a principal stress exceeds rock strength in the stress direction. For engineering and other purposes, it is important to be able to predict the conditions under which failure will occur. We can make laboratory measurements, but such measurements are only approximations to the real world; it is impossible to determine the strength of all materials under all conditions. And although, as discussed above, rocks and other materials are generally classified as exhibiting brittle failure or ductile failure, there is much variation not captured by these two descriptions. Furthermore, depending on temperature, pressure, confining stresses, load rate, and other factors, rocks may fail brittlely, fail ductilely, or not fail at all. Consequently, materials scientists and others have developed models to predict rock failure. These models are generally called *failure criteria*.

Figure 16.16a shows the Mohr-Coulomb criterion for rock failure. It is depicted as a straight line, called the *failure envelope*, that intersects the τ axis above 0. The value of the intercept, termed *cohesion*, or *unconfined compressive strength*, is the shear strength of the rock when unconfined. The line showing the failure envelope moves up (strong rock) or down (weaker rock) depending on the kind of rock being considered. Additionally, the slope of the failure envelope also varies depending on rock type.

The circles on the diagrams in Figure 16.16 are Mohr's circles depicting the state of stress of a rock. If the circle does not intersect the failure envelope, the

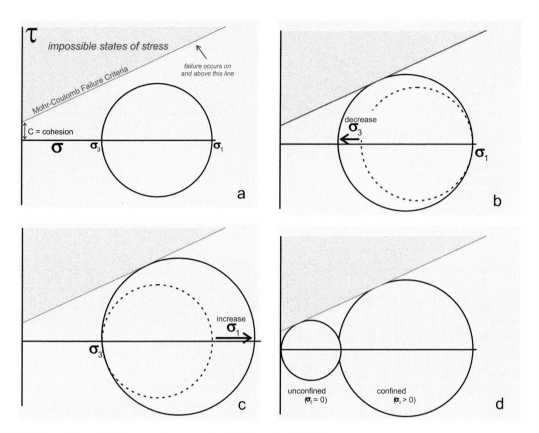

Figure 16.16   Mohr's circles and different conditions under which a rock will fail.

rock will be stable. Note that if confining stress (σ_3) is great, the Mohr's circle will be a long way from the failure envelope; the rock will have great strength. But, if the confining stress (σ_3) is decreased, as shown in Figure 16.16b, the size of the Mohr circle will increase, and eventually it will intersect the failure envelope. Similarly, if the confining stress remains constant, but the maximum stress (σ_1) increases (Fig. 16.16c), the Mohr circle will become larger and eventually intersect the failure envelope.

As shown in Figure 16.16, confining pressure makes rocks stronger; if grains within a rock are being squeezed together, failure is less likely to occur. Figure 16.16d depicts the importance of confining pressure—the size of the Mohr's circle, and thus the stable conditions for a rock, are much less when the rock is unconfined ($\sigma_3 = 0$), shown by the smaller circle, than when it is confined (larger circle).

The Mohr-Coulomb criterion is, like any model, an approximation, and other models have been developed to correct some deficiencies in Mohr-Coulomb theory. For example, the Hoek-Brown failure crieterion more accurately predict strength of rock under low confining

pressures. Unlike the Mohr-Coulomb model, the Hoek-Brown failure envelope curves and approaches the origin at low stress values. Other models have been developed that work better for rocks that are very ductile; failure envelopes for such rocks exhibit a great deal of curvature. The principles illustrated in Figure 16.16, however, are the same for all models and all rocks: increasing maximum stress and decreasing confining pressure leads to failure. Beyond that, the various models all have strengths and limitations, and the key is to use the one that is most appropriate for whatever it is going to be used for.

16.3.3.4   *Effective stress*

Most rocks contain pore fluids, and pore fluid pressure can affect rock strength. If pressure is applied to a rock from the outside, and pore pressure is acting from the inside, the pore pressure offsets some of the external force. Thus, rock strain is a product of the *effective stress* (usually denoted σ') on the rock, which is the difference between the total stress (σ_a) and the effects of pore pressure (u). When, in the previous part of this chapter, we talked about stress that caused strain, the stress being considered was the effective stress:

$$\sigma' = \sigma_a - u$$
$$\text{or } \sigma' = \sigma_a - \beta P_p$$

P_p is the pore pressure, and β, the *effective stress coefficient*, sometimes called *Biot's coefficient*, is different for different kinds of rock. Values of β may be 400% greater for weak sedimentary rocks compared with stronger igneous or metamorphic rocks. Thus, the concept of effective stress (the importance of pore fluids) is much more significant for sedimentary rocks than for igneous and metamorphic rocks. It is even more significant for soils and other unconsolidated materials. The equations above are sometimes called *Terzaghi's principle*, named after Karl Terzaghi who first recognized, in 1936, the importance of pore pressure when studying stresses in soils.

Figure 16.17 shows, using a Mohr's circle diagram, why pore pressure can have implications for rock failure. If pore pressure is great, Mohr's circle moves to the left, closer to zero as pore pressure pushes rock grains apart, decreasing strength. This notion is consistent with Terzaghi's principle, which says that if pore pressure within a rock is significant, it subtracts from the externally applied stresses. With enough pore pressure, the Mohr's circle may move far enough to intersect the failure envelope, leading to brittle fracture. Alternatively, if pore pressure decreases, it increases effective stress, which may lead to failure due to rock contraction or grain crushing. Petroleum engineers know about the effects of pore pressure and sometimes induce brittle fracturing on purpose by applying controlled increases in pore pressure, a practice called hydraulic fracturing ("fracking"), to create cracks that enhance oil and natural gas recovery.

16.4 On a larger scale

Figure 16.18 shows a quarry wall in Dorset, England. We can most easily visualize and model rock properties if the rock is a homogenous mass in which all properties are the same throughout. But, as seen in this photo, such is not always the case. The quarry wall shown consists of limestone that has well-developed vertical jointing with sympathetic horizontal jointing between. These joints, more than anything else, control the strength of this rock formation. Additionally, any measurements made on a fist-sized sample from this quarry would not reflect the overall rock mass properties.

Early in the development of rock mechanic theory, scientists and engineers recognized that most rocks contain joints, fault traces, bedding planes, dikes, and other discontinuities that are planes of weakness that subtract from rock strength. Some discontinuities are *primary*—they developed when the rocks were first formed. Others are *secondary*, perhaps resulting from tectonic stresses. On a very small scale, cleavage of individual minerals can introduce directions of weakness. On a

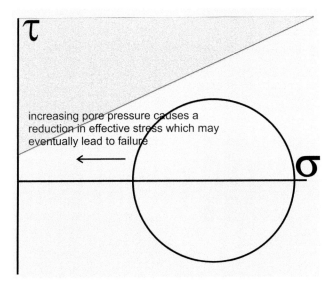

increasing pore pressure causes a reduction in effective stress which may eventually lead to failure

Figure 16.17 The effect of pore pressure on rock stability.

Figure 16.18 Limestone in the Swanworth Rock Quarry, Dorset, England.

Photo from Ian West.

larger scale, bedding planes and metamorphic foliations can be planes of weakness.

Studies of the strength of coal (Fig. 16.19) have shown that strength depends on sample size. Tests on small samples revealed that the material had high axial strength, but tests on larger samples showed lower strength. The difference is because the larger samples contained more planes of weakness that affected the measurements. For those in the coal mining business, this distinction has major implications. Thus, coal mining engineers and others concerned with outcrop-scale or larger phenomena know that instead of modeling a rock unit as a homogeneous, isotropic, perfectly elastic material, they must consider differences caused by rock fabric and structure.

Mapping discontinuities—planes of weakness—and cataloging their spacing, orientation, roughness, and continuity provides a more comprehensive picture of the overall strength of a rock mass than just doing measurements on small samples. *Spacing* is a measure of the distance between continuities, *continuity* is a measure of their size, and *roughness* deals with friction that hinders movement. Rough or jagged surfaces, for example, are stronger than smooth surfaces. Additionally, secondary features, such as the in-filling of fractures and joint-surface weathering, are important. For example, fractures filled with cement or that are mineralized are stronger than mud-filled fractures are. Thus, measurements of rock properties can vary greatly depending on the size of the sample studied. Rock masses with few discontinuities overall have a strength similar to intact rock. Rock masses with many discontinuities have lower strength, and sometimes a single plane of weakness can control the strength of an entire rock mass. So, for engineering purposes, a thorough investigation of discontinuities is crucial.

16.4.1 *Rock quality designation (RQD)*

Drill core, when brought to the surface, may be fractured to various degrees. First described by Deere (1962), the *rock quality designation* (RQD) provides a *fracture index* (a percent value that describes the extent to which a rock core is fractured) that is used in many rock classification systems. Drill cores typically break along joints or fractures, and the RQD is based on evaluating the percentage of the drill core that is recovered in hard/sound lengths of 10 cm (4 inch) or more, measured along the centerline of the core. Sections of fractured or missing core count "against" the index, and hence lower RQD value. High-quality rock has an RQD greater than 75%, fair to poor rock has a value less than 50%:

RQD (%)	Rock quality
90–100	Very good
75–90	Good
50–75	Fair
50–25	Poor
< 25	Very poor

The RQD system is generally applicable to N-size (75.7 millimeters) and larger diameter core, and is sometimes problematic for smaller core and low-strength, easily broken, rocks. Additionally, fissile and foliated rocks will separate during coring and transport. Thus, it is sometimes difficult to evaluate whether fractures were primary or secondary artifacts of sampling. The distinction is more easily made for large diameter cores.

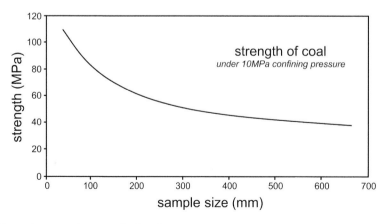

Figure 16.19   The effects of sample size on the strength of coal.
Data from Hoek and Brown (1997).

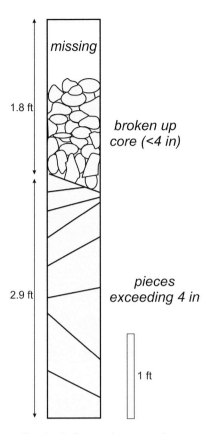

Figure 16.20 Sketch of a fractured core sample.

As an example, consider the sketch of an example core shown in Figure 16.20. It is a sample drilled using a 4.7-foot long-core barrel. Figure 16.20 shows the result of inspecting the core and approximate size of the broken pieces recovered. The pieces in the bottom 2.9 ft exceed 10 cm in size; the top 1.8 ft is missing or is highly fractured. The RQD is, thus, calculated as follows:

$$RQD(\%) = 100 \times \frac{\sum \text{Length of pieces} > 100\text{mm}\,(4")}{\text{Total Expected Length}}$$

$$= \frac{2.9 \text{ feet}}{4.7 \text{ feet}} \times 100 = 61\%$$

RQD allows cores to be compared, but does not provide a description of the properties of a large rock mass. Part of this deficiency is because RQD does not account for rock mass characteristics such as joint orientation, tightness, continuity, and other characteristics.

16.4.2 The geological strength index (GSI)

Completely characterizing the strength of a heterogeneous rock mass requires, among other things, measuring, describing, and cataloging discontinuities in the rock. This endeavor can be tedious, especially if the rock contains many discontinuities and discontinuity sets with different orientations.

The Geological Strength Index (GSI), created in 1955, was developed to be a faster method of assessing the strength of a heterogeneous or jointed rock mass. It is efficient and works well in many applications. Geologists or engineers conduct the process in the field. They make qualitative evaluations of the internal structure and surface conditions of a rock mass, considering the lithology and any discontinuities, such as joints and fractures, that are present. The evaluations allow a value for GSI to be read from charts such as the simplified one shown in Figure 16.21.

The GSI has possible values between 0 and 100. Heavily broken rock masses receive lower values, while fresh, unweathered surfaces tend to have higher values.

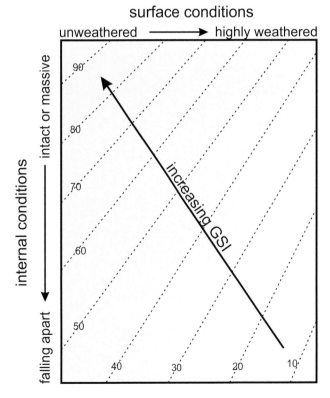

Figure 16.21 Schematic diagram showing how the Geological Strength Index (GSI) of a rock mass may be estimated. Drawing based on diagrams of Marinos and Hoek (2001).

This process of assigning a strength index is subjective but provides generally consistent values and helps to put all known discontinuities into context for a geologic site. Because the GSI uses in-field observations of the entire rock mass, it may better reflect the true strength of a rock mass than RQD determinations made for a single drill core in the laboratory.

The GSI was developed to help visually classify rock masses, and the classification assumes that discontinuities are uniform across the mass. Thus, it does not work well if random and highly persistent discontinuities are present, because a single discontinuity that extends through an entire rock's exposure can greatly weaken the rock mass. However, this limitation of the GSI need not create difficulty if careful observations are made during the field inspection. If, for example, a joint plane creates a single failure block, the use of the GSI is not warranted. Additional limitations on the use of the GSI involve structurally controlled failures, such as fault places and fold limbs.

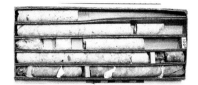

Figure 16.22   Drill core from Flynn Creek (top) and students examining core samples

Photo from the USGS (top) and A. Simon (bottom).

16.5   Determining values for rock strength

We can measure rock strength and other mechanical properties both in the field and in laboratories Laboratory measurements generally give reliable results, but field measurements add other values. Although some field tests are not as precise or accurate as laboratory tests, conducting measurements in a natural geologic setting is valuable because relationships with surrounding rocks can be seen and noted if significant. Additionally, some testing methods require *in situ* testing. *In situ* testing has an advantage because it eliminates any errors introduced during sampling that might affect rock fabric and introduce discontinuities not present in the original rock mass. Sampling for laboratory measurements is frequently done from outcrops due to ease of access and low acquisition cost. Outcrop sampling can be done with a hammer or with a core drill that produces round core samples, typically two to several centimeters in diameter. However, outcrop surfaces have often experienced weathering that affects their properties, and collecting outcrop samples can fracture or in other ways alter a rock. So, often, special drilling equipment is used to collect *cores* from intact rock at depth. These cores are brought to the surface, cataloged, described, and tested for their properties. The top photo in Figure 16.22 shows examples of drill core from Tennessee. The box is about 1 m across. The bottom photo shows two students examining core using hand lenses.

For applications of most interest to people, compressive strength is more significant than tensile strength, and a rock's compressive strength can be 5–25 times greater than the tensile strength. Most commonly reported as unconfined compressive strength (q_u or UCS), this property is routinely measured in laboratory experiments. Additionally, other laboratory and field techniques provide indirect measurements. Like tensile strength, compressive strength is different for different kinds of rock. Lower-strength rocks, such as volcanic tuff and shale, have UCS values of 11–40 MPa (1,600–5800 psi). Intact limestone can have a range of 50–250 MPa (7,250–36,260 psi), while some quartzite and basalt formations have UCS values greater than 320 MPa (46,400 psi) and 350 MPa (50,760 psi), respectively.

When studying rocks in an outcrop, an estimation of strength may be possible using well-established criteria. Table 16.2 presents one system that correlates field observations with laboratory measurements on typical rocks. Details of the laboratory measurements are discussed later in this chapter.

16.5.1   *Uniaxial compressive strength tests*

Figure 16.23 shows a typical device used to measure uniaxial compressive strength (UCS) of a rock, cement, concrete, or other solid material. UCS, discussed briefly earlier in this chapter, is the most widely used metric

Table 16.2   Classification nomenclature for rock strength and comparison to rock strength.

Description	Field identification of sample	Unconfined uniaxial compressive strength q_u (MPa)	Point-load strength (MPa)
Extremely strong	Can only be chipped with geological hammer	> 250	> 10
Very strong	Requires many blows of geological hammer to break it	100–250	5–10
Strong	Requires more than one blow of geological hammer to fracture it	50–100	2–5
Moderately strong	Cannot be scraped or peeled with a pocket knife; can be fractured with single firm blow of geological hammer	20–50	1–2
Weak	Can be peeled by a pocket knife with difficulty; shallow indentations made by firm blow with point of geological hammer	5–20	< 1
Very weak	Crumbles under firm blows with point of geological hammer; can be peeled by a pocket knife	1–5	
Extremely weak (also needs additional description in soil terminology)	Indented by thumb nail or referred to by other lesser strength terms used for soils	< 1	

Note: No correlation is implied between q_u and I_{s50}.

From Williams *et al.* (2005).

Figure 16.23   Example of a device used for measuring uniaxial compression strength of a rock cylinder.
Photo from Xb-70, Wikimedia Commons.

Table 16.3   Classification terms used to describe uniaxial compressive strength.

Class	Description	Uniaxial compressive strength (lb/in²)(psi)
A	Very high strength	> 32,000
B	High strength	16,000–32,000
C	Medium strength	8,000–16,000
D	Low strength	4,000–8,000
E	Very low strength	< 4,000

From Deere and Miller (1966).

and reliable way to learn the strength of an intact material and the material's response to stress. The measurements allow calculation of important rock properties, including axial strain, Young's modulus (the modulus of elasticity), Poisson's ratio, and ultimate strength.

UCS tests use a core—a cylindrical piece of rock, or other material, with a diameter on the scale of centimeters. Such cores are commonly obtained directly during drilling of boreholes. The core is put in a hydraulic press such as the one in Figure 16.23, and pressure is slowly applied to the core ends by two platens. No force is applied in any other direction. Because rocks sometimes burst, wire mesh or some other shield generally surrounds the apparatus to provide protection from potential flying rock fragments. As stress increases, a detector—which can be seen on the left side of the cylinder—measures the amount of deformation in response to stress until, eventually, the cylinder fractures.

for rock strength. It is a numerical value that suggests whether a rock or other material is likely to have problematic strength characteristics. Measuring the UCS, using a device such as the one shown, is a generally straightforward

Deere and Miller (1966) provide a simple classification for intact rock based on UCS strength (Table 16.3). It is a helpful way to talk about small homogeneous rock samples but does not apply to larger and heterogeneous rock masses.

16.5.2 Point-load tests

Figure 16.24 shows a rock cylinder undergoing a point-load test. The UCS may be the standard when it comes to evaluating rock strength, but point-load tests (PLTs) provide an attractive alternative, because they can provide similar data at a lower cost. Additionally, the PLT is sometimes preferred because it requires smaller sample sizes and less sample preparation, as well as allowing samples to be tested in different orientations. PLT tests are conducted by subjecting a rock to load applied by two conical platens; in contrast with UCS tests, force is applied at points, not across flat surfaces. Pressure is increased until the sample ultimately fractures. The point-load test provides an indirect measure of the sample's uniaxial compressive strength, which is calculated using well-calibrated algorithms.

16.5.3 Uniaxial tension strength tests

Uniaxial tension tests (UTS) and uniaxial compression tests are most appropriate for testing isotropic or near-isotropic materials. (For anisotropic materials, stresses must be applied in more than one direction for complete characterization.) Uniaxial measurements of tensile strength are made using devices similar to the one shown in Figure 16.23. The difference is that, for tensile tests, the sample is subjected to increasing tension until failure, instead of subjected to increasing compression. Additionally, sample preparation is more complicated for uniaxial tensile tests than for compression tests, because samples for tensile stress are generally shaped so that they have lips or shoulders that can be easily gripped.

16.5.4 Brazilian test

In practice, direct measurements of tensile strength require gluing (adhering platens to rock surfaces) or shaping rock samples, as described above for UTS tests, and then applying tension. So, making direct tests poses special challenges. However, the Brazilian test is the commonly and widely used alternative that provides, simply and inexpensively, an indirect measurement of tensile strength.

Figure 16.25 shows a rock cylinder undergoing a Brazilian test; two plates are applying pressure to the surfaces of a rock cylinder. The compressive forces have squeezed

Figure 16.25 Example of a rock failing during a Brazilian test.
Photo from Brianne Mackinnon, The University of Queensland, School of Mechanical and Mining Engineering.

Figure 16.24 Example of a device used to make point-load tests.
Photo courtesy of Franz Campero, MEC International LLC.

the sample, causing it to bulge (become wider). This bulging, which stressed and strained the sample horizontally, created tensional stress within the sample, creating the tensile crack that can be seen in the photo. The crack is oriented parallel to the force being applied, but is normal to the tensional stress. The yield strength—the amount of stress that causes cracking—is used to estimate the tensile strength of the sample.

16.5.4.1 *Shear strength testing*

The standard method for measuring the shear strength of rock involves applying a normal stress to a specimen and then applying shear forces. Figure 16.26 shows schematically how this is done. While the normal force remains constant, an increasing amount of shear stress is applied until the sample fractures. The values of the applied forces are then used to calculate shear strength. Figure 16.27 shows one commercially available device used to measure shear strength.

16.5.5 *The Schmidt hammer*

Figure 16.28 shows a geologist measuring the strength of a specimen using a *Schmidt hammer*. Ernst Schmidt created this hammer in the 1950s, and these hammers are currently marketed by the BBR Network in Zurich, Switzerland, hence the alternative name of "Swiss hammer." The device is a small, hand-held instrument, originally developed to test the elastic properties of concrete, but it may also be used on rocks. A user presses the device against the surface of the rock he or she wants to test. The device releases a small, spring-loaded piston that bounces off the rock surface, and the amount of rebound is an indirect measure of the compressive strength of the material. The Schmidt hammer provides quick, reliable results in the field. There is no need to obtain samples or go through the costly time of preparing samples for uniaxial compression testing. However, the hammer does have limitations, and results may need to be adjusted to account for non-ideal situations.

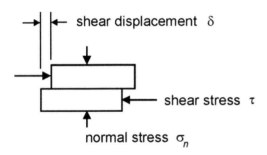

Figure 16.26 Diagram showing how measurements of shear strength are made.

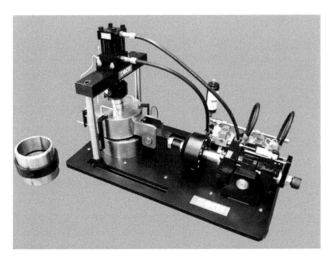

Figure 16.27 Example of a device used for measuring shear strength.
Photo from Zachary Phillips, GCTS Testing Systems.

Figure 16.28 Making measurements using a Schmidt hammer.
Photo Arjuncm3, Wikimedia Commons.

16.5.6   *Triaxial rock testing*

Figure 16.29 shows a typical device used for conducting triaxial strength tests. Triaxial testing involves applying axial compression to a specimen that is already under confining pressure. The name implies three directions of force, but most triaxial devices, such as the one shown in this photo, use a pressurized fluid (typically mineral oil) as the source of confinement, so σ_2 and σ_3 are equal. The use of a fluid "jacket" around the core provides continuous stress along the entire surface of the sample. The axial compression is applied to the test specimen using the same kind of device used for the uniaxial compressive test.

For triaxial testing, the device (pressure vessel) that contains the confining fluid and allows the sample to be simultaneously compressed is called a *cell*. Researchers use many kinds of cells, and they can hold cubic or cylinder-shaped rock samples of varying sizes. One of the simplest, more effective cell designs is the Hoek cell; the cylindrical silver unit in the center of Figure 16.29 is an example. The metal cylinder seen in this figure contains oil that applies confining pressure to a rock specimen while the vertical press slowly compresses the sample. Stress and strain are measured simultaneously, yielding data such as depicted in Figures 16.9 and 16.10, which allow calculation of rock strength and other important rock properties.

16.5.7   *Durability tests*

The durability of weak, intact rock is an important consideration in engineering applications. So, experimental laboratory tests with devices, like the one shown

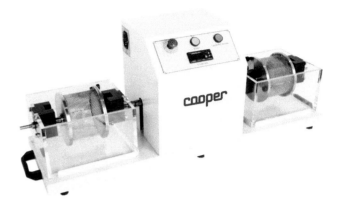

Figure 16.30   Example of a device used to measure a rock's slake durability.

Photo courtesy of Zameer Syed, Cooper Technology.

in Figure 16.30, are commonly used to determine the *slake durability* (the durability of a rock exposed to cycles of wetting and drying) and indirectly evaluate the weathering resistance of shales, mudstones, silt-stones, and other clay-bearing rocks. There are several standard protocols used to measure durability, but all include wetting and stressing rock samples and measuring the results. In most experiments, small samples of rock are placed in baskets and slowly tumbled as they pass through liquid. Pieces of the samples fall off and out of the basket, and what remains is weighed to determine loss of material.

Questions for thought—chapter 16

1. Where is the Coeur d'Alene Mining District? What is mined there? And, what causes the earthquakes in the Coeur d'Alene Mining District? Are the earthquakes dangerous? Why have the number of earthquakes increased at the Coeur d'Alene Mining District in recent years?
2. What is the difference between body and surface forces? What are some examples of each?
3. What is stress and how is it related to strain?
4. What is the difference between elastic and plastic deformation? Can either of them be reversed by removing stress?
5. What three different kinds of stress might a rock experience? How do they differ? Describe a situation where each of the three may occur.
6. The ratio of shear stress to normal stress often determines whether a rock fails. What happens when a rock fails? And why is the ratio of shear to nor-

Figure 16.29   A device for conducting triaxial rock strength tests.

Photo courtesy of Medeo Olivares, CONTROLS Group, Italy.

mals stress a key? And, how does lithostatic pressure affect rock strength?

7. Describe what happens to a rock that undergoes compressional, tensional, and shear strain?

8. How does uniaxial stress tests differ from triaxial tests? What purpose does each serve?

9. Would a very rigid rock have a high or low Poisson's ratio? Why would multiple samples of a rock be analyzed for Poisson's ratio rather than just one sample?

10. What is the difference between ductile and brittle deformation? Name one kind of geological material that you expect would be prone to ductile deformation. And name one that would be prone to brittle deformation.

11. What happens when a rock reaches its elastic limit? What happens when a rock reaches its point of failure?

12. If a rock is at some shallow depth and experiences a high level of stress, it may fail. But suppose the same rock is considerably deeper in Earth. Would it fail at the same level of stress, at a lower level of stress, or at a greater level of stress?

13. How do temperature and pressure affect the fracturing of a rock? Why do earthquakes generally not occur at depths greater than 750 km?

14. What is Poisson's ratio? Why does a shale generally have a much lower Poisson's ratio than halite? (You may wish to look at Table16.1.)

15. Are most rocks stronger or weaker when they are water saturated compared with when they are dry. Explain. How does this relate to effective stress?

16. What is the strain rate? How does it affect the behavior of most rocks? Explain your answer.

17. Sometimes just determining the strength of a particular kind of rock can be very misleading. Even small hand samples may not yield reliable strength measurements. It is often more appropriate to consider and entire rock mass. Why? Explain.

18. What are the Geological Strength Index (GSI) and Rock Quality Designation (RQD)? How are values for each determined? Why may evaluating rocks strength using the GSI instead of the RQD be advantageous?

19. What is the purpose of the uniaxial compressive strength (UCS) test? How is such a test made? What kinds of information are obtained from UCS tests? What are some limitations of the tests?

20. What is the purpose of uniaxial tension strength (UTS) tests? Why is the Brazilian test sometimes preferred instead?

21. What is a triaxial rock test? It is not as commonly done as other kinds of testing. Why do you think this is? What kind of strength information does triaxial testing give that is not obtainable using the other testing methods described? Do triaxial tests provide values for compressive strength, tensional strength, shear strength, or for all three?

References

In general, we have tried not to clutter the text with many references to specific publications, because much of the information in *Earth Materials* is available in other places. We have, however, combined and connected information in original ways to create new knowledge by presenting different ideas and ways of thinking. Besides a small number of references given in the text, some other important sources of information are listed below, chapter by chapter.

This book contains figures with many photographs and graphics. Half of the figures are original to this publication and most of the others came from public sources with no copyright restrictions. Many figures came from Wikimedia Commons, a robust and fantastic resource. Whether copyrighted or not, we have given credit in figure captions for anything that was not created by us.

Chapter 1—The origin of the elements and Earth

General References

Much of this information comes from a number of different NASA web sites. Some general geochemical ideas came from:

White, W.M. (2013) *Geochemistry*. Wiley-Blackwell, Hoboken.

The data in Table 1.1 come from:

Abbott, P.L. (2004) *Natural Disasters*. McGraw Hill Higher Education, Boston. p. 14.
Faure, G. (1998) *Principles and Applications of Geochemistry*, 2nd ed. Prentice Hill, Upper Saddle River, NJ. p. 24.
Levy, D.H. (1994) *Skywatching*. Fog City Press, San Francisco, CA.
Solar System, NASA. Available from: https://solarsystem.nasa.gov
Spudis, P.D. (1999) The moon. In: Beatty, J.K. Petersen, C.C. & Chaikin, A. (eds) *The New Solar System*, 4th ed. Sky Publishing Corp. and Cambridge University Press, Cambridge. pp. 125–140.
Stern, S. A., Bagenal, F., Ennico, K., Gladstone, G. R., Grundy, W. M., McKinnon, W. B., Moore, J. M., Olkin, C. B., Spencer, J. R., Weaver, H. A., and Young, L. A. (2015) The Pluto system: Initial results from its exploration by new horizons. *Science*, 350(6258).
Hamilton, C.J. (1995–2015) *Views of the Solar System*. Available from: http://solarviews.com/eng/index.htm
Vilas, F. (1999) Mercury. In: Beatty, J.K. Petersen, C.C. & Chaikin, A. (eds) *The New Solar System*, 4th ed. Sky Publishing Corp. and Cambridge University Press, Cambridge. pp. 87–96.

Opening information on the Orion Nebula comes mostly from various NASA websites and from Wikipedia.

The data in Table 1.2 come from:

Yanagi, T. (2011) *Arc Volcano Japan Generation of Continental Crust from the Mantle*. Springer-Verlag, Berlin.

Chapter 2—Earth systems and cycles

General references

Berner, E.K. & Berner, R.K. (2012) *Global Environment: Water, Air, and Geochemical Cycles*, 2nd ed. Princeton University Press, Princeton and Oxford.
Langmuir, C.H. & Broeker, W. (2012) *How to Build a Habitable Planet—The Story of Earth from the Big Bang to Humankind*. Princeton University Press, Princeton and Oxford.

The opening vignette is based on:

Leopold, A. (1949) *A Sand County Almanac, and Sketches Here and There*. Oxford University Press, New York.

The data in Table 2.1 come from:

Freeze, R.A. & Cherry, J.A. (1979) *Groundwater*. Prentice Hall, Englewood Cliffs. p. 5.

Chapter 3—Minerals

General References

Dyar, M.D., Gunter, M.E. & Tasa, D. (2008) *Mineralogy and Optical Mineralogy*. Mineralogical Society of America, Chantilly.

Perkins, D. (2011) *Mineralogy*, 3rd ed. Pearson Education, Upper Saddle River, NJ.

Additional references in the chapter

Fuhrman, M.L. & Lindsley, D.H. (1988) Ternary feldspar modeling and thermometry. *American Mineralogist*, 73, 201–215.

Nickel, E.H. & Grice, J.D. (1998) The IMA commission on new minerals and mineral names: Procedures and guidelines on mineral nomenclature. *Canadian Mineralogist*, 36, 14.

Switzer, G.S. (1975) Composition of garnet xenocrysts from three kimberlite pipes in Arizona and New Mexico. *Smithsonian Contributions to Earth Science*, 19(1), 21.

The IMA definition of a mineral given on the top of page 72 comes from Nickel and Grice (1998).

Chapter 4—Mineral crystals

General references

Dyar, M.D., Gunter, M.E. & Tasa, D. (2008) *Mineralogy and Optical Mineralogy*. Mineralogical Society of America, Chantilly.

Perkins, D. (2011) *Mineralogy*, 3rd ed. Pearson Education, Upper Saddle River, NJ.

Information about Cuevo de los Cristales in the opening vignette comes mostly from:

Lovgren, S. (2007) Giant crystal cave's mystery solved. *National Geographic News*. Available from: www.nationalgeographic.com/science/2007/04/giant-crystal-cave-mexico-mystery-solved/

A general reference in the text to Shannon and Prewitt refers to several papers, including:

Shannon, R.D. (1976) Revised effective ionic radii and systematic studies of interatomic distances in halides and chalcogenides. *Acta Crystallographica*, A32(5), 751–767.

Shannon, R.D. & Prewitt, C.T. (1969) Effective ionic radii in oxides and fluorides. *Acta Crystallographica*, B25, 925–946.

Chapter 5—Igneous petrology and the nature of magmas

General References

Huang, H-H., Lin, F.C., Schmandt, B., Farrell, J., Smith, R.B. & Tsai, V.C. (2015) The Yellowstone magmatic system from the mantle plume to the upper crust. *Science*, 348(6236), 773–776.

Raymond, L.A. (2007) *Petrology: The Study of Igneous, Sedimentary, and Metamorphic Rocks*, 2nd ed. Waveland Press, Long Grove.

Zimmer, M.M, Plank, T, Hauri, E.H., Yogodzinski, G.M., Stelling, P., Larsen, J., Singer, B., Jicha, C., Mandeville, C. & Nye, C.J. (2010) The role of water in generating the calc-alkaline trend: New volatile data for Aleutian magmas and a new tholeiitic index. *Journal of Petrology*, 51(12), 2411–2444.

Williams, H.H. (1942) The geology of Crater Lake National Park, Oregon with a reconnaissance of the Cascade Range southward to Mount Shasta. *Carnegie Institute of Washington Pub*, 540, 162.

Winter, J.D. (2009) *Principles of Igneous and Metamorphic Petrology*, 2nd ed. Pearson Education, Upper Saddle River, NJ.

The data in Table 5.1 come from:

Raymond, L.A. (2007) *Petrology: The Study of Igneous, Sedimentary, and Metamorphic Rocks*, 2nd ed. Waveland Press, Long Grove.

The data in Table 5.3 come from:

Duncan, R.A., Backman, J. & Peterson, L.C. *et al.* (1990) *Proceedings of the Ocean Drilling Program. Volume: 115, Scientific Results: Mascarene Plateau*. Texas AM University, College Station, TX (Ocean Drilling Program).

The data in Figure 5.53 come from:

Rollinson, H.R. (1993) *Using Geochemical Data: Evaluation, Presentation, Interpretation*. Routledge Longman Geochemistry Series, London.

Chapter 6—Plutonic rocks

General references

Raymond, L.A. (2007) *Petrology: The Study of Igneous, Sedimentary, and Metamorphic Rocks*, 2nd ed. Waveland Press, Long Grove.

Winter, J.D. (2001) *An Introduction to Igneous and Metamorphic Petrology*, 1st ed. Pearson Education, Upper Saddle River.

Winter, J.D. (2009) *Principles of Igneous and Metamorphic Petrology*, 2nd ed. Pearson Education, Upper Saddle River.

Information about the geology of Yosemite National Park comes from:

Harris, A.G., Tuttle, E. & Tuttle, S.D. (1997) *Geology of National Parks*, 5th ed. Kendall-Hunt Publishing, Dubuque.

Huber, N.K. (1987) The geologic story of Yosemite National Park. *USGS Bulletin*, 1595.

Kiver, E.P. & Harris, D.V. (1999) *Geology of U.S. Parklands*, 5th ed. John Wiley & Sons, New York.

National Park Service. *Geology of Yosemite National Park*. Available from: www.nps.gov/yose/learn/nature/geology.htm

Schaffer, J.P. (1997) *The Geomorphic Evolution of the Yosemite Valley and Sierra Nevada Landscapes: Solving the Riddles in the Rocks*. Wilderness Press, Berkeley.

Table 6.2 references the IUGS rock classification system. Although it has been modified several times, the seminal version of the IUGS rock classification system is found at:

Le Bas, M.J. & Streckeisen, A.L. (1991) The IUGS systematics of igneous rocks. *Journal of the Geological Society*, 148, 825–833.

Table 6.3, details about the petrology of El Capitan, and the geologic map in Figure 6.18 derive from:

Putnam, R., Glazner, A., Coleman, D. & Ingalls, M. (2015) Plutonism in three dimensions: Field and geochemical relations on the southeast face of El Capitan, Yosemite National Park, California. *Geosphere*, 11(4). Available from: http://dx.doi.org/10.1130/GES01133.1

The map in Figure 6.20 is based on a map by Winter (2009).

Chapter 7—Volcanoes and volcanic materials

Much of the information in this chapter comes from various USGS websites, and from Volcano World (Oregon State University; http://volcano.oregonstate.edu/).

General references

Lockwood, J.P. & Hazlett, R.W. (2010) *Volcanoes: Global Perspectives*. Wiley-Blackwell, Hoboken.

Nicholson, E. (2017) *Volcanology*. Larsen and Keller Education, New York.

Information about Tambora and Toba volcanoes comes from many sources on the internet and also:

Evans, R. (2002) Blast from the past—the eruption of Mount Tambora killed thousands, plunged much of the world into a frightful chill and offers lessons for today. *Smithsonian Magazine*, July.

Gibbons, A. (1993) Pleistocene population explosions. *Science*, 262(5130), 27–28.

Klemetti, E. (2015) Tambora 1815: Just how big was the eruptions. *Wired*, 4 October. Available from: www.wired.com/2015/04/tambora-1815-just-big-eruption/

Rampino, M.R. & Self, S. (1993) Climate—volcanism feedback and the Toba eruption of ~74,000 years ago. *Quaternary Research*, 40, 269–280.

Although it has been modified several times, the seminal version of the IUGS rock classification system is found at:

Le Bas, M.J. & Streckeisen, A.L. (1991) The IUGS systematics of igneous rocks. *Journal of the Geological Society*, 148, 825–833.

Chapter 8—Sediments and sedimentary rocks

General references

Boggs, S. Jr. (2012) *Principles of Sedimentology and Stratigraphy*, 5th ed. Pearson Education, Upper Saddle River.

Hewitt, K., Gosse, J., & Clague, J.J. (2011) Rock avalanches and the pace of late Quaternary development of river valleys in the Karakoram Himalaya. *Geological Society of America Bulletin*, 123, 1836–1850.

Nichols, G. (2009) *Sedimentology and Stratigraphy*, 2nd ed. Wiley-Blackwell, Hoboken.

Prothero, D.R. (2003) *Sedimentary Geology: An Introduction to Sedimentary Rocks and Stratigraphy*. W.H. Freeman & Co., New York.

Tucker, M.E. (ed.) (2012) *Sedimentary Petrology: An Introduction to the Origin of Sedimentary Rocks*, 3rd ed. Wiley-Blackwell, Hoboken.

Information about the Queenston and Catskill deltas comes mostly from:

Gibson, R. (2014) *History of the Earth: The Queenston Delta*, March 21. Available from: http://historyoft

heearthcalendar.blogspot.com/2014/03/march-21-queenston-delta.html

Reference for stream migration in Indiana is a USGS publication at https://pubs.usgs.gov/sir/2013/5168/pdf/sir2013-5168.pdf

Chapter 9—Stratigraphy

General references

Boggs, S. Jr. (2012) *Principles of Sedimentology and Stratigraphy*, 5th ed. Pearson Education, Upper Saddle River.

Nichols, G. (2009) *Sedimentology and Stratigraphy*, 2nd ed. Wiley-Blackwell, Hoboken.

Prothero, D.R. (2003) *Sedimentary Geology: An Introduction to Sedimentary Rocks and Stratigraphy*. W.H. Freeman & Co.

Information about the Geology of the Grand Canyon is widely available, but the opening vignette was based in large part on a National Park Service website at www.nps.gov/grca/learn/nature/grca-geology.htm

Table 9.1, listing depositional environments, is based on a table by R. Dawes that can be found at https://commons.wvc.edu/rdawes/g101ocl/basics/bscstables/depenv.html

Chapter 10—Metamorphic rocks

General References

Burnham, C.W. (1959) Contact Metamorphism of Magnesian Limestones at Crestmore, California. *Geological Society of America Bulletin*, 70, 879–920.

Perkins, D. (2011) *Mineralogy*, 3rd ed. Pearson Education, Upper Saddle River.

Raymond, L.A. (2007) *Petrology: The Study of Igneous, Sedimentary, and Metamorphic Rocks*, 2nd ed. Waveland Press, Long Grove.

Winter, J.D. (2009) *Principles of Igneous and Metamorphic Petrology*, 2nd ed. Pearson Education, Upper Saddle River.

Information about wollastonite and metamorphism in the Adirondacks comes from:

Bohlen, S.R., Valley, J.W. & Essene, E.J. (1985) Metamorphism in the Adirondacks: Petrology, pressure, and temperature. *Journal of Petrology*, 26, 971–992.

Clechenko, C.C. & Valley, J.W. (2003) Oscillatory zoning in garnet from the Willsboro wollastonite skarn, Adirondack Mts, New York: A Record of Shallow Hydrothermal Processes Preserved in a Granulite Facies Terrane. *Journal of Metamorphic Geology*, 21, 771–784.

Clechenko, C.C., Valley, J.W. & McLelland, J. (2002) Timing and depth of intrusion of the Marcy anorthosite massif: Implications from field relations, geochronology, and geochemistry at Woolen Mill, Jay Covered Bridge, Split Rock Falls, and the Oak Hill Wollastonite Mine. *New England Intercollegiate Geological Conference, New York State Geological Association*, C1-1–C1-17.

Chapter 11—Weathering and soils

General references

Boggs, S. (2012) *Principles of Sedimentology and Stratigraphy*, 5th ed. Pearson Education, Upper Saddle River.

Eash, N.S., Sauer, T.J., O'Dell, D. & Odoi, E. (2016) *Soil Science Simplified*, 6th ed. Wiley-Blackwell, Hoboken.

Schaetzl, R. & Thompson, M.L. (2015) *Soils: Genesis and Geomorphology*, 2nd ed. Cambridge University Press, New York.

Tucker, M.E. (ed). (2012) *Sedimentary Petrology: An Introduction to the Origin of Sedimentary Rocks*, 3rd ed. Wiley-Blackwell, Hoboken.

Weil, R.R. & Brady, N.C. (2016) *The Nature and Properties of Soils*, 15th ed. Pearson Education, Upper Saddle River.

Information about Mesopotamia comes mostly from:

Ancient History Encyclopedia: Fertile Crescent. Available from: www.ancient.eu/Fertile_Crescent/

History of the World.Net: Mesopotamia. Available from: www.historyworld.net/wrldhis/PlainTextHistories.asp?historyid=aa53

Learner.org: Collapse—Why Do Civilizations Fall? Mesopotamia. Available from: www.learner.org/exhibits/collapse/mesopotamia.html

The information in Table 11.2 is from:

Birkeland, P. (1999) *Soils and Geomorphology*. Oxford University Press, New York.

The reference to Simonson and pedogenesis is:
Simonson, R.W. (1959) Outline of a generalized theory of soil genesis. *Soil Science Society of America Journal*, 23(2), 152–156.

The seminal Whittaker biome diagram (that was the basis for Figure 11.40) can be found at:
Whittaker, R.H. (1975) *Communities and Ecosystems*, 2nd ed. Macmillan, New York.

Table 11.7 is from:
U.S. Department of Agriculture. (1999) *Soil Taxonomy, A Basic System of Soil Classification for Making and Interpreting Soil Surveys*, 2nd ed. United States Department of Agriculture, Washington, DC, Agriculture Handbook No. 436.

Chapter 12—Water and the hydrosphere

The Wikipedia "water cycle" reference is https://en.wikipedia.org/wiki/Water_cycle.

Information about the Salton Sea can be found in countless recent publications and websites. For this book, most of the information comes from publications of the USGS available from: https://www2.usgs.gov/saltonsea/.

The data in Table 12.2 come from unattributed sources in Wikipedia and from:
Dai, A. & Trenberth, K.E. (2002) Estimates of freshwater discharge from continents: Latitudinal and seasonal variations. *Journal of Hydrometeorology*, 3, 660–687.
Dai, A., Qian, T., Trenberth, K.E. & Milliman, J.D. (2009) Changes in continental freshwater discharge from 1948–2004. *Journal of Climate*, 22, 2773–2791.
Syvitski, J.P.M., Kettner, A.J., Overeem, I., Hutton, E.W.H., Hannon, M.T., Brakenridge, G.R., Day, J., Vörösmarty, C., Saito, Y., Giosan, L. & Nicholls, R.J. (2009) Sinking deltas due to human activities. *Nature Geoscience*, 2, 681–686.

The data in Table 12.3 and Figure 12.23 come from many different sources on the internet.

Chapter 13—Mineral deposits

General references

Guilbert, J.M. & Park, C.F. (1986) *The Geology of Ore Deposits*. W.H. Freeman and Company, New York.
Kesler, S.E. & Simon, A.C. (2015) *Mineral Resources, Economics and the Environment*. Cambridge University Press, Cambridge.
Misra, K. (2000) *Understanding Mineral Deposits*. Springer, Amsterdam.
Ridley, J.A. (2013) *Ore Deposit Geology*. Cambridge University Press, Cambridge.
Robb, L. (2005) *Introduction to Ore-Forming Processes*. Blackwell Science Limited, Oxford.

Information about the history and geology of Bingham Canyon comes from the sources listed above and from Wikipedia.

Chapter 14—Energy resources

General references

Balasubramanian, V. (2017) *Energy Sources—Fundamentals of Chemical Conversion Processes and Application*. Elsevier, Amsterdam.
Dhupper, R. (2015) *Textbook on Energy Resources and Management*. CBS Publishers and Distributers, Hyderabad.
Ghosh, T. & Prelas, M. (2009) *Energy Resources and Systems. Volume 1: Fundamentals and Non-Renewable Resources*. Springer, Amsterdam, The Netherlands.
Ghosh, T. & Prelas, M. (2011) *Energy Resources and Systems. Volume 2: Renewable Resources*. Springer, Amsterdam, The Netherlands.
Kesler, S.E. & Simon, A.C. (2015) *Mineral Resources, Economics and the Environment*, 2nd ed. Cambridge University Press, Cambridge.
Singh, S.N. (2015) *Non Conventional Energy Resources*. Pearson Education India, Delhi.

Information about the Gemasolar Thermosolar Plant was gathered by one of the authors (DP) during a visit and tour in 2017. Information about the Crescent Dunes Solar Energy Project comes from Wikipedia and:

Dietrich, R. (2018) 24-Hour solar energy: Molten salt makes it possible, and prices are falling fast. *Inside Climate News*, January 16. Available from: https://insideclimatenews.org/news/16012018/csp-concen

trated-solar-molten-salt-storage-24-hour-renewable-energy-crescent-dunes-nevada

Overton, T.W. (2016) *Top Plant: Crescent Dunes Solar Energy Project*, January 12. Tonopah, Nevada. Power. Available from: www.powermag.com/crescent-dunes-solar-energy-project-tonopah-nevada-2/

Chapter 15—Soil mechanics

General references

Das, B.M. (2002) *Soil Mechanics Laboratory Manual.* Oxford University Press, New York.

de Vallejo, L.G. & Ferrer, M. (2011) *Geological Engineering*. CRC Press, Leiden.

Fredlund, D.G., Rahardjo, H. & Fredlund, M.D. (2012) *Unsaturated Soil Mechanics in Engineering Practice*. John Wiley & Sons, New York.

Rahn, P.H. (1996) *Engineering Geology: An Environmental Approach*, 2nd ed. Prentice Hall, Englewood Cliffs.

Terzaghi, K., Peck, R.B. & Mesri, G. (1996) *Soil Mechanics in Engineering Practice*. John Wiley & Sons, New York.

Information about the Leaning Tower of Pisa comes from many articles on the web: references can be found on Wikipedia at https://en.wikipedia.org/wiki/Leaning_Tower_of_Pisa.

Chapter 16—Rock mechanics

General references

Bieniawski, Z.T. (1984) *Rock Mechanics Design in Mining and Tunneling*. Balkema Publishers, Leiden.

de Vallejo, L.G. & Ferrer, M. (2011) *Geological Engineering*. CRC Press, Leiden.

Goodman, R.E. (1989) *Introduction to Rock Mechanics*, vol. 2. John Wiley & Sons, New York.

Jaeger, J.C., Cook, N.G. & Zimmerman, R. (2009) *Fundamentals of Rock Mechanics*. John Wiley & Sons, New York.

Rahn, P.H. (1996) *Engineering Geology: An Environmental Approach*, 2nd ed. Prentice Hall, Englewood Cliffs.

U.S. Bureau of Reclamation. (1998) *Engineering Geology Field Manual*, vol. 1, 2nd ed. U. S. Government Printing Office, Washington, DC.

Much of the information on rock burst in the Coeur D'Alene mining district comes from:

Whyatt, J.K., Blake, W., Williams, T.J. & White, B.G. (2002) *60 Years of Rockbursting in the Coeur D'Alene District of Northern Idaho, USA: Lessons Learned and Remaining Issues*. The National Institute for Occupational Safety and Health, Mining Publication, Centers for Disease Control and Prevention, Washington.

Reference in Table 16.2:

Williams, A., Burns, D., Farquhar, G. & Mills, M. (2005) *Field Descriptions of Soil and Rock*. New Zealand Geotechnical Society, Wellington.

Reference in Table 16.3:

Deere, D. & Miller, R. (1966) *Engineering Classification and Index Properties for Intact Rock*. Tech. Report No AFWL—TR-65–116, Air Force Weapons Lab, Kirtland Air Base, New Mexico.

Index